AF751262

AIRO Springer Series

Volume 1

Editor-in-chief

Daniele Vigo, Dipartimento di Ingegneria dell'Energia Elettrica e dell'Informazione "Guglielmo Marconi", Alma Mater Studiorum Università di Bologna, Bologna, Italy

Series editors

Alessandro Agnetis, Dipartimento di Ingegneria dell'Informazione e Scienze Matematiche, Università degli Studi di Siena, Siena, Italy

Edoardo Amaldi, Dipartimento di Elettronica, Informazione e Bioingegneria (DEIB), Politecnico di Milano, Milan, Italy

Francesca Guerriero, Dipartimento di Ingegneria Meccanica, Energetica e Gestionale (DIMEG), Università della Calabria, Rende, Italy

Stefano Lucidi, Dipartimento di Ingegneria Informatica Automatica e Gestionale "Antonio Ruberti" (DIAG), Università di Roma "La Sapienza", Rome, Italy

Enza Messina, Dipartimento di Informatica Sistemistica e Comunicazione, Università di Milano-Bicocca, Milan, Italy

Antonio Sforza, Dipartimento di Ingegneria Elettrica e delle Tecnologie dell'Informazione, Università degli Studi di Napoli Federico II, Naples, Italy

The **AIRO Springer Series** focuses on the relevance of operations research (OR) in the scientific world and in real life applications.

The series publishes contributed volumes, lectures notes, and monographs in English language resulting from workshops, conferences, courses, schools, seminars, and research activities carried out by AIRO, Associazione Italiana di Ricerca Operativa - Optimization and Decision Sciences: http://www.airo.org/index.php/it/. The books in the series will discuss recent results and analyze new trends focusing on the following areas: Optimization and Operation Research, including Continuous, Discrete and Network Optimization, and related industrial and territorial applications. Interdisciplinary contributions, showing a fruitful collaboration of scientists with researchers from other fields to address complex applications, are welcome. The series is aimed at providing useful reference material to students, academic and industrial researchers at an international level.

More information about this series at http://www.springer.com/series/15947

Patrizia Daniele · Laura Scrimali
Editors

New Trends in Emerging Complex Real Life Problems

ODS, Taormina, Italy, September 10–13, 2018

AIRO
ASSOCIAZIONE ITALIANA DI RICERCA OPERATIVA
OPTIMIZATION AND DECISION SCIENCE

Springer

Editors
Patrizia Daniele
Department of Mathematics
and Computer Science
University of Catania
Catania, Italy

Laura Scrimali
Department of Mathematics
and Computer Science
University of Catania
Catania, Italy

ISSN 2523-7047 ISSN 2523-7055 (electronic)
AIRO Springer Series
ISBN 978-3-030-00472-9 ISBN 978-3-030-00473-6 (eBook)
https://doi.org/10.1007/978-3-030-00473-6

Library of Congress Control Number: 2018958352

© Springer Nature Switzerland AG 2018
This work is subject to copyright. All rights are reserved by the Publisher, whether the whole or part of the material is concerned, specifically the rights of translation, reprinting, reuse of illustrations, recitation, broadcasting, reproduction on microfilms or in any other physical way, and transmission or information storage and retrieval, electronic adaptation, computer software, or by similar or dissimilar methodology now known or hereafter developed.
The use of general descriptive names, registered names, trademarks, service marks, etc. in this publication does not imply, even in the absence of a specific statement, that such names are exempt from the relevant protective laws and regulations and therefore free for general use.
The publisher, the authors and the editors are safe to assume that the advice and information in this book are believed to be true and accurate at the date of publication. Neither the publisher nor the authors or the editors give a warranty, express or implied, with respect to the material contained herein or for any errors or omissions that may have been made. The publisher remains neutral with regard to jurisdictional claims in published maps and institutional affiliations.

This Springer imprint is published by the registered company Springer Nature Switzerland AG
The registered company address is: Gewerbestrasse 11, 6330 Cham, Switzerland

Preface

This book contains the contributions of the International Conference on Optimization and Decision Science (ODS2018), held in Taormina (Messina), Italy, from September 10 to 13, 2018. ODS2018 was the 48th annual meeting of AIRO, the Italian Operations Research Society, and was organized in cooperation with the Department of Mathematics and Computer Science (DMI) of the University of Catania. ODS2018 was addressed to the entire Operations Research community working in the field of optimization, problem-solving, and decision-making methods. Its aim was to bring together scholars and decision-makers from both the academic and the industrial domain in order to present results with the potential to solve concrete problems and to provide new insights, bridging the researcher–practitioner gap.

This book presents state-of-the-art knowledge relating to optimization, decisions science, and problem-solving methods, as well as a large variety of applications of extreme importance in relation to computer science, mathematical physics, engineering, statistics, and economics. It constitutes a fine collection of methodological and application-oriented papers that characterize the current research in challenging and worthwhile areas. The papers included in this volume present new developments in topics of great interest, such as scheduling, routing, heuristics and metaheuristics, logistics, health care, Bayesian optimization, supply chain management, and game theory. The selection of papers also provides novel applications in the optimal design of photovoltaic installations, parking pricing problems, industrial IoT networks, cybersecurity investments, and autonomous driving. Combining new methodological advances with a wide variety of real applications, this volume is certainly of great value not only to researchers and practitioners working in these areas but also to the Operations Research community.

The work includes international contributions attesting to the success and global dimension of the project. Some papers were selected for publication as full papers, and some were accepted as short presentations. The peer-review process was conducted by experts in Operations Research and related fields. All the abstracts can be found at the website: http://www.airoconference.it/ods2018/.

Moreover, we are to host three outstanding contributions given by internationally distinguished professors, namely Nicholas Hall (Ohio State University, USA), Panos Pardalos (University of Florida, USA), and Maria Grazia Speranza (University of Brescia, Italy), which increase the overall quality of the book and provide a deeper understanding of the fields of interest faced here.

Finally, we would like to express our thanks to the invited speakers for their invaluable contributions, to the authors for their work and dedication, and to all members of the Program Committee and auxiliary reviewers who helped by offering their expertise and time. Special gratitude should be addressed to Springer for strongly supporting us during the publishing process.

Catania, Italy
September 2018

Patrizia Daniele
Laura Scrimali

Contents

About the Editors

Patrizia Daniele is a Full Professor of Operations Research at the University of Catania. In 2006, she was a Visiting Scholar at the Division of Engineering and Applied Science (DEAS), Harvard University. Her areas of research include global optimization, variational inequality theory, infinite-dimensional duality, humanitarian logistics, equilibrium problems, cybersecurity, and computational procedures. She is the author or editor of six books and has published several papers in national and international journals. She is also director of various international workshops and has served on the scientific committees of several national and international conferences.

Laura Scrimali is an Associate Professor of Operations Research at the University of Catania. Her areas of research include convex optimization, variational inequality theory, numerical methods, sustainable development, and equilibrium problems in transportation, economic, and finance. She has published several papers in national and international journals and has served on the scientific committees of various national and international conferences.

INFORMS, Analytics, Research and Challenges

Nicholas G. Hall

Abstract The profession of operations research and analytics is experiencing a period of exceptionally rapid changes. These changes are being stimulated by the availability of much greater data, increased competition for publication space, and a business environment that expects greater technical skills. This paper reviews these significant changes and various resulting challenges for the profession. It also discusses possible responses to these challenges by INFORMS, the world's leading academic and professional society for operations research and analytics, and its members.

Keywords Operations research and analytics · Emerging trends · Challenges INFORMS

1 Introduction

This article and the related presentation at AIRO Conference, Taormina, Sicily, in September 2018 provide an overview of (a) the current status and activities of INFORMS, the world's leading academic and professional society for operations research and analytics (OR&A), (b) the impact of analytics, (c) three research areas that will generate significant intellectual and career opportunities over the next 10 years, and (d) several challenges currently faced by the field and profession of OR&A.

N. G. Hall (✉)
The Ohio State University, Columbus, USA
e-mail: hall.33@osu.edu

© Springer Nature Switzerland AG 2018

P. Daniele and L. Scrimali (eds.), *New Trends in Emerging Complex Real Life Problems*, AIRO Springer Series 1,
https://doi.org/10.1007/978-3-030-00473-6_1

2 INFORMS

A review of INFORMS' current situation reveals that it stands at a time of unique opportunity. Financially, the society is healthier than ever, with a net asset position exceeding $22 million. Membership is also strong, with an expected total number of members of about 12,500 by year end 2018, including regular, student and retired members. A key component of INFORMS' success is its publications. The variety of publications has expanded in recent years and now includes 16 refereed journals and two magazines, *OR/MS Today* and *Analytics*. The impact of INFORMS' publications can be measured by the number of interested readers, and the annual number of fulltext downloads is expected to pass 2 million for the first time in 2018.

INFORMS' current and future activities are closely aligned with its four strategic goals, as follows:

- INFORMS will identify, recognize, and promote the work of our members.
- Decision makers will use innovative technologies and methodologies to achieve better outcomes.
- Organizations will identify OR&A as core components of success.
- OR&A will advance society and make the world a better place.

Two major and interrelated activities of INFORMS in 2018 are Visibility and Awareness [7]. Visibility refers to a structured live event, "Government and Analytics Summit", which was developed for the purpose of improving outreach to policymakers. As a U.S.-based organization, our first focus is naturally on reaching U.S. federal agencies. The first event of this type was held on Capitol Hill in Washington on May 21, 2018. Keynote addresses were delivered by former Director of the National Security Agency and the Central Intelligence Agency General Michael Hayden, and former Secretary of Transportation Anthony Foxx. Their presentations about the importance of OR&A for leaders in government were followed by three concurrent panels featuring leading INFORMS experts on the topics of transportation, healthcare, and national security, each of which is currently a key focus area for the U.S. government. A review of the successful Summit event appears in Tucker [13].

Awareness refers to an ongoing advocacy campaign that is also addressed to policymakers. The main components of this campaign are:

- Legislative Advocacy: The purpose of legislative advocacy is to build awareness, interest, relationships and support for the value proposition of OR&A among policymakers and their staff.
- Federal Administration Advocacy: Spreading the word to senior career officials in federal agencies about the impact of OR&A is an important component of building allies and advocates for INFORMS and the OR&A profession. This advocacy provides us with an opportunity to leverage the expertise of members who have direct experience with various agencies and with the problems and issues they face.

- Indirect Advocacy: We have a specific focus on working with those in the media who study and write about public policy issues and the operations of government. We need to ensure that they understand at a high level what OR&A is and why it matters, and that they are familiar with compelling examples of the work OR&A professionals do.

Together, the above advocacy initiatives represent the foundations of INFORMS' efforts to raise interest and awareness about the value which OR&A can bring to societal and government problems, how its use can provide greater efficiencies, effective use of resources, and "Save money, Save lives, and Solve problems". We expect that these initiatives can serve as a model for similar important initiatives in other countries.

INFORMS is also moving forward strongly with other initiatives. These initiatives will affect Publications, Continuing Education, and International Activities. For Publications, we are initiating a greatly improved process for delivering already accepted journal papers to publication. We estimate that the new process will reduce the average time between acceptance and online publication from 184 to 36 days! See Hall [8] for additional details. For Continuing Education, we are conducting a review of our current programs for the purposes of enhancing them and evaluating the potential to develop new programs. For International Activities, a big focus of the current year has been the INFORMS International Conference held in Taipei, June 17–20. This conference attracted more than 800 participants from 25 countries. Meanwhile, the international Teaching Effectiveness Colloquium (TEC) program started by Jim Cochran (University of Alabama) has run workshops to help colleagues in 14 developing nations offer effective OR&A courses, with more TEC programs planned for 2019 and 2020. INFORMS currently has members in 80 countries, and there are plans to increase this number with outreach programs including TEC.

3 Analytics

This section contains several informal comments about analytics. INFORMS uses the following definition of analytics: "Analytics is the scientific process of transforming data into insight for making better decisions." An immediately relevant question is whether OR is a subset of analytics or *vice versa*. However, this question can be answered by counterexamples. First, a high school student counting cars at a tollbooth is doing analytics, but not operations research. Whereas, the development of a polynomial time algorithm for a special class of integer programs is operations research, but it is not analytics.

Closely associated with, and supportive of, the analytics movement is the availability of much larger quantities of data and for a broader range of applications than was previously the case. There are some important positive effects of greater data availability. First, it brings models and applications closer together by enabling calibration of modeling parameters. Second, it enables and enhances new modeling approaches and the study of more applications. Both these effects enable OR&A to deliver much greater value.

However, some negative effects of the increased focus on data can also be observed. First, some seminars and conference presentations have become less interesting, due to lengthy explanations of data acquisition and preparation. A further concern is that the data requirements of research projects are not homogeneous. For the study of a specific urban transport or power system, where the only user of the study is the utility that operates the system, it is certainly appropriate to expect the use of real data to validate research results. However, for a research study on the fundamental planning methodology of project management, there are many thousands of projects for which to solicit data, with a wide variety of possible research outcomes depending on that choice. Hence, a generic requirement to "provide real data" in support of a research study is specious. A third concern is that data availability may distract Ph.D. students to work on applications with convenient data availability, without regard to their level of fundamental, long-run interest in the application being studied.

Some confusion has arisen about whether the development of analytics research provides additional justification for empirical research that is predominantly regression-based and provides results that are descriptive or at best predictive. However, this is a difficult case to make, for two reasons. First, the analytics maturity curve [2] classifies the "competitive advantage" of value delivered by analytics into three levels; importantly, the highest level, "prescriptive", uses optimization and stochastic optimization rather than traditional empirical methodology. Secondly, analytics typically expects contributions to multiple steps in the decision making life cycle that consists of: problem identification, data preparation, data exploration, model creation, model testing and validation, model deployment, and monitoring and assessment of models.

Finally, the impact of analytics is strongly problem dependent. There are some business applications where the decisions to be made are very simple, and hence the predictive and prescriptive problems are almost identical. IBM [10] provides an example of using an analytics study of the relationship between the weather and customer demand to plan inventory in a retail bakery, which apparently resulted in a 20% increase in profit. In this case, if demand can be accurately predicted from the weather forecast, then the inventory planning decisions follow trivially. However, there are many planning problems where knowledge of a data pattern still leaves a complex decision problem to be solved, hence the benefit of analytics is less obvious and harder to estimate.

4 Research

This section discusses three research areas where important work is expected over the next 10 years for applications with multiple billions of dollars in annual value. The first area is the sharing economy, for which a huge increase in impact is predicted within the next decade. The second area is personalized health care, which offers the potential for greatly improved health outcomes and cost reductions. The third is project management, the business process that—along with supply chain management—dominates the world's economy.

4.1 The Sharing Economy

Regarding the sharing economy, Federal Trade Commission [3] predicts an increase in annual value from $15 billion at that time to $335 billion in 2025. Very few marketplaces have the potential for such dramatic expansion. The fundamental motivation for the sharing economy comes from underutilization of individually owned, expensive resources. A revealing perspective on this new environment is provided by Goodwin [5]: "Uber, the world's largest taxi company, owns no taxis. Facebook, the world's most popular media owner, creates no content. Alibaba, the world's most valuable retailer, has no inventory. Airbnb, the world's largest accommodation provider, owns no real estate. Something interesting is happening." See Hu [9] for several examples of recent research on the sharing economy.

The main features of the sharing economy marketplace include:

- Two-sided markets with externalities among users on both sides.
- A decentralized system where the intermediate platform cannot directly control either side.
- On-demand operational decisions and incentives that allocate resources in real time.
- The availability of massive data.

All these features, and especially the dramatic growth that is projected for the sharing economy, will make this research topic a highly important one over the next 10 years. Specific research opportunities include:

- The design of matching mechanisms that provide the most efficient way to match service providers and customers.
- The evaluation of social impact for different stakeholders.
- How social preferences or behavioral biases affect users' decision making and the dynamics of the market.
- Data-driven operational decision making, i.e. how to use data to provide guidance to the operational decisions of policymakers and platforms.

4.2 Personalized Health Care

Chronic diseases are a leading cause of death in most developed countries. They account for 80% of health care expenditures in the U.S. A feature of chronic diseases is that they are often asymptomatic in their early stages, which results in late detection and poor health outcomes. From an OR perspective, since disease progression occurs over protracted time periods, it is necessary to make complex sequential decisions under uncertainty. Steimle and Denton [12] provide an overview of research on Markov decision processes for chronic disease treatment. Because of the seriousness of many chronic diseases and the high cost of medical care, especially in the U.S., substantial resources are being invested in personalized health care. The concept of personalizing health care is a potentially valuable one, since biomarkers of an individual patient may suggest which forms of treatment are most likely to be successful, thereby improving treatment and reducing costs.

Some examples of important recent research studies follow.

- For cardiovascular disease, ongoing research is developing stochastic models of disease progression using health records and robust Markov methods for optimization of treatment policies under model ambiguity.
- Research is underway on whether biomarker-based screening strategies increase lifespan and improve quality of life through early detection of prostate cancer, for example through Markov model estimation using longitudinal health records with hidden and missing information, or through the optimization of screening decisions using partially observable Markov decision processes.
- Regarding the use of predictive models for disease diagnosis, research is studying whether observational data from medical records can predict future patient outcomes, for example by using optimization and machine learning to develop predictive models, or using robust optimization methods and approximation algorithms to account for prediction errors.

As populations in developing nations continue to age, the prevalence of chronic diseases is likely to increase. This will make research in the area of personalized health care of great importance for the foreseeable future.

4.3 Project Management

Project Management Institute [11] reports that the annual value of economic activity managed as projects was \$12 trillion, or approximately one-fifth of the economic activity of the world. Moreover, the value annually at risk from a deficiency of well-trained project managers was \$4.5 trillion. More problematically, as much as 30% of the current project management workforce is estimated to become eligible for retirement during the next 10 years. Hence, the forecast availability of 15.7 million new jobs in project management between 2016 and 2020. According to Project Management Institute [11], a survey of senior managers by Economist Intelligence Unit established that 97% of the skills needed for a successful business career fall within the scope of project management. The expansion of professional interest in project management is impressive, with membership in Project Management Institute growing from about 50,000 in 1996, to 300,000 in 2008, to more than 500,000 today. Also impressive is the increase in the number of degree programs in project management in China from one in 2003 to 103 in 2008.

An unusual feature of project management is that innovation from the business side has outpaced academic research over the last 20 years. For example, the development of critical chains [4] occurred in response to observed poor performance of projects. Similarly, agile methodology [1] was developed to address the different needs of a new class of projects with less well determined deliverables. Important applications within this new class include software development, pharmaceutical development, and new product and service development. As a result of research lagging behind practice in this area, many high value research opportunities now present themselves. These include:

- How to design a work breakdown structure to create an efficiently executable project.
- How to manage incentives to mitigate the effect of Parkinson's Law, without the loss of control implied by dropping task deadlines as is recommended within critical chain project management.
- How to optimize learning within a project.
- How to estimate the net present value of a project that is subject to failure.
- For a given project, how to decide whether to use waterfall or agile methodology, or even some hybrid approach.
- How to overcome a lack of scalability for agile methodology.

An overview of current research issues in project management is provided by Hall [6]. Since such a large part of the world's economy is managed using projects, the economic value that is realizable through addressing any of these issues is potentially great. Moreover, further expansion of the range of applications that are managed as projects will generate additional research issues of similar importance.

5 Challenges

This section provides a discussion of some challenges facing INFORMS, its members, and the profession of OR&A. The first challenge is a decline in public funding of research. A recent restructuring of programs at the National Science Foundation, for example, has apparently made it more difficult to obtain grants for some types of research. These grants are not only essential to support research expenditures, but are also an important validating component of promotion and tenure cases, especially in U.S. engineering schools.

Some 15 years ago, more than 70% of INFORMS' revenue came from publications. Although this number has declined to slightly below 50% due to improved meetings and analytics revenues, publications are still a large component of INFORMS' financial viability. Two potential challenges are evident here. The first is reductions in library budgets and increased bundling of publications. The second is open access publication, a trend which is developing slowly but in the long run may become very significant. For authors to be able to afford open access charges for their publications, which are typically around $3,000 per paper, a new funding model for faculty research would be needed.

The competition for publications space has apparently intensified in recent years. Essentially, the number of pages that are available in refereed journals viewed as "tier one" by typical business and engineering schools has not increased with the number of high quality submissions. As a result, acceptance rates have become very low in some of INFORMS' premier publications, especially *Management Science* and *Operations Research*. Nevertheless, universities' expectations about the number and quality of publications required for promotion and tenure are in many cases slow to adjust. This has placed a lot of pressure on faculty, especially those in the early stages of their careers.

Many graduate business programs in the U.S., in particular MBA, have been reporting declining enrollment for several years. Some nationally ranked business schools have reported class size reductions approaching one third over a period of less than five years. Since MBA programs have been a major source of revenue for U.S. business schools over the last 40 years, this leaves a significant revenue gap. This issue has generated a number of innovative responses, including alternative Masters degree programs with a more specialized focus, for example in analytics, finance, and operational excellence. In many cases, this has not yet fully alleviated the revenue shortfall, but on the positive side these innovative new programs should have potential for substantial growth.

Looking at the long run health of academic careers in OR&A, another challenge has emerged over the last decade. The growth of analytics has led to the creation of many technically interesting jobs within industry for which a Ph.D. degree is not required. These positions not only offer generous salaries relative to those available in academia, but also a variety of technical challenges and intellectual development opportunities that previously may have led talented young people into academic careers. The full significance of this increased competition from industry for talent, for example on Ph.D. application numbers and on faculty retention, has yet to be observed.

Acknowledgements This work is supported by the Berry Professorship at the Fisher College of Business, The Ohio State University. Permission to reproduce content from INFORMS' publications is appreciated. Content for the discussion of research on the sharing economy was provided by Prof. Guangwen Kong, University of Minnesota. Content for the discussion of research on personalized health care was provided by Prof. Brian Denton, University of Michigan.

References

1. Agilemanifesto.org: Manifesto for agile software development (2001). http://agilemanifesto.org

2. Davenport, T.H., Harris, J.G.: Competing on Analytics: The New Science of Winning. Harvard Business Review Press, Boston, MA (2017)
3. Federal Trade Commission: The "Sharing" Economy: Issues Facing Platforms, Participants and Regulators. Federal Trade Commission, Washington, D.C. (2016)
4. Goldratt, E.M.: The Critical Chain. North River Press, Great Barrington, MA (1997)
5. Goodwin, T.: The battle is for the customer interface (2015). https://www.linkedin.com/pulse/battle-customer-interface-tom-goodwin
6. Hall, N.G.: Research and teaching opportunities in project management. In: Tutorials in Operations Research, INFORMS, Catonsville, MD, pp. 329–388 (2016). https://doi.org/10.1287/educ.2016.0146
7. Hall, N.G.: INFORMS goes to Washington: policy days and advocacy. President's Desk. OR/MS Today **45**(1), 8–9 (2018)
8. Hall, N.G.: Moving INFORMS' publications forward. President's Desk. OR/MS Today **45**(4), 8 (2018)
9. Hu, M.: Sharing economy: making supply meet demand. In: Tang, C. (ed.) Springer Series in Supply Chain Management (2018)
10. IBM: Smarter analytics: businesses use analytics to find hidden opportunities (2012). https://www.youtube.com/watch?reload=9&v=TBjNzJEWpCI
11. Project Management Institute: Should You Be Teaching Project Management?. PMI Publications, Newton Square, PA (2008)
12. Steimle, L.N., Denton, B.T.: Markov decision processes for screening and treatment of chronic diseases. In: van Dijk, N.M. (ed.) Markov Decision Processes in Practice, pp. 189–222. Springer International Publishing (2017)
13. Tucker, K.: Government & analytics summit. OR/MS Today **45**(3), 14–16 (2018)

On the Limits of Computation in Non-convex Optimization

Panos M. Pardalos

1 Extended Abstract

Large-scale problems in engineering, the design of networks and energy systems, biomedicine, and finance are modeled as optimization problems. Humans and nature are constantly optimizing to minimize costs or maximize profits, to maximize the flow in a network, or to minimize the probability of a blackout in a smart grid.

Due to new algorithmic developments and the computational power of machines (digital, analog, biochemical, quantum computers, etc. …), optimization algorithms have been used to "solve" problems in a wide spectrum of applications in science and engineering.

But what do we mean by "solving" an optimization problem? What are the limits of what machines (and humans) can compute? In the first part of the talk we are going to address some fundamental questions about the limits of computation. In particular we will discuss these questions:

What are the limits of what humans can compute?

What are the limits of what machines can compute?

Are these limits the same?

What are the physical foundations and limitations of computation?

After a brief discussion of different computational machines such as analog computers, DNA computers, and quantum computers, I will present a short history of optimization. In the second part of the talks we will address the question: "What is Global Optimization" and we will focus on these questions:

How can we find a globally optimal solution (and how we can provide a certificate of optimality?)

P. M. Pardalos (✉)
Center for Applied Optimization, University of Florida, Gainesville, Florida, USA
e-mail: p.m.pardalos@gmail.com
URL: http://www.ise.ufl.edu/pardalos

© Springer Nature Switzerland AG 2018

P. Daniele and L. Scrimali (eds.), *New Trends in Emerging Complex Real Life Problems*, AIRO Springer Series 1,
https://doi.org/10.1007/978-3-030-00473-6_2

How do we compute "good" locally optimal solutions? (or points that satisfy the optimality conditions?)

Do we compute "better" solutions than "known" solutions?

How do we address in-feasibility?

Next I will summarize results on the complexity of the following problems:

Checking existence of a point satisfying the optimality conditions.

Complexity of local optimization and the fundamental question of how hard is to check the convexity of a function.

In addition I will discuss phase transition problems and will discuss issues of evaluation heuristic algorithms for hard optimization problems.

Operations Research in Transportation and Supply Chain Management

M. Grazia Speranza

Abstract The developments in digital technologies are creating new challenges and opportunities to Operations Research. In this paper, research trends in transportation and supply chain management will be discussed and some examples briefly presented.

Keywords Digital technologies · Trends · Research opportunities Transportation · Supply chain management · Logistics

1 Introduction

Information and communication technologies (ICT) are having a continuously increasing impact on our daily life and on all economic and social activities. With ICT we generally mean the infrastructure and components that enable modern computing. Although there is no single, universal definition of ICT, the term is generally accepted to mean all devices, networking components, applications and systems that combined allow people and organizations (i.e., businesses, nonprofit agencies, governments) to interact in the digital world.

Computers have been part of our daily life since the early eighties and the digitalization processes have proceeded at a regular pace over time since then. However, recently new phenomena, enabled by developments in digital technologies, are taking place and changing the economy and the society. Internet and all the technological devices that can receive and transmit data and information, including the mobile cellular phones, have enabled the design and introduction of new products and services, and new management styles. Companies with a long successful history

M. G. Speranza (✉)
University of Brescia, Brescia, Italy
e-mail: grazia.speranza@unibs.it

© Springer Nature Switzerland AG 2018

P. Daniele and L. Scrimali (eds.), *New Trends in Emerging Complex Real Life Problems*, AIRO Springer Series 1,
https://doi.org/10.1007/978-3-030-00473-6_3

have disappeared, replaced by new companies because businesses have completely changed. The behaviour of citizens, in the daily activities, keeps changing.

Expressions like *Internet of things*, *big data*, *machine learning*, *sharing economy* may not be clearly defined but indicate global phenomena. The fact that such expressions are regularly used in the media and by everyone is a sign of the high impact they are having. The Internet of things is the interconnection via the Internet of computing devices embedded in everyday objects, enabling them to send and receive data. Big data refers to data sets that are so big and complex that traditional data-processing application software are inadequate to deal with them. Machine learning is a subset of artificial intelligence in the field of computer science that often uses statistical and optimization techniques to give computers the ability to 'learn' (i.e., progressively improve performance on a specific task) with data. Sharing economy is an umbrella term with a range of meanings, often used to describe economic activities involving online transactions.

Transportation and logistics are two fields where Operations Research (OR) has substantially contributed in the last 50 years, especially improving the efficiency of operations but also providing support to decision making in strategic and tactical phases of the decision processes. The meaning of 'logistics' is slightly different from that of 'supply chain management'. Logistics refers to the planning, execution, and control of the procurement, movement, and stationing of personnel, material, and other resources to achieve the objectives of a plan, project, or strategy. Logistics may be defined as the management of inventory in motion and at rest. Supply chain management (SCM) is the broad range of activities required to plan, control and execute a product flow, from acquiring raw materials and production through distribution to the final customer. While logistics does not usually include the production phases that transform raw materials into a final product, supply chain management does not include the logistic operations in different areas such as the organization of events (for example, conferences or concerts). In this paper, we will refer to contributions to supply chain management but part of the discussion can be extended to logistics.

ICT is changing the processes in supply chain management and the way goods and people are transported, and this in turn is changing the problems where OR can contribute. May be it is also, at least partially, changing the domain of OR, pushing its boundaries.

2 Supply Chain Management

Supply chain management aims at integrating the various components of a supply chain and has been a research topic for OR for already some decades. Four major directions for research can be derived from the technological changes: a systemic, a collaborative, a dynamic and a data-driven direction. In the following we briefly overview the research opportunities associated with each of these directions.

2.1 Systemic Direction

Better solutions to problems can be identified when parts of a supply chain are jointly modeled and optimized. The digital technologies have enabled the implementation of integrated management styles. Research efforts in this direction have already been made recently. For example, in the area of vehicle routing, several papers have studied combined problems previously studied separately. Integrated vehicle routing problems (VRPs) is the expression increasingly used to denote the class of problems where routing decisions are taken jointly with other decisions (as outlined by the special issue edited by Bektaş et al. [9]). Location-routing problems jointly optimize location and routing. Inventory-routing problems combine routing and inventory management. Production-routing problems integrate production, routing, and inventory decisions. Multi-echelon routing problems optimize the routes of vehicles in distribution systems comprising two or more echelons. Routing problems with loading constraints simultaneously optimize the routing of vehicles and the loading of goods.

Integrated VRPs combine problems that are usually NP-hard when treated individually. However, solving the individual problems sequentially, even by means of exact methods, leads to a sub-optimal solution for the integrated problem, even if solved with a heuristic. The cost reduction achieved through the integration may be an order of magnitude greater than the cost reduction usually obtained by an exact method with respect to a heuristic. One of the first papers that showed the benefits of integrated decisions is due to Chandra and Fisher [11]. More recently, Archetti and Speranza [6] compared the heuristic solution of an inventory-routing problem with the solution obtained by sequentially and optimally solving the inventory management and the routing problems. The sequential solution models a traditional management style where customers of a supply chain control their inventory and decide order times and quantities. Then, the supplier organizes the distribution that, however, has to take order times and quantities as constraints. The inventory-routing problem models a more recent integrated policy, called Vendor Managed Inventory (VMI), where the supplier is responsible for the distribution and for the inventory at its customers. In [6] the results of computational tests show that solutions of the inventory-routing problem allow average savings of 10%. The integration of production and inventory-routing is studied in [1]. The value of integrating loading and routing is analyzed in [13].

Integrated optimization problems model integrated management styles adopted in supply chain management, contribute to exploit the advantages of the integration and can quantify the benefits. The research opportunities in this direction are endless.

2.2 Collaborative Direction

Different forms of collaboration in supply chain management have been discussed, internal and external, horizontal and vertical. When dealing with road transportation, horizontal cooperation among carriers can be further classified according to the operational collaboration mode in order sharing and capacity sharing (see, for instance, the recent survey [20]).

Obviously, partners of a collaboration initiative aim at improving the performance of their own business. Each partner will be focused on its own business rather than on global performance. The collaboration initiative may bring benefits to some of the partners only or the benefits may be distributed in a way that is unacceptable to some of the partners. Thus, integration must be mediated with individual interests to make the collaboration initiative successful. This concept makes models for decision support in collaboration initiatives often different from models for global optimization.

Several statistics show that approximately 90% of freight travels on road. In European countries between 15 and 30% of trucks travel empty and contribute to traffic, pollution, accidents. The average load of a truck is much lower than its capacity, especially in city distribution. As a consequence, the number of trucks on road is much higher than it could be. Collaboration among carriers may improve the statistics and generate economic benefits for the carriers involved, a sector which lives on very low margins, as well as social and environmental benefits. A pilot European project, called 'Collaboration Concepts for Co-modality', has been implemented by two companies with production sites in Belgium and with large quantities of goods delivered to Greece. One of the two companies transports light large-volume goods, the other heavy low-volume goods. The project has achieved a reduction of 150,000 km travelled, and a reduction of 17% of the routing costs.

As an example, in Fernández et al. [14] a collaboration scheme is adopted by a consortium of carriers. Each carrier can decide which of its customers to serve and which ones to share with the other carriers. A shared customer can be served by any of the carriers. A carrier will tend to share customers not conveniently located and/or with low demand. The revenue from a shared customer is partly collected by the 'owner' of the customer and partly by the one actually serving the customer. The optimization model proposed assigns the shared customers to carriers and builds the routes for all the carriers to serve all the customers in such a way that the total cost is minimized and each carrier has a profit at least as high as the profit it would have without collaboration. The computational results show that the profit increase in the collaborative setting strongly depends on the location and demand of customers, and ranges from small values to up to 85%.

The research opportunities include the modelling and the solution of collaborative schemes. While imposing that the individual participants in a collaborative initiative gain with respect to a non collaborative scheme is a necessary condition for the individuals to join the initiative, it may not be a sufficient condition, as the profit may end up being distributed in an unfair way among the participants.

2.3 Dynamic Direction

The flow of data about customers, purchases, deliveries, locations, inventories, combined with the frequency of the update of the information, gives rise to a number of new dynamic problems. The phenomenon of e-commerce is in constant growth and is challenging all commercial companies and their logistics partners. The last-mile delivery problem remains a challenging problem to solve and logistic companies keep exploring new ways to offer a service to their customers that is up to their expectations. Additional transportation needs are set by the flow of on-line ordered goods that are returned, and in general by the reverse logistics.

Classical optimization models are based upon the assumption that information is available, that a model is solved and the solution implemented. This is a less and less realistic assumption. While until few years ago a delivery time of 5–10 days was acceptable to customers, today on-line customers wish and expect to receive the goods within the same day of the order or the day after. Moreover, they wish to choose the day, the time, the delivery point. The increased frequency and the reduction in quantities of the deliveries to shops and retailers make the urban logistics even more complex. The urban city logistics keep changing because new and disruptive ways to tackle the problem are being experienced by the logistic companies, sometimes under constraints or incentives set by the public authorities.

Dynamic problems in transportation have been discussed for a long time (for example, by Psaraftis [18]), but research on dynamic and stochastic VRPs received increasing interest only in the last decade (see the recent survey by Ritzinger et al. [19] where the importance of appropriately modeling dynamic events and simultaneously incorporating information about the uncertainty of future events is outlined). Research opportunities related to the dynamism of supply chains are discussed in [21].

Another challenge to the OR community implied by the dynamism of processes is related to the computational times needed to find a solution to a problem. Also, long computational times may not be justified when the solution may be only partially implemented and the data is likely to change shortly after.

2.4 Data-Driven Direction

The quantity of data available is huge because of the devices that can collect and transmit data and because of the capacity of the systems that can store the data. Data creates enormous learning opportunities in all sectors. The amount of research in machine learning (see, for example, [3]) has been growing over the last years and various machine learning techniques have been successfully applied to a number of real life situations. Machine learning is related to mathematical optimization because several learning problems are modelled as optimization problems where the cost function, to be minimized, measures the discrepancy between the forecasts of the model and the real observations over time. A different relation with mathematical optimization consists in using the machine learning techniques in optimization problems. Presentations on the use of machine learning techniques in transportation and supply chain management have started to appear in conferences but this line of research is still in its infancy.

3 People Transportation

OR has contributed to the solution of many problems with big impact in people transportation, especially problems related to modes, air, railways, buses, for example, timetabling and crew scheduling. The way people move is, however, changing dramatically. Cars are becoming connected and, especially in the United States, huge research resources are invested on the development of autonomous vehicles. People expect to have flexible transportation services, with different levels of cost and comfort. Bike and car sharing are services already offered in most towns. In this field, we hear more and more the expression *transportation as a service*. Someone predicts that private cars will be owned in the future for the pleasure of driving and not for the need of people to be transported.

Public authorities need to study solutions that protect the well-being of citizens, challenged by high-frequency deliveries of small quantities of goods and cars with one passenger only. Space for the parking of cars and for the loading/unloading operations of vehicles is a scarce resource. The evolution towards electric vehicles will reduce pollution in urban areas but will not reduce the number of traveling vehicles, the need for space and the level of congestion.

From the OR perspective, the only optimization models OR contributed to private vehicles have been those aimed to optimize the shortest path from origin and destination, embedded nowadays in the several sat-nav systems available. The evolution of the way people are transported is opening new research directions to OR.

3.1 On-demand Services

The number of traveling vehicles can be reduced only by reducing the number of people in need of travel and/or by increasing the number of people transported in the same vehicle. While OR can hardly contribute to the former option, contribution of OR may be relevant in supporting the latter.

The main reason that leads people to use their own vehicle is the lack of flexibility of mass transit systems. Such systems work on fixed itineraries and fixed schedules. Nowadays, for most potential customers the frequency is too low and the travel time is too high, with changes of transportation mean and waiting times. Moreover, those systems do not provide a transportation service from origin to destination. Taxis, on the other hand, are too expensive for regular use for most travelers.

Demand Responsive Transit (DRT) systems (also called dial-a-ride systems) are flexible services that provide 'door-to-door' transportation. DRT systems have been mainly implemented as services for small groups of people (see [12] for a survey).

Martínez et al. [17] have suggested a classification of DRT systems:

- with fixed itineraries and stops, with pre-booking;
- with fixed itineraries and stops with possible detours;
- with unspecified itineraries and predefined stops;
- with unspecified itineraries and unspecified stops.

An implementation of a DRT system in Maryland is presented by Marković et al. [16] and the benefits, with 450 trip requests daily, of a computerized routing and scheduling system are estimated with annual savings of $0.82 million, or about 18% of the total annual expense, with respect to manual operations.

In [7] a simulation study is performed where a conventional mass transit system, say buses, is offered together with an on-demand service without fixed itineraries and schedules, provided through minibuses. A minibus, if acceptable to the user in terms of arrival time to destination, will pickup the user at the origin of the trip and deliver him/her to the destination. In case neither the conventional bus nor the on-demand minibus provide an acceptable service to the user, he/she will use a private car. A recent paper [8] introduced a Flexible Mobility On Demand (FMOD) system that offers different services, taxis, shared taxis and minibuses, where the minibus service works as a regular bus service with fixed schedules.

The so called dynamic ride-share systems share with the DRT systems the goal of increasing the number of people sharing the same vehicle. Such systems aim to bring together travelers with similar itineraries and time schedules on short-notice. Optimization methods that match drivers and riders in real-time are necessary for a successful implementation of such systems (see [2] for a review of dynamic ride-sharing systems).

The need for parking space can be also reduced through car sharing systems, where a car is pre-booked, used and returned to a parking station. One-way, with respect to two-way, systems provide more flexibility to users since cars can be dropped-off at any station (see, for instance, [10]). Research opportunities include the location of stations and cars, car relocation problems, coordination of reservations.

3.2 New Shortest Path problems

As previously mentioned, sat-nav systems nowadays embed shortest path algorithms. Real-time information on traffic conditions imply the need of different algorithms that consider time-dependent uncertain travel times (see [15] for a recent survey).

Moreover, coordination opportunities among the routes of vehicles exist because of their capability of receiving and transmitting information. In [4] a model is proposed for the problem of coordinating the routes of vehicles in such a way that congestion is eliminated or reduced to a minimum level by keeping the inconvenience, measured as the increase of the traveling time, with respect to the minimum possible, for cars below a given threshold.

3.3 *People and Freight*

Collaboration opportunities are being explored by companies in all directions, including collaboration of people to transport goods with the goal of reducing the costs of a distribution process that is extremely expensive. The line between freight and passenger transportation, which was clearly defined until some time ago, is not any longer. An example is related to regular customers of shops who may become available, for a small compensation, to deliver goods to on-line customers on their way home. These customers are called occasional drivers (see, for example, [5]) and can contribute to make the delivery of goods to on-line customers more efficient.

4 Conclusions

New research opportunities keep arising because of new and disruptive options for transporting goods and people, enabled by the digital technologies. Lack of data has been an obstacle to the implementation of OR models and algorithms. Nowadays, data is huge in volume, variety, velocity and OR can exploit its potential, provided it will be capable of adapting to the new research challenges.

References

1. Absi, N., Archetti, C., Dauzère-Pérès, S., Feillet, D., Speranza, M.G.: Comparing sequential and integrated approaches for the production routing problem. Eur. J. Oper. Res. (to appear). https://doi.org/10.1016/j.ejor.2018.01.052
2. Agatz, N., Erera, A., Savelsbergh, M., Wang, X.: Optimization for dynamic ride-sharing: a review. Eur. J. Oper. Res. **223**, 295–303 (2012)
3. Alpaydin, E.: Introduction to Machine Learning. The MIT Press (2014)
4. Angelelli, E., Arsik, I., Morandi, V., Savelsbergh, M., Speranza, M.G.: Proactive route guidance to avoid congestion. Transp. Res. B **94**, 1–21 (2016)
5. Archetti, C., Savelsbergh, M., Speranza, M.G.: The vehicle routing problem with occasional rrivers. Eur. J. Oper. Res. **254**, 472–480 (2016)
6. Archetti, C., Speranza, M.G.: The inventory routing problem: the value of integration. Int. Trans. Oper. Res **23**, 393–407 (2016)
7. Archetti, C., Speranza, M.G., Weyland, D.: On-demand public transportation. Int. Trans. Oper. Res. (2018) (to appear)
8. Atasoy, B., Ikeda, T., Song, X., Ben-Akiva, M.E.: The concept and impact analysis of a flexible mobility on demand system. Transp. Res. Part C: Emerg. Technol. **56**, 373–392 (2015)

9. Bektaş, T., Laporte, G., Vigo, D.: Integrated vehicle routing problems. Comput. Oper. Res. **55**, 126 (2015)
10. Bruglieri, M., Colorni, A., Lu, A.: The relocation problem for the one-way electric vehicle sharing. Networks **64**, 292–305 (2014)
11. Chandra, P., Fisher, M.L.: Coordination of production and distribution planning. Eur. J. Oper. Res. **72**, 503–517 (1994)
12. Cordeau, J.-F., Laporte, G.: The dial-a-ride problem: models and algorithms. Ann. Oper. Res. **153**, 29–46 (2007)
13. Côté, J.-F., Guastaroba, G., Speranza, M.G.: The value of integrating loading and routing. Eur. J. Oper. Res. **257**, 89–105 (2017)
14. Fernández, E., Fontana, D., Speranza, M.G.: On the collaboration uncapacitated arc routing problem. Comput. Oper. Res. **67**, 120–131 (2016)
15. Gendreau, M., Ghiani, G., Guerriero, E.: Time-dependent routing problems: a review. Comput. Oper. Res. **64**, 189–197 (2015)
16. Marković, N., Nair, R., Schonfeld, P., Miller-Hooks, E., Mohebbi, M.: Optimizing dial-a-ride services in Maryland: benefits of computerized routing and scheduling. Transp. Res. Part C: Emerg. Technol. **55**, 156–165 (2015)
17. Martínez, L.M., Viegas, J.M., Eiró, T.: Formulating a new express minibus service design problem as a clustering problem. Transp. Sci. **49**, 85–98 (2014)
18. Psaraftis, H.N.: Dynamic vehicle routing: status and prospects. Ann. Oper. Res. **61**, 143–164 (1995)
19. Ritzinger, U., Puchinger, J., Hartl, R.F.: A survey on dynamic and stochastic vehicle routing problems. Int. J. Prod. Res. **54**, 215–231 (2016)
20. Verdonck, L., Caris, A., Ramaekers, K., Janssens, G.K.: Collaborative logistics from the perspective of road transportation companies. Transp. Rev. **33**, 700–719 (2013)
21. Waller, M.A., Fawcett, S.E.: Data science, predictive analytics, and big data: a revolution that will transform supply chain design and management. J. Bus. Logist. **34**, 77–84 (2013)

Energy Optimization of a Speed-Scalable and Multi-states Single Machine Scheduling Problem

MohammadMohsen Aghelinejad, Yassine Ouazene and Alice Yalaoui

Abstract This study deals with the single-machine scheduling problem to minimize the total energy consumption costs. The considered machine has three main states (OFF, ON, Idle), and the transitions between states OFF and ON are also considered (Turn-on and Turn-off). Each of these states as well as the processing jobs consume different amount of energy. Moreover, a speed scalable machine is addressed in this paper. So, when the machine performs a job faster, it consumes more units of energy than with a slower speed. In this study, two new mathematical formulations are proposed to model this problem, and their efficiency are investigated based on several numerical experiments.

Keywords Energy efficiency · Single machine scheduling · Time-dependent energy cost · Speed-scalable multi-states system

1 Introduction

In the last few years, the increase of the electricity price, which is one of the main type of energy used in manufacturing industry, attracted the attention of many researchers all around the world. They studied minimization of manufacturing system's energy consumption or total energy cost to reduce the total production costs and environmental impacts simultaneously. A comprehensive review of previous researches

M. Aghelinejad (✉) · Y. Ouazene · A. Yalaoui
Industrial Systems Optimization Laboratory (ICD, UMR 6281, CNRS),
Université de Technologie de Troyes, Troye, France
e-mail: mohsen.aghelinejad@utt.fr

Y. Ouazene
e-mail: yassine.ouazene@utt.fr

A. Yalaoui
e-mail: alice.yalaoui@utt.fr

© Springer Nature Switzerland AG 2018

P. Daniele and L. Scrimali (eds.), *New Trends in Emerging Complex Real Life Problems*, AIRO Springer Series 1,
https://doi.org/10.1007/978-3-030-00473-6_4

shows that, the minimization of energy consumption in a manufacturing system can be applied in different sectors such machine-level, product-level, and system-level. Unlike the two other levels that need enormous financial investments, in the system-level, manufacturers may reduce the system's energy consumption by using the existing decision models and optimization techniques in production planning and scheduling. Based on the literature review analysis, the studies which deal with the energy efficiency of a manufacturing system generally consider decreasing energy consumption value or energy consumption cost (operational cost) as the main objective. In the following, at first, the papers which deal with energy consumption are analyzed and then, a summary of papers on the minimization of total energy costs are presented.

In practical cases, the energy consumption of a production system is composed of the amount of energy consumed during the non-processing states (NPE) (start-up, transition between different states, shut down and idle states), and the amount of energy consumed during the processing state (PE). So, the energy consumption of the machine is state-dependent. Moreover, several characteristics may change the energy consumption of any machine during the processing (ON) states (PE) such as: the type of the machine (machine-dependent), the type of the processing job (job-dependent) and the processing speed of the machine (speed-dependent). Investigating the NPE consumption, and using a scheduling method to change the processing job's order and machine's state during a production shift is one of the easiest and most popular ways.

The complexity of the non-preemptive classical deadline-based scheduling problems under a variable processing speed to minimize the total energy consumptions is analyzed in [4]. The minimization of energy consumption and total completion time of a single machine is addressed in [11] using a multi-objective genetic algorithm and dominance rules. A framework is presented in [8] to minimize total energy consumption and total tardiness simultaneously when idle and setup energy are considered.

In practice, electricity suppliers in different countries propose variable pricing to balance the electricity supply and demand to improve the reliability and efficiency of electrical power grids. The most common categories of time-varying rates which are used previously to investigate the total energy cost of a system are Time-Of-Use (TOU), Critical Peak Pricing (CPP) and Real Time Pricing (RTP). The variation of electricity prices during the time period may impact the total energy costs of any production system. In some papers, the time-dependent energy cost is assumed to compute the total energy consumption cost. While, in the others, the other characteristics which may change the energy consumption of the machine are also considered. For example, an energy-conscious single machine scheduling problem when each processing job has its power consumption, and electricity prices may vary from hour to hour throughout a day is assumed in [5]. Several single machine scheduling problems with arbitrary power demands for the jobs and uniform or variable speeds for processing, are studied in [6] to minimize total electricity cost under a time of use electricity tariffs. They also investigated the complexity of these two problems in preemptive and non-preemptive cases. A preemptive scheduling problem with energy constraint in each period, different energy consumption for each job, and the electricity time-varying prices is investigated in [7] to minimize the total electricity consumption costs and operation's postponement penalty costs.

A mathematical model is proposed in [10], to minimize total energy consumption costs when variable energy prices, different possible states and energy consumption are considered for the machine. Aghelinejad et al. [1] studied the same problem as [10] to improve the previous mathematical model. They also presented a new mathematical model to obtain the optimal schedule for the machine state and job's sequence simultaneously. Then, a new heuristic algorithm and a genetic algorithm are proposed in [2] to solve this problem. The complexity of the preemptive version of this problem, using a dynamic programming approach, is analyzed in [3].

Based on the literature review analysis, some papers addressed the energy efficiency of a multi-states single machine system and some other papers addressed the energy efficiency of a multi-speeds single machine system. So, to the best of our knowledge, there is no previous work which studies the energy efficiency of a multi-states and multi-speeds single machine system with the time-dependent electricity cost. This paper aims to fill this gap within the literature.

The remainder of this work is organized as follows. Section 2 introduces the considered problem with different notations and assumptions. Section 3 describes two new mathematical models which are presented for this problem. Moreover, Sect. 4 represents the numerical experiments' results to evaluate the efficiency of the proposed models. Finally, Sect. 5 draws the conclusions of this paper, as well as the future directions for this study.

2 Problem Definition

This paper deals with the energy efficient scheduling of n jobs on a single machine system during T periods. The TOU electricity pricing is considered for each period. The studied machine has 3 main states (ON, OFF and Idle), and the possible transitions between OFF and ON states (Ton and Toff) are considered (Fig. 1). The machine consumes a different amount of energy (e_s) in each state ($s \in \{ON, OFF, Idle, Ton, Toff\}$), and must be in the same state for a specific number of periods (d_s). Moreover, the energy consumption of the machine during state ON is depends on the processing job. Since a speed scalable machine is addressed in this work, there are different possibilities for processing each job. That means for each job $j = 1, \ldots, n$ with v_j possible speeds, there are different values for the processing time as $P_j = \{p_{j,1}, \ldots, p_{j,v_j}\}$, and for each $p_{j,i}$, a corresponding energy consumption $q_{j,i}$ is associated. $Q_j = \{q_{j,1}, \ldots, q_{j,v_j}\}$ is the set of the different energy consumptions of job $j = 1, \ldots, n$. Without loss of generality, the following relations are assumed:

$$p_{j,1} > p_{j,2} > \cdots > p_{j,v_j}; \qquad \forall j \in \{1, \ldots, n\} \tag{1}$$

$$q_{j,1} < q_{j,2} < \cdots < q_{j,v_j}; \qquad \forall j \in \{1, \ldots, n\} \tag{2}$$

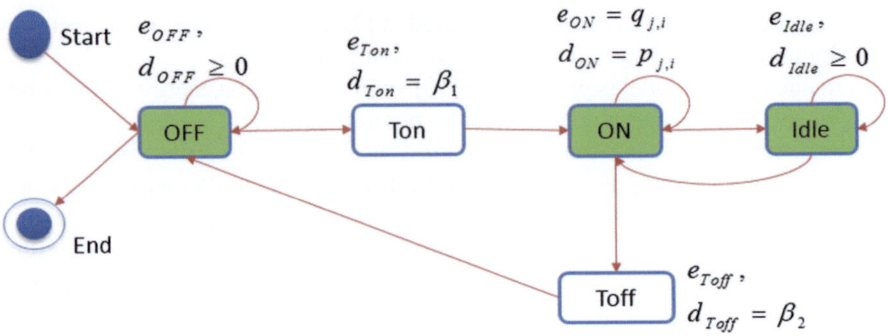

Fig. 1 Machine states and possible transitions

The machine must be in OFF state during the initial and final periods. The machine's energy consumption in state OFF is equal to 0 ($e_{OFF} = 0$). When it is decided to turn on the machine, this transition takes β_1 periods and it consumes e_{Ton} units of energy. Then the machine is in ON state and is ready to process a job $j = 1, \ldots, n$ with speed $i = 1, \ldots, v_j$, that takes $p_{j,i}$ periods and consumes $q_{j,i}$ units of energy per period. Once the job is processed completely, there are three possibilities for the machine: it may stay in ON state and process another job; it can go to Idle state for one period or more; the machine can go to OFF state. Regarding to the energy consumption and the unit of energy price, any of these possibilities may be selected. Note that, in this study the transition time between Idle and ON state, and its energy consumption are neglected. In addition, the transition between Idle and OFF states is not allowed. Therefore, when the machine is in Idle state, for the next period, it may stay in Idle state or go to ON state (Fig. 1).

The objective of this study is to find the most economical production scheduling for the jobs and the machine's states in terms of energy consumption costs during the horizon time. In the next section, two new mathematical models are proposed for this problem.

3 Mathematical Formulations

3.1 First Model

To describe this programming model, first of all, the parameters and decision variables are defined and then, the mathematical model including objective functions and constraints is presented.

Parameters:

n: Number of jobs
T: Total number of periods
c_t: Unit of energy price in period t
v_j : Number of possible processing speed for job j
$p_{j,i}$: Processing time of job j with speed $i = \{1, \ldots, v_j\}$ (in number of periods)
$q_{j,i}$: Energy consumption of job j per period which is associated with $p_{j,i}$
s: States of the machine (s = $\{1, 2, 3\}$ for ON, OFF and idle states, respectively)
E_s: Energy consumption of the machine during state s
$E_{ss'}$: Energy consumption of the machine in transiting between s and s'
$d_{ss'}$: Required number of periods for switching from state s to s' ($s \neq s'$).

Decision variables:

$\alpha_{s,t}$ = 1 if machine is in state s during period t, 0 otherwise.
$\beta_{ss',t}$ = 1 if machine is in transition from state s to s' in period t, 0 otherwise.
$x_{j,i}$ = 1 if job j is processed with speed $i = \{1, \ldots, v_j\}$, 0 otherwise.
$y_{j,i,t}$ = 1 if job j is processed with speed i in period t, 0 otherwise.

Mathematical model:

$$Min \sum_{t=0}^{T} c_t \left(\sum_{j=1}^{n} \sum_{i=1}^{v_j} q_{j,i} \cdot y_{j,i,t} + \sum_{s=2}^{3} E_s \cdot \alpha_{s,t} + \sum_{s=1}^{3} \sum_{s'=1}^{3} E_{ss'} \cdot \beta_{ss',t} \right) \tag{3}$$

$$\sum_{s=1}^{3} \alpha_{s,t} + \sum_{s=1}^{3} \sum_{s'=1}^{3} \beta_{ss',t} = 1; \qquad \forall\, t \in \{0, \ldots, T\} \tag{4}$$

$$\alpha_{s,t} \leq \sum_{s'=1|d_{ss'}=0}^{3} \alpha_{s',t+1} + \sum_{s=1|d_{ss''}\geq 1}^{3} \sum_{s''=1}^{3} \beta_{ss'',t+1}; \qquad \forall\, t \in \{0, \ldots, T-1\};\ \forall\, s \in \{1, 2, 3\} \tag{5}$$

$$\beta_{ss',t} \leq \beta_{ss',t+1} + \alpha_{s',t+1}; \qquad \forall\, t \in \{0, \ldots, T-1\};\ \forall\, s, s' \in \{1, 2, 3\}|\ d_{ss'} \geq 1 \tag{6}$$

$$\sum_{t'=t+1}^{t+d_{ss'}} \beta_{ss',t'} \geq (\alpha_{s,t} + \beta_{ss',t+1} - 1) \cdot d_{ss'}; \qquad \forall\, t \in \{0, \ldots, T-1\};\ \forall\, s, s' \in \{1, 2, 3\}|\ d_{ss'} \geq 1 \tag{7}$$

$$\beta_{ss',t} + \beta_{ss',t+d_{ss'}} \leq 1; \qquad \forall\, t \in \{0, \ldots, T - t_{ss'}\};\ \forall\, s, s' \in \{1, 2, 3\}|\ d_{ss'} \geq 1 \tag{8}$$

$$\sum_{j=1}^{n} \sum_{i=1}^{v_j} y_{j,i,t} = \alpha_{1,t}; \qquad \forall\, t \in \{1, \ldots, T\} \tag{9}$$

$$\sum_{j=1}^{n}\sum_{i=1}^{v_j} y_{j,i,t} \leq 1; \qquad \forall\, t \in \{0, \ldots, T\} \tag{10}$$

$$\sum_{t'=0}^{t-p_{j,i}} y_{j,i,t'} + \sum_{t'=t+p_{j,i}}^{T} y_{j,i,t'} \leq p_{j,i} \cdot (1 - y_{j,i,t});$$
$$\forall\, t \in \{p_{j,i}, \ldots, T - p_{j,i} - 1\};\ \forall\, j \in \{1, \ldots, n\};\ \forall\, i \in \{1, \ldots, v_j\} \tag{11}$$

$$\sum_{i=1}^{v_j} x_{j,i} = 1; \qquad \forall\, j \in \{1, \ldots, n\} \tag{12}$$

$$\sum_{t=0}^{T} y_{j,i,t} \geq p_{j,i} \cdot x_{j,i}; \qquad \forall j \in \{1, \ldots, n\};\ \forall\, i \in \{1, \ldots, v_j\} \tag{13}$$

$$\alpha_{2,t} = 1; \qquad \forall\, t \in \{0, T\} \tag{14}$$

In this model, the objective value depends on the machine states, the processing job, the processing speed (energy consumption of the machine), and the unit of electricity price in each period (Eq. (3)). Equation (4) expresses that in each period the machine must be in one of the possible states (ON, OFF, Idle, Ton, and Toff). Equations (5) and (6) limit the machine's state in each period regarding to the machine's state in previous period. Equations (7) and (8) identify lower and upper number of periods that takes for Ton and Toff states. Equation (9) indicates that the machine may process at most one job per period, and if the machine processes job j during period t, it must be in ON state ($s = 1$). Equation (10) imposes the constraint that the machine can process at most one job per period. Equation (11) demonstrates the non-preemption constraints of the jobs. Equation (12) imposes the constraint that each job must be processed with only one speed. Equation (13) specifies the processing time of each job, regarding to its processing speed. Equation (14) identifies that the machine is in OFF state during the initial and final periods.

3.2 Second Model

In the first formulation, two decision variables ($y_{j,i,t}$ and $x_{j,i}$) and three indexes (j, i and t) are used to model the problem. In this section, it is attempted to propose a more efficient model to formulate the problem by using just one decision variable and two indexes. This formulation is inspired by the approach proposed by [9].

Let define M jobs such that $M = \sum_{j=1}^{n} v_j$, and $J = \{J_1, J_2, \ldots, J_n\}$ the set of all the job such that $J_1 = \{1, 2, \ldots, v_1\}$, $J_j = \{v_{j-1} + 1, \ldots, v_{j-1} + v_j\}$; $\forall j = 2, \ldots, n$. Let also consider the set $P = \{p_1, \ldots, p_M\}$ of the processing time of each job

$k \in J$ and the set $Q = \{q_1, \ldots, q_M\}$ of the energy consumptions of the jobs $k \in J$ such that q_k is the unit of energy consumption corresponding to the processing time p_k. $R_{k,t} = \sum_{\tau=t}^{t+p_k-1} c_\tau$ is also defined to compute the sum of energy unit cost if the machine performs job $k = 1, \ldots, M$ from period t to period $t + p_k - 1$. Then, $q_k * R_{k,t}$ represents the total energy consumption for performing job k from period t. Moreover, $x_{k,t}$ is used as a decision variable to formulate the problem:
$x_{k,t} = \begin{cases} 1 \text{ ; If job} k \in \{1, \ldots, M\} \text{begins to be processed at period} \, t \\ 0 \text{ ; Otherwise} \end{cases}$
So, in this model, the objective function can be written as follow:

$$Min \sum_{t=0}^{T} \sum_{k=1}^{M} q_k \cdot x_{k,t} \cdot R_{k,t} + \sum_{t=0}^{T} c_t \cdot \left(\sum_{s=2}^{3} E_s \cdot \alpha_{s,t} + \sum_{s=1}^{3} \sum_{s'=1}^{3} E_{ss'} \cdot \beta_{ss',t} \right) \quad (15)$$

Also, the constraints which consist of $y_{j,i,t}$ and $x_{j,i}$ (Eqs. (9)–(13)) must be changed. Therefore, Eq. (16) is replaced (9), (11) and (13). Moreover, equations (17) and (18) respectively replace equations (10) and (12). The new equations are given in the following:

$$\sum_{s=2}^{3} \alpha_{s,t} + \sum_{s=1}^{3} \sum_{s'=1}^{3} \beta_{ss',t} = 1 - \sum_{k=1}^{M} \sum_{t'=t-p_k+1}^{t} x_{k,t'}; \qquad \forall\, t \in \{0, \ldots, T\} \quad (16)$$

$$\sum_{k=1}^{M} x_{k,t} \leq 1; \qquad \forall\, t \in \{0, \ldots, T\} \quad (17)$$

$$\sum_{k \in J_j} \sum_{t=0}^{T} x_{k,t} = 1; \qquad \forall\, j \in \{1, \ldots, n\} \quad (18)$$

It must be mention that, in the second model all the related constraints for defining the machine's state in each period are kept the same as in the first model (Eqs. (4)–(8)).

Note that, the second model integrates a pre-treatment phase that computes the value of parameter $R_{k,t}$ for any k and t.

4 Numerical Experiments

The performance of these two mathematical models have been examined by several randomly generated instances. For this purpose, CPLEX 12.6.1 software is used to solve instances with the exact method (Branch and Cut). Five different examples are randomly generated for each instance by changing the processing times and the energy consumptions of the jobs among [1, 8], as well as the unit of energy price

Table 1 Comparison between Model1 and Model2

(n-v-T)	$cons_1$	$cons_2$	GAP (%)	var_1	var_2	GAP (%)	$CPU_1(s)$	$CPU_2(s)$
(2-3-15)	578	517	11.80	295	289	2.08	0.36	0.29
(3-3-25)	1030	858	19.95	556	547	1.65	1.23	0.49
(4-3-30)	1318	1029	28.03	757	745	1.61	2.00	0.82
(5-2-40)	1677	1370	22.39	913	903	1.11	3.08	1.07
(5-3-40)	1827	1370	33.36	1123	1108	1.35	4.46	1.07
(5-5-40)	2150	1370	56.89	1543	1518	1.65	4.12	1.62
(10-2-80)	4149	2735	51.69	2613	2593	0.77	375.67	2.58
(10-3-80)	4857	2735	77.57	3433	3403	0.88	188.57	2.56
(10-5-80)	6277	2735	129.48	5073	5023	1.00	390.72	3.22
(15-5-120)	12609	4100	207.54	10603	10528	0.71	1748.52	4.32
Average			63.87			1.28	271.87	1.80

in each period among [1, 10]. It must be mention that these generations are inspired from the literature [10]. The computation time for all the experiments was set to 1 h or 3600 s. By using the first model, CPLEX was able to find the optimal solutions for the problems smaller than 15 jobs, 5 speeds, and 120 periods. Therefore, the results of the models in terms of the number of constraints and variables, as well as the computation time are compared together. These results are presented in Table 1. The second model is very faster than the first one (in average, it takes 271.87 s for the first model, and 1.80 s for the second one), and it decreased the number of constraints and variables of about 64% for the number of constraints and 1.3% for the number of variables in average.

5 Conclusion

A speed scalable and multi-states single machine scheduling problem is addressed in this paper to minimize the total energy consumption costs. The main contribution of this paper consists of the proposition of two new mathematical formulations to model the problem. In the first formulation, two decision variables and three indexes are used, while, just one decision variable and two indexes are used in the second one. The performance of these models are evaluated on several randomly generated instances. The numerical results demonstrates that, the second model is very faster than the first one, and it decreased the number of constraints and variables.

For the future works, first of all, we are interested to solve the problem by using the other exact methods. Then, since this problem is NP-hard, it could be also interesting to propose some heuristic and meta-heuristic algorithms to solve the large size instances.

References

1. Aghelinejad, M., Ouazene, Y., Yalaoui, A.: Machine and production scheduling under electricity time varying prices. In: 2016 IEEE International Conference on Industrial Engineering and Engineering Management (IEEM), pp. 992–996. IEEE (2016)
2. Aghelinejad, M., Ouazene, Y., Yalaoui, A.: Production scheduling optimisation with machine state and time-dependent energy costs. Int. J. Prod. Res. 1–18 (2017). https://doi.org/10.1080/00207543.2017.1414969
3. Aghelinejad, M., Ouazene, Y., Yalaoui, A.: Preemptive Scheduling of a Single Machine with Finite States to Minimize Energy Costs, pp. 591–599. Springer International Publishing (2017)
4. Antoniadis, A., Huang, C.-C.: Non-preemptive speed scaling. J. Schedul. **16**(4), 385–394 (2013)
5. Che, A., Zeng, Y., Lyu, K.: An efficient greedy insertion heuristic for energy-conscious single machine scheduling problem under time-of-use electricity tariffs. J. Clean. Prod. (2016)
6. Fang, K., Uhan, N.A., Zhao, F., Sutherland, J.W.: Scheduling on a single machine under time-of-use electricity tariffs. Ann. Operat. Res. 1–29 (2014)
7. Mikhaylidi, Y., Naseraldin, H., Yedidsion, L.: Operations scheduling under electricity time-varying prices. Int. J. Product. Res. **53**(23), 7136–7157 (2015)
8. Mouzon, G., Yildirim, M.B.: A framework to minimise total energy consumption and total tardiness on a single machine. Int. J. Sustain. Eng. **1**(2), 105–116 (2008)
9. Nguyen, N.-Q., Yalaoui, F., Amodeo, L., Chehade, H., Toggenburger, P.: Solving a malleable jobs scheduling problem to minimize total weighted completion times by mixed integer linear programming models. In: Asian Conference on Intelligent Information and Database Systems, pp. 286–295. Springer (2016)
10. Shrouf, F., Ordieres-Meré, J., García-Sánchez, A., Ortega-Mier, M.: Optimizing the production scheduling of a single machine to minimize total energy consumption costs. J. Clean. Prod. **67**, 197–207 (2014)
11. Yildirim, M.B., Mouzon, G.: Single-machine sustainable production planning to minimize total energy consumption and total completion time using a multiple objective genetic algorithm. IEEE Trans. Eng. Manag. **59**(4), 585–597 (2012)

Constrained Job Rearrangements on a Single Machine

Arianna Alfieri, Gaia Nicosia, Andrea Pacifici and Ulrich Pferschy

Abstract In several scheduling applications, one may be required to revise a pre-determined plan in order to meet a certain objective. This may happen if changes in the scenario predicted beforehand occur (e.g., due to disruptions, breakdowns, data values different from the expected ones). In this case costly reorganization of the current solution impose a limit on the allowed number of modifications. In our work, we address a single-machine scheduling problem where we need to alter a given (original) solution, by re-sequencing jobs with constraints on the number and type of allowed job shifts. For different objectives and rearrangement types, we propose mathematical programming models and possible solution approaches.

Keywords Scheduling · Integer linear programming · Re-sequencing · Dynamic programming

---

A. Alfieri
Dipartimento di Ingegneria Gestionale e della Produzione, Politecnico di Torino, Corso Duca degli Abruzzi 24, 10129 Turin, Italy
e-mail: arianna.alfieri@polito.it

G. Nicosia (✉)
Dipartimento di Ingegneria, Università Roma Tre, Via della Vasca Navale 79, 00146 Rome, Italy
e-mail: nicosia@ing.uniroma3.it

A. Pacifici
Dipartimento di Ingegneria Civile e Ingegneria Informatica, Università di Roma "Tor Vergata", Via del Politecnico 1, 00133 Rome, Italy
e-mail: andrea.pacifici@uniroma2.it

U. Pferschy
Department of Statistics and Operations Research, University of Graz, Universitaetsstrasse 15, 8010 Graz, Austria
e-mail: pferschy@uni-graz.at

© Springer Nature Switzerland AG 2018

P. Daniele and L. Scrimali (eds.), *New Trends in Emerging Complex Real Life Problems*, AIRO Springer Series 1,
https://doi.org/10.1007/978-3-030-00473-6_5

1 Introduction

Scheduling decisions in real world applications are often taken in multiple phases. This is the case, for instance, in multi-echelon supply chain processes where, at each stage, the solution computed in the previous phase may be altered in order to account for possible different criteria and constraints (see e.g., [1]). Hence, as the orders proceed from one level to the next—since it is unlikely that the same schedule gives a good performance value for all the stages of the supply chain—re-sequencing operations may be needed at intermediate inventories to consider new requirements; unless one decides to opt for suboptimal solutions and considers robust schedules instead [2].

Also in other settings the need of rearrangements may occur: Consider a system with a single objective function and subject to events that lead to a perturbation of the originally given or estimated data, e.g., changes in the expected job durations or availabilities as in [3]. In this case, the given, previously optimal, solution sequence is not performing so well anymore and re-scheduling should be carried out.

Re-scheduling could be of course very expensive in terms of both costs and operational difficulties associated to unstable solutions (as in the the so-called nervousness phenomenon [6]). As a consequence, when changing a solution schedule, the cost and stability trade-off becomes of primary importance. On these grounds, we aim at balancing the cost for re-positioning the jobs and the improvement in the objective. In the following, we specifically refer to this problem as *scheduling with rearrangements*.

2 Problem Definition

Consider a deterministic single-machine environment where n jobs with given processing times have to be scheduled according to a regular objective function $C(\sigma)$ (e.g., minimization of the total completion time, or minimization of the maximum tardiness, etc.), which depends on the job sequence σ.

Without loss of generality, it is assumed that an initial sequence $\sigma_0 = \langle 1, 2, \dots, n \rangle$, of the n jobs (representing an optimal or satisfactory solution for the original data) is given and that job j is the job placed in the j-th position of σ_0. Suppose now that, due to changed conditions (e.g., altered processing times, different objective at the current supply chain echelon, etc.), σ_0 is no more adequate in terms of the current performance indicator $C(\cdot)$. In order to get back to a satisfactory solution, it is possible to rearrange jobs, so that a new sequence $\sigma = \langle \sigma_1, \sigma_2, \dots, \sigma_n \rangle$ is obtained which achieves a better performance. The problem we address in this work is to determine a new job sequence σ such that:

(i) $C(\sigma)$ is minimum (or, at least, $C(\sigma) \leq C(\sigma_0)$) and
(ii) the value of a "distance function" between the new sequence σ and the initial sequence σ_0 is bounded.

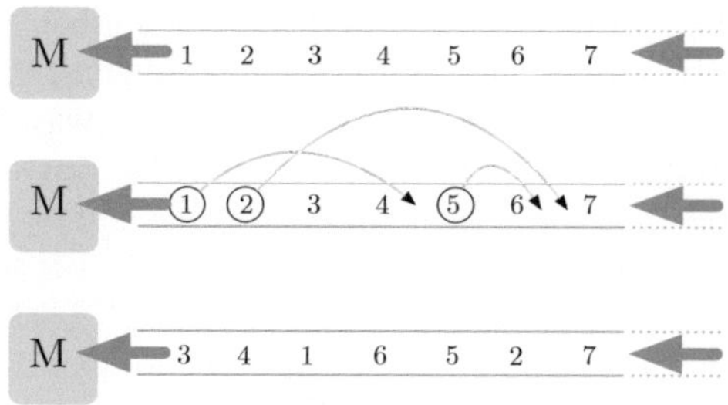

Fig. 1 As the conveyor belt proceeds leftward, three moves are performed

Several different metrics are used in the literature to define distances between *rankings*. Hereafter, the ranking of a job denotes its position in the sequence (e.g., the ranking of job j in σ_0 is j) and hence a permutation or sequence is completely described by the rankings of its jobs. In [4] a number of classical metrics are surveyed and new ones are proposed.

The metric we adopt in this paper is the number of *moves* necessary to reach σ starting from σ_0, where a move is the extraction of a job from the current sequence and its insertion in a *successive* position. By "successive position" we mean a position with larger index number, i.e., further "to the right" of the sequence. This restricted notion of moves is motivated by the real-world example of a machine that receives its input by a conveyor belt. Since the conveyor keeps moving forward, it is not possible to insert an extracted job in an earlier position in the sequence, i.e., in front of the queue. An illustration is given in Fig. 1, where the new sequence σ is obtained in three moves. As the conveyor belt proceeds to the left, we can remove jobs 1 and 2 to place them after 4 and 6, respectively. Doing so, 3 becomes the first job of the new sequence. Note that, in this setting, we would not be allowed to move jobs 3 and 4 before job 1.

Observe that even under our restricted definition of moves, every permutation of the jobs can be obtained from σ_0 by a suitable sequence of feasible moves. Note also that in our metric we do not care about the distance (i.e., number of positions) a job is moved (as it is done in the classical Kendall-Tau or Kemeny distance, where the number of neighbour interchanges is counted), but we focus on the number of moving operations. This is motivated by the practical situation where a job is associated with an item (work piece, part of a machinery) and moving such an item entails, for instance, a forklift to pick up the item and drive it to the new position. In this case, driving distance is mostly negligible, while the pickup and placing operations are costly.

The following Lemma determines the number of moves necessary to reach a given sequence σ, starting from the initial sequence σ_0. (Recall that, in σ_0, the name of a job indicates its position/ranking in the sequence.)

Lemma 1 *Transforming a starting sequence σ_0 into a new, given sequence σ by moves to successive positions, a job j destined for a position i in σ must be moved from its original position j in σ_0, if and only if, there is a job $k > j$ placed before the i-th position in σ.*

Proof ($\Leftarrow$) Assume in σ there is a job $k > j$ placed at a smaller position number than job j. Since in the original sequence σ_0, j was placed before k and only "rightward" moves are allowed, there is no way for k to reach a position "left of" j by a move, but σ can be obtained only by moving j "behind" k.
($\Rightarrow$) Suppose that, contrary to the assumption, in σ there is no such job k but any job ℓ placed in positions $1, 2, \ldots, i-1$ is such that $\ell \leq j$. Thus, by the pigeon-hole principle, i cannot exceed j, i.e., $i \leq j$. This means that the new position i of j in σ does not have a larger position number than j in σ_0. Hence it is not meaningful to move job j "to the right" in order to reach position i.

For instance, in Fig. 1, $\sigma = \langle 3, 4, 1, 6, 5, 2, 7 \rangle$, therefore 1, 5, and 2, have been re-positioned, since they are preceded by higher ranking jobs, while 3, 4, 6 and 7 have not.

Let $\mathfrak{S}_n$ indicate the set of all sequences (i.e. permutations) of the n jobs. We are now in the position to give a formal definition of our problem.

> SINGLE-MACHINE SCHEDULING WITH REARRANGEMENTS (SSRP)
> *Given:* a set $J = \{1, \ldots, n\}$, a sequence of jobs $\sigma_0 = \langle 1, 2, \ldots, n \rangle$ with nonnegative processing times p_j, $j = 1, \ldots, n$, a regular objective function $C : \mathfrak{S}_n \longrightarrow \Re_+$, and an integer k;
> *Find:* a sequence $\sigma \in \mathfrak{S}_n$ that can be reached in at most k moves from σ_0 such that $C(\sigma)$ is minimum.

The job in position i in the sequence $\sigma \in \mathfrak{S}_n$ is denoted by $\gamma_\sigma(i) \in \{1, 2, \ldots, n\}$ (e.g., $\gamma_{\sigma_0}(i) = i$), while $\pi_\sigma(j)$ denotes the position of job j. If it is clear from the context which sequence we are talking about, we omit the subscript σ and write, e.g., $\pi(\gamma(i)) = i$ or $\gamma(\pi(j)) = j$.

3 Solution Approaches and Preliminary Results

We propose integer linear programming models for three cases of our problem, corresponding to three different objective functions. In addition, we present efficient procedures to solve restricted versions of SSRP. These can be of use, as building blocks, in devising heuristic solution algorithms for the general case.

3.1 Mathematical Programming Model

Hereafter we introduce MIP models for our problem. Since the mathematical programs for the three objective functions are very similar, we first introduce a MIP for a generic objective function $C(\sigma)$ and then describe the linear expressions for the three different objectives.

In all programs, we use the assignment variables $x \in \{0, 1\}^{n \times n}$, where $x_j(i) = 1$ indicates that job j is placed in position i in σ and variables $z \in \{0, 1\}^n$ to count the number of jobs that are to be repositioned. The quantities $\gamma(i)$ and $\pi(j)$ introduced above may be easily expressed in terms of the x variables. In particular, in the following models, we use $\gamma(i) = \sum_{j=1}^{n} j\, x_j(i)$.

The SINGLE-MACHINE SCHEDULING WITH REARRANGEMENTS problem can be modelled as follows:

$$\min C(\sigma) \tag{1}$$

$$\text{s.t.} x \in \mathscr{A} \tag{2}$$

$$z_i \geq \frac{1}{n}(\gamma(h) - \gamma(i)) \qquad h = 1, \ldots, i-1;\ i = 1, \ldots, n \tag{3}$$

$$\sum_{i=1}^{n} z_i \leq k \tag{4}$$

$$x_j(i),\ z_i \in \{0, 1\} \tag{5}$$

where Eq. (2) are the usual assignment constraints ($\sum_{i=1}^{n} x_j(i) = 1$ for all j and $\sum_{j=1}^{n} x_j(i) = 1$ for all positions i), (3) enforce the counting of moves in z_i: Variable z_i must assume value 1 when there is a job with index greater than i at position h before the job in position i. Finally, (4) bounds the total number of moves by k.

We consider the minimization of three objective functions, *total completion time* C^1, *maximum lateness* C^2, and *number of tardy jobs* C^3:

$$C^1 = \sum_{i=1}^{n} \sum_{h=1}^{i} \sum_{j=1}^{n} p_j x_j(h) \tag{6}$$

$$C^2 = L_{\max} \text{ with } L_{\max} \geq \sum_{h=1}^{i} \sum_{j=1}^{n} p_j x_j(h) - \sum_{j=1}^{n} d_j x_j(i) \quad \forall i = 1, \ldots, n. \tag{7}$$

$$C^3 = \sum_{i=1}^{n} U_i \text{ with } U_i \geq \frac{1}{M}\left(\sum_{h=1}^{i} \sum_{j=1}^{n} p_j x_j(h) - \sum_{j=1}^{n} d_j x_j(i)\right) \quad \forall i = 1, \ldots, n. \tag{8}$$

Note that the RHS of (7) becomes positive if the completion time of *the job in position i exceeds its due date* and hence $U_i = 1$ in (8) (for suitably large M).

We performed a number of preliminary computational experiments on randomly generated instances using the commercial ILP solver Gurobi v7.5.0 on a 3.6 GHz Quad Core i7, 16GB RAM PC. These tests show that Gurobi is able to solve quite efficiently instances with up to $n = 40$ jobs and $k = 3$. As soon as n or k grows, the time required to solve the problem becomes prohibitively large. So, we also implemented a few simple greedy rules to detect the k jobs to be repositioned and on which positions to schedule them. Such greedy rules are extremely fast (a few milliseconds for quite large instances), but poorly effective when the number of jobs is small. However, their performance improves when the number of jobs becomes larger.

A few experiments were also performed to test how the value of k influences the objective function values. In most cases, it turns out that it is possible to find the optimal solution as soon as k reaches $\approx \frac{n}{3}$.

3.2 Restricted Problems

We consider a restricted variant of SSRP in which the set $\mathcal{K}$ of k jobs that we are allowed to move is given as an input. For this restriction we are able to provide exact and efficient solution algorithms for the three objectives (6)–(8).

The following algorithms can also be used to find a solution for the general (unrestricted) SSRP. In fact, one may run them for all possible choices of $\mathcal{K}$, i.e., for all k-tuples of jobs, and return the best solution found. This also shows that, when k is fixed, SSRP can be solved in polynomial time for all three objective functions.

3.2.1 Minimizing the Total Completion Time

It is not hard to see and to prove by an exchange argument that the jobs in $\mathcal{K}$ induce a partitioning of σ_0 into segments in each of which the jobs in $\mathcal{K}$ will be positioned in SPT-order among themselves.

Lemma 2 *Given the set of movable jobs $\mathcal{K}$ for SSRP with total completion time minimization, if $i, j \in \mathcal{K}$ and $p_i < p_j$, then either i precedes j in any optimal schedule σ, or $j < i$ and $\pi(j) < i$.*

The first case $p_i < p_j$ refers to the SPT-order of jobs in the same segment while the second case means that j and $\pi(j)$ are in a segment separated from a subsequent segment containing i by same (large) jobs.

This gives rise to a simple algorithm: Consider the jobs in $\mathcal{K}$ in non-increasing order of processing times. For each job $j \in \mathcal{K}$, find the best new position for j by moving it right and evaluating the objective function for all possible slots. Place j in that position $\pi(j)$ (permanently) and remove it from $\mathcal{K}$, iterate. Note that jobs in different segments will not interact with each other.

3.2.2 Minimizing the Maximum Lateness

In this case, we may easily adapt Lawler's algorithm for $1 \mid prec \mid L_{\max}$ to express the precedences induced by movable and unmovable jobs.

Figure 2 illustrates the idea: Digraph (a) represents the original sequence where the grey nodes are the movable jobs $\mathcal{K}$, while digraph (b) illustrates the input precedence-constraints that must be considered when applying Lawler's algorithm.

Note that this result easily extends to the minimization of regular functions $C(\sigma) = f_{\max} = \max_{j \in J}\{f_j(C_j(\sigma))\}$ where $f_j(\cdot)$ are nondecreasing functions.

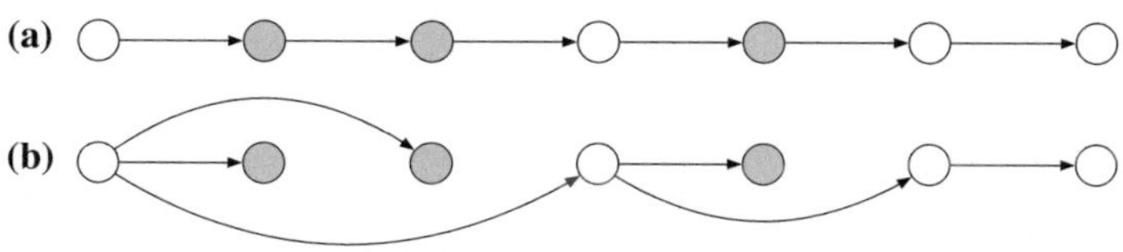

Fig. 2 Example of precedence for application of Lawler's algorithm

3.2.3 Minimizing the Number of Tardy Jobs

Minimizing the number of tardy jobs on a single machine can be done optimally by Moore's algorithm [5]. Unfortunately, Moore's algorithm is not useful for our SSRP in its general form for unknown $\mathcal{K}$. The following example shows that—even if the original starting sequence is EDD—the same algorithm does not provide any insight on the jobs that one should move in order to minimize the number of tardy jobs. In fact, the best solution consists in moving jobs that Moore's algorithm would not move. Consider an instance with the following data sorted by EDD.

Job	1	2	3	4	5	6	7
p_j	1	2	5	6	20	20	20
d_j	1	2	6	12	30	50	70

If we apply Moore's algorithm it is quite easy to show that it would move jobs 2 and 5 and find an optimal solution in which 2 jobs are late.

If we restrict ourselves to move only one job, i.e. $k = 1$, then we can observe that if we move job 2 we get a solution with 4 late jobs (2, 5, 6, and 7). The same solution value is attained when job 5 is delayed (so that jobs 2, 3, 4, and 5 are late). However, if we consider the solution in which job 3 is moved at the end of the schedule, we can see that the number of tardy jobs (jobs 2 and 3) is 2, so it is optimum.

Turning back to our restricted version of SSRP, i.e., if we assume that the set $\mathcal{K}$ of k jobs which are allowed to be moved *is known*, determining a schedule that minimizes the number of late jobs can be done by a two-dimensional, strictly polynomial, dynamic programming scheme which we sketch hereafter. The set of jobs J is partitioned into movable jobs $\mathcal{K}$ and the complement set $\mathcal{U} := J \setminus \mathcal{K}$. The algorithm is based on the following three structural properties (whose easy proofs are omitted). Here, *removing* a job $j \in \mathcal{K}$ means that we extract j from the schedule and append it at the end of the schedule in the final step of the algorithm.

1. Jobs $j \in \mathcal{K}$ are either scheduled on time or they are removed.
2. All jobs $j \in \mathcal{K}$ which are on time are scheduled in EDD order among themselves (disregarding all jobs in $\mathcal{U}$).
3. If a job $j \in \mathcal{K}$ is on time, it is always scheduled at the latest possible time.

Assuming the k jobs in $\mathcal{K}$ are numbered in EDD order; the procedure starts from an empty schedule and proceeds by iteratively inserting jobs in $\mathcal{K}$. Whenever a job $j \in \mathcal{K}$ is inserted, all jobs in $\mathcal{U}$ positioned before j in σ_0 are scheduled as well.

Let $C_j(\ell, m)$, $\ell = 0, 1, \ldots, n$, $m = 0, 1, \ldots, |\mathscr{U}|$, $j = 0, 1, \ldots, k$, indicate the minimum makespan—not considering the removed jobs—of a schedule in which exactly ℓ jobs in J are late, the first m jobs of $\mathscr{U}$ are scheduled, and each of the first j jobs in $\mathscr{K}$ is scheduled or removed. In each partial schedule represented by $C_j(\ell, m)$, the last scheduled job is always a job from $\mathscr{K}$. $C_j(\ell, m) := +\infty$ if no such solution exists. The optimal solution of SRRP, minimizing the number of late jobs, is given by the minimum ℓ such that $C_k(\ell, |\mathscr{U}|)$ is finite (and the $\mathscr{U}$-jobs placed after the latest on time $\mathscr{K}$-job have to be inserted).

$C_j(\ell, m)$ can be determined by adding job j to every finite previously generated dynamic programming entry $C_{j-1}(\ell, m)$. To do so, new *candidate* entries $CC_j(\ell, m)$ are computed. At the end of each iteration, $C_j(\ell, m)$ is calculated as the minimum over all generated candidates $CC_j(\ell, m)$ (if any). Two cases are to be distinguished:

Case (i): j is removed and classified as late job, set $CC_j(\ell + 1, m) := C_{j-1}(\ell, m)$.

Case (ii): j is scheduled on time (if this is possible). Thus, all unscheduled jobs in $\mathscr{U}$ with positions after m, but before j in σ_0, must also be inserted preceding j in the resulting sequence. Moreover, due to the above property 3, j is possibly moved to the right, by inserting some other jobs in $\mathscr{U}$, as long as j remains on time. This implies the placement for a subset $U \subseteq \mathscr{U}$ of jobs out of which u' are late. Eventually, a new candidate solution is obtained by appending jobs U and j at the end of the current sequence given by $C_j(\ell, m)$ (which ends with a job in $\mathscr{K}$):

$$CC_j(\ell + u', m + |U|) := C_{j-1}(\ell, m) + \sum_{i \in U} p_i + p_j.$$

4 Conclusions

In this paper, we address SSRP, a task rearrangement scheduling problem on a single machine. Similar decision problems are relevant in dynamic scenarios, where a pre-computed optimal schedule becomes sub-optimal due, e.g., to parameter changes, or errors in the data used in the computations.

We propose an integer programming approach and report about its performance on some preliminary test sets. In addition, we propose some efficient methods for the solution of restricted versions of SSRP. We believe, those may be effectively adapted as subroutines in a heuristic algorithm for solving the original (general) version of our problem. This, together with a computational complexity characterization of SSRP, is also our main future direction for this study.

References

1. Agnetis, A., Hall, N.G., Pacciarelli, D.: Supply chain scheduling: sequence coordination. Discret. Appl. Math. **154**(15), 2044–2063 (2006)
2. Detti, P., Nicosia, G., Pacifici, A., Zabalo Manrique de Lara, G.: Robust single machine scheduling with external-party jobs. IFAC-PapersOnLine **49**(12), 1731–1736 (2016)

3. Hall, N.G., Potts, C.N.: Rescheduling for job unavailability. Oper. Res. **58**(3), 746–755 (2010)
4. Kumar, R., Vassilvitskii, S.: Generalized distances between rankings. In: Proceedings of the 19th International Conference on World Wide Web, pp. 571–580. ACM (2010)
5. Moore, J.M.: An n job, one machine sequencing algorithm for minimizing the number of late jobs. Manag. Sci. **15**(1), 102–109 (1968)
6. Wang, X., Disney, S.M.: The bullwhip effect: progress, trends and directions. Eur. J. Oper. Res. **250**(3), 691–701 (2016)

Reducing Overcrowding at the Emergency Department Through a Different Physician and Nurse Shift Organisation: A Case Study

Roberto Aringhieri, Giovanni Bonetta and Davide Duma

Abstract Overcrowding is a widespread problem affecting the performance of an Emergency Department (ED). In this paper we deal with the overcrowding problem at the ED sited at *Ospedale Sant'Antonio Abate di Cantù*, Italy. Exploiting the huge amounts of data collected by the ED, we propose a new agent-based simulation model to analyse the real impact on the ED overcrowding of a different physicians and nurses shift organisations. The proposed simulation model demonstrates its capability of analysing the ED performance from a patient-centred perspective.

Keywords Emergency department · Overcrowding · Agent based simulation

1 Introduction

An Emergency Department (ED) is a medical treatment facility inside of a hospital or another primary care centre and is specialised in emergency medicine providing a treatment to unplanned patients, that is patients who present without scheduling. The ED operates 24 h a day, providing initial treatment for a broad spectrum of illnesses and injuries with different urgency. Such treatments require the execution of several activities, such as visits, exams, therapies and intensive observations. Therefore human and medical resources need to be coordinated to efficiently manage the flow of patients, that varies over time for volume and characteristics.

R. Aringhieri · G. Bonetta · D. Duma (✉)
Dipartimento di Informatica, Università degli Studi di Torino, Turin, Italy
e-mail: davide.duma@unito.it

R. Aringhieri
e-mail: roberto.aringhieri@unito.it

G. Bonetta
e-mail: giovanni.bonetta@edu.unito.it

© Springer Nature Switzerland AG 2018

P. Daniele and L. Scrimali (eds.), *New Trends in Emerging Complex Real Life Problems*, AIRO Springer Series 1,
https://doi.org/10.1007/978-3-030-00473-6_6

A phenomenon that affects EDs all over the world reaching crisis proportions is the overcrowding [12]. It is manifested through an excessive number of patients in the ED, long waiting times and patients Leaving Without Being Seen (LWBS); sometimes patients being treated in hallways and ambulances are diverted [2, 3, 10]. Consequently, the ED overcrowding has a harmful impact on the health care: when the crowding level raises, the rate of medical errors increases and there are delays in treatments, that is a risk to patient safety. Not only overcrowding represents a lowering of the patient outcomes, but it also entails an increase in costs [9]. Moreover, the ED overcrowding causes stress among the ED staff, patient dissatisfaction and episodes of violence [6].

Simulation is widely used to test what-if scenarios to deal with overcrowding [12], analysing the use of different resources, setting or policy within the care planning process. Although most of the solutions proposed in literature foresee the use of new additional resources, often they are scarce and there is no economic possibility of new investments [5]. Then human and equipment resources available should be used as efficiently as possible in order to improve the ED processes. For this reason, research addressing short-term decision problems are becoming increasingly in recent years [1]. Placing in the perspective to alleviate the ED overcrowding without changing the ED resources and settings, there are two way to act: (i) changing the human resources planning [13] or (ii) adopting different policies in the allocation of the human and equipment resources [8, 11].

In this paper we deal with the problem of the management of the staff (physicians and nurses) operating at the ED sited at *Ospedale Sant'Antonio Abate di Cantù*, which is a medium size hospital in the region of Lombardy, Italy. Exploiting the huge amounts of data collected by the ED, we propose a new agent-based simulation model to analyse the real impact of a different physicians and nurses shift organisations on the ED overcrowding.

2 Modelling Approach

After a brief description of the case study under consideration (more details available in [7]), we report the pre-processing procedure to make the huge amounts of data usable within the simulation model.

The case study. The resources available within the ED are: 4 beds for the medical visits placed in 3 different visit rooms, in addition to one bed within the shock-room and another one in the Minor Codes Ambulatory (MCA), one X-ray machine, 5 Short-Stay Observation (SSO) units (beds), 10 stretchers and 10 wheelchairs to transport patients with walking difficulties. The medical staff is composed of 4–6 nurses and 1–3 physician(s), depending on the time of day and the day of week, in addition to the X-ray technician.

A patient is interviewed and registered as soon as possible by a triage-nurse on his/her arrival in the ED, recording personal data, the main symptom and the urgency code c from 1 (most urgent) to 5 (less urgent). Table 1 reports all the activities that

Table 1 Activities in a patient path

Id	Description	Class	EDc	TS	Id	Description	Class	EDc	TS
A	Triage	Triage	✓	t_E	B	Medical Visit	Visit	✓	t_E
C	Shock-Room	Visit	✓	t_E	D	MCA Visit	Visit	✓	t_E
E	Paediatric Fast-Track	Discharge		t_P	F	Therapy	Tests and Care	✓	t_P, t_E
G	Laboratory Exams	Tests and Care	✓	t_P, t_R	H	X-Ray Exams	Tests and Care	✓	t_P, t_R
I	Comp. Tomography (CT)	Tests and Care		t_P, t_R	J	Echography	Tests and Care		t_P, t_R
K	Specialist Visit	Tests and Care		t_P, t_R	L	SSO	Tests and Care	✓	t_S, t_E
M	Pre-hosp. SSO	Tests and Care	✓	t_S, t_E	N	Revaluation Visit	Revaluation	✓	t_E
O	Hospitalisation	Discharge	✓	t_E	P	Discharge (ordinary)	Discharge	✓	t_E
Q	Interruption	Discharge		t_E					

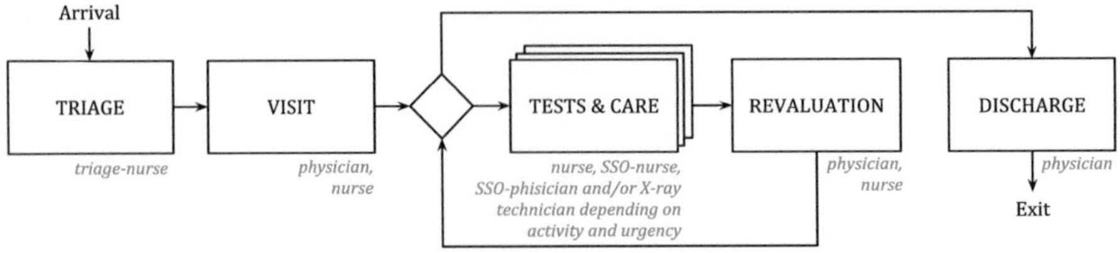

Fig. 1 A simplified path for a patient within the ED and the human resources associated

could be performed by a patient within the ED, each one classified into 5 classes (*Triage*, *Visit*, *Tests and Care*, *Revaluation* and *Discharge*). Columns EDc indicate those that are competence of the ED. Columns TS report the timestamps available from the data, that is the start time t_S, the prescription or request time t_P, the report time t_R and the end time t_E.

Figure 1 depicts a simplified version of the patient path also reporting the human resources associated to each activity: after the triage, a Visit class activity is always provided except for a LWBS patient. Then the patient can be discharged or continue with a sequence of Tests and Care class activities, that is always followed by a revaluation visit, after which the patient can be discharged or go on with other Tests and Care class activities.

Data analysis. In order to supply our simulation model, we are required to pre-process the ED dataset to create an event log, which consists of a set of traces (i.e. ordered sequences of events of a single case), their multiplicity and other information about the single events, such as timestamps and/or durations, resources, case attributes and event attributes. The event log has been generated taking into account the 27 039 accesses of the year 2015. Each case of the event log consists in an access and events consist in activities, which has been classified into 17 event classes corresponding to the activities reported in Table 1.

We should take into account several *noises* in the dataset to estimate start and end time of each activity. In the following, we consider the following 7 noises, that is missing timestamps ($\mathfrak{N}_o$), timely execution ($\mathfrak{N}_1$), forgetfulness in therapies recording ($\mathfrak{N}_2$), multiple recording ($\mathfrak{N}_3$), fake or missing revaluation visit ($\mathfrak{N}_4$), fake medical visit ($\mathfrak{N}_5$), tests reported after discharge ($\mathfrak{N}_6$).

In accordance with Table 2, the pre-processing algorithm has been implemented as follows with parameters τ and δ fixed to 10 and 30 min:

1. Start time and end time of each activity are estimated in accordance with Table 2 ($\mathfrak{N}_o$).
2. A sorting time $\bar{t}$ is fixed for each activity in order to avoid overlapping of activities (because of $\mathfrak{N}_o$); we chose the more reliable time, that is $\bar{t} = t_S$ for activities F, L and M, $\bar{t} = t_E$ for the other ones.
3. If activity E occurs, all the other activities are removed, except the triage ($\mathfrak{N}_5$).
4. The activities of the same path are sorted in chronological order of $\bar{t}$ composing the trace.

Table 2 Average duration in min of the activities according to the ED staff and estimation of the missing timestamps (✓ stands for available)

Id	Activity duration d	Reporting duration r	Start time t_S	End time t_E	Sorting time $\bar{t}$	Id	Activity duration d	Reporting duration r	Start time t_S	End time t_E	Sorting time $\bar{t}$
A	5	n.a.	$t_E - d$	✓	t_E	B	15	n.a.	$t_E - d$	✓	t_E
C	15	n.a.	$t_E - t_E^{\text{triage}}$	✓	t_E	D	15	n.a.	$t_E - d$	✓	t_E
E	0	n.a.	t_E	✓	t_E	F	2	n.a.	$t_E - d$	✓	t_S
G	3	15	$t_R - r - d$	$t_R - r$	t_E	H	3	30	$t_R - r - d$	$t_R - r$	t_E
I	10	45	$t_R - r - d$	$t_R - r$	t_E	J	15	45	$t_R - r - d$	$t_R - r$	t_E
K	15	n.a.	$t_E^{\text{last before K}}$	t_R	t_E	L	✓	n.a.	✓	✓	t_S
M	✓	n.a.	✓	✓	t_S	N	1	n.a.	$t_E - d$	✓	t_E
O	1	n.a.	$t_E - d$	✓	t_E	P	1	n.a.	$t_E - d$	✓	t_E
Q	0	n.a.	t_E	✓	t_E						

5. For each trace, let $\bar{t}_{\text{exit}}$ be the sorting time of the discharge (one among activities O, P and Q) and let $\tau > 0$ be a parameter denoting the amount of time before the discharge in which the forget recording of therapies is remedied. If $\bar{t}_{\text{exit}} - \bar{t}_F < \tau$, then $\bar{t}_F = \max\{\bar{t}_F, t_R^F + 1 \text{ min}\}$, where t_R^F is the prescription time of that therapy ($\mathfrak{N}_2$).
6. For each trace, let $\bar{t}_Y$ be the sorting time of a certain Tests and Care class activity. If $\bar{t}_Y > \bar{t}_{\text{exit}}$, then $\bar{t}_Y$ is fixed one minute before the first revaluation visit after the prescription time of that activity ($\mathfrak{N}_6$).
7. For each activity of each trace, if it precedes the triage time, then it is moved one minute after the triage time ($\mathfrak{N}_1$); if it is not a triage and it precedes the visit time, then it is moved one minute after the visit time ($\mathfrak{N}_1$).
8. For each trace, if there is no revaluation visit between a Tests and Care activity and the discharge, then a fake revaluation visit is inserted a minute before the discharge ($\mathfrak{N}_4$).
9. For each trace, consecutive Tests and Care activities of the same type such that the time between them is less than δ are merged keeping the start time of the first one and the end time of the last one ($\mathfrak{N}_3$).

The simulation model. We propose an Agent-Based Simulation (ABS) model to represent the interactions among patients and medical personnel inside the patient path within the ED by replicating in detail the progress of ED activities. ABS is well suited for modelling such a type of interactions [4]. Five type of agents are implemented as follows:

Decision-maker. A unique agent is used to manage the resource allocation and to assign tasks to the medical staff. When a patient require the execution of an activity, such a request is inserted in a prioritised queue recording the patient ID, the request timestamp, the set of resources needed, the urgency code c and the priority class γ initially equals to c. Every time a new request is made or a resource is released, the agent scans the queue considering the priority, that is (i) patients with lower values of γ first, and (ii) patients with the same value of γ are sorted in decreasing order of the waiting time. When the whole set of resources to provide an activity is available, the agent send a message to the agents representing the patient and the human resources involved in the activity.

Patient. The patient population is reproduced from the event log: an agent is created for each access to the ED from the dataset and relevant information for the replication of its path (i.e. urgency code c, trace, arrival time and several activity durations) are represented by agent attributes. Each agent progresses in their path within the ED replying that in Fig. 1 in accordance with its trace.

Physician. Each physician shift is represented by an agent with an attribute that indicates its competence (visit rooms, SSO units, etc.). This agent is characterised by two relevant aspects determined by the end of the shift and the handover between physicians: (i) λ_1 min before the shift end, the agent can be assigned only to urgent patients with $c \leq 2$, or taken over previously; (ii) the last ρ min of the shift models the handover to the next physician (whose agent arrives ε min before starting the shift) in which the agent is no more available for any tasks.

Before the handover, the agent waits for a task assignment from the decision-maker agent.

Nurse. The agent is implemented as well as the physician, omitting the handover of the medical records. We denote with λ_2 the min before the shift end in which the agent deal with only urgent patients with $c \leq 2$.

X-ray technician. The agent is implemented as well as the nurse, omitting the limitation on serving non-urgent patients close to the shift end. Since at nighttime no technician is working in the ED, we model the on demand technician availability by adding a delay of 30 min to represent the time needed to reach the ED.

The ABS approach allows us to model the continuity of the care process, which is allowed by the ability to identify individual resources (i.e., single physician and nurses) and to simulate their interactions: the same physician is always assigned to a patient for the activities that follow its first medical visit, that is revaluation visits and discharge; furthermore, if the assigned physician ends its shift before the completion of the care process, the activities are performed by the physician to which the medical record has been passed.

Another important aspect represented by the ABS model is the simulation of the behaviour of the human resources during the beginning and the ending of their shift, which are the critical moments that cause a slowdown in the flow of patients. To this end, the parameters ε, ρ and $\lambda_{1,2}$ have been introduced: the arrival of the physician ε min before the beginning of the shift is a common practice to ease the handover made in the last ρ min of the current physician shift; the assignment of non-urgent patients to physicians and nurses $\lambda_{1,2}$ min before the end of the shift is usually avoided to guarantee continuity in the process of care. Finally, to avoid the *starvation* of the less urgent patients, a re-prioritisation of such patients has been planned each υ hours of stay increasing the urgency code c by 1, without going beyond 2 if $c \geq 3$ and 1 otherwise.

3 Quantitative Analysis

The dataset of all the accesses to the ED of the year 2015 is used for the analysis in this section. The shifts of the medical staff replicate those of the real case, as reported in Table 3. In last 4 columns the competence of the ED staff are indicated with a check symbol; an asterisk means that the competence is limited to a time slot in which SSO units do not have a dedicated physician. Furthermore, a X-ray technician is available in the ED from 8:00 to 20:00 every day. The model is implemented in AnyLogic 7.3.7. The average computational time for a simulation run is 10 s.

To validate the model, we compare the obtained average waiting times of the patients belonging to 4 urgency classes and the overall, with those computed from the real data. To this purpose we fix $\varepsilon = 15$ min and range υ in [0.5, 6] with step 0.5 h and both λ_1 and λ_2 in [15, 30] with step 5 min. The best fitness has been obtained for the values $\upsilon = 2$ h, $\lambda_1 = 20$ min, $\lambda_2 = 10$ min.

Table 3 Shifts of the ED staff (real case)

Resource		Shift				Competence			
Type	Number	Start	Duration (h)	Weekday	Weekend	Triage	Visits	MCA	SSO
Physician	2	8:00	8	✓	✓		✓		✓*
	2	16:00	7	✓	✓		✓		✓*
	1	23:00	9	✓	✓		✓		✓
	1	8:00	8	✓				✓	✓*
	1	8:00	7	✓					✓
Nurse	2	7:00	7	✓	✓	✓			
	2	14:00	7	✓	✓	✓			
	1	21:00	10	✓	✓	✓			
	2	7:00	7	✓	✓		✓		
	3	14:00	7	✓	✓		✓		
	2	21:00	10	✓	✓		✓		
	1	7:00	7	✓	✓				✓
	1	14:00	7	✓	✓				✓
	1	21:00	10	✓	✓				✓
	1	8:00	8	✓				✓	

Starting from a request of the ED, a what-if analysis is conducted with the aim of reducing the waiting times of the patients without adding new resources but only changing the time span of the shifts. Table 4 reports the structure of the shifts considered: from the real case (base), we obtain other physician shifts by moving them of 30 and 60 min forward and backward. After selecting the best shift structure for the physicians (phase 1), we repeat the same experiment for the nurse shifts (phase 2). We measure the average waiting times and the average ED Length of Stay (EDLOS) of the patients to evaluate the best configuration: the former is the time elapsed between the triage and the beginning of the first visit, while the latter starts with the first visit and ends with the discharge.

From the results of the phase 1 reported in Table 4, the base configuration seems to be the best for minimising both waiting times and EDLOS. However, most urgent codes could take a slight advantage when the physician shifts are postponed of 30 min. On the contrary, all the other shift configuration worse the waiting times up to 26 min and the the EDLOS up to 30 min in the worst case compared with the base configuration. Regarding the phase 2, the nurse shifts postponed of 30 min can give slightly improvements, that is 2 min for the waiting times and 8 min for the EDLOS on average. On the contrary, the preponing of the nurse shifts of 30 min causes a significant worsening of the indices. These results demonstrate the urgent need of additional resources for a significant reduction of the overcrowding. For this reason, we provide a further analysis, in which we evaluate the impact of adding one physician. Starting from the best shift configuration in accordance with Table 4, we

Table 4 Analysing the impact of existing resources: what-if analysis

	Shift		Average waiting time (min)					Average EDLOS (h)				
	Physician	Nurse	1	2	3	4–5	Overall	1	2	3	4–5	Overall
Phase 1	Base	Base	13	34	76	98	70	8.5	7.5	5.0	4.1	5.4
	+30 min	Base	12	34	75	104	71	8.7	7.4	5.0	4.1	5.4
	−30 min	Base	37	41	87	117	82	8.7	7.6	5.0	4.1	5.5
	+60 min	Base	50	57	95	118	90	9.1	8.2	5.4	4.4	5.9
	−60 min	Base	52	61	99	126	96	8.9	8.2	5.5	4.5	5.9
Phase 2	Base	+30 min	10	31	72	101	68	8.4	7.5	4.9	4.1	5.3
	Base	−30 min	36	46	91	120	86	8.6	8.0	5.3	4.4	5.8
	Base	+60 min	16	36	77	99	71	8.5	7.6	5.0	4.1	5.5
	Base	−60 min	16	35	75	99	70	8.5	7.5	4.9	4.0	5.5

Table 5 Adding one physician: what-if analysis

Additional shift	Average waiting time (min)					Average EDLOS (h)				
	1	2	3	4–5	Overall	1	2	3	4–5	Overall
None	10	31	72	101	68	8.4	7.5	4.9	4.1	5.3
S1	23	33	74	95	68	8.6	7.4	4.8	3.9	5.3
S2	10	24	58	84	55	8.2	6.9	4.2	3.5	4.7
S3	6	22	56	82	53	8.3	6.6	4.2	3.4	4.7
S4	6	26	65	91	60	8.5	6.9	4.5	3.7	5.0

analyse the performance inserting a physician shifts with competence on the visit rooms in four different way: (S1) from 8:00 to 16:00 in weekdays, (S2) from 15:00 to 23:00 in weekdays, (S3) from 12:00 to 20:00 in weekdays, and (S4) from 10:00 to 16:00 in weekdays and from 11:00 to 16:00 in weekend.

The resulting average waiting times and the EDLOS are reported in Table 5 proving that an inadequate addition of physician shifts would be useless for the overall performance or even worse, as shown by S1. The best configuration is S3, for which patients wait 15 min less and have an EDLOS 40 min shorter, on average.

4 Conclusions

The proposed ABS model demonstrates its capability of analysing the ED performance from a patient-centred perspective. The change of the existing physician and nurse shifts seems to be insufficient to get a significant alleviation of the ED overcrowding. However the insertion of one physician can reduce the average waiting times of the 24% and the EDLOS of the 14% compared to the current ones.

Acknowledgements The authors wish to thank Alessandra Farina, Elena Scola and Filippo Marconcini of the ED at *Ospedale Sant'Antonio Abate di Cantù* for the fruitful collaboration and for providing us the data set and allowing their use in this paper.

References

1. Aboueljinane, L., Sahin, E., Jemai, Z.: A review on simulation models applied to emergency medical service operations. Comput. Ind. Eng. **66**, 734–750 (2013)
2. Aringhieri, R., Bruni, M., Khodaparasti, S., van Essen, J.: Emergency medical services and beyond: addressing new challenges through a wide literature review. Comput. Oper. Res. **78**, 349–368 (2017)

3. Aringhieri, R., Dell'Anna, D., Duma, D., Sonnessa, M.: Evaluating the dispatching policies for a regional network of emergency departments exploiting health care big data. In: Lecture Notes in Computer Science, vol. 10710, pp. 549–561. Springer (2018)
4. Aringhieri, R., Duma, D., Fragnelli, V.: Modeling the rational behavior of individuals on an e-commerce system. Oper. Res. Perspect. **5**, 22–31 (2018)
5. Derlet, R.: Overcrowding in emergency departments: increased demand and decreased capacity. Ann. Emerg. Med. **39**(4), 430–432 (2002)
6. Derlet, R., Richards, J.: Overcrowding in the nation's emergency departments: complex causes and disturbing effects. Ann. Emerg. Med. **35**(1), 63–68 (2000)
7. Duma, D., Aringhieri, R.: Mining the Patient Flow Through an Emergency Department to Deal with Overcrowding, vol. 210, pp. 49–59. Springer, New York LLC (2017)
8. Feng, Y.-Y., Wu, I.-C., Chen, T.-L.: Stochastic resource allocation in emergency departments with a multi-objective simulation optimization algorithm. Health Care Manag. Sci. **20**(1), 55–75 (2017)
9. George, F., Evridiki, K.: The effect of emergency department crowding on patient outcomes. Health Sci. J. **9**(1), 1–6 (2015)
10. Hwang, U., Concato, J.: Care in the emergency department: how crowded is overcrowded? Acad. Emerg. Med. **11**(10), 1097–1101 (2004)
11. Luscombe, R., Kozan, E.: Dynamic resource allocation to improve emergency department efficiency in real time. Eur. J. Oper. Res. **255**(2), 593–603 (2016)
12. Paul, S., Reddy, M., Deflitch, C.: A systematic review of simulation studies investigating emergency department overcrowding. Simulation **86**(8–9), 559–571 (2010)
13. Sinreich, D., Jabali, O., Dellaert, N.: Reducing emergency department waiting times by adjusting work shifts considering patient visits to multiple care providers. IIE Trans. (Inst. Ind. Eng.) **44**(3), 163–180 (2012)

Integrating Mental Health into a Primary Care System: A Hybrid Simulation Model

Roberto Aringhieri, Davide Duma and Francesco Polacchi

Abstract Depression and anxiety appear to be the most frequently encountered psychiatric problems in primary care patients. It has been also reported that primary care physicians under-diagnose psychiatric illness in their patients. Although collaborative care has been shown to be a cost-effective strategy for treating mental disorders, to the best of our knowledge few attempts of modelling collaborative care interventions in primary care are known in literature. The main purpose of this paper is to propose a hybrid simulation approach to model the integration of the collaborative care for mental health into the primary care pathway in order to allow an accurate cost-effectiveness analysis. Quantitative analysis are reported exploiting different and independent input data sources in order to overcome the problem of the data appropriateness. The analysis demonstrates the capability of the collaborative care to reduce the usual general practitioner overcrowding and to be cost-effective when the psychological treatments have a success rate around the 50%.

Keywords Mental health · Collaborative care pathway · Cost effectiveness
Discrete event · Agent based · Hybrid simulation

1 Introduction

World Health Organization (WHO) estimates that as much as 24% of all patients contacting general health services suffer from well-defined psychological disorders and that another 10% have psychological problems which may not meet the criteria for a formal diagnosis of mental disorder, but diminish the quality of life and cause

R. Aringhieri (✉) · D. Duma
Dipartimento di Informatica, Università degli Studi di Torino, Turin, Italy
e-mail: roberto.aringhieri@unito.it

D. Duma
e-mail: davide.duma@unito.it

F. Polacchi
Università degli Studi di Torino, Turin, Italy

© Springer Nature Switzerland AG 2018

P. Daniele and L. Scrimali (eds.), *New Trends in Emerging Complex Real Life Problems*, AIRO Springer Series 1,
https://doi.org/10.1007/978-3-030-00473-6_7

disability [13]. Nowadays such disorders are often diagnosed and treated in primary care settings adopting the *collaborative care* approach [9], in which specialised staff, i.e., psychologists, support the primary care practices. In Italy, collaborative care experiences are reported in [7] confirming also the association between medical-psychiatric co-morbidity and frequent utilisation of primary care resources, that is *frequent attenders* of a primary care service are usually affected by some mental disorders.

Cost-effectiveness analysis plays an important role in the economic evaluation of such interventions [14]. A critical review of model-based economic studies of depression argued hat little attention has been paid to issues around modelling studies with a focus on potential biases [1]. To the best of our knowledge, few attempts of modelling collaborative care interventions in primary care are known in literature [2]. None of these works analyse the impact of the frequent attenders to the usual general practitioner overcrowding. Further, the problem of the appropriateness of data sources used to estimate input parameters is discussed in [12].

The main purpose of this paper is to propose a hybrid simulation approach to model the integration of the collaborative care for mental health into the primary care pathway in order to allow an accurate cost-effectiveness analysis. The hybridisation of different methodologies is a way to deal with challenging problem arising in health care analysis [4]. In this problem, the challenge is to model the behaviour of the population pertaining the general practitioner, which differs from their frequency of attendance, and the patient flow within the collaborative care pathway. To this end, the proposed hybrid approach exploits the Agent Based Simulation (ABS) and the Discrete Event Simulation (DES) methodologies to model the population behaviour and the collaborative care pathway, respectively. Instead of using our own data, we use different and independent data sources in order to overcome the problem of their appropriateness.

2 The Mental Care Pathway: A Case Study

In the last years, several trials has been carried out in Piedmont Region.[1] Among them, we selected the trial carried out at the local health unit ASLTO3 in Turin. The trial consists in offering mental care for 10 h per week within a general practitioner office. Further, phone and home support is also provided. Such a trial defines a *mental primary care pathway*, that is the the step-by-step patient flow within the collaborative care pathway.

The mental primary care pathway implemented at the ASLTO3 is depicted in Fig. 1. The general practitioner (GP) meets a patient that could suffer from a mental disorder. If the GP recognises a mental disorder and the patient is willing to accept

[1]Regione Piemonte, *Indirizzi e raccomandazioni per l'implementazione dell'assistenza psicologica nelle cure primarie nella rete sanitaria territoriale del Piemonte*, Scheda P.A.S. 2012 - n. 4.1.7, 2013.

psychological care, the collaborative care process starts with a counselling between the GP and the psychologist in order to define the level (*low*, *medium* or *high*) of the mental disorder. In the case of indecision, the patient can follow the usual primary care pathway or can accept a consultation with the psychologist before starting the collaborative care process. Each treatment ends up with a success or a failure. A *success* consists in a remission of the patient' symptoms with a decrease in the number of visits to the GP: in our case study, the success of a treatment is measured adopting the CORE-OM self-evaluation scale [6], which evaluates the improvement of the patient's life quality. Note that only patients with a low or medium level of mental disorder—in accordance with ICD-10 criteria—are treated, while high level patients will be in charge of the specialised service.

3 The Hybrid Model

We report the proposed hybrid (DES and ABS) simulation model to represent the integration of the collaborative care for mental health into the primary care pathway in order to allow an accurate cost-effectiveness analysis.

Modelling the mental primary care pathway with DES. Inspired by the case study discussed in Sect. 2, we propose a model for the mental primary care pathway. The DES model is a straightforward implementation of the pathway depicted in Fig. 1. As mentioned in the introduction, one of the main characteristics of our approach is to different and independent data sources to animate our model. From this perspective, the main interesting part of the DES model is its parametrization. According to the daily schedule of an Italian GP (the studio is open at least 4 h per day) and to the national statistics [11], the number of daily patients is 20 (5 per hour) and, if the patient requires a consultation, it lasts around 30 min. The inter-arrival time of a patient is modelled using an exponential distribution while the duration is a triangular

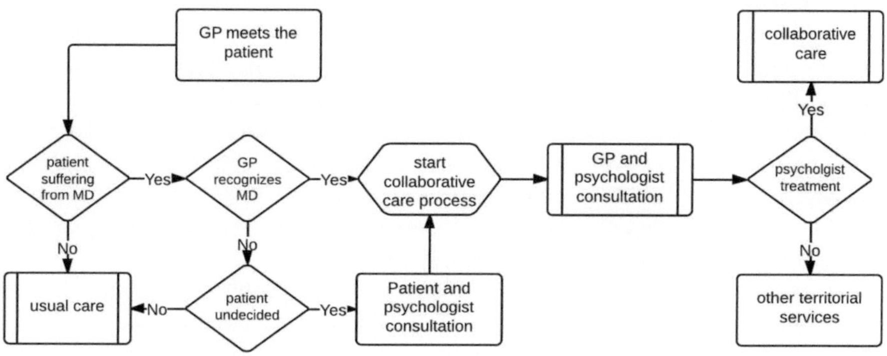

Fig. 1 Collaborative care pathway at the ASLTO3

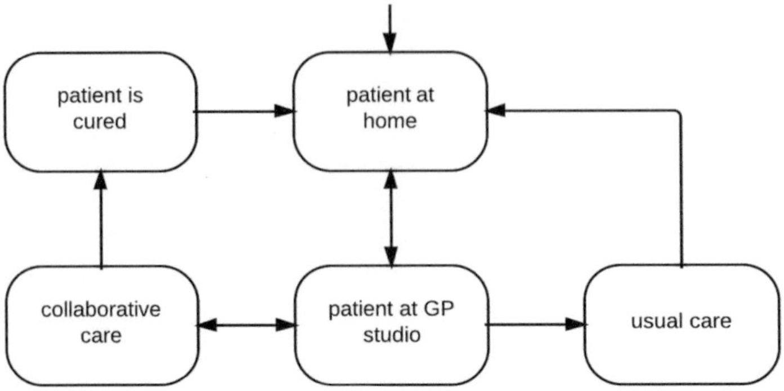

Fig. 2 Behavioural model: state-chart representing all the possible status of a patient

in the [20, 40] interval. Among all the patient entering to the GP, the percentage of the those suffering from a mental disorder is 24% as reported in [13].

According to the case study trial, the psychologist can work with no more than 3 patients per week for no more than 10 h per week. The treatment duration for low and medium level patients is, on average, 3 (in the case of the treatment is a simple consultation) and 8 h (in the case of the treatment is a psychotherapy). The percentage of high level patients is 7.3% while the remaining have low or medium level. Among them, the 42.1% follows a psychotherapy while the remaining the simple consultation. Finally, the success rate of a treatment is set to 80%, as in the case study.

Modelling the population behaviour with ABS. Different studies [7] confirm that primary care frequent attenders are usually affected by mental disorders. Thus we should pay a particular attention when modelling the population pertaining a GP since we have to take into account their frequency. As in [3], we would exploit the ABS methodology to describe a behavioural model representing the different patient status. Figure 2 describes the proposed behavioural model of a patient.

The patient usually stays at home, that is he/she is not going to the GP; when needed, the patient goes to the GP to have a consultancy; after the consultancy, two possible paths are available: the former is the usual care pathway while the latter is the collaborative care one. On the usual care pathway will flow all the patients not affected by a mental disorder and those affected but not recognised and/or not accepting the collaborative care pathway. Starting the collaborative care pathway, after a while a patient could be cured (depending on the treatment success) becoming a *routine attender*. We recall that the remission of the patient' symptoms implies also a decrease in the number of visits to the GP.

In our ABS model, a population of patients pertaining a given GP is modelled by generating a number of agents, each one following the behavioural model depicted in Fig. 2. The behavioural model of each agent is characterised by a specialised setting determining, for instance, if that patient is a routine or a frequent attenders.

Such settings are defined according to the parameters reported in Table 1, which also report the source of each value.

The hybrid model. The proposed hybrid model is composed of a population of n patients ($n \in [1300, 1350]$ [11]) pertaining of a given GP. Each patient is modelled by an agent following the behavioural model depicted in Fig. 2. When the agent have a state transition from “patient at home” to “patient at GP studio”, the hybrid model generates an item to represent the patient flowing within the usual care or collaborative care pathways following the DES model. If the patient will follow the usual care pathway then a state transition will be enabled from “patient at GP studio” to “usual care”; otherwise, the model will enable a state transition from “patient at GP studio” to “collaborative care”. When the patient finishes his/her “collaborative care” a state transition to “patient is cured” is enabled. When this item exits from the DES, the agent/patient makes a state transition from the “usual care” or “patient is cured” to the “patient at home”.

4 Quantitative Analysis

We provide an overview of the results that can be obtained using the proposed hybrid model. We recall that the reported results are obtained running the model using different and independent data sources in order to overcome the problem of the data sources appropriateness. The hybrid model is implemented using AnyLogic [5]. The running time required for each experiment is negligible.

Validation. Although the validation of a simulation model usually requires a quite complex analysis, in our case it is quite easy due to the simplicity of our DES model. Our validation experiment consists in a repeated test (30 times) to evaluate the output and the outcome of the model. The validation experiment is performed by forcing that the number of patients entering in the collaborative care process is around the number of patients actually participated in the trial, that is 41. Further, the time horizon is the same of the trial, that is two years.

The output, at the end of the time horizon, complies with the trial, that is the number of low/medium patients treated with a consultation is 17, those treated with a psychotherapy is 11 and those taken in charge by the specialised territorial service is 3. The remaining 10 patients have not yet finished their pathway. The outcome of the collaborative care is measured in term of successes and failures of the treatments. Again, the results at the end of the time horizon complies with the trial: the consultation has 14 successes and 3 failures while the psychotherapy 9 and 2.

On the basis of these considerations, the comparison is satisfactory with respect to our objective, that is the validation of the logical correctness of the proposed hybrid model.

Tuning of the parameters. The tuning of the model consists in determining a suitable parametrization of the resources involved in the DES model: actually, the validation

Table 1 Values of the parameters determining the patient population (MD = mental disorder)

Description	Value (%)	Source	Description	Value	Source
Number of frequent attenders (FA)	15.0	[11]	Number of high level among FAs	11.8%	[8]
Number of routine attenders (RA)	85.0	[11]	Number of high level among RAs	1.85%	[8]
Number of FAs suffering from MD	30.2	[8]	Number of accesses to the GP by FA	>12 per year	[11]
Number of RAs suffering from MD	10.8	[8]	Number of accesses to the GP by RA	<5 per year	[11]

of the DES model has been performed on the case study which involved a limited number of resources, that is 1 psychologists operating for 10 h per week (from Monday to Friday). We refer to this case as scenario S_1.

We introduce a further scenario, say S_2, in which 2 psychologists operate for 20 h per week for a grand total of 40 h which seems closer to the real needs of a mental primary care pathway. Both scenarios shares the same settings about service times within the DES model while the patient population has been generated according to the Table 1. We recall that the success rate of the treatments is estimated at the 80% as in the case study reported in Sect. 2.

Table 2 reports the results regarding the two scenario S_1 and S_2. The table is divided in three main sections: population, accesses to the GP and mental primary care pathway. Regarding the first section, it is worth noting that the composition of the population complies with the values reported in Table 1. Regarding the number of accesses, the number of saved accesses is an estimate computed with respect to the instant in which a frequent attender with mental disorder is cured becoming a routine attender. This estimate is higher in scenario S_2 due to the higher number of resources available. Anyway, it shows that the mental primary care pathway can effectively reduce the GP overcrowding. The third and last section shows the number of patient treated and those waiting for a treatment. Note that the increase of the number of patient treated in scenario S_2 is proportional to the increased number of hours offered by the psychologists.

Cost-effectiveness analysis In this section, we would provide an analysis in order to evaluate the cost effectiveness of the integration of the collaborative care for mental health into the primary care pathway, In this analysis, we will consider a third scenario, say S_3, which is the same of S_2 but the time horizon is extended from 2 to 5 years. This scenario is introduced to evaluate the economic sustainability in the medium/long term.

Table 2 Tuning the hybrid model: results on the scenario S_1 and S_2

Description	S_1	S_2
Population		
Total number of patients	1328	1329
Number of FAs with mental disorder	61	62
Number of RAs with mental disorder	123	123
Number of patients with low/medium mental disorder	174	176
Accesses to the GP		
Number of accesses by FAs	375	380
Number of accesses by RAs	1879	1895
Number of saved accesses by FAs	154	640
Mental primary care pathway		
Number of patients treated	30	120
Number of patients waiting for their treatment	141	53

While it is quite easy to identify the cost of a psychologist treatment within the mental primary care pathway, the cost of the usual care pathway is more difficult since it is concerned with different way of treating a not recognised mental disorder. Considering the Italian NHS, to the best of our knowledge the most accurate estimation is reported in [11] in which a range of yearly costs per patient is reported: the usual care costs from 2100 to 2500 € while the mental primary care costs from 900 to 1100 €. This difference is due to the fact that the latter is essentially the hourly cost of a psychologist multiplied the number of hours of the treatment while the former should considers not only the cost of the GP but also of the examinations usually prescribed by the GP himself.

Table 3 reports a cost comparison between the two pathways, that is the mental primary care and the usual care pathways. The results show the cost effectiveness of the mental primary care pathway proving also the capability of reducing the cost of the whole NHS system. Further, results for scenario S_3 show the sustainability of the system in the medium/long term.

The treatment success rate is the crucial parameter: actually, a lower success rate can determine the cost-ineffectiveness of the mental care pathway. To provide more insights to our analysis, we repeated our cost analysis varying the success rate of the psychologist treatment so far set to the 80%. We report the results only for the scenario S_2 since those for S_3 are almost the same.

Table 3 Cost analysis: ranges of results (in Euro) on the 3 scenarios

Description	S_1	S_2	S_3
Estimated costs (€)			
Mental primary care pathway overall cost	26850–32816	107880–131853	148350–181316
Usual care pathways overall cost	43470–51750	179340–213500	264030–314416
Variations	16860–18707	71460–81527	115010–133170

Table 4 Cost analysis: ranges of results (in Euro) varying the success rate on scenario S_2

Description	80%	40%	50%
Estimated costs			
Mental primary care pathway overall cost	107880–131853	108270–132330	108000–132000
Usual care pathways overall cost	179340–213500	88130–104917	108920–129667
Variations	71460–81527	−20140 to −27413	900 to −2334

Table 4 reports the results of the analysis. It is worth noting that the mental primary care pathway become cost effective as soon as the treatment success rate is around the 50%. Note that the same result is reported in the well-known Depression Report [10] providing also a further ex-post model validation.

5 Conclusions

We proposed a hybrid simulation model to evaluate the integration of collaborative care for mental health into a primary care system. In line with the current trends, the hybrid approach allows us to face properly the challenging modelling issues, that is how to model and integrate the behaviour of the population pertaining the general practitioner and the patient flow within the collaborative care pathway overcoming the the main weakness of the previous analysis, that is the appropriateness of data sources. To the best of our knowledge, this is the first attempt in the health care management literature. The quantitative analysis demonstrates the capability of the collaborative care to reduce the usual GP overcrowding and to be cost-effective when the psychological treatments have a success rate around the 50%, as reported in [10]. The proposed model could be extended to evaluate (i) the net social benefits in terms of quality-adjusted life years, and (ii) several resource sharing strategies and their impact on the GPs overcrowding.

References

1. Afzali, H., Karnon, J., Gray, J.: A critical review of model-based economic studies of depression: modelling techniques, model structure and data sources. PharmacoEconomics **30**(6), 461–482 (2012)
2. Afzali, H.H.A., Karnon, J., Merlin, T.: Improving the accuracy and comparability of model-based economic evaluations of health technologies for reimbursement decisions: a methodological framework for the development of reference models. Med. Decis. Mak. **33**(3), 325–332 (2013)
3. Aringhieri, R., Duma, D., Fragnelli, V.: Modeling the rational behavior of individuals on an e-commerce system. Oper. Res. Perspect. **5**, 22–31 (2018)
4. Aringhieri, R., Tànfani, E., Testi, A.: Operations research for health care delivery. Comput. Oper. Res. **40**(9), 2165–2166 (2013)
5. Borshchev, A.: The Big Book of Simulation Modeling. Multimethod Modeling with AnyLogic, vol. 6 (2013). ISBN 978-0-9895731-7-7
6. Evans, C., Connell, J., Barkham, M., Margison, F., McGrath, G., Mellor-Clark, J., Audin, K.: Towards a standardised brief outcome measure: psychometric properties and utility of the core-om. Br. J. Psychiatry **180**, 51–60 (2002)
7. Ferrari, S., Galeazzi, G., Mackinnon, A., Rigatelli, M.: Frequent attenders in primary care: impact of medical, psychiatric and psychosomatic diagnoses. Psychother. Psychosom. **77**(5), 306–314 (2008)
8. Gili, M., Luciano, J., Serrano, M., Jimnez, R., Bauza, N., Roca, M.: Mental disorders among frequent attenders in primary care: a comparison with routine attenders. J. Nerv. Ment. Dis. **199**(10), 744–749 (2011)
9. Glied, S., Herzog, K., Frank, R.: Review: the net benefits of depression management in primary care. Med. Care Res. Rev. **67**(3), 251–274 (2010)
10. Layard, R., Bell, S., Clark, D., Knapp, M., Meacher, M., Priebe, S., et al.: The depression report: a new deal for depression and anxiety disorders. LSE London (2006)
11. Marino, M., Gnavi, R., Rusciani, R., Spadea, T., Migliardi, A., Costa, G.: Caratteristichc dell'utenza del MMG in Italia e determinanti dell'accesso. I dati dell'indagine Multiscopo "Salute 2005". In: Prima Conferenza Italiana sulle Cure Domiciliari, Roma, maggio 2011
12. Nuijten, M.: The selection of data sources for use in modelling studies. PharmacoEconomics **13**(3), 305–316 (1998)
13. Üstün, T., Sartorius, N. (eds.): Mental Illness in General Health Care: An International Study. Wiley (1995)
14. Watkins, K., Cuellar, A., Hepner, K., Hunter, S., Paddock, S., Ewing, B., de la Cruz, E.: The cost-effectiveness of depression treatment for co-occurring disorders: a clinical trial. J. Subst. Abus. Treat. **46**(2), 128–133 (2014)

Cost Minimization of Library Electronic Subscriptions

Laura Bigram, Patrick Hosein and Jonathan Earle

Abstract Many libraries, particularly those at Universities in developing countries, are facing challenging financial times. This has led to the need for budget cuts and more efficient management of limited resources. One of the major costs of an academic library are the fees paid for subscriptions to electronic journals, databases, conference proceedings and for costs associated with downloads of individual papers if there is no subscription to the corresponding resource. Typically the decision as to whether or not a particular subscription is acquired is done based on faculty member requests, information about the resource (such as cost) and policies of the library. However, with the availability of a wide range of collected statistics, (number of downloads, impact factors, etc.) one can make better informed decisions. In this paper we provide a decision support system in which we define a metric for the value obtained per access to a resource and then determine the minimum budget required to achieve a given total value of this metric.

Keywords Optimisation · Data analysis · Decision support system

1 Introduction

Due to the recent economic downturn in some countries, many academic institutions are having to slash their budgets resulting in a reduction in financial allocations to their libraries. Many academic libraries have been left with the challenge of providing the same value and experience to faculty and students despite the significant decrease in funding and spiraling increases in resource costs over the past few years [4].

L. Bigram · P. Hosein (✉) · J. Earle
The University of the West Indies, St. Augustine, Trinidad
e-mail: patrick@lab.tt; patrick.hosein@sta.uwi.edu

L. Bigram
e-mail: laura@lab.tt

J. Earle
e-mail: jonathan@lab.tt

© Springer Nature Switzerland AG 2018

P. Daniele and L. Scrimali (eds.), *New Trends in Emerging Complex Real Life Problems*, AIRO Springer Series 1,
https://doi.org/10.1007/978-3-030-00473-6_8

Acquisitions of resources can account for several millions of dollars in a budget and so even small percentage reductions can result in significant savings. For example, in one particular library, approximately eighty two percent of the library's budget was spent on e-resource acquisitions for the year 2015. Decision support systems have been successfully used in the past to manage human resources in libraries (e.g., see [1]) with considerable savings. We focus on library cost savings for electronic subscriptions.

In this paper we provide a model for optimizing the selection of resources to subscribe to at the beginning of the academic year. We formulate the associated mathematical model and develop an algorithm for obtaining the solution. We then use the approach to determine the optimal subscriptions using real data and compare this solution with the actual performance in prior years.

We first develop a metric that measures the value of electronic resources to an academic community. For example, a simple metric is the number of downloads; more downloads implies more value to the community because of the increased knowledge provided. However, we go further and add other factors that affect value, such as impact factors and the importance of the field to the general goals of the university or country. Given this metric we can now measure the total value achieved for the given subscriptions made. What we then do is minimize the total cost while keeping the total value fixed. To our knowledge, this particular model is new to the area. The optimization approach is also one of our contributions. Note that, the work in [1–3] assumes material acquisition is budgeted by categories (books, journals etc.) whereas we optimize over all categories. The reason being that e-resources are becoming the preferred form of acquisitions and hence categorization can contribute to underutilization.

Although the model can take into account any of the resources acquired by the library, we only had access to a subset of the data, that of Science Direct e-journals. Since this formed a significant percentage (approximately 18%) of the resources, we believe that the results reflect what can be obtained if all resources are taken into account. We obtained statistics on e-journals in the Science Direct database for the years 2011–2015. This dataset consists of 1739 individual journals and for each of these it includes the title, a unique identifier, year, number of downloads for the year, annual subscription cost, cost per download and subject areas (keywords). One of our objectives is to check how closely the subscribed journals relate to the research goals of a university. We therefore also produced a list of the keywords that reflect such goals. Journals that have more common subject areas with the set of keywords in the research goals should be given more value.

2 Mathematical Model

In this section we provide the optimization objective and a mathematical model that will be used to achieve this objective. Note that the library pays for a service (access to resources) and the associated community receives some value from this service.

The service cost is easily obtained but the associated value of a downloaded resource depends on various factors. In this paper, we take into account two of the factors that contribute to the value but others can be added in the formulation in the future. Note that in this paper we are focused only on subscription or pay-per-download services of electronic journals and not acquisition of books.

Let $\mathcal{J}$ denote the set of resources (e.g., Journals, Conference Proceedings, etc.) accessible through the library, either through subscriptions or pay-per-download. For any resource $j \in \mathcal{J}$ we use the following notation:

$$\begin{aligned} S_j &= \text{the annual subscription rate for the resource} \\ P_j &= \text{the price per publication download for the resource} \\ I_j &= \text{the Impact Factor of the resource} \\ D_j &= \text{the number of completed downloads for the year} \\ R_j &= \text{the number of requested downloads for the year.} \end{aligned}$$

Note that some requests for a paper may not be satisfied because there is no journal subscription and no pay-per-download option or the cost for the download is considered too much or the budget for downloads has been exhausted and hence $R_j \geq D_j$.

We next take into account the relevance of the journal/conference resource to the goals of the university. In many developing states the Government has indicated specific areas that need to be developed as high priority as part of an overall vision for the country. For example, in a small island state there may be less interest in Quantum Physics research than in ICT since the latter will help in development of the country. If this is the case, we take this factor into account as follows. For any resource j, let K_j denote the set of keywords for the resource. Let O denote the set of keywords that represent the development objectives of the country. We assume that members of O are contained within the union of the sets K_j. We will weight each resource by the number of its keywords that are included in O.

2.1 Resource Value and Cost

The cost of a subscribed resource j is simply the annual cost of the subscription, S_j. For other resources the cost depends on the number of downloads and is given by $P_j D_j$. Naturally if $S_j < P_j R_j$ then one should subscribe to the resource otherwise it is better to pay for each download. Therefore we define the cost of resource j as a function of the number of satisfied requests x as:

$$C_j(x) = \min\{x P_j, S_j\} \tag{1}$$

Since we do not know the number of downloads in advance we will use historical data (with linear regression) to estimate the number of requests in the upcoming year.

For the value V_j of a resource we may take into account the Impact Factor as well as the goals of the university. This is given by,

$$V_j = 1 + \alpha I_j + \beta |K_j \cap O| \tag{2}$$

where α and β determine the importance of each factor. The third component is the number of common keywords between the resource and goals. For our numerical results we do not have Impact Factors and so we set $\alpha = 0$. For Governmental research goals we use $\beta = 0.1$.

2.2 *Problem Formulation*

The objective of this problem is to minimize the total cost for a given total value. However, as we mentioned before, this optimization requires knowledge of the number of requests per resource which is not known in advance. This will be obtained by using historical data to predict the expected number of requests for each resource. Let x_j represent the number of requests satisfied for resource j. This is the decision variable over which we must optimize. Note that the number of requests is integer but we relax this constraint and assume that x_j is continuous. We will find that, in the optimal solution of the relaxed problem, all but one of the optimal variables x_j will in fact be integral and hence we may only need to round one decision variable with negligible loss in optimality. We use $F(\mathbf{x})$ to denote the objective function value for a given decision vector $\mathbf{x}$. The optimization problem can be stated as follows:

$$\min_{\mathbf{x}} F(\mathbf{x}) \equiv \sum_{j \in \mathcal{J}} C_j(x_j) \tag{3}$$

$$\text{s.t.} \quad \sum_{j \in \mathcal{J}} V_j x_j = T$$

$$\text{and} \quad 0 \leq x_j \leq R_j \quad \forall\, j \in \mathcal{J}$$

where T denotes the total value to be achieved. We first provide an important property of the optimal solution.

Lemma 1 *The optimal solution has the property that, for all except maybe one resource, either all requests are satisfied or none is satisfied.*

Proof This is a well known property of the solution obtained when taking the minimization of a sum of concave functions over a polytope. □

Although this property is helpful it is still difficult to check all possible solutions to find the optimal one. Next, we consider a modified version which can be solved and has a solution close to optimal. Consider the following lower bound to the objective function:

$$C'_j(x) = \min\left\{P_j, \frac{S_j}{R_j}\right\}x \equiv G_j x \tag{4}$$

For convenience we use the representation $G_j x$ for some constant G_j for each j. Note that for all feasible x we have $C'(x) \leq C(x)$. Consider the problem (3) but with $C(x)$ replaced with $C'(x)$ or equivalently Gx since for a given resource the function is linear in x. Furthermore we convert the optimization problem to a maximization problem by considering the negative of the objective function value. This modified problem can be solved using Lagrange Multiplier methods by introducing λ, μ and γ. The Lagrangian is given by,

$$\mathcal{L}(\mathbf{x}, \lambda, \boldsymbol{\mu}, \boldsymbol{\gamma}) = -\lambda T + \sum_{j\in\mathcal{J}} -G_j x_j + \lambda V_j x_j + \mu_j x_j - \gamma_j(x_j - R_j) \tag{5}$$

$$\text{s.t } \mu_j x_j = 0, \quad \gamma_j(x_j - R_j) = 0, \quad \mu_j \geq 0, \quad \gamma_j \geq 0 \quad \forall\, j \in \mathcal{J} \tag{6}$$

Taking partial derivatives and setting to zero we obtain:

$$\frac{\partial \mathcal{L}}{\partial x_j} = -G_j + \lambda V_j + \mu_j - \gamma_j = 0 \tag{7}$$

$$\frac{\partial \mathcal{L}}{\partial \lambda} = -T + \sum_{j\in\mathcal{J}} V_j x_j = 0 \tag{8}$$

Let us consider the various cases:

$$\mu_j > 0, \gamma_j > 0 \Rightarrow x_j = 0, \quad x_j = R_j \quad \text{not possible}$$

$$\mu_j > 0, \gamma_j = 0 \Rightarrow x_j = 0, \quad \lambda < \frac{G_j}{V_j}$$

$$\mu_j = 0, \gamma_j > 0 \Rightarrow x_j = R_j, \quad \lambda > \frac{G_j}{V_j}$$

$$\mu_j = 0, \gamma_j = 0 \Rightarrow 0 \leq x_j \leq R_j, \quad \lambda = \frac{G_j}{V_j}$$

Therefore, we just need to find $\lambda = \lambda^*$ that satisfies these conditions. This can be accomplished by starting with the resource with the smallest ratio $\frac{G_j}{V_j}$ and satisfying all of the requests for that resource. We then go to the resource with the next smallest ratio and repeat. We keep track of the total value and stop when this total reaches T. In the case in which $V_j = 1$ we note that all optimal decision variables are integers and hence the optimal solution for the relaxed (approximate) problem is the same as that of the integer (approximate) problem. The solution of the modified problem will be used for our numerical results and so we next show that the error is bounded and in fact quite small. We start with the following Lemma.

Lemma 2 *Let* $\mathbf{x}^*$ *denote the optimal solution to the original problem (3) and let* $\mathbf{y}^*$ *denote the optimal solution to the modified problem then* $F(\mathbf{y}^*) - F(\mathbf{x}^*) \leq \max_{j\in\mathcal{J}} S_j$.

Proof Note that the solution of the modified problem also has the property that for all but one resource, requests are totally satisfied or not satisfied. Suppose that some resource i is not at an extreme point. Since $C'(x) < C(x)$ then the optimal solution for the modified problem is less than that of the original problem so,

$$F(\mathbf{x}^*) \geq F'(\mathbf{y}^*) \tag{9}$$

where F' is used to represent the objective function value of the modified problem. Now note that for all resources j except i we have $C'(y_j) = C(y_j)$ since both functions are equal at extreme points. Hence, we can write $F(\mathbf{y}^*) = F'(\mathbf{y}^*) + C_i(y_i) - C_i'(y_i)$. Also, note that the maximum difference between $C_i(y_i)$ and $C_i'(y_i)$ is S_i and hence we can write $F(\mathbf{y}^*) \leq F'(\mathbf{y}^*) + S_i$. Inserting this into (9) we obtain $F(\mathbf{x}^*) \geq F(\mathbf{y}^*) - S_i$ and hence,

$$F(\mathbf{y}^*) - F(\mathbf{x}^*) \leq S_i \leq \max_{j\in\mathcal{J}} S_j \tag{10}$$

□

Note that, if in the optimal solution all resources are at extreme points then the optimal solution for the modified problem is also optimal for the original one. Since subscriptions for hundreds of resources are acquired and these contribute to F then the error is small when compared to the optimal function value. We will compute this error for our numerical results to illustrate this point.

3 Numerical Results

In this section, we provide numerical results to illustrate the performance of the proposed model as well as present our prediction model. We focus on the year 2014. The subscription decisions for 2014 must be made in 2013 and so to make a fair comparison we assume that only information available in 2013 can be used for the proposed algorithm. Using the publication download statistics for years prior to 2014 we predict download values for 2014. We use simple linear regression for this prediction. In addition to the journals that are active in 2014 (i.e., had at least one download), we also include in our model journals that are potential candidates for downloads. To do this, we look at historical data and if a journal previously had a subscription we predict what the number of downloads would have been had it been kept as an option. We use the Symmetric Mean Absolute Percent Error (sMAPE) to ensure that our predictions were sufficiently accurate.

Note that our model also requires the 2014 target value which is also not known in 2013. We again use historical data with linear regression to estimate this value for 2014. Given the predicted downloads and predicted target value we can run the optimization algorithm to determine the set of journals that should be subscribed, $\mathcal{J}_s^*$ and the set of journals for which the library should pay per download $\mathcal{J}_d^*$.

In order to compare the proposed approach with the actual 2014 subscriptions, we have to evaluate the resulting cost for both approaches given the actual downloads in 2014. Let $\mathcal{J}_s$ and $\mathcal{J}_d$ denote the subscription and non-subscription journal decisions made by the library for 2014 and $\hat{x}_j$ denote the number of pay-per-downloads for journal j. The total cost to the library would therefore be:

$$C_{lib} = \sum_{j \in \mathcal{J}_s} S_j + \sum_{j \in \mathcal{J}_d} \hat{x}_j P_j \tag{11}$$

Let us now consider the cost that would have resulted if the proposed approach was used. Different subscription decisions made in 2013 would have affected the resulting downloads. There may have been journals that the optimal solution decided not to purchase (either by subscription or pay-per-download) but the library solution had allowed it. In such cases the cost would be included for the library solution but not for the optimal solution. There may also be journals that the optimal solution decided to include (i.e. they were included in the past and the optimal solution includes them) but there are no downloads for these in 2014. For such cases we include the cost for the predicted values instead. For each such journal j, let x_j^* denote the optimal number of requests for the journal. The 2014 cost for the optimal solution can therefore be written as:

$$C_{opt} = \sum_{j \in \mathcal{J}_s^*} S_j + \sum_{j \in \mathcal{J}_d^*} x_j^* P_j \tag{12}$$

So C_{lib} provides the cost incurred based on the library's decisions while C_{opt} denotes the cost incurred if the optimal algorithm is used.

Let us now consider the target values. The proposed algorithm uses a predicted target value but this can be different to the value actually obtained in 2014. Let T_{opt} be used to denote the value achieved in 2014 if the optimal solution was used and let T_{lib} denote the actual value achieved in 2014. Since we have costs for different total values then we need to normalize them and so we compute the cost ratio as,

$$Q = \frac{C_{opt}/T_{opt}}{C_{lib}/T_{lib}} \tag{13}$$

This ratio determines the cost gain obtained with the optimization approach.

We also repeated this computation with perfect prediction. In other words, we ran the algorithm using the downloads actually experienced in 2014. This allows us to see the gain that is possible without including prediction errors. We performed

Table 1 Performance Results

Year	Prediction	Value function	Cost ratio Q	Error bound
2014	Perfect	Downloads	0.036	0.012
2014	Perfect	Downloads + goals	0.038	0.092
2014	Regression	Downloads	0.051	0.077
2014	Regression	Downloads + goals	0.091	0.075

these evaluations for two use cases $V = 1$ (i.e. just using downloads) and also for the value function for the keyword component using a value of $\beta = 0.1$. These are all included in Table 1. Note the significant cost savings that are achievable by simply making better subscription choices. For the case of imperfect predictions and including research goals in our objective the cost for the optimized approach is approximately 10% of that of the present approach. The source of these savings can easily be verified by inspection. For example, for the journal *Surgery (Oxford)*, the library paid per download while the optimal solution decision was to pay for a subscription which resulted in a savings of $157,423. On the other hand in the case of *Journal of Sound and Vibration* the library subscribed to the journal but there were few downloads and so the optimal solution decision was to pay for each download rather than to subscribe and this resulted in a savings of $7,122. We also find that the error bounds are relatively small and will become even tighter when the complete dataset is considered.

References

1. Chorba, R.W., Bommer, M.R.W.: Developing academic library decision support systems. J. Am. Soc. Inf. Sci. **34**(1), 40–50 (1983)
2. Enger, K.B.: Using citation analysis to develop core book collections in academic libraries. Libr. Inf. Sci. Res. **31**(2), 107–112 (2009)
3. Ho, T.F., Shyu, S.J., Wu, Y.L.: Material acquisitions in academic libraries. In: 2008 Asia-Pacific Services Computing Conference, APSCC'08, pp. 1465–1470. IEEE (2008)
4. Jotwani, D.: Trends in acquisition and usage of electronic resources at indian institutes oftechnology libraries. Ann. Libr. Inf. Stud. (ALIS) **61**(1), 33–40 (2014)

The Mahalanobis Distance for Feature Selection Using Genetic Algorithms: An Application to BCI

Maria Elena Bruni, D. Nguyen Duy, Patrizia Beraldi and Antonio Violi

Abstract High dimensionality is a big problem that has been receiving a lot of interest from data scientists. Classification algorithms usually have trouble handling high dimensional data, and Support Vector Machine is not an exception. Trying to reduce the dimensionality of data selecting a subset of the original features is a solution to this problem. Many proposals have been applied and obtained positive results, including the use of Genetic Algorithms that has been proven to be an effective strategy. In this paper, a new method using Mahalanobis distance as a fitness function is introduced. The performance of the proposed method is investigated and compared with the state-of-the-art methods.

Keywords Mahalanobis distance · Genetic algorithm · Feature selection

1 Introduction

Recent technological developments such as the Internet of Things, microarrays and the advent of Big Data, have facilitated the emergence of enormous amounts of complex multivariate data in various applications involving thousands of attributes. In the machine learning context, an individual measurable property of the process

M. E. Bruni (✉) · P. Beraldi · A. Violi
Department of Mechanical, Energy and Management Engineering, University of Calabria, Cosenza, Rende, Italy
e-mail: mariaelena.bruni@unical.it

P. Beraldi
e-mail: patrizia.beraldi@unical.it

A. Violi
e-mail: antonio.violi@unical.it

D. Nguyen Duy
Department of Mathematics and Computer Science, University of Calabria, Cosenza, Rende, Italy
e-mail: dund.hust@gmail.com

© Springer Nature Switzerland AG 2018

P. Daniele and L. Scrimali (eds.), *New Trends in Emerging Complex Real Life Problems*, AIRO Springer Series 1,
https://doi.org/10.1007/978-3-030-00473-6_9

being observed is called feature. Using a set of features is necessary to perform classification, however, many of them could be highly correlated with other variables leading to an extra classification burden and unwanted noise that might introduce bias and reduce the classification performance. By eliminating the redundant variables with little or no predictive information, the relevant information content can be obtained from fewer unique features, that have enough discrimination power. To remove irrelevant features, without significantly degrading the performance of the system, several techniques have been developed to identify and select an informative subset of features. The aims of these techniques (called feature selection methods) are multifold, since they try to avoid over-fitting, reduce data dimension, improve classification and clustering performance and produce models which can interpret data better (see [2] and the references therein).

There are two main groups of feature selection methods proposed in the literature: filters and wrappers. Filter approaches use only intrinsic properties of the data, regardless the chosen classifier. Some popular methods from this group are Information Gain, ReliefF, Correlation-based Feature Selection and Consistency-based Feature Selection [11]. In the second approach, the selection criterion of the features depends on the learning algorithm used to construct the classifier [5]. In particular, the predictor is used as a black box and its performance as the objective function to evaluate the feature subset. Since exhaustive search methods can become computationally intensive for large datasets, heuristic search algorithms have been employed to find a subset of variables which maximizes the objective function. These can be broadly classified as sequential search, which start with an empty set (full set) and add features (remove features) until the maximum objective function is obtained, and evolutionary algorithms [6]. Among them, genetic algorithms (GAs, for short), which are popular meta-heuristics inspired by Darwin's theory about evolution, are the most used [1, 3, 4]. In this heuristics, a few individuals belonging to a generation are selected, their genetic information recombined and randomly mutated through the application of genetic operators, to produce a new generation, hopefully better than the previous one. The selection is guided by a so-called fitness function, which has been commonly considered as the classification accuracy, usually obtained by applying a classic linear SVM method [12]. The straightforward approach of applying GA in the feature selection process has a drawback, since evolution, guided by classification accuracy, will favour individuals with many features and hence high classification accuracy. To overcome this problem, a GA with Aggressive Mutation (GAAM) was proposed by Rejer [9], where individuals contain a specified number of features. This number of features, set by the user, can be still too high and running the algorithm several times using different number of genes, time consuming. To overcome this weakness, a variant of the GAAM has been proposed in [10], called by the authors GA with melting individuals (GAMI, for short). The optimization process not only maximizes the classifier accuracy but also minimizes the number of features, by randomly removing from each individual one feature in successive iterations of the algorithm. In the selection step of both the GAAM and the GAMI, the fitness function was evaluated as the classifier accuracy, tested with tenfold cross-validation. Even though classification accuracy has been commonly used, and it has been proven to be a good method, its main drawback is the slow performance, due to the fact that the classifier has to be applied to each single feature subset.

This paper proposes a new method that uses the separation between classes of data (i.e. the Mahalanobis distance) as the fitness function within the GAAM and GAMI approaches reviewed above. This new approach will be referred to as GAMI with Mahalanobis distance (GAMD, for short).

2 The Proposed Approach

In order to be applied, each GA needs a proper encoding of the genetic information of the individuals, a fitness function for evaluating individuals in each generation, and genetic operators used for mutation and reproduction. The initial population is usually randomly generated and one or more individuals are selected from the current population, randomly mutated and then recombined to form a new generation, that will be used in the next iteration of the algorithm.

In the GAAM algorithm, each individual is represented by a chromosome with a fixed number of genes N (set at the beginning of the algorithm), corresponding to a subset of N features out of P. Hence, M individuals are created randomly by choosing values from the set $\{0, 1, 2, \ldots P\}$, where a value of 0 means that any feature has been selected. When this occurs, the number of features of the individual is reduced by one. The aggressive mutation of the GAAM is performed on each of the M individuals of the population, by randomly mutating a gene for each children. Therefore, a total of NM new individuals will be created. Then, the classic one-point crossover (with a probability equal to one) is applied on the individuals from the previous population hence creating M new individuals during the crossover operation (the population now has $M + NM + M$ individuals). On the basis of the fitness function, the worst $NM + M$ individuals will be discarded, and only the best M individuals will remain in the population. In the variant proposed later on, the GAMI, the number of features is iteratively decremented during the algorithm and a threshold θ for the fitness value of the best individual ϕ is used to eventually terminate the algorithm, stuck in a local optimum. In Fig. 1, the scheme of the GAMI is reported.

In this paper, we present our proposal which is based on a modification of the GAMI approach reviewed above, on which we embed the Mahalanobis distance function, used as a fitness function (that will be indicated in the following with ϕ^{MAL}). The Mahalanobis distance is a well known statistical distance function, between two or more multidimensional points in the space defined by correlated variables. Since it accounts for correlations between groups of features, it will adequately evaluate the distance between classes of data points. Only when the features are uncorrelated, the distance under a Mahalanobis distance metric is identical to that under the Euclidean distance metric. In particular, if we have two groups of data with mean $\hat{x}^i$ and $\hat{x}^j$, respectively, and a covariance matrix Σ, the Mahalanobis distance is given by

$$\phi^{Mal} = \sqrt{(\hat{x}^i - \hat{x}^j)^\top \Sigma^{-1} (\hat{x}^i - \hat{x}^j)}.$$

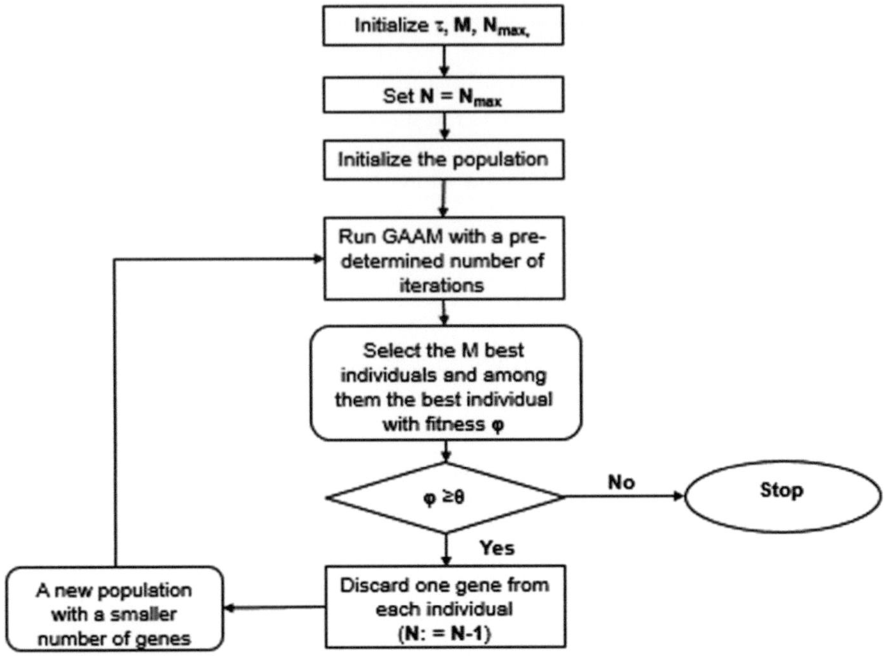

Fig. 1 The work-flow of the GAMI

Whilst the commonly used Euclidean distance metric assumes that each feature of data point is equally important and independent from others, the Mahalanobis distance is able to identify important features and discriminate relevant and irrelevant features. This claim is supported by a previous research [7], where an experiment was conducted to asses the relationship between the Mahalanobis distance and the performance of the SVM. The obtained results indicated a noticeable decline in the performance when the Mahalanobis distance decreases. This suggests that if we could reduce the dimensionality of feature vectors, while increasing or even just not deteriorating the Mahalanobis distance, the performance of the SVM would be improved as well. Moreover, distance metric is a key issue in many machine learning algorithms [13]. One family of algorithms is developed with known class labels of training data points. Also in our case, the threshold τ is derived by using class label information brought by training examples and, in particular, it equals the Mahalanobis distance between the feature vectors of the positive samples and those classified as negative samples in training data. Then, each subset of features is evaluated by considering the distance between the positive and the negative data and by comparing it with the threshold τ. The higher is the Mahalanobis distance, the more distinct two groups are, the better is.

The algorithm starts with a random population of M individuals coding different subsets of features. Then, the GAAM is applied for a given number of iteration IT_{max}. In each iteration the individuals are crossed-over and mutated. Since the individuals from the mother population P^{curr}, as well as from the mutated population P^{mut} and from the new generation obtained by applying one-point crossover crossover P^{cross} are selected in Step 15, the last population is the best one. Then, the individual that has the highest fitness function is selected from the current population and its fitness value compared with the threshold τ. If $\phi^{MAL}_{currbest} \geq \tau$ the number of genes is reduced by randomly selecting and discarding one gene from each individual, otherwise, the algorithm terminates. The workflow of the GAMD is reported in Algorithm 1.

Algorithm 1: The workflow of the GAMD

---

1 **Input:** The maximum number of iterations IT_{max}, the maximum number of of genes of an individual N_{max} and the threshold τ.
2 **Initialization:** Set the number of genes $N := N_{max}$, $best, currentbest := null$; P^{curr}, P^{cross}, P^{mut}, $P^{best} = \emptyset$.
3 Create an initial population of M individuals randomly choosing N values from the set $\{0, 1, 2, \ldots, P\}$ and store them in P^{curr}
4 **repeat**
5     $it = 0$
6     $P^{it} = P^{curr}$
7   **while** $it \leq IT_{max}$ **do**
8   P^{cross}, $P^{mut} = \emptyset$
9   **for** $i \in P^{it}$ **do**
10   Apply 1 -point crossover operator to i, obtaining anew individual and store it in P^{cross}
11   **end**
12   **for** $i \in P^{it}$ **do**
13   **for** $g = 1, \ldots, N$ **do**
14   Assign a random value from the interval $\{0, 1, 2, \ldots, P\}$ to the gene g and save the individual i as a new individual in P^{mut}
15   **end**
16   **end**
17   Select the best M individuals from $P^{mut} \cup P^{it} \cup P^{cross}$ according to the evaluated fitness function ϕ^{MAL} and store them in P^{curr}
18   it:=it+1
19   **end**
20     $currentbest = argmin_{i \in P^{curr}} \phi^{MAL}_i$
21   **if** $\phi^{MAL}_{currbest} > \phi^{MAL}_{best}$ **then**
22   best:=currentbest
23   **end**
24   **for** $i \in P^{curr}$ **do**
25   Randomly select one gene to discard
26   **end**
27   Set the number of genes $N := N - 1$
28 **until** $\phi^{MAL}_{currentbest} < \tau$;
29 Return $best$

---

3 Experimental Phase

In this Section, we empirically evaluate the GAMD for classification of Brain-Computer Interface (BCI, for short) data. BCI technology represents a highly growing field of research with applications in many fields, ranging from prevention to neuronal rehabilitation for serious injuries. BCIs are systems that measure and convert electro-physiological measures from the brain into external devices, enabling communication between humans and computers. From a data mining point of view, this is a challenging task for several reasons, among which the presence of noisy, correlated and highly dimensional data.

The experiments were performed considering the data set III motor imagery, taken from the second BCI Competition provided by the Department of Medical Informatics, Institute for Biomedical Engineering, Graz University of Technology. This dataset was recorded from a female, 25 years old, sat in a relaxing chair with armrests. The task was to control a feedback bar by means of imagery left or right hand movements. The whole data set contains data from 280 trials, but target values (1—left hand, 2—right hand), are known only for 140 trials. These trials have been then used to test the algorithms, and divided into two subsets, the first (with cardinality 50) for classifier training and the second for external classifier testing. Ten-cross fold validation was used to measure the fitness function of individuals in the GAAM and GAMI and to assess the accuracy of the solution provided by the GAMD. In both cases,

$$\phi_i = \sum_{k=1}^{10} \frac{\frac{R_k}{U_k}}{10}$$

where ϕ_i is the fitness value of ith individual (or the best one), R_k ($k = 1, \ldots, 10$) is the number of correctly classified samples and U_k is the number of all samples in the kth validation subset. The simulations have been performed on a personal computer with a CPU Intel(R) Core(TM) i5-2450M @2.50 GHz, 8 GB DDR II RAM and 500 GB of HDD Hitachi. Matlab (The Mathworks, USA; version 2016a) has been used to implement the heuristics as well as for running the SVM classifier.

In the first set of experiments the GAAM, the GAMI and the GAMD were run 10 times with the following parameter setting. $N_{max} = 6$, the number of generations for the GAAM (also equal to IT_{max} in the GAMI and GAMD) was set to 100, the number of individuals in each population M was set to 10, $\tau = 0.8$ for the GAMI and $\tau = 2.34$ for the GAMD. Since Provost and Fawcett [8] have pointed out that accuracy is not always a suitable assessment metric, when, for instance, the class distribution is imbalanced which is our case, the AUC has been used as an alternative metric for assessing classifiers. The larger the AUC, the better the classifier is. Table 1 shows the best solutions and the corresponding fitness value and AUC values obtained after running GAAM, GAMI and GAMD 10 times each.

In comparison with other methods, GAMI always obtains solutions with a smaller number of features. Although the best subset of features came from the combination of GAMI and SVM, in general the GAMD achieves better performance than others.

Table 1 The results obtained after running the algorithms 10 times

GAAM			GAMI			GAMD		
Indexes of features	Fitness	AUC	Indexes of features	Fitness	AUC	Indexes of features	Fitness	AUC
24. 104. 156. 240. 258	0.86	0.80	23. 86. 263	0.84	0.87	95. 104 263	2.41	0.74
23. 30. 55. 173. 185. 263	0.84	0.70	27. 66. 97. 104. 242. 263	0.78	0.71	5. 23. 125	2.35	0.89
5. 23. 29. 78. 227. 264	0.86	0.82	24. 77. 250. 282	0.82	0.70	5. 104. 213	2.46	0.86
24. 47. 61. 104. 135. 278	0.82	0.74	23. 86. 263	0.84	0.87	86. 104. 263	2.42	0.88
179. 207. 247. 248. 255. 266	0.8	0.70	23. 86. 263	0.84	0.87	5. 104. 213	2.46	0.86
102. 105. 173. 203. 320	0.8	0.68	23. 24. 303	0.8	0.71	86. 104. 263	2.42	0.88
57. 86. 104. 185. 234. 288	0.82	0.87	5. 24. 195. 246. 269	0.84	0.77	5. 23. 125	2.35	0.89
3. 167. 266. 288. 309. 310	0.84	0.69	5. 23. 156	0.84	0.64	5. 23. 125	2.35	0.89
5. 22. 104. 287. 289	0.84	0.92	24. 87. 266	0.84	0.88	6. 86. 104. 105. 140. 282	2.29	0.89
23. 24. 104. 140. 146. 151	0.8	0.73	16. 24. 55. 169. 175. 273	0.82	0.64	14. 23. 289. 300	2.59	0.72
Avg		0.76			0.77			0.85

The average CPU is, respectively, 792.26, 2593.2 and 5.09 for the GAAM, the GAMI and the GAMD.

In order to provide a statistical comparison between the performances of different genetic algorithms, a t-test was employed. Each algorithm was run 100 times. For the data sets, it is revealed that there is no significant difference between the classification accuracy of GAAM versus GAMI (h = 0 with p = 0.9948). The null hypothesis at the 0.05 significance level is rejected instead if we compare GAMD versus GAMI.

The second set of experiments is devoted to the comparison of the GAMD approach with the approaches GAAM and GAMI, in terms of classification accuracy. In order to guarantee a fair comparison, the algorithms were run 50 times (except SVM because it is deterministic) with the parameters used in the papers [9, 10] for the GAAM and the GAMD, respectively. The first 112 samples were used for training the algorithms. The resulting threshold for the GAMD was $\tau = 3.538$. The classification accuracy of the final classifier calculated over the whole set is reported in Table 2. The conclusion that could be drawn after observing the results is that GAMD returned feature sets with higher classification accuracy with a maximum vale of 0.96 and an average improvement of at least 7% in classification accuracy with respect to the other methods.

Table 2 Classification accuracy obtained from 50 runs

	GAAM	GAMI	GAMD		GAAM	GAMI	GAMD		GAAM	GAMI	GAMD
1	0.71	0.82	0.82	21	0.71	0.71	0.86	41	0.82	0.68	0.89
2	0.79	0.71	0.93	22	0.68	0.71	0.71	42	0.68	0.75	0.93
3	0.64	0.79	0.86	23	0.71	0.79	0.79	43	0.68	0.71	0.93
4	0.86	0.75	0.79	24	0.71	0.75	0.86	44	0.68	0.86	0.79
5	0.89	0.71	0.96	25	0.82	0.82	0.82	45	0.82	0.71	0.75
6	0.71	0.82	0.61	26	0.75	0.79	0.93	46	0.75	0.75	0.89
7	0.71	0.71	0.75	27	0.71	0.71	0.93	47	0.71	0.71	0.79
8	0.75	0.79	0.93	28	0.93	0.82	0.79	48	0.82	0.86	0.79
9	0.64	0.75	0.79	29	0.75	0.86	0.86	49	0.57	0.79	0.93
10	0.75	0.71	0.82	30	0.68	0.79	0.82	50	0.75	0.89	0.93
11	0.68	0.82	0.79	31	0.82	0.86	0.79	Avg	0.74	0.76	0.83
12	0.71	0.71	0.79	32	0.68	0.71	0.96				
13	0.86	0.79	0.79	33	0.68	0.82	0.93				
14	0.68	0.75	0.93	34	0.86	0.82	0.93				
15	0.79	0.71	0.79	35	0.71	0.79	0.86				
16	0.79	0.82	0.79	36	0.86	0.68	0.89				
17	0.71	0.71	0.93	37	0.64	0.71	0.79				
18	0.68	0.79	0.79	38	0.68	0.86	0.68				
19	0.68	0.75	0.93	39	0.75	0.82	0.68				
20	0.71	0.71	0.79	40	0.79	0.75	0.89				

4 Conclusions

In this paper, we have investigated and solved the problem of features selection of small or high dimension data and proposed a new approach for this problem using GA and the Mahalanobis distance as fitness function. In general, the proposed method worked well in terms of accuracy and CPU time. As future work, a more comprehensive study on the relationship between the separation of data, the number of features and classification accuracy should be conducted.

References

1. Brester, C., Semenkin, E., Sidorov, M., Minker, W.: Self-adaptive multi-objective genetic algorithms for feature selection. In: Proceedings of International Conference on Engineering and Applied Sciences Optimization, pp. 1838–1846 (2014)
2. Chandrashekar, G., Sahin, F.: A survey on feature selection methods. Comput. Electr. Eng. **40**, 16–28 (2014)
3. Eads, D., Hill, D., Davis, S., Perkins, S., Ma, J., Porter, R., Theiler, J.: Genetic algorithms and support vector machines for time series classification. In: 5th Conference on the Application and Science of Neural Networks, Fuzzy Systems and Evolutionary Computation, pp. 74–85 (2002)
4. Garcia-Nieto, J., Alba, E., Jourdan, L., Talbi, E.: Sensitivity and specificity based multiobjective approach for feature selection: application to cancer diagnosis. Inf. Process. Lett. **109**(16), 887–896 (2009)
5. Kohavi, R., John, G.H.: Wrappers for feature subset selection. Artif. Intell. **97**(1–2), 273–324 (1997)
6. Mukhopadhyay, A., Maulik, U., Bandyopadhyay, S., Coello, C.A.C.: A survey of multiobjective evolutionary algorithms for data mining: part I. IEEE Trans. Evol. Comput. **18**(1), 4–19 (2014)
7. Nguyen Duy, D., Nguyen Hoang, H., Nguyen Xuan, H.: The impact of high dimensionality on SVM when classifying ERP data-a solution from LDA. In: Proceedings of the Sixth International Symposium on Information and Communication Technology, pp. 32–37. ACM, New York, NY, USA (2015). https://doi.org/10.1145/2833258.2833290
8. Provost, F., Fawcett, T.: Robust classification for imprecise environments. Mach. Learn. **42**(3), 203–231 (2001)
9. Rejer, I.: Genetic algorithm with aggressive mutation for feature selection in BCI feature space. Pattern Anal. Appl. **18**, 485–492 (2015)
10. Rejer, I.: Genetic algorithms for feature selection for brain–computer interface. Int. J. Pattern Recogn. Artif. Intell. **29**(5), 1559008-1–1559008-27 (2015) (World Scientific Publishing Company)
11. Sanchez-Marono, N., Alonso-Betanzos, A., Tombilla-Sanroman, M.: Filter methods for feature selection-a comparative study. In: Yin, H., Tino, P., Corchado, E., Byrne, W., Yao, X. (eds.) Intelligent Data Engineering and Automated Learning—IDEAL 2007. Lecture Notes in Computer Science, vol. 4881. Springer, Berlin, Heidelberg (2007)
12. Vapnik, V.: Statistical Learning Theory. Wiley-Interscience, NY (1998)
13. Yang, L., Jin, R.: Distance metric learning: a comprehensive survey, Technical Report, Michigan State University (2006)

An Iterated Local Search Algorithm for the Pollution Traveling Salesman Problem

Valentina Cacchiani, Carlos Contreras-Bolton, John W. Escobar, Luis M. Escobar-Falcon, Rodrigo Linfati and Paolo Toth

Abstract Motivated by recent works on the Pollution Routing Problem (PRP), introduced in Bektas and Laporte (Transp Res Part B: Methodol 45(8):1232–1250, 2011) [1], we study the Pollution Traveling Salesman Problem (PTSP). It is a generalization of the well-known Traveling Salesman Problem, which aims at finding a Hamiltonian tour that minimizes a function of fuel consumption (dependent on distance travelled, vehicle speed and load) and driver costs. We present a Mixed Integer Linear Programming (MILP) model for the PTSP, enhanced with sub-tour elimination constraints, and propose an Iterated Local Search (ILS) algorithm. It first builds a feasible tour, based on the solution of the Linear Programming (LP) relaxation of the MILP model, and then loops between three phases: perturbation, local search and acceptance criterion. The results obtained by the ILS on instances with up to 50 customers are compared with those found by a Cut-and-Branch algorithm based on the enhanced MILP model. The results show the effectiveness of the ILS algorithm, which can find the best solution for about 99% of the instances within short computing times.

V. Cacchiani (✉) · C. Contreras-Bolton · P. Toth
University of Bologna, Viale Risorgimento 2, 40136 Bologna, Italy
e-mail: valentina.cacchiani@unibo.it

C. Contreras-Bolton
e-mail: carlos.contrerasbolton@unibo.it

P. Toth
e-mail: paolo.toth@unibo.it

J. W. Escobar
Pontificia Universidad Javeriana, Cl. 18 #118-250, Cali, Colombia
e-mail: jwescobar@javerianacali.edu.co

L. M. Escobar-Falcon
Technological University of Pereira, Cl. 27 #10-02 Barrio Álamos, Pereira, Colombia
e-mail: luismescobarf@utp.edu.co

R. Linfati
University of Bío Bío, Av. Collao 1202, Concepción, Chile
e-mail: rlinfati@ubiobio.cl

© Springer Nature Switzerland AG 2018

P. Daniele and L. Scrimali (eds.), *New Trends in Emerging Complex Real Life Problems*, AIRO Springer Series 1,
https://doi.org/10.1007/978-3-030-00473-6_10

Keywords Pollution Traveling Salesman Problem · Iterated Local Search MILP model

1 Introduction

Nowadays, environmental issues are becoming more important. Emissions from vehicles traveling on roads are one of the main causes of pollution. Therefore, reducing carbon emissions is one of the most important goals that have to be taken into account in vehicle routing problems [4, 7]. In [1], the Pollution Routing Problem (PRP) was introduced: it is a variant of the Vehicle Routing Problem (VRP) in which the goal is not only to minimize the travel distance, but also the amount of green-house emissions, fuel, travel times and their costs. The authors proposed a Mixed Integer Linear Programming (MILP) model, and analyzed trade-offs between various performance measures of vehicle routing, such as distance, load, emissions and costs. In [3], the MILP model was extended to allow for low travel speeds, and an effective adaptive large neighborhood search heuristic was proposed for the PRP. A matheuristic approach was proposed in [5] for PRP and other green VRP variants.

Motivated by these recent works on the PRP, we study the Pollution Traveling Salesman Problem (PTSP), i.e. the problem of determining a Hamiltonian tour that minimizes a function of fuel consumption (dependent on vehicle speed and load) and driver costs. More precisely, we refer to the PRP as modelled in [3] and consider the single vehicle case. The PTSP is formally described in Sect. 2, where we also present a MILP model, enhanced with explicit subtour elimination constrains, for its exact solution. The main contribution of this work is an Iterated Local Search (ILS) algorithm [6], presented in Sect. 3, that is able to find good solutions for PTSP instances in very short computing times. In Sect. 4, we report computational experiments on instances with up to 50 customers, proposed in [3] and adapted here for the single vehicle case: in particular, we compare the results obtained by the ILS algorithm with those found by a Cut-and-Branch algorithm, based on the enhanced MILP model, we developed for the PTSP. Finally, we draw some conclusions in Sect. 5.

2 Problem Description and Formulation

The PTSP is defined on a complete directed graph $G = (\mathcal{N}, \mathcal{A})$ where $\mathcal{N} = \{0, \ldots, n\}$ is the set of nodes, 0 is a depot and $\mathcal{A}$ is the set of arcs. The distance from node i to node j is denoted by d_{ij}. The set $\mathcal{N}_0 = \mathcal{N} \setminus \{0\}$ is the customer set. Each customer $i \in \mathcal{N}_0$ has a non-negative demand q_i, and a service time t_i. We define

Table 1 Parameters used in the PTSP model

Notation	Description	Typical values
w	Curb-weight (kg)	6350
ξ	Fuel-to-air mass ratio	1
k	Engine friction factor (kJ/rev/L)	0.2
N	Engine speed (rev/s)	33
V	Engine displacement (L)	5
g	Gravitational constant (m/s^2)	9.81
C_d	Coefficient of aerodynamic drag	0.7
ρ	Air density (kg/m^3)	1.2041
A	Frontal surface area (m^2)	3.912
C_r	Coefficient of rolling resistance	0.01
η_{tf}	Vehicle drive train efficiency	0.4
η	Efficiency parameter for diesel engines	0.9
f_d	Driver wage per (£/s)	0.0022
κ	Heating value of a typical diesel fuel (kJ/g)	44
ψ	Conversion factor (g/s to L/s)	737
v^l	Lower speed limit (m/s)	5.5 (or 20 km/h)
v^u	Upper speed limit (m/s)	25 (or 90 km/h)

$D = \sum_{i \in \mathcal{N}_0} q_i$ as the capacity of the vehicle, and f_d as the driver wage per unit time. We consider a discretized speed function defined by $|\mathcal{R}|$ non-decreasing speed levels $\bar{v}^r$ ($r \in \mathcal{R}$), where each $r \in \mathcal{R}$ corresponds to a speed interval $[v^l, v^u]$, and $\bar{v}^r$ is set as $(v^l + v^u)/2$ (where v^l and v^u are, respectively, the lower and upper speed limits).

We adopt the fuel consumption expression proposed in [3], which extends the one presented in [1] to allow for speeds lower than 40 km/h, and refer the interested reader to these papers for explanations of how it is derived. For a given arc $(i, j) \in \mathcal{A}$ of length d_{ij}, traversed at speed v by a vehicle carrying load $M = w + f_{ij}$ (where w is the weight of the empty vehicle a.k.a. curb weight, and $f_{ij}$is the load carried by the vehicle on this arc), the fuel consumption can be expressed as:

$$F(v) = \lambda k N V d_{ij}/v + \lambda \beta \gamma d_{ij} v^2 + \lambda w \gamma \alpha_{ij} d_{ij} + \lambda \gamma \alpha_{ij} f_{ij} d_{ij} \tag{1}$$

where $\lambda = \xi/\kappa\psi$ and $\gamma = 1/1000\eta_{tf}\eta$ are constants, $\beta = 0.5C_d\rho A$ is a vehicle specific constant, $\alpha_{ij} = \tau + g \sin\theta_{ij} + gC_r \cos\theta_{ij}$ is an arc specific constant depending on the road angle θ_{ij}, and all other parameters and values, taken from [3], are reported in Table 1. In particular, the first two terms of (1) represent the speed-induced energy requirements, while the last two terms represent the load-induced energy requirements.

The PTSP calls for determining the minimum cost tour that departs from the depot and visits each customer exactly once by serving its demand, where the cost is given by the sum of the fuel consumption and driver wage. We introduce the following decision variables: (i) binary variables x_{ij} assuming value 1 if arc $(i, j) \in \mathscr{A}$ is traversed; non-negative variables f_{ij} representing the amount of flow (i.e. the load on the vehicle) on arc $(i, j) \in \mathscr{A}$; (iii) binary variables z^r_{ij} assuming value 1 if arc $(i, j) \in \mathscr{A}$ is traversed at speed level $r \in \mathscr{R}$. The MILP model for the PTSP reads as follows:

$$\text{Minimize} \sum_{(i,j)\in\mathscr{A}} \lambda kNVd_{ij} \sum_{r\in\mathscr{R}} z^r_{ij}/\bar{v}^r + \sum_{(i,j)\in\mathscr{A}} \lambda\beta\gamma d_{ij} \sum_{r\in\mathscr{R}} z^r_{ij}(\bar{v}^r)^2 \quad (2)$$

$$+ \sum_{(i,j)\in\mathscr{A}} \lambda w\gamma\alpha_{ij}d_{ij}x_{ij} + \sum_{(i,j)\in\mathscr{A}} \lambda\gamma\alpha_{ij}d_{ij}f_{ij} \quad (3)$$

$$+ \; f_d\Big(\sum_{(i,j)\in\mathscr{A}} \sum_{r\in\mathscr{R}} (d_{ij}/\bar{v}^r)z^r_{ij} + \sum_{i\in\mathscr{N}_0} t_i\Big) \quad (4)$$

subject to

$$\sum_{j\in\mathscr{N}_0} f_{0j} = D \quad (5)$$

$$\sum_{j\in\mathscr{N}_0} f_{j0} = 0 \quad (6)$$

$$\sum_{j\in\mathscr{N}} x_{ij} = 1, \; \forall i \in \mathscr{N} \quad (7)$$

$$\sum_{i\in\mathscr{N}} x_{ij} = 1, \; \forall j \in \mathscr{N} \quad (8)$$

$$\sum_{j\in\mathscr{N}} f_{ji} - \sum_{j\in\mathscr{N}} f_{ij} = q_i, \; \forall i \in \mathscr{N}_0 \quad (9)$$

$$q_j x_{ij} \le f_{ij} \le (D - q_i)\, x_{ij}, \forall (i, j) \in \mathscr{A} \quad (10)$$

$$\sum_{r\in\mathscr{R}} z^r_{ij} = x_{ij}, \; \forall (i, j) \in \mathscr{A} \quad (11)$$

$$x_{ij} \in \{0, 1\}, \; \forall (i, j) \in \mathscr{A} \quad (12)$$

$$f_{ij} \ge 0, \; \forall (i, j) \in \mathscr{A} \quad (13)$$

$$z^r_{ij} \in \{0, 1\}, \; \forall (i, j) \in \mathscr{A}, \forall r \in \mathscr{R} \quad (14)$$

The objective function consists of three main components to be minimized: (2) and (3) represent the fuel consumption, as defined in (1), by taking into account, respectively, the energy required by speed variations and that used to curry the curb weight and the load on the vehicle, while (4) corresponds to the driver wage, where the term in the external brackets is the total tour duration which depends on the speed and service times. Constraints (5) and (6) ensure, respectively, that the vehicle leaves full and returns empty at the depot. Constraints (7) and (8) guarantee that each node is visited exactly once. Constraints (9) and (10) define the load of the vehicle on each visited arc. Finally, constraints (11) link the x and z variables by imposing that exactly one speed level is chosen for each arc $(i, j) \in \mathscr{A}$, and constraints (12)–(14) define the variable domains.

We enhance model (2)–(14) with the explicit subtour elimination constraints (SECs), as proposed in [2] for the Asymmetric TSP:

$$\sum_{i \in \mathscr{S}} \sum_{j \in \mathscr{N} \setminus \mathscr{S}} x_{ij} \geq 1, \quad \mathscr{S} \subset \mathscr{N}, \quad \mathscr{S} \neq \emptyset. \tag{15}$$

Model (2)–(15) is used as benchmark in the computational experiments to evaluate the performance of the proposed ILS algorithm. In particular, we solve model (2)–(15) by a Cut-and-Branch algorithm, in which the SECs are separated, at the root node, by using the separation procedure proposed in [8], and the general purpose solver CPLEX is used to derive integer solutions.

3 Iterated Local Search Algorithm

The pseudo-code of the ILS algorithm is reported in Algorithm 1. The first step (lines 1–7) consists of iteratively solving the Linear Programming (LP) relaxation of model (2)–(15), by applying the separation procedure proposed in [8] to derive a set $\mathscr{S}$ containing a sub-tour: if one exists, then the corresponding cut (15) is added to the model and the LP-relaxation is solved again. Once the optimal solution x of the LP-relaxation has been derived, it is used to build a feasible tour, as follows. Initially, we define the depot 0 as the starting node h. Then, iteratively, we select the node j such that $x_{hj} + x_{jh}$ has the highest value: arc (h, j) is added to the tour, and j becomes the new starting node h. If, for all nodes j connected to h, $x_{hj} + x_{jh} = 0$, then we choose the arc with the smallest d_{hj}. The procedure is repeated until we obtain a complete feasible tour. We call s^* the locally optimal solution and s^{**} the current best solution. The following loop (lines 9–29) is made of three phases: *perturbation*, *local search* and *acceptance criterion*.

To perturb the current best solution s^{**} (lines 10–14) we apply, with probability 80% a double-bridge move, and a scramble tour move otherwise. The former move consists of randomly removing four edges (A, B), (C, D), (E,F), (G, H) and reconnecting them as (A, F), (G, D), (E, B), (C, H). The latter move corresponds to randomly choose a path of the tour and randomly mixing its nodes. After perturbation, we obtain solution s'.

The following phase (lines 15–24) is local search that is applied to the locally optimal solution s^*: with probability 80% we apply a 2-opt move, otherwise we perform an exchange improvement. The former move consists of executing the 2-opt procedure by using as arc costs only d_{ij} $(i, j) \in \mathscr{A}$, and the procedure is stopped at the first improvement. The latter move requires exchanging two nodes of the tour: if an improvement is obtained, then the exchange is performed, otherwise the original tour is kept. This procedure is executed $|\mathscr{N}|$ times. After applying local search, we obtain solution s''. Then, we choose to store in s^* the best solution between s' and s'', by considering function ϕ that gives the value of the PTSP objective function (2)–(4).

The last phase (lines 25–27) is the acceptance criterion (check-history): if s^{**} has not been improved for 10 iterations, then we apply an additional local search step to s^{**} by executing the 2-opt procedure. It uses as arc costs d_{ij} $(i, j) \in \mathscr{A}$, but each time an improvement is possible, it checks if the PTSP objective function value improves too, and accepts the change only in this case.

Finally, the best solution between s^* and s^{**} is stored in s^{**}. The termination condition is reached when I iterations are performed ($I = 5000$ in our computational experiments).

Algorithm 1 Iterated Local Search

1: **repeat**
2: $x \leftarrow$ solve LP-relaxation of (2)–(15)
3: $\mathscr{S} \leftarrow$ separation-procedure(x)
4: **if** $\mathscr{S} \neq \emptyset$ **then**
5: LP-relaxation of (2)–(15) $\leftarrow$ add-cut($\mathscr{S}$)
6: **end if**
7: **until** $\mathscr{S} = \emptyset$
8: $s^*, s^{**} \leftarrow$ build-feasible-tour(x)
9: **repeat**
10: **if** $rnd(0, 1) < 0.8$ **then**
11: $s' \leftarrow$ double-bridge-move(s^{**})
12: **else**
13: $s' \leftarrow$ scramble-subtour(s^{**})
14: **end if**
15: **if** $rnd(0, 1) < 0.8$ **then**
16: $s'' \leftarrow$ 2-opt-move(s^*)
17: **else**
18: $s'' \leftarrow$ exchange-improvement(s^*)
19: **end if**
20: **if** $\phi(s') < \phi(s'')$ **then**
21: $s^* \leftarrow s'$
22: **else**
23: $s^* \leftarrow s''$
24: **end if**
25: **if** check-history($\phi(s^{**})$) **then**
26: $s^* \leftarrow$ 2-opt-improvement(s^{**})
27: **end if**
28: $s^{**} \leftarrow$ store if $\phi(s^*) < \phi(s^{**})$
29: **until** termination condition

4 Computational Experiments

We used instances with 10–25, and 50 customers, proposed in [3] for the PVRP and adapted them to the PTSP. In particular, to make the instances feasible for a single vehicle, we removed the time window constraints for every customer, and used a vehicle with capacity $D = \sum_{i \in \mathscr{N}_0} q_i$. In addition, we generated instances with 30–45 customers. Each set of instances contains 20 instances. The ILS and model (2)–(15) were implemented in C++, and all experiments were executed on an

Intel Core i7-6900 K with 16-Core 3.20 GHz and 66 GB RAM (single thread). We used CPLEX 12.7.1 as LP and MILP solver, and set a time limit of two hours for the Cut-and-Branch algorithm.

In Table 2, we report the computational results obtained on instances with up to 50 customers by the Cut-and-Branch C&B and ILS algorithms. We wish to mention that we also solved model (2)–(14) with CPLEX, but the Cut-and-Branch is able to find some additional optimal solutions and has a smaller average gap and computing time. Therefore, we only report the results obtained by the C&B. Each row of Table 2 corresponds to a set of instances and shows average results over the 20 instances in the set. For the C&B algorithm, we report the integer solution value (UB) and the lower bound (LB) obtained at the end of the C&B solution process, the final percentage gap (Gap%) between UB and LB, the number of obtained optimal solutions (#Opt), and the computing time (Time) in seconds. For the ILS algorithm, we executed 10 runs on each instance in every set, and we report the average (Avg) and the minimum (Min) results obtained over 10 runs. In particular, we show the solution value (Val), the percentage gap (Gap%) with respect to UB (if negative, then an improvement has been obtained by ILS), the number of best solutions (#B) found (i.e. solutions having the same value or a smaller value than that of the solutions found by the C&B), and the computing time (Time) in seconds. Obviously, the computing time for obtaining the minimum out of 10 runs is ten times the value reported in column Time for Avg. Finally, the last row reports, for each column, the average value over all the sets of instances.

We observe that the C&B is capable of deriving the optimal solution for all the instances with up to 25 customers in very short computing times (on average about 22 s). All but one instance with 30 customers are solved even though the computing time increases (on average about 911 s). As expected, instances that contain more customers are more difficult: in particular, no instance with 45 or more customers can be solved to optimality within the time limit. The average percentage gap is rather small for instances with up to 40 customers, while it increases up to 5.55% for instances with 50 customers.

When considering the average results obtained, for each instance, by the ILS algorithm over 10 runs, we can see that several best solutions can be found (for more than half of the instances on average), and the average gap from UB is very small: only 0.083% on average. The computing time is very short, being 0.50 s on average. Given the very short computing time, the ILS algorithm can be executed 10 times for every instance to select the solution found in the best of the 10 runs. In this case, the best solutions can be found, on average, for 19.78 instances in 5 s, and the gap is slightly negative (−0.004%). I.e., the ILS algorithm obtains the best solution for 98.9% of the instances. In addition, we notice that for instances with 40 and 50 customers, the average gap is negative, meaning that the ILS algorithm is able to derive better solutions than C&B.

Table 2 Comparison between the Cut-and-Branch and the ILS algorithms

	C&B					ILS						
						Avg				Min		
#Cust	UB	LB	Gap%	#Opt	Time	Val	Gap%	#B	Time	Val	Gap%	#B%
10	150.64	150.64	0.00	20	0.29	150.64	0.00	19	0.02	150.64	0.00	20
15	215.69	215.69	0.00	20	0.58	215.69	0.00	18	0.06	215.69	0.00	20
20	288.56	288.56	0.00	20	2.61	288.60	0.02	17	0.11	288.56	0.00	20
25	311.45	311.45	0.00	20	22.19	311.92	0.12	14	0.21	311.45	0.00	20
30	417.52	417.23	0.05	19	911.08	417.85	0.08	10	0.42	417.52	0.00	20
35	493.21	488.17	1.00	12	4226.34	493.76	0.11	10	0.58	493.31	0.02	19
40	563.13	548.05	2.67	3	6495.88	563.48	0.06	6	0.79	563.01	−0.02	20
45	644.22	615.61	4.40	0	7200.00	645.47	0.23	6	0.99	644.05	0.01	19
50	720.66	680.46	5.55	0	7200.00	721.05	0.14	5	1.28	719.77	−0.04	20
Avg.	422.79	412.87	1.518	12.67	2895.44	423.16	0.083	11.67	0.50	422.67	−0.004	19.78

5 Conclusions

We proposed an Iterated Local Search (ILS) algorithm for the Pollution Traveling Salesman Problem (PTSP), a generalization of the TSP in which fuel consumption and driver wage are objectives to be minimized. The ILS algorithm starts by building a feasible tour: it is computed by using the Linear Programming (LP) solution of a Mixed Integer Linear Programming (MILP) model for the PTSP, that contains exponentially many sub-tour elimination constraints. Then, the ILS algorithm loops between three phases: perturbation, local search and acceptance criterion.

We tested the ILS on instances with up to 50 customers adapted from those proposed in [3] for the Pollution Routing Problem. To evaluate the performance of the ILS algorithm we developed a Cut-and-Branch algorithm, in which subtour elimination constraints are added at the root node. The obtained results show that the Cut-and-Branch is able to derive the optimal solution for instances with up to 30 customers. However, no instance with 45 or more customers can be solved to optimality within two hours of time limit. The ILS algorithm is very effective, as it is able to derive, in 5 s, the best solution for about 99% of the instances. Future work will be devoted to develop exact methods that combine the ILS with the Cut-and-Branch algorithm in order to solve instances with a larger number of customers.

Acknowledgements This material is based upon work supported by the Air Force Office of Scientific Research under award number FA9550-17-1-0025. We also acknowledge project CONICYT FONDECYT 11150370.

References

1. Bektaş, T., Laporte, G.: The pollution-routing problem. Transp. Res. Part B: Methodol. **45**(8), 1232–1250 (2011)
2. Dantzig, G.B., Fulkerson, D.R., Johnson, S.M.: Solution of a large-scale traveling salesman problem. Oper. Res. **2**, 393–410 (1954)
3. Demir, E., Bektaş, T., Laporte, G.: An adaptive large neighborhood search heuristic for the pollution-routing problem. Eur. J. Oper. Res. **223**(2), 346–359 (2012)
4. Demir, E., Bektaş, T., Laporte, G.: A review of recent research on green road freight transportation. Eur. J. Oper. Res. **237**(3), 775–793 (2014)
5. Kramer, R., Subramanian, A., Vidal, T., Lucídio dos Anjos, F.C.: A matheuristic approach for the pollution-routing problem. Eur. J. Oper. Res. **243**(2), 523–539 (2015)
6. Lourenço, H.R., Martin, O.C., Stützle, T.: Iterated local search: framework and applications. In: Handbook of Metaheuristics, pp. 363–397. Springer, Boston, MA (2010)
7. Lin, C., Choy, K.L., Ho, G.T., Chung, S.H., Lam, H.Y.: Survey of green vehicle routing problem: past and future trends. Expert Syst. Appl. **41**(4), 1118–1138 (2014)
8. Padberg, M., Rinaldi, G.: An efficient algorithm for the minimum capacity cut problem. Math. Program. **47**(1), 19–36 (1990)

Data Throughput Optimization for Vehicle to Infrastructure Communications

Angela Sara Cacciapuoti, Marcello Caleffi, Adriano Masone, Antonio Sforza and Claudio Sterle

Abstract The ultra-high bandwidth available at millimeter (mmWave) and Terahertz (THz) frequencies can effectively realize short-range wireless access links in small cells enabling potential uses such as driver-less cars, ultra-high-definition infotainment services and data backhauling. In this context, in alternative to fiber-based and legacy wireless-based backhauling, vehicles can be used as digital mules to increase the data throughput of a region served by a software defined network (SDN) transmitting data to the Software Defined Base Station (SD-BS), equipped with only one mmWave/THz transceiver. In real applications, multiple vehicles may concurrently pass through the region and related data throughput depends on the relative distance with respect to the transceiver. For technological reasons, the SD-BS transceiver can be used by just one vehicle at each time instant (time-slot). Hence, an operational decision problem arises consisting in determining the assignment of the vehicles to the time-slots of the SD-BS maximizing the data throughput. The problem can be conceived as a variant of different combinatorial optimization problems like scheduling and assignment problems. An original mixed integer linear programming formulation of the problem is presented and tested on real-like instances generated from a case study.

A. S. Cacciapuoti · M. Caleffi · A. Masone (✉) · A. Sforza · C. Sterle
Department of Electrical Engineering and Information Technology,
University "Federico II" of Naples, Via Claudio 21, 80125 Naples, Italy
e-mail: adriano.masone@unina.it

A. S. Cacciapuoti
e-mail: angelasara.cacciapuoti@unina.it

M. Caleffi
e-mail: marcello.caleffi@unina.it

A. Sforza
e-mail: sforza@unina.it

C. Sterle
e-mail: claudio.sterle@unina.it

© Springer Nature Switzerland AG 2018

P. Daniele and L. Scrimali (eds.), *New Trends in Emerging Complex Real Life Problems*, AIRO Springer Series 1,
https://doi.org/10.1007/978-3-030-00473-6_11

Keywords Millimeter wave and TeraHertz communications · Data throughput optimization · Generalized assignment · Scheduling

1 Introduction

The proliferation of outdoor small cell deployments, already started in 2017, gives rise to spectrum scarcity and congestion problems in the sub-6 GHz bands [11]. Among the different solutions proposed to tackle these problems, one of the most promising is the use of millimeter-wave (mmWave) (around 30–100 GHz) and Terahertz (THz) (around 0.1–10 THz) frequencies.

Unfortunately, the severe path loss for a traveling wave in the mmWave and THz Bands limits the range of communications to 200 m for mmWave and few meters for THz [2, 8], respectively. In this context, an alternative architecture, based on the use of vehicles as digital mules, i.e., mobile small cells and data caches, has been proposed with particular reference to data backhauling. Indeed, using the vehicles as digital mules, it is possible to reduce the load on the backhaul and core network [13, 14]. Moreover, the vehicles during their trip have the possibility to get in the range of communication of mmWave and THz so allowing the use of these frequencies.

In this work we tackled the problem presented in literature [3], where a fleet of vehicle passes through a region served by a software defined network (SDN) and the vehicles have to transmit data to a Software Defined Base Station (SD-BS), equipped with only one mmWave/THz transceiver. The arising optimization problem consists in finding the schedule of the connections between the vehicles and the SD-BS that maximizes the transmitted data, satisfying also several operational constraints. To this aim, an original mixed integer linear programming model has been formulated and tested on instances generated from the real problem.

The paper is organized as follows: in Sect. 2 a detailed description of the real problem is given. In Sect. 3 different combinatorial optimization models, which can represent the case study, are discussed and an original mixed integer linear programming model is proposed. In Sect. 4 we report the computational results of the tests performed on instances generated from the real problem. Finally, in Sect. 5, conclusions and research perspectives are given.

2 Problem Description

In real applications, multiple vehicles may concurrently pass through a region, served by a SDN, transmitting data in a given time horizon to a SD-BS. The vehicles are characterized by specific communication needs, i.e. the amount of data to transfer (e.g. a content related to the specific part of the region which the vehicle passes through), that bounds the maximum amount of transmittable data by a vehicle. Moreover, each vehicle is supposed to be provided with a GPS technology and, since the SD-BS transceiver position is fixed, it is possible to assume that the localization of both the vehicles and the transceiver is known in each time instant.

Since the data throughput depends on the relative distance between the vehicles and the transceiver of the SD-BS, the full knowledge of the localization of the elements of the network in each time instant, together with the assumption that the routes and the speed of the vehicles are known in advance, allows us to compute off-line the value of the data shower bulk (data throughput) transmittable by adopting the presented architecture. To this aim, it would be necessary to formulate the analytic derivation of the capacity. The approach is out of scope of this work, hence only the fundamental elements needed to present the expression of the data throughput will be provided and, for further information, the interested reader is addressed to [3]. Let d_v be the relative distance between a generic vehicle and the transceiver of the SD-BS. Let also $R_{mm} = (d^{THz}, d^{mm}]$ ($R_{THz} = (0, d^{THz}]$) be the distance interval in which a mmWave (THz) communication is established. It is possible to define a function $L_{mm}(d_v)$ ($L_{THz}(d_v)$) that is equal to 1 if $d_v \in R_{mm}(R_{THz})$, 0 otherwise. Moreover, let $C_{mm}(d_v)$ be the capacity in mmWave and $C_{THz}(d_v)$ the capacity in THz. On this basis, it is possible to express the capacity available for transmitting data $C(d_v)$ by a generic vehicle v at a given distance d_v, as follows:

$$C(d_v) = C_{mm}(d_v)L_{mm}(d_v) + C_{THz}(d_v)L_{THz}(d_v) \tag{1}$$

The capacity $C(d_v)$ is expressed in bit/s. Since each vehicle has a different route to follow, the relative distance d_v is not fixed but is a function of the time $d_v(t)$. Hence, also the capacity is a function of the time $C(d_v(t))$ and so the expression of the data transmittable by a vehicle v for a given capacity function $C(d_v(t))$ is:

$$z_v = \int_{t_{start}}^{t_{end}} C(d_v(t))dt \tag{2}$$

where t_{start} and t_{end} are the time instants when the data transmission starts and ends, respectively. If all the vehicles could connect to the SD-BS simultaneously, the total data throughput transmitted would be the sum of the data transmittable by each vehicle.

As previously said, the SD-BS is equipped with only one mmWave/THz transceiver so allowing to connect one vehicle at a time.

Therefore, the SDN controller has to schedule the vehicles at different time-instants, i.e. it chooses when and which vehicle has to connect to the SD-BS in such a way that the total data throughput is maximized. Moreover, it decides which physical layer between the mmWave or THz one should be used for a given SD-BS to vehicle communication.

The problem in this form could be very hard to solve, so we have assumed that the system operates in a discrete time horizon and so the time interval $(0, t]$ is segmented into a sequence of t/h (where h is an integer) time slots each of duration $h > 0$. In Fig. 1 an example of the capacity over time horizon for two vehicles and the related discretization is represented.

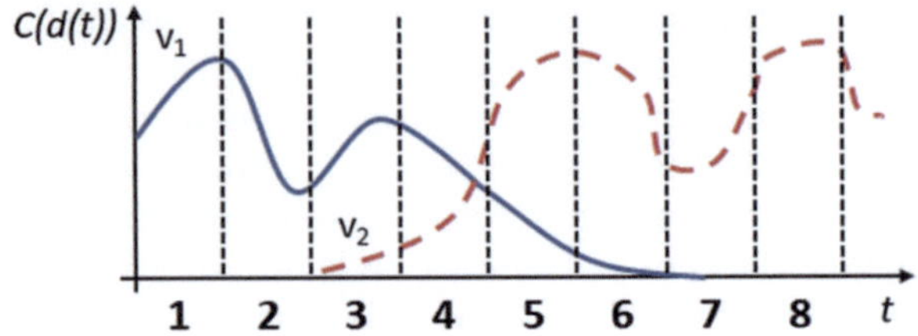

Fig. 1 Example of the capacity over time of two vehicle and related discretization

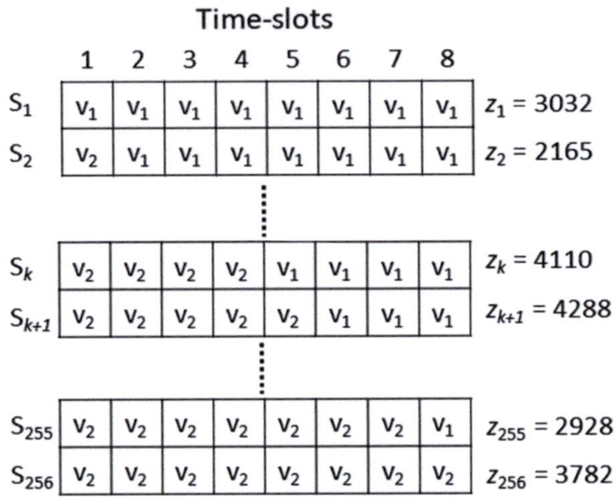

Fig. 2 All possible solutions with 2 vehicles and 8 time-slots

On this assumption, the SDN controller has to schedule the vehicles to connect to the SD-BS throughout the different time-slots. It is important to underline that the vehicles during the considered time horizon follow their routes, so we have different amounts of transmittable data for each vehicle in each time-slot, since they depend on the position of a given vehicle in a given time-slot. Hence, at each schedule of the vehicles will correspond a different data throughput. For example, if we consider a problem with 2 vehicles and 8 time-slots, supposing that the SD-BS is always connected to one vehicle, all the possible schedules are $2^8 = 256$, from the one where the first vehicle is assigned to every time-slots to the one where the second vehicle is assigned to every time-slots (a schematic illustration of all the possible schedules and the corresponding data throughputs is reported in Fig. 2). Therefore, the SDN controller has to determine the optimal schedule that maximizes the data throughput among all the possible ones. Moreover, in the determination of the optimal scheduling, the SDN controller will have to take into account also the overhead time caused by the beamsteering complexity which affects the data throughput. Indeed, before an assigned vehicle can connect to the SD-BS, a beam alignment between vehicle and SD-BS is required. During the beam alignment data cannot be transmitted, so resulting in a loss of transmittable data in a time-slot when the alignment is required [5, 15].

3 Problem Formulation

To determine the schedule that maximizes the total data throughput, the problem can be formulated referring to different combinatorial optimization problems, like scheduling and assignment problems.

Indeed, scheduling is concerned with the allocation of limited resources over time. Scheduling problems involving multiple customers (jobs) competing for a common processing resource. In our problem the limited resource is represented by the SD-BS while the jobs are represented by the vehicles. In particular, the problemunder investigation can be classified as a scheduling problem on a single machine (the SD-BS) with setup times (overhead time), release dates and deadlines (the beginning and the end of each time-slot), maximizing the data transmitted without taking into account the processing and completion times of the jobs [1, 4, 12].

Instead, assignment problems arise when we have to assign resources to several activities in such a way that a cost function is maximized/minimized. In our case, the problem is a particular variant of the generalized assignment problem where resource is represented by the time-slots to be assigned to vehicles, the activities are represented by the vehicles and the objective function is the total transmitted data [6, 9, 10].

Conceiving the problem under investigation as an assignment problem, we propose an original mixed integer linear programming model. To this aim, let $V = \{i, \ldots, v\}$ be the set of vehicles, each one characterized by the communication needs k_i. Let $T = \{i, \ldots, t\}$ be the set of time-slots of predefined fixed duration h. Each time-slot $t \in T$ can be considered to be composed of two parts: the first part is reserved for the beam alignment between the vehicle and the base station (synchronization part), while the second one is reserved for the data transmission. The durations of synchronization and transmission parts are fixed and known in advance. It is important to highlight that, if a vehicle is assigned to two consecutive time-slots, it can transmit the data in the synchronization part of the second time-slot since it has already performed the alignment with the base station in the first time-slot. Hence, being e_i^j and n_i^j the maximum number of bits that the vehicle i can transmit during the synchronization part and transmission part of the time-slot j, respectively, we compute them by:

$$e_i^j = \int_{s_{start}^j}^{s_{end}^j} C(d_i(t))dt \quad and \quad n_i^j = \int_{t_{start}^j}^{t_{end}^j} C(d_i(t))dt \tag{3}$$

where $d_i(t)$ is the relative distance between the i-th vehicle and the SD-BS at a given time t, and s_{start}^j and s_{end}^j (t_{start}^j and t_{end}^j) are the starting and ending time instant of the synchronization part (transmission part) of the j-th time-slot. Finally, we have to take also into account the operational constraints related to the communication needs, which impose that if a vehicle transmits more than a predefined threshold the exceeding transmitted data are unuseful and can be neglected. Therefore, we have to distinguish two kinds of data: the total data transmitted and the meaningful data transmitted. To represent the meaningful data transmitted, we can introduce two kind of continuous variables: $x_i^j, i \in V, j \in T$ and $y_i^j, i \in V, j \in T$, where x_i^j are the meaningful data transmitted by the vehicle i during the transmission part of the time-slot j and y_i^j are the meaningful data transmitted by the vehicle i during the synchronization part of the time-slot j.

Moreover, we define a binary variable $\phi_i^j, i \in V, j \in T$, that is equal to 1 if the vehicle i is assigned to the time-slot j, 0 otherwise. The products $n_i^j \phi_i^j$ and $n_i^j \phi_i^j \phi_i^{j-1}$ are the data transmitted by the vehicle i in the transmission and synchronization part of the time-slot j, respectively.

On the basis of this notation, it is possible to formulate the following mixed integer linear programming model:

$$\max z = \sum_{i \in V} \sum_{j \in T} (y_i^j + x_i^j) \tag{4}$$

$$\sum_{i \in V} \sum_{j \in T} (y_i^j + x_i^j) \leq k_i \qquad \forall\, i \in V \tag{5}$$

$$\sum_{i \in V} \phi_i^j \leq 1 \qquad \forall\, j \in T \tag{6}$$

$$y_i^j \leq e_i^j \phi_i^j \qquad \forall\, i \in V, \forall\, j \in T \tag{7}$$

$$y_i^j \leq e_i^j \phi_i^{j-1} \qquad \forall\, i \in V, \forall\, j \in \{2, \ldots, T\} \tag{8}$$

$$\phi_i^0 = 0 \qquad \forall\, i \in V \tag{9}$$

$$x_i^j \leq n_i^j \phi_i^j \qquad \forall\, i \in V, \forall\, j \in T \tag{10}$$

$$\phi_i^j \in \{0,\ 1\} \qquad \forall\, i \in V, \forall\, j \in T \tag{11}$$

$$0 \leq y_i^j \leq e_i^j \qquad \forall\, i \in V, \forall\, j \in T \tag{12}$$

$$0 \leq x_i^j \leq n_i^j \qquad \forall\, i \in V, \forall\, j \in T \tag{13}$$

The objective function (4) maximizes the total meaningful transmitted data. Constraints (5) ensure that the meaningful data transmitted by the vehicle i cannot exceed its threshold k_i. Constraints (6) guarantee that, during each time slot of the considered time horizon, there exists at most one vehicle in connection with the SD-BS. Constraints (7)–(9) ensure that a vehicle i needs to synchronize with the base station in the time-slot j if another vehicle l, $l \neq i$ has been assigned to the time-slot $j - 1$. Constraints (10) guarantee that a data transmission is possible only if the vehicle and SD-BS are connected. Requirements for the nature of the variables are given by (11)–(13). The difference $\sum_{i \in V} \sum_{j \in T} (y_i^j + x_i^j - (n_i^j \phi_i^j + e_i^j \phi_i^j \phi_i^{j-1}))$ represents the data transmitted which are not meaningful, since the vehicle that has transmitted the data has already satisfied its communication needs. It is possible to find the optimal solution of instances of hundreds vehicles and hundreds of time-slots in few seconds solving the proposed model by a commercial optimization solver. In our experiments we have noted that if the values of the communication needs of the vehicles are similar, the time needed to solve the model increases. Indeed, in this case there are many similar solutions in terms of data throughput.

4 Computational Results

In this section we report the computational results of several instances based on the real case study, presented in [3], where SD-BS is located along a straight road. In our instances we have considered a road traveled by three hundreds vehicles and a time horizon of 1280 s grouped in time-slots of different durations. In particular, we have generated 5 instances considering the duration of each time-slot equal to 1.25, 2.5, 5, 10, 20 s. Moreover, each instance is characterized by a fixed synchronization time equal to 1 s. We consider the range of the transceiver of the SD-BS equal to 200 m and so a vehicle i can transmit data to the SD-BS in a time-slot j if the relative distance between vehicle and SD-BS is lower or equal to 200 m. In particular, we have computed the data transmittable by each vehicle in each time-slot simulating the relative distance of each vehicle in each time-slot. To this aim, we have assumed that each vehicle is characterized by a certain speed and an initial time-slot during which the vehicle enters in the range of the SD-BS transceiver. The speeds of the vehicles are randomly generated in the range (10–30 Km/h) that are coherent with the speed of a vehicle that travels along a city road. The times-slot during which a vehicle enters in the communication range is randomly chosen between the time-slots of the time horizon. On this basis, it is possible to compute the relative distance and so the matrix of transmittable data, i.e. the maximum number of bits that each vehicle can transmit during the synchronization part and transmission part of each time-slot . Moreover, each vehicle is characterized by certain communication needs randomly generated in the range ($0.5e13$–$1.5e13$ bites) coherently with the flood of data generated by an autonomous vehicle [7]. The tests were run on an Intel Core i7-4750HQ, 2.00 GHz, 8 GB RAM, Windows 10 (64 bit). The details of the instances and the corresponding results are given in Table 1 where:

Name	Instance name.
$\|V\|$	Number of vehicles.
h	Duration of each time-slot in seconds.
$\|T\|$	Number of time-slots.
Cn	Sum of the communication needs of all the vehicles in bit.
Dt	Total amount of transmitted data in bit.
MDt	Total amount of meaningful transmitted data in bit.
Nv	Number of vehicles that transmit the data to the SD-BS.
Nsv	Number of vehicles that have satisfied their communication needs.
Ts	Number of time-slots during which the synchronization with the SD-BS has been avoided.
Time	Computation time needed to solve the instance in seconds.

Table 1 Case study—instances details and computational results

Name	$\|V\|$	h	$\|T\|$	$Cn(E+15)$	$Dt(E+15)$	$MDt(E+14)$	Nv	Nsv	Ts	Time
IS1	300	20	64	3.03	2.05	5.58	55	40	4	0.7
IS2	300	10	128	3.03	1.59	8.12	101	41	23	1.2
IS3	300	5	256	3.03	1.14	9.64	146	37	87	2.6
IS4	300	2.5	512	3.03	1.15	9.79	174	20	301	6
IS5	300	1.25	1024	3.03	1.11	9.92	180	8	774	33

We can observe that, for all the instances, the upper bound, represented by the sum of the communication needs of each vehicle, on the total meaningful data transmitted, and so on the objective function value, has not been exceeded by the total data transmitted. Moreover, the results show that a greater number of time-slots of shorter duration lead to a decrease of the total data transmitted and an increase of the meaningful data transmitted. The reduction of the difference between the data transmitted and the meaningful data transmitted shows a better management of the connection between SD-BS and vehicles. These results are explainable considering that if more time-slots of shorter duration are available the SDN controller in the same period of time can choose between different vehicles, so avoiding that a vehicle exceeds its communication needs. This explanation is supported observing that, when a greater number of time-slots are available, the number of the vehicles connecting to the SD-BS increases and the number of vehicles exceeding their communication needs decreases.

Hence, it is clear that, the duration of each time-slots represents a key parameter of the problem. On this basis, let us also remark that for duration smaller than a certain value the variation of the objective function value become less significant. This value of the duration of the time-slots depends on the synchronization time and on the particular configuration of the matrix of the transmittable data. Hence, a trade-off arises between the effectiveness of the assignment and the level of discretization of the problem.

5 Conclusions

In this paper we have presented and analyzed a real problem arising in communication networks when vehicles can be used as digital mules in alternative to fiber-based and wireless backhauling. Different combinatorial optimization problems can be used to schematize the problem and an original assignment based mixed integer linear programming model has been presented. Some computational results on instances based on a real case study have been presented.

Future research perspectives may include: an extensive analysis on the choice of the duration time of each time-slots; an integration/adaptation of the proposed model to the case with more than one SD-BS; the development of exact and heuristic approaches able to effectively tackle real large size instances with low computation time given the real time requirement of the problem under investigation.

References

1. Allahverdi, A., Ng, C.T., Cheng, T.C.E., Kovalyov, M.Y.: A survey of scheduling problems with setup times or costs. Eur. J. Op. Res. **187**, 985–1032 (2008)
2. Akyildiz, I.F., Jornet, J.M., Hana, C.: Terahertz band: Next frontier for wireless communications. Phys. Commun. **12**, 1632 (2014)
3. Cacciapuoti, K.A.S., Subramanian, R., Caleffi, M., Chowdhury, K.C.: Software-defined network controlled switching between millimeter wave and terahertz small cells, Feb 2017. arXiv:1702.02775
4. Drexl, A., Kimms, A.: Lot sizing and scheduling—survey and extensions. Eur. J. Op. Res. **99**, 221–235 (1997)
5. Gonzalez-Prelcic, N., Mendez-Rial, R., Heath, R. Jr.: Radar aided mmWave beam alignment in V2I communications supporting antenna diversity. In: Information Theory and Applications Workshop (ITA), Feb 2016
6. Morales, D.R., Romeijn, H.E.: The generalized assignment problem and extensions, handbook of combinatorial optimization, Du, D.Z., Pardalos, P.M. (eds.), vol. 5, pp. 259–311 (2004)
7. Nelson, P.: Just one autonomous car will use 4,000 GB of data/day, Network World. https://www.networkworld.com/article/3147892/internet/one-autonomous-car-will-use-4000-gb-of-dataday.html
8. Niu, Y., Li, Y., Jin, D., Su, L., Vasilakos, A.V.: A survey of millimeter wave communications (mmWave) for 5G: opportunities and challenges. Wireless Netw. **21**, 2657–2676 (2015)
9. Oncan, T.: A survey of the generalized assignment problem and its applications. INFOR **45**(3), 123–141 (2007)
10. Shmoys, B., Tardos, E.: An approximation algorithm for the generalized assignment problem. Math. Program. **62**, 461–474 (1993)
11. SCF050. Market status statistics Feb 2016 mobile experts, Small Cell Forum document number SCF050
12. Sousa, J.P., Wolsey, L.A.: A time indexed formulation of non-preemptive single machine scheduling problems. Math. Program. **54**, 353–367 (1992)
13. Sugihara, R., Gupta, R.K.: Path planning of data mules in sensor networks. ACM Trans. Sensor Netw. **8**(1) (2011)
14. Vigneri, L., Spyropoulos, T., Barakat, C.: Storage on wheels: offloading popular contents through a vehicular cloud. In: IEEE 17th International Symposium on a World of Wireless, Mobile and Multimedia Networks (WoWMoM), Jun 2016, Coimbra, Portugal
15. Xia, Q., Hossain, Z., Medley, M., Jornet, J.M.: A link-layer synchronization and medium access control protocol for terahertz-band communication networks. In: Proceedings of GLOBECOM, Dec 2015

Evaluation of Cascade Effects for Transit Networks

Antonio Candelieri, Ilaria Giordani, Bruno G. Galuzzi and Francesco Archetti

Abstract This paper presents a network analysis approach to simulate cascading effects in a transit network with the aim to assess its resilience and efficiency. The key element of a cascade is time: as time passes by, more locations or connections of the transit network which are nodes and edges of the associated graph can be affected consecutively as well as change their own condition. Thus, modifications in terms of efficiency and resilience are also dynamically evaluated and analysed along the cascade. Results on the two case studies of the RESOLUTE project (i.e., Florence, in Italy, and the Attika region, in Greece) are presented. Since the two case studies are significantly different, important differences are reflected also on the impacts of the relative cascades, even if they were started in both the two cases from the node with the highest betweenness centrality.

Keywords Network analysis · Resilience · Cascading effects
Urban transport system

A. Candelieri · I. Giordani · B. G. Galuzzi (✉) · F. Archetti
Department of Computer Science, Systems and Communications,
University of Milano-Bicocca, viale Sarca 336, 20125 Milan, Italy
e-mail: bruno.galuzzi@gmail.com; brunogaluzzi@gmail.com

A. Candelieri
e-mail: antonio.candelieri@unimib.it

I. Giordani
e-mail: ilaria.giordani@disco.unimib.it

F. Archetti
e-mail: francesco.archetti@unimib.it

A. Candelieri · I. Giordani · F. Archetti
Consorzio Milano-Ricerche, via Roberto Cozzi, 53, 20126 Milan, Italy

© Springer Nature Switzerland AG 2018

P. Daniele and L. Scrimali (eds.), *New Trends in Emerging Complex Real Life Problems*, AIRO Springer Series 1,
https://doi.org/10.1007/978-3-030-00473-6_12

1 Introduction

In complex Urban Transport System (UTS), such as urban city or regional transports, the nodes individually experience a load, such as the volume of passengers passing through that node, and in normal circumstances, this load does not exceed the capacity of that node [1]. Cascades failures are initiated when a heavily loaded node is lost for some reason (an accident, an infrastructure collapses or an attack) and a load of that node (i.e. the flow passing through it) must be redistributed to other nodes of the UTS. This redistribution may cause other nodes to exceed their capacity causing them also to fail. Then, the number of failed nodes increase, propagating throughout the network.

The analysis of **cascades** in complex networks [2–4] is an important topic towards modelling and assessment of network resilience. In this case, the term **resilience** indicates the capability of a UTS to resist to a node failure, and the possible resulting cascade. Resilience has come to define a set of properties of a much broader socio-technical framework to cope with infrastructure threats and disruptions including preparedness, response, recovery and adaptation. All these concepts run across several application domains like ecology, economics and networked infrastructures.

Methods of representation and analysis of resilience come from several different communities like statistical physics, graph theory, optimization, network science and engineering design. An interesting tool of quantitative analysis of recovery capability of a system is reported in [5], which considers the London underground where a new resilience measure is proposed according to the speed with which the passenger count time series return to normal condition. This information is taking as an indicator of how quickly the underground transport system is able to recover from the shock and, after that, resume normal operations.

In the extensive literature about cascades and resilience, a primary distinction must be made between percolation cascades and capacity cascades [6]. In the former, e.g. epidemiological networks, the nodes change their status due to interaction with their neighbourhood. In the latter, e.g. water distribution networks [7, 8] and transit networks, cascades occur when, due to failure in edges/nodes, the flow can no longer be carried out by the edges with their capacities or when some of the nodes fail. The failure in a capacity cascade can jump to nodes that are many hops away from the initial failure also skipping the neighbourhoods.

This paper is focused on the capacity cascades in UTSs. Recent works about it can be found, for example, in [9] and [10]. The main goal is to characterize the level of efficiency and resilience that ensure the persistence of key functions even in the presence of cascading failures. This is coherent with the activities of the EU project RESOLUTE (http://www.RESOLUTE-eu.org), whose general aim is the operationalization of the resilience guidelines and development of software tools for resilience assessment and support to a quick recovery of the service.

Two main lines of analysis have been pursued and are represented in the literature. When demand and supply data are available, the interaction between demand and supply is simulated through mathematical modelling and/or software simulation and generates the flow within the network. Simulation allows to capture the operational and economic aspects, such as the flow-induced costs, and the behaviour of users both prior and post any disruptions, as presented in [11].

When, as in most cases, data are not available, one is left to work with the network topology; traffic flows are not explicitly modeled, and the number of shortest paths between any two points passing through a node/link is taken as a proxy of the traffic demand in that node/link. This approach is adopted, for instance, in [12] and [13]. The topological structure of a network provides critical information and enables the computation of efficiency/resilience measure, which has been applied to study the Boston subway network and the transit networks of major cities worldwide [14]. Typically, only information about the interconnections is needed to create the graph associated to the transit infrastructure and still they can provide fundamental insights about the structural weakness of a transport network. Therefore, to operationalize resilience management in a wide set of conditions, the topological approach has been adopted in this paper, based on the description of a transit network and a possible capacity cascade using the graph theory.

2 Graph-Based Modelling of a Transit Network

The main elements of a networked infrastructure, such as a UTS, can be easily mapped into elements of a directed graph $G = (V, E)$, where V is a set of n nodes and E is a set of edges, which are ordered 2-element subsets (i, j), with i and j elements of the V set. In case of a UTS, the nodes represent the locations of interest on the transportation network, such as towns, bus/rail stops, road intersections, etc. whereas the edges represent connections/links between locations, such as roads, rail lines, bus line sections, etc. Furthermore, another relevant concept is the route, that can be mapped into a series of connected edges of the graph, and with a specific label, to distinguish by other routes.

Considering connectivity information at node levels, a measure of the network organization is the **betweenness centrality** [15], computed for each node i as:

$$g(i) = \sum_{j,k \in V, j \neq i, k \neq i} \frac{\sigma_{jk}(i)}{\sigma_{jk}}, \quad (1)$$

where σ_{jk} is the total number of shortest paths from the node j to node k and $\sigma_{jk}(i)$ is the number of those paths passing through node i. Two different parameters are considered to measure the efficiency and the resilience of a network, during and after a cascade. The first is the relative size of the **largest connected component** (S) [12, 13], defined as

$$S = \frac{N'}{N}, \quad (2)$$

where $N^{'}$ and N are, respectively, the number of nodes in the **largest connected component** after and before the cascade. The second is the **network efficiency** (E) [16], defined as

$$E = \frac{1}{n(n-1)} \sum_{i,j \in V, i \neq j} \frac{1}{d_{ij}}, \tag{3}$$

where $d_{ij} = d(i, j)$ represent the **length of shortest path** between nodes i and j, named **distance**. Normalization by $n(n-1)$ ensures that $E \leq 1$, where 1 is obtained for a complete graph. These two quantities can be computed before, during and after the cascade event.

3 Capacity Cascades Caused from a Station Closure

Disruptions of a transportation network can be of different types (accidents, infrastructure collapses, attacks, etc.) and can lead to impacts with different severities: injuries, fatalities. Common disruptions, such as a road link blocked, a rail service interruption, a strike, etc., have an impact with low severity. In this work, we simulate a capacity cascade caused by the **closure of a station or stop**, that means the access to that station/stop is disabled but transport lines/routes passing through it are not interrupted. In this case, the node k of a graph, corresponding to the station of the UTS, is removed from the graph but with the possibility that all the paths passing through it are still maintained. The resulting graph after this event is $G^{'} = (V^{'}, E^{'})$ where: $V^{'} = V - \{k\}$ and

$$E^{'} = E - \{(i,k) \in E, (k, j) \in E\} \cup \{(i, j) : \exists(i,k) \in E \wedge \exists(k, j) \in E\} \tag{4}$$

To select the station to close, the g value is computed for every node of the network and the node with the highest g value is selected as target node. The intentional capacity cascade in the network is simulated starting from the removal of the target node. As in [1], we definite $L_i(t)$ the total number of shortest paths, or load, passing through node i at a certain iteration t of the cascade, whereas ψ_i is the maximum load that the node i can handle, or capacity. This value corresponds to the initial load at iteration $t = 0$, multiplied by a tolerance parameter $\alpha \geq 1$:

$$\psi_i = \alpha\, L_i(0) \;\; i = 1, 2, \ldots, n \tag{5}$$

To start the cascading effect, the node with higher (initial) load is removed, and a new graph is re-computed using Eq. 4. Then, the g value for each of the remained nodes is re-computed: if any node has a g value exceeding its own capacity, then that node is removed from the network. The process iterates until no more node must be removed from the network, that is the termination of the cascade. Note that the load for each node is updated along the iterations of the cascade simulation, whereas the capacities are set at the beginning. Logically, capacity could be increased with the aim to contrast cascades of failures in the network.

4 Experimental Setting and Results

In this section, we report the results obtained from network analysis applied to two real UTSs. These systems are modelled through a directed multi-graph, because more than one route/line may connect two stations; moreover the graph is direct to model direction of each line from one stop to the next.

The first one consists of the public bus transportation in Florence. The number of bus stops is 999, whereas the number of directed edge is 3226. Figure 1a shows the associated graph of the network. To improve the visualization, we did not draw multiple edges. The different colors and sizes of the nodes indicate the different values of their g-value. Passing from the yellow color to the blue one, and from the lower size to the greater size of the nodes, we have an increase of the g-value. The bigger red point indicates the node with the highest g-value (35517.7).

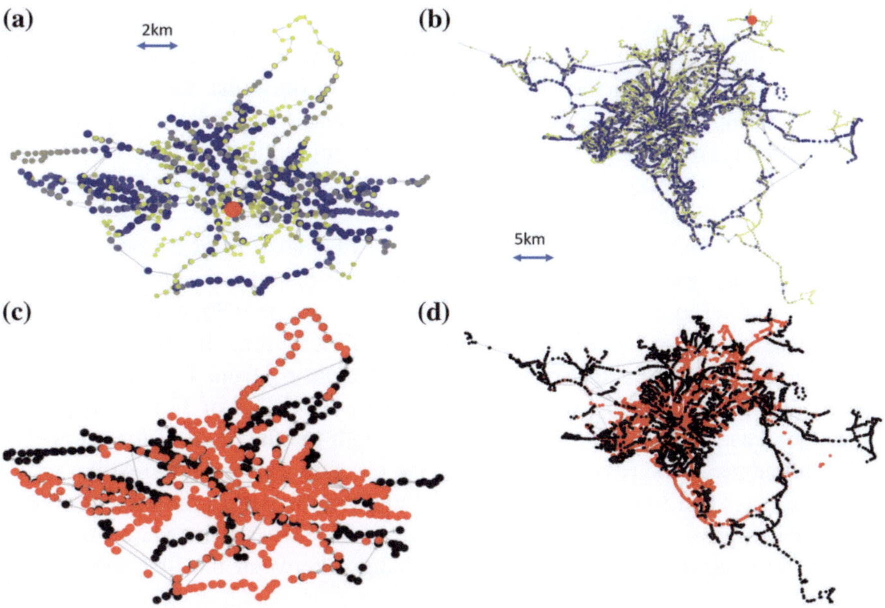

Fig. 1 **a** Graph of the bus transport network in Florence before the cascade. **b** Graph of the transport network in Attika region before the cascade. **c** Graph of the bus transport network in Florence after the cascade. **d** Graph of the transport network in Attika after the cascade effect. In (**a**) and (**b**) the red points represent the nodes from which the cascades begin (the ones with the highest g-value), whereas in (**c**) and (**d**) the red points represent the node failed along the cascade

The second network consists of the public transportation network (bus, tramway, …), in the Attika region. The number of stops is 7681, whereas the number of directed edges is 18.128. Figure 1b shows the associated graph of the network. Again, passing from the green color to the blue one we have an increase of the g-value, and the bigger red point indicates the node with the highest g-value (438585.6). The peculiar location of the node with highest betweenness in Attika depends on different factors. First of all, the Attika's UTS has more branches towards peripheral regions. This leads to having some sub-graphs (i.e. clusters) associated with the peripheral areas. Nodes—as well as links—connecting clusters in a graph, are characterized by high values of betweenness since all the paths between two clusters pass through them. Thus, several nodes with high betweenness values are located on the branches connecting peripherals. Secondly, there are a lot of nodes with very high betweenness also in the center of the network, however the rule we have adopted to choose the "triggering" node is just to select the one with highest node betweenness value, even if the difference with the second or the third ones—which could be less peripheral—is very small.

To simulate a possible capacity cascade for the two networks we simulate the removal of the node with the largest g-value of the graphs. The re-computation of the g-value for each node permits to identify the new failing nodes in the cascade (i.e. nodes with the capacity lower than the current load). These nodes are removed, and the process iterated until no more nodes fail. Figure 1c, d show the two final networks at the end of the cascade, respectively, in which the black points represent the nodes removed along the cascade. Figure 2a, b show the values of E and S computed during the cascade, for Florence (black) and Attika (red), respectively. In Fig. 2c, d we represent also the number of remaining nodes and edges of the two networks, during cascades. Both these quantities are divided by their corresponding values before the cascade.

For both the cases we note that the removal of a such critical nodes generate an important disruption of the network, causing the decrease of both E and S. In particular, in the Florence UTS S and E decrease from 1 to 0.35, and from 0.044 to 0.007, respectively, with a cascade consisting of 17 iterations. In the second Attika UTS case we also note a decrease of S and E, decreasing from 1 to 0.32 and from 0.023 to 0.002, respectively, with a longer cascade consisting of 37 iterations.

The decrease of such quantities is coherent with the decrease of the number of nodes and edges of the two graphs, during the cascade. However, the Attika's UTS shows a greater resilience to the cascading failure compared to the Florence's one. Indeed, it requires a largest number of iteration to reach similar values of E and S during the cascade, with no significant changes between the first and the fifth iteration.

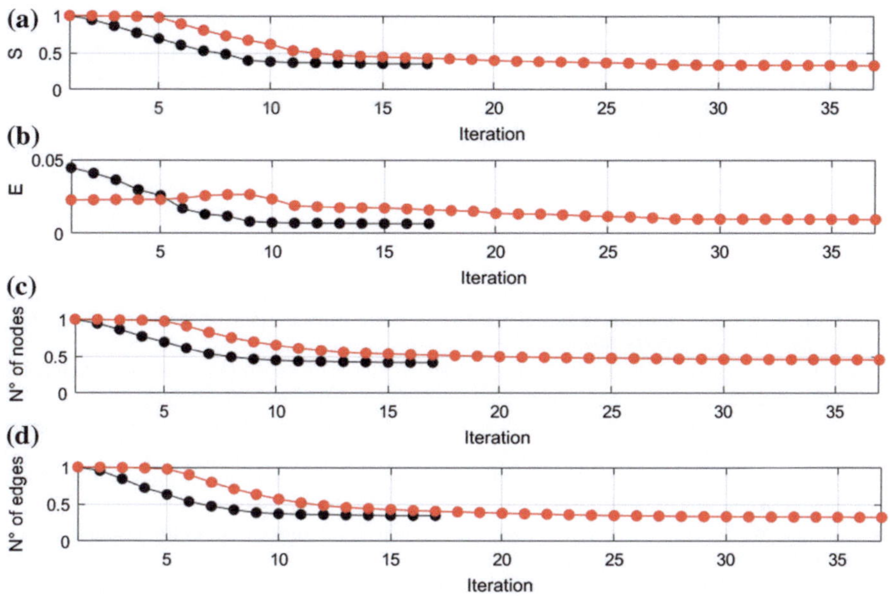

Fig. 2 Values of S (**a**), E (**b**), the relative number of remaining nodes (**c**), and the relative number of remaining edges (**d**), along the cascade. Black curves refer to the Florence UTS; red curves refer to the Attika UTS

5 Conclusions

The experimental results have demonstrated that a set of analytical functionalities can be used simulate a cascading failure and assess, dynamically along the cascade, the re-organization of flows into the network. A software tool has been developed to dynamically analyse the graph associated to the UTS, even under changing conditions, to identify the new failing components along the cascade and, therefore, the critical components which could potentially be empowered to block or at least mitigate the impact of the cascade. This analytical tool is important for assessing the resilience, as well as the efficiency, of a UTS even with respect to disruptive events starting from different locations (i.e. selecting any node as the target) and to support decisions about the capacity increase on critical nodes (i.e. those which may guarantee a lower impact).

Finally, the analytical software tool has been validated on two real UTSs, the bus transport network in Florence and the transport system in the Attika region, respectively. The analysis allowed to identify important differences between the two UTSs with respect to the impact of the corresponding cascades, even if, in both the two cases, the cascade was started by considering, as starting node, the node with highest betweenness centrality in the two networks.

Acknowledgements This work has been supported by the RESOLUTE project (http://www.RESOLUTE-eu.org) and has been funded within the European Commissions H2020 Programme under contract number 653460. This paper expresses the opinions of the authors and not necessarily those of the European Commission. The European Commission is not liable for any use that may be made of the information contained in this paper.

References

1. Ash, J., Newth, D.: Optimizing complex networks for resilience against cascading failure. Phys. Stat. Mech. Appl. (2007). https://doi.org/10.1016/j.physa.2006.12.058
2. Buldyrev, S., Parshani, R., Paul, G., Stanley, H., Havlin, S.: Catastrophic cascade of failures in interdependent networks. Nature (2010). https://doi.org/10.1038/nature08932
3. Motter, A.: Cascade control and defense in complex networks. Phys. Rev. Lett. **93** (2004). https://doi.org/10.1103/PhysRevLett.93.098701
4. Phadke, A., Thorp, J.: Expose hidden failures to prevent cascading outages [in power systems]. Comput. Appl. Power IEEE (1996). https://doi.org/10.1109/67.526849
5. D'Lima, M., Medda, F.: A new measure of resilience: an application to the London Underground. Transp. Res. Part A Policy Practise (2015). https://doi.org/10.1016/j.tra.2015.05.017
6. Gutfraind, A.: Optimizing network topology for cascade resiliance. In: Handbook of Optimization in Complex Networks (2012). https://doi.org/10.1007/978-1-4614-0857-4_2
7. Candelieri, A., Archetti, F.: Detecting events and sentiment on twitter for improving urban mobility. ESSEM AAMAS **1351**, 106–115 (2015)
8. Soldi, D., Candelieri, A., Archetti, F.: Resilience and vulnerability in urban water distribution networks through network theory and hydraulic simulation. Proc. Eng. **2015**(119), 1259–1268 (2015). https://doi.org/10.1016/j.proeng.2015.08.990
9. Crucitti, P., Latora, V., Marchiori, M.: Model for cascading failures in complex networks. Phys. Rev. E. **69**(4), 045–104 (2004). https://doi.org/10.1103/PhysRevE.69.045104
10. Lai, Y., Motter, A., Nishikawa, T.: Attacks and cascades in complex networks. Complex Netw. Lect. Notes Phys. (2004). https://doi.org/10.1007/978-3-540-44485-5_14
11. Nagurney, A., Qiang, Q.: A network efficiency measure for congested networks. EPL (Europhysics Letters) (2007). https://doi.org/10.1209/0295-5075/79/38005
12. Zou, Z., Xiao, Y., Gao, J.: Robustness analysis of urban transit network based on complex networks theory. Kybernetes **42**(3), 383–399 (2013). https://doi.org/10.1108/03684921311323644
13. Von Ferber, C., Holovatch, T., Holovatch, Y., Palchykov, Y.: Public transport networks: empirical analysis and modeling. Eur. Phys. J. B (2009). https://doi.org/10.1140/epjb/e2009-00090-x
14. Freeman, L.C.: A set of measures of centrality based on betweenness. Sociometry (1977). https://doi.org/10.2307/3033543
15. Latora, V., Marchiori, M.: Efficient behavior of small-world networks. Phys. Rev. Lett. (2001). https://doi.org/10.1103/PhysRevLett.87.198701
16. Latora, V., Marchiori, M.: Is the Boston subway a small-world network? Phys. A Stat. Mech. Appl. (2002). https://doi.org/10.1016/S0378-4371(02)01089-0

Maximizing Lifetime for a Zone Monitoring Problem Through Reduction to Target Coverage

F. Carrabs, R. Cerulli, C. D'Ambrosio and A. Raiconi

Abstract We consider a scenario in which it is necessary to monitor a geographical region of interest through a network of sensing devices. The region is divided into subregions of regular sizes (zones), such that if a sensor can even partially monitor the zone, the detected information can be considered representative of the entire subregion. The aim is to schedule the sensor active and idle states in order to maximize the lifetime of the network. We take into account two main types of scenarios. In the first one, the whole region is partitioned into zones. In the second one, a predefined number of possibly overlapping zones are randomly placed and oriented inside the region. We discuss how to transform any problem instance into a target coverage one, and solve the problem through a highly competitive column generation-based method.

Keywords Wireless sensor networks · Maximum lifetime problem · Zone monitoring · Area coverage · Target coverage

F. Carrabs · R. Cerulli · C. D'Ambrosio
Department of Mathematics, University of Salerno, Fisciano, Italy
e-mail: fcarrabs@unisa.it

R. Cerulli
e-mail: raffaele@unisa.it

C. D'Ambrosio
e-mail: cdambrosio@unisa.it

A. Raiconi (✉)
Department of Computer Engineering, Electrical Engineering and Applied Mathematics, University of Salerno, Fisciano, Italy
e-mail: araiconi@unisa.it

© Springer Nature Switzerland AG 2018

P. Daniele and L. Scrimali (eds.), *New Trends in Emerging Complex Real Life Problems*, AIRO Springer Series 1,
https://doi.org/10.1007/978-3-030-00473-6_13

1 Introduction

The issue of monitoring efficiently geographical regions through sensor networks has been intensively studied in the literature. Given the limited amount of energy provided by the battery of each device, it is indeed of great relevance to optimize their usage in order to prolong the working time (or *lifetime*) of the network for as long as possible. This is particularly relevant in vast or hardly accessible areas, where frequent substitutions of the sensors could be not practical or impossible. In order to face this issue, many researchers have proposed approaches for the Maximum Lifetime Problem (MLP). The underlying idea is to activate at any given time only a subset of sensors, capable of performing the required monitoring task, while the others are kept idle in order to preserve their batteries. Such a subset of sensors is called *cover*. Formally, the problem consists in finding a family of covers and in determining for how long each of them should be activated (activation time). The aim is maximize the sum of these activation times, while respecting the battery duration constraints of each sensor. The MLP problem is usually studied in terms of target coverage, meaning that we consider the existence of some special points of interest inside the area, called *targets*. A subset of sensors is then a cover if all targets fall within the sensing range of at least a sensor. The most effective resolution approaches proposed in the literature for MLP are based on the Column Generation (CG) technique. In such methods, the master problem is an LP formulation that individuates the optimal solution given a set a covers, while the pricing subproblem identifies new covers that could be introduced into the set considered by the master in order to improve the incumbent solution. The main differentiating factor among such CG based approaches is represented by the method used to solve the subproblem, that is NP-Hard. A simple greedy heuristic is proposed in [1]. A genetic algorithm (GA) was instead proposed in [2]. To the best of our knowledge, this algorithm represents to date the most effective resolution approach for the target coverage MLP problem. CG based approaches have also been proposed to study several MLP variants with additional requirements, see for instance [3–9]. An alternate proposed definition of the problem considers area coverage, that is, the case in which we are interested to observe the entire area, rather than single points of it. Hence, the region resulting from the union of the sensing ranges for each cover should correspond to the whole area. It was however shown ([10, 11]) that any area coverage instance can be reduced to an equivalent target coverage one, by identifying in pre-processing specific targets, such that their coverage would induce coverage for the whole area.

In this work, we aim to solve MLP within a context that does not strictly correspond neither to target nor to area coverage. We start by observing that, in an area coverage context, the requisite to cover the whole area can usually be realistically relaxed in real-world applications. Suppose that, for instance, we are interested to collect average temperatures or monitor the occurrence of fires. Within small distances, a sensor would not detect significantly different values. We can then discretize the original area into sub-areas of appropriate size (*zones*), and guarantee a partial coverage for each of them in each cover. Analogously, in a target coverage case, we can imagine that collecting information in the close proximities of the chosen target location can generally be sufficient. Again, in this case we can define a zone around each target. By relaxing in both cases the coverage requirement, such an approach can bring improvements in the network lifetime, without decreasing the quality of the solution, given appropriately chosen zone sizes. As will be shown, this new problem can be reduced to an equivalent target coverage one as well.

The specific considered scenarios are described in Sect. 2. In Sect. 3 we discuss the reduction to target coverage, and resume the algorithm presented in [2] that we use to solve it. Finally, computational results are presented in Sect. 4.

2 Considered Scenarios

We consider zone monitoring in the context of two different test scenarios, that we call Type 1 and Type 2, respectively. The Type 1 scenarios are meant to model area coverage. Given a square area, with L being the length of its side, we partition it into $(L/l)^2$ square zones with side l. An example of Type 1 instance (with $L/l = 5$) is shown in Fig. 1a, along with an example of cover (only active sensors are shown). As can be seen, each zone is at least partially within a sensing range, even if a relevant portion of the area is uncovered.

Type 2 scenarios model the target coverage case instead. To build these instances, we first randomly dispose a predefined number of targets within the area. Then, we consider for each target a square zone, such that the target is the center point of it. To further generalize this case, each zone in a Type 2 instance is rotated by a

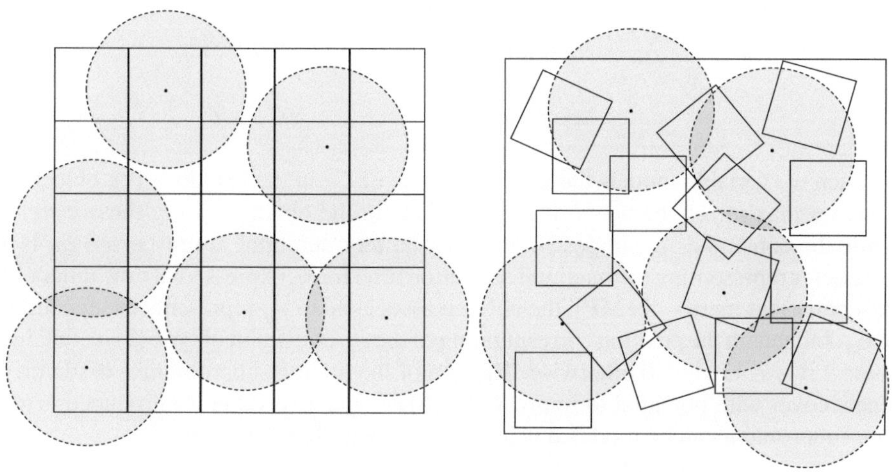

(a) Type 1 scenario and cover (b) Type 2 scenario and cover

Fig. 1 Example instances and covers for the two considered scenario types

randomly chosen angle. Note that, differently from Type 1 instances, the zones may present overlaps, and their union does not necessarily correspond to the whole area. An example of Type 2 instance and cover is shown in Fig. 1b.

3 Reduction to Target Coverage and CG Solution Approach

In order to reduce our problem to a target coverage one, we first need to perform a preprocessing on each input instance. For each sensor s_i and zone z_k, we use the position of s_i and of the center point of z_k, the radius of s_i and the area size of z_k to determine whether z_k intersects the sensing range of s_i. Hence, we derive a binary parameter δ_{ik} equal to 1 if s_i can keep z_k under observation, and 0 otherwise. We note that any instance of our problem can then be reduced to an equivalent instance of the target coverage MLP problem, in which there exists a target t_k for each zone z_k, and such that any given sensor s_i covers t_k if and only if $\delta_{ik} = 1$.

In order to solve the problem, we apply the highly effective CG-based exact algorithm proposed in [2]. Let $\mathcal{C} = \{C_1, \ldots, C_h\}$ be a set of covers. Note that the overall number of covers can be exponential, hence the aim of the CG algorithm is to find the optimal solution while avoiding to generate most of them. The master problem is defined as follows:

$$\textbf{[MP]} \max \sum_{C_j \in \mathcal{C}} w_j \tag{1}$$

s.t.

$$\sum_{C_j \in \mathcal{C} : s_i \in C_j} w_j \leq b_i \qquad \forall s_i \in S \tag{2}$$

$$w_j \geq 0 \qquad \forall C_j \in \mathcal{C} \tag{3}$$

Each w_j variable models the activation time of C_j in the solution. The objective function maximizes the network lifetime that can be obtained using these covers, while the constraints (2) impose that sensor battery durations are respected (b_i is a parameter representing the maximal activation time for s_i, expressed in time units). In the coefficient matrix of **[MP]**, the column associated to w_j represents the encoding of C_j. Indeed, in the position corresponding to the ith constraint of type (2) it contains value 1 if $s_i \in C_j$, and 0 otherwise. The aim of the pricing subproblem is to identify a new cover with potential to improve the **[MP]** objective value if introduced in $\mathcal{C}$. The subproblem can be modeled using the following ILP formulation:

$$\textbf{[SP]} \qquad \min \sum_{s_i \in S} \pi_i x_i - 1 \tag{4}$$

s.t.

$$\sum_{s_i \in S:\delta_{ik}=1} x_i \geq 1 \qquad \forall z_k \in Z \tag{5}$$

$$x_i \in \{0, 1\} \qquad \forall s_i \in S \tag{6}$$

Each binary variable x_i represents the choice related to the inclusion of sensor $s_i \in S$ in the new cover. The constraints (5) make sure that at least a sensor is chosen among the ones that can monitor each zone. The π_i values are the shadow prices associated to constraints (2) after solving **[MP]**. The objective function identifies the cover with minimum reduced cost, where the constant value 1 corresponds to the coefficient of each variable in (1). If the optimal **[SP]** solution value is greater or equal than 0, the incumbent solution found by **[MP]** is also optimal, otherwise the new cover (which is said to be *attractive*) is added to $\mathcal{C}$ and **[MP]** is solved again. The CG procedure can be iterated until a proven optimal solution is found. The **[SP]** is an NP-hard covering problem, hence the algorithm proposed in [2] also integrates a GA to solve it heuristically. In more detail, after each **[MP]** resolution, the algorithm first calls the GA. If the final population of the GA contains one or more attractive covers, they are all added to $\mathcal{C}$. Otherwise, **[SP]** is used to solve the subproblem; clearly, the exact resolution of the subproblem is needed at least once, in order to certify the optimality of the solution. In the following, we briefly resume the GA procedure; for additional details, see [2]. Each chromosome corresponds to a solution for **[SP]**, and is therefore a binary string of size $|S|$, representing the encoding of a cover. The fitness function coincides with the objective function value of **[SP]**. The GA starts from a fixed-size population P, composed of randomly generated chromosomes. For a predefined number of iterations it, two parent chromosomes are chosen through binary tournament selection, and the following operators are applied in sequence to produce a child chromosome c:

- *Crossover*: The crossover builds c by applying a bitwise AND operation on p_1 and p_2. Each sensor then belongs to c if and only if it belongs to both p_1 and p_2.
- *Mutation*: Given the chromosome c obtained after applying the crossover, the mutation operator switches the value of one of its elements, chosen randomly among those that have identical value among the two parents. If the parents are completely different, a random element of c is mutated.
- *Feasibility*: The c chromosome resulting from crossover and mutation could be unfeasible, that is, some zones could be uncovered. Iteratively, the feasibility operator choses at random a sensor that can cover additional zones, and sets the related bit to 1 in c, until feasibility is obtained.
- *Redundancy*: The redundancy operator is used to remove from c sensors that are not needed for feasibility. Iteratively, it computes a list of such redundant sensors, choses a random element of it and sets the related bit to 0 in c.

Finally, if the new produced chromosome c is not already in P, it is introduced in the population, and replaces an older chromosome, chosen among the $|P|/2$ ones with worse fitness function. Otherwise, it is discarded. The described GA is also used to generate the starting columns belonging to $\mathscr{C}$, used to first solve **[MP]**. In this case, random values are used for the shadow prices π_i, and the entire final population P is assigned to $\mathscr{C}$.

4 Computational Results

In this section, we report and comment the results obtained by the CG algorithm used to solve the two different proposed scenarios. The algorithm was coded in C++ on an OSX platform, and was run on an 3.1 Ghz Intel Core i5 processor with 8 GB of RAM. The **[MP]** and **[SP]** formulations within the CG approach have been solved using the Concert library of IBM ILOG CPLEX 12.7, running in single thread mode.

For Type 1 instances, we considered a square area with side length $L = 500$, partitioned into 25, 100 or 400 zones; hence, the side length l of each zone is equal to either 100, 50 or 25. For each of these cases, we considered instances containing 500, 750 or 1000 randomly deployed sensors ($b_i = 1$ for each sensor), generating 10 different instances for each combination of parameters. Average results for Type 1 instances are reported in Table 1.

The first three columns in the table contain instance characteristics, while the remaining two contain the lifetime value (Lifetime) in time units and the computational time (Time) required by the CG algorithm in seconds. From the Lifetime column we can note that, as expected, by reducing the number of zones (and thus increasing their size) we can prolong the lifetime of the network. Within instances containing the same number of zones, it is also predictably possible to obtain a longer lifetime by and the same occurs by increasing the number of deployed sensors. In more detail, we can observe that reducing the number of zones from 400 to 25 brings to a lifetime that is about three times longer, while by increasing the number of sensors from 500 to 1000 the lifetime can be roughly doubled. With respect to computational times, all the instances are optimally solved in less than 1 minute, showing

Table 1 Computational results for Type 1 instances

	Zones	l	Sensors	Lifetime	Time
L=500	25	100	500	67.80	3.36
			750	102.10	7.72
			1000	135.70	13.22
	100	50	500	34.00	3.38
			750	53.90	8.39
			1000	71.90	17.74
	400	25	500	21.40	8.97
			750	31.40	24.91
			1000	42.10	48.31

Table 2 Computational results for Type 2 instances

	d	Sensors	Lifetime	Time		*d*	Sensors	Lifetime	Time
L = 500 - 15 zones	0	500	43.10	1.83	L = 1000 - 15 zones	0	500	27.30	1.32
		750	58.80	3.71			750	42.10	3.22
		1000	79.60	6.21			1000	58.50	5.76
	50	500	48.10	1.91		50	500	28.70	1.35
		750	71.30	4.44			750	42.70	3.20
		1000	88.10	6.53			1000	59.50	5.67
	100	500	54.70	2.11		100	500	29.50	1.24
		750	84.30	5.13			750	43.10	2.90
		1000	97.60	6.74			1000	60.80	5.50
	150	500	63.40	2.32		150	500	30.00	1.23
		750	97.40	5.83			750	46.10	3.06
		1000	107.60	7.76			1000	63.40	5.47
L = 500 - 30 zones	0	500	23.90	1.41	L = 1000 - 30 zones	0	500	12.30	0.80
		750	40.80	3.91			750	20.50	1.95
		1000	66.90	8.43			1000	28.10	3.37
	50	500	29.20	1.64		50	500	12.90	0.91
		750	51.70	4.36			750	21.40	2.06
		1000	87.50	11.32			1000	30.00	3.49
	100	500	33.00	2.06		100	500	14.20	0.83
		750	65.30	5.29			750	23.30	1.96
		1000	104.80	11.99			1000	33.70	3.83
	150	500	38.70	2.34		150	500	16.50	1.01
		750	79.50	7.19			750	25.30	2.24
		1000	121.60	13.42			1000	36.50	4.03

the efficiency of the CG algorithm. We can observe that the more complex instances are the ones containing more zones. In the worst case, corresponding to 400 zones and 1000 sensors, 48.31 s are required on average.

For the Type 2 instances, extending target coverage problems, we considered areas with side length L equal to either 500 or 1000. We randomly disposed in the area either 15 or 30 targets and, again, 500, 700 or 1000 sensors. The diagonal length d of each zone was chosen equal to either 0, 50, 100 or 150. Note that the case $d = 0$ corresponds to a classical target coverage problem. Again, we generated 10 instances for each combination of the parameters, reporting average values in Table 2, where columns report instance characteristics and solution values, like in Table 1.

We can note that the average computational time required to solve each scenario never exceeds 14 s. As for Type 1 instances, both increasing the zone sizes and the number of sensors bring to lifetime improvements. In particular, in the instances with $L = 500$ and 1000 sensors, moving from the target coverage case to the one with $d = 150$ brings a 26% improvement (15 zones) or a 45% improvement (30 zones). The improvement is much less noticeable for the instances with $L = 1000$, given that the same zones are scattered over a larger area, and therefore the overlap of sensors with multiple zones is less likely. Overall, given instances with the same number of sensors and zones, the lifetimes obtained for $L = 1000$ is usually about half of the ones for $L = 500$.

5 Conclusion

We presented the problem of lifetime maximization for zone monitoring by means of wireless sensor networks. In this variant, the coverage requirement of specific points that is usually considered by target coverage problems is relaxed to allow partial coverage of a wider surrounding portion of the area. We considered two types of scenarios, aimed at modeling both classical area coverage and target coverage problems, and discussed how to reduce the new problem to a classical target coverage one. Computational results obtained by using a competitive algorithm available in the literature validate the idea that the approach can be used to obtain a longer network lifetime, in particular when applied to area coverage or to target coverage scenarios with a large number of targets.

References

1. Deschinkel, K.: A column generation based heuristic for maximum lifetime coverage in wireless sensor networks. In: SENSORCOMM 11, 5th International Conference on Sensor Technologies and Applications, vol. 4, pp. 209–214 (2011)
2. Carrabs, F., Cerulli, R., D'Ambrosio, C., Raiconi, A.: A hybrid exact approach for maximizing lifetime in sensor networks with complete and partial coverage constraints. J. Netw. Comput. Appl. **58**, 12–22 (2015)
3. Alfieri, A., Bianco, A., Brandimarte, P., Chiasserini, C.F.: Maximizing system lifetime in wireless sensor networks. Eur. J. Oper. Res. **181**(1), 390–402 (2007)
4. Carrabs, F., Cerrone, C., D'Ambrosio, C., Raiconi, A.: Column generation embedding carousel greedy for the maximum network lifetime problem with interference constraints. In: Sforza, A., Sterle, C. (eds.) Optimization and Decision Science: Methodologies and Applications. ODS 2017, Springer Proceedings in Mathematics & Statistics, vol. 217, pp. 151–159. Springer, Cham (2017)
5. Carrabs, F., Cerulli, R., D'Ambrosio, C., Raiconi, A.: Extending lifetime through partial coverage and roles allocation in connectivity-constrained sensor networks. IFAC-PapersOnline **49**(12), 973–978 (2016)
6. Carrabs, F., Cerulli, R., D'Ambrosio, C., Raiconi, A.: An exact algorithm to extend lifetime through roles allocation in sensor networks with connectivity constraints. Optim. Lett. **11**(7), 1341–1356 (2017). https://doi.org/10.1007/s11590-016-1072-y
7. Carrabs, F., Cerulli, R., D'Ambrosio, C., Raiconi, A.: Exact and heuristic approaches for the maximum lifetime problem in sensor networks with coverage and connectivity constraints. RAIRO—Oper. Res. **51**(3), 607–625 (2017)
8. Castaño, F., Bourreau, E., Velasco, N., Rossi, A., Sevaux, M.: Exact approaches for lifetime maximization in connectivity constrained wireless multi-role sensor networks. Eur. J. Oper. Res. **241**(1), 28–38 (2015)
9. Rossi, A., Singh, A., Sevaux, M.: Lifetime maximization in wireless directional sensor network. Eur. J. Oper. Res. **231**(1), 229–241 (2013)
10. Berman, P., Calinescu, G., Shah, C., Zelikovsky, A.: Power efficient monitoring management in sensor networks. Proc. Wirel. Commun. Netw. Conf. **4**, 2329–2334 (2004)
11. Slijepcevic, S., Potkonjak, M.: Power efficient organization of wireless sensor networks. IEEE Int. Conf. Commun. **2**, 472–476 (2001)

Mathematical Formulations for the Optimal Design of Resilient Shortest Paths

Marco Casazza, Alberto Ceselli and Andrea Taverna

Abstract We study a Resilient Shortest Path Problem (RSPP) arising in the literature for the design of communication networks with reliability guarantees. A graph is given, in which every edge has a cost and a probability of availability, and in which two vertices are marked as source and destination. The aim of our RSPP is to find a subgraph of minimum cost, containing a set of paths from the source to the destination vertices, such that the probability that at least one path is available is higher than a given threshold. We explore its theoretical properties and show that, despite a few interesting special cases can be solved in polynomial time, it is in generalbreak NP-hard. Computing the probability of availability of a given subgraph is already NP-hard; we therefore introduce an integer relaxation that simplifies the computation of such probability, and we design a corresponding exact algorithm. We present computational results, finding that our algorithm can handle graphs with up to 20 vertices within minutes of computing time.

Keywords Network reliability · Shortest path problem · Branch and price
Column generation

M. Casazza (✉) · A. Ceselli
Dipartimento di Informatica, Università degli Studi di Milano, Milan, Italy
e-mail: marco.casazza@unimi.it

A. Ceselli
e-mail: alberto.ceselli@unimi.it

A. Taverna
RSE - Ricerche sul Sistema Energetico, Milan, Italy
e-mail: andrea.taverna@outlook.com

© Springer Nature Switzerland AG 2018

P. Daniele and L. Scrimali (eds.), *New Trends in Emerging Complex Real Life Problems*, AIRO Springer Series 1,
https://doi.org/10.1007/978-3-030-00473-6_14

1 Introduction

In a world where everything is connected, the value of network reliability can be hardly overestimated. Everyday more and more companies rely on the availability of their online platforms, and the trend is to push services to the cloud in order to reduce costs. Within such a context, network reliability is a key factor for the success of a business: physical links are subject to failures, entire portions of a network can fail due to local disruptions, and the overall probability of success of a communication, called *availability*, is often subject to Service Level Agreements (SLA) between network operators and clients. Therefore, simple shortest path structures are too fragile, since the failure of a single link induces the overall connection failure. At the same time, there might be alternative ways to re-establish connectivity, ensuring *path protection* by considering one or more backup paths [3]. However duplicating flows through multiple paths increases also the cost of the communications.

We address what we call the *Resilient Shortest Path Problem (RSPP)*. It is given a network where each edge has both a cost and a probability of being available. Two vertices of the network are marked as endpoints of connections. The RSPP is the problem of finding a minimum cost subgraph, whose probability of containing an available path between the endpoints is not lower than a given threshold.

The problem of establishing reliable communications has already been investigated in several works, such as [4], where the authors investigate on the complexity of computing the probability of failure of a network, and [2], where the authors proposed heuristic methods and an exact model based on an enumeration of all the possible states of the network.

In this paper we formally model the RSPP and investigate its theoretical properties. In Sect. 2 we formalize the RSPP, we provide its mathematical programming formulation, and we highlight theoretical properties. However, computing the availability of a solution to the RSPP is already a NP-hard task that can be accomplished through complex methodologies such as Markov Chains and Bayesian Networks. Therefore, in Sect. 3 we focus on solving a noticeable integer relaxation, whose simplifying idea is to consider path failures as independent events. We name such a relaxation *Independent-paths RSPP (IRSPP)*: we provide a branch-and-price algorithm that solves it to proven optimality. In Sect. 4 we present the results of our experimental analysis and include some brief conclusions

2 Formal Description

An undirected graph $G = (V, E)$ is given, where V is the set of vertices and E is the set of edges. A source vertex $s \in V$ and a destination vertex $t \in V$ are also given. For each edge $e \in E$ we are given a cost c_e and its availability, that is the probability $0 \leq p_e \leq 1$ that such an edge *will not fail*. In this paper we assume that failures of single edges are independent events. A feasible path $r = (\sigma_1, \sigma_2, \ldots, \sigma_k)$

is a sequence of vertices σ_m having $\sigma_1 = s$, $\sigma_k = t$, and such that it exists an edge between each pair of vertices σ_m and σ_{m+1}, for $m = 1, \dots, k-1$. An RSPP solution is a subset of edges $\bar{E} \subseteq E$; the solution cost is the sum of the costs of edges in $\bar{E}$. Given a SLA availability $0 \leq \mathscr{A} \leq 1$, an RSPP solution $\bar{E}$ is feasible if the probability of finding a path from s to t in $\bar{E}$ which is not failing is greater than or equal to $\mathscr{A}$. A feasible RSPP solution is optimal when its cost is minimum.

Intuitively, the RSPP is the problem of selecting a set of feasible paths such that $\mathscr{A}$ is met and the sum of the costs of the selected edges is minimum. When an edge e is traversed by at least one path, its cost c_e is paid. The cost of an edge is paid only once even if it is shared by many paths.

2.1 Modelling

Let R be the set of all feasible paths. Each path $r \in R$ is modelled as a pattern $\bar{\mathbf{z}}^r \in \mathbb{B}^{|E|}$ having $\bar{z}^r_e = 1$ if edge e is selected, and 0 otherwise. We model the RSPP as follows:

$$\min \sum_{e \in E} c_e \cdot x_e \tag{1}$$

$$\text{s.t.} \quad P(\mathbf{y}) \geq \mathscr{A} \tag{2}$$

$$\sum_{r \in R} \bar{z}^r_e \cdot y^r \leq |\{r' \in R : \bar{z}^r_e = 1\}| \cdot x_e \qquad \forall e \in E \tag{3}$$

$$x_e \in \mathbb{B} \qquad \forall e \in E \tag{4}$$

$$y^r \in \mathbb{B} \qquad \forall r \in R \tag{5}$$

where each y^r is a binary variable associated to a path r and it is set to 1 if path r is selected, 0 otherwise, each x_e is a binary variable that is set to 1 if an edge e belongs to at least one path in the solution, and 0 otherwise, and $P(\mathbf{y})$ is a function computing the probability that at least one of the paths having $y^r = 1$ succeeds. The objective function (1) aims at minimizing the overall cost of the selected edges. Constraint (2) imposes that the set of selected paths ensures the SLA on availability, that is the overall probability of availability is not below the given threshold. Constraints (3) impose that each variable x_e is set to 1 if edge e belongs to at least a path.

2.2 Theoretical Properties

Let us denote as $\bar{p}^r = \prod_{e \in E} p_e \cdot \bar{z}^r_e$ the probability of each path $r \in R$ to be available. We can first observe that:

Observation 1 *If (1)–(5) is feasible, then there always exists an optimal solution where no path r having* $\bar{p}^r = 0$ *is selected.*

Therefore we assume w.l.o.g. that:

Remark 1 An edge having $p_e = 0$ can be removed from E.

We also prove that:

Observation 2 *If an optimal solution exists, in which a path r with* $\bar{p}^r = 1$ *is selected, then also the solution selecting only r is optimal.*

In fact, by selecting a path having probability equal to 1, any availability target $\mathscr{A}$ is satisfied. It also follows that:

Corollary 1 *If a path r having* $\bar{p}^r = 1$ *is not representing an optimal solution, then no optimal solution exists in which such a path is selected.*

These observations lead to the following:

Theorem 1 *When* $\mathscr{A} = 1$ *any instance of RSPP can be solved in polynomial time.*

A simple proof is to solve a Shortest Path Problem on a graph having edges with $p_e = 1$ only. Therefore, since when $\mathscr{A} = 0$ the problem admits a trivial *null* solution, in the following we assume w.l.o.g. $0 < \mathscr{A} < 1$.

We remark that, in this general case, even the problem of deciding if a RSPP solution is feasible is NP-hard [1].

However, Theorem 1 leads to the following:

Observation 3 *A path having* $\bar{p}^r = 1$*, if any, can be found in polynomial time.*

We therefore conclude by observing the following.

Observation 4 *Given an instance of RSPP, an optimal solution is the best between the shortest path found on the subset of edges having* $p_e = 1$ *as in Theorem 1, and the solution on a RSPP where paths having* $\bar{p}^r = 1$ *are forbidden.*

In the following, we then assume that all paths having $\bar{p}^r = 0$ and $\bar{p}^r = 1$ are excluded from set R, since the former are never selected in an optimal solution, while the best of the latter can be found in a preprocessing phase.

2.3 Integer Relaxation

Computing $P(\mathbf{y})$ is hard in the general case, because the paths in R could share edges, and thus, their failure events would not be independent. Therefore, we build our computational methods around the following relaxation:

Theorem 2 *When the failures of the paths are assumed to be independent, Inequality (2) becomes:*

$$\tilde{P}(\mathbf{y}) = 1 - \prod_{r \in R}(1 - \bar{p}^r \cdot y^r) \geq \mathscr{A} \tag{6}$$

thereby obtaining an integer relaxation of the RSPP.

Indeed Equation (6) is nonlinear, but by means of logarithmic mapping and since y^r are binary variables, we obtain:

$$\sum_{r \in R} \log(1 - \bar{p}^r) \cdot y^r \leq \log(1 - \mathscr{A}). \tag{7}$$

Formally, we indicate as Independent-paths RSPP (IRSPP) the relaxed model obtained by (1)–(5), replacing (2) with (7).

3 Solving the IRSPP

To compute our relaxation, that is solving the IRSPP, we propose a branch-and-price algorithm. In fact, the set R grows exponentially in the size of the graph, and we recur to column generation techniques to solve the continuous relaxation of the IRSPP, which we refer to as *Master Problem (MP)*: we solve to optimality a *Restricted MP (RMP)* involving a small set of columns $\bar{R} \subseteq R$, and we iteratively search for negative reduced cost variables solving a pricing problem. If no negative reduced cost variable is found, the optimal RMP solution is optimal for the MP as well, and the corresponding value is a valid lower bound for both the IRSPP and RSPP. If the final RMP solution is integer, then it is also optimal for the IRSPP, otherwise we enter a search tree by performing branching operations to find a proven global optimum. In particular, integrality conditions on y^r variables are relaxed as follows:

$$y^r \leq 1 \quad \forall r \in \bar{R} \tag{8}$$

and non-negativity conditions. While these variable upper bounds are often disregarded in column generation algorithms, in our case the following holds.

Theorem 3 *By dropping (8), an optimal MP solution always exists, in which a single path having best ratio between cost and availability is selected (possibly in multiple copies).*

In our preliminary tests, this yielded very poor lower bounds. On the other hand, constraints (8) are *exponential in number*: their handling requires a special procedure, which is discussed in the following.

3.1 Pricing Problem

We initialize the RMP by including both the shortest path and the highest probability path in R. Let $\lambda_e \leq 0$, $\mu_r \leq 0$, and $\eta \leq 0$ be the dual variables corresponding to Eqs. (3), (8), and (7), respectively. The pricing problem can be stated as follows:

$$\min \quad -\sum_{e \in E} \lambda_e \cdot z_e - \sum_{r \in \bar{R}} \mu_r \cdot u_r - \eta \cdot \log\left(1 - \prod_{e \in E} p_e \cdot z_e\right) \tag{9}$$

$$\text{s.t.} \quad \mathbf{z} \text{ is a path from } s \text{ to } t \tag{10}$$

$$u_r = 1 \text{ if } \mathbf{z} = \bar{\mathbf{z}}^r \qquad \forall r \in \bar{R} \tag{11}$$

$$u_r \in \mathbb{B} \qquad \forall r \in \bar{R} \tag{12}$$

$$z_e \in \mathbb{B} \qquad \forall e \in E \tag{13}$$

where each variable z_e is set to 1 if edge e is selected, and 0 otherwise, each variable u_r is set to 1 if the path is already in the set $\bar{R}$, and 0 otherwise.

In our pricing problem we pay a cost $-\lambda_e$ to select edge e and a cost $-\mu_r$ when the generated path r is already in $\bar{R}$. Instead, we gain a profit η that is proportional to the logarithm of the probability of failure of the path. The objective function (9) minimizes the difference between costs and profit.

To solve our pricing problem we design a label correcting algorithm: let $l = (i, c, p, \rho)$ be a label defining the cost c and the probability p of a partial path $\rho = (s, \ldots, i)$. Our algorithm is implemented as follows:

initialization: we start by creating a single label $l = (s, 0, 1, (s))$ in a label queue;

extension: at each iteration we select a label $l = (i, c, p, \rho)$ from the queue and create a new label $l' = (j, c', p', \rho')$ for each neighbour j of i, such that $c' = c - \lambda_e$ and $p' = p \cdot p_e$, where e is the edge connecting i and j, and $\rho' = (s, \sigma_1, \sigma_2, \ldots, i, j)$;

dominance: for each new label $l = (i, c, p, \rho)$ we perform a dominance check: we define μ_ρ^{min} as the minimum between all the μ_r values such that ρ is a partial path contained in r. Therefore $-\mu_\rho^{min}$ is the maximum cost of generating a path containing ρ; if it exists a label $l' = (i, c', p', \rho')$ such that $c \geq c'$, $p \leq p'$, $c - \eta \cdot \log(1-p) \geq c' - \eta \cdot \log(1-p') - \mu_{\rho'}^{min}$, and at least one inequality is strict, then label l cannot lead to an optimal solution, and therefore can be deleted. Otherwise, we add l to the labels queue;

stopping criteria: when the labels queue is empty, we stop and select the label $l = (t, c, p, r)$ corresponding to a path r having minimum $c - \eta \cdot \log(1-p) - \mu_r$.

We remark that paths not in $\bar{R}$ have $-\mu_r$ cost equal to 0: in fact such a cost is the dual variable corresponding to Inequality (8), but for paths not in $\bar{R}$ such inequality is not binding.

Set of generated paths.

One of the key elements of our label correcting algorithm is how we manage the set of generated paths, in order to pay μ_r and μ_ρ^{min} costs. To overcome such a problem we designed an enumeration tree $T^R = (N^R, A^R)$, where N^R is the set of nodes of the tree and A^R the set of the edges. The tree starts with $N^R = \{s\}$ and $A^R = \emptyset$. When a new path $r = (\sigma_1, \sigma_2, \dots, \sigma_k)$ is discovered, the tree is updated: first, we search for the longest subpath $(\sigma_1, \dots, \sigma_m)$ of r in the tree T^R; if no subpath is found, r is inserted into the tree as it is, eventually creating copies of its nodes, otherwise $(\sigma_m, \dots, \sigma_k)$ is appended to such a subpath in the tree. It follows that our tree has a root node that is the source node s, while each leaf is a copy of the destination node t. At each leaf corresponds a path r, to which is attached the cost μ_r. The cost of each other node of the tree is computed as follows:

$$c^T((\sigma_1, \dots, \sigma_m)) = \begin{cases} \mu_r, \text{ if } \sigma_m = t \\ \min_{\sigma_{m+1} \in nodeChildren}\{c^T((\sigma_1, \dots, \sigma_m, \sigma_{m+1}))\} \end{cases} \tag{14}$$

3.2 Branching Strategy

When we find an optimal solution to MP that is fractional, we perform branching to enforce integrality. We use a single branching rule to fix edges in the integer solution: we search for the most fractional edge $\hat{e}$, such that $\hat{e} \in \text{argmin}_{e \in E} |x_e - 0.5|$, and then create two new branching nodes: one having $x_{\hat{e}} = 1$ and one where $x_{\hat{e}} = 0$. In both nodes we add a constraint to the MP: $\sum_{r \in \bar{R}} \bar{z}_e \cdot y^r \geq 1$ for the first branching node and $\sum_{r \in \bar{R}} \bar{z}_e \cdot y^r \leq 0$ for the second one.

Unfortunately our branching strategy changes the nature of our pricing algorithm, because setting $x_{\hat{e}} = 1$ may induce a profit for selecting edge $\hat{e}$ in a solution to the pricing problem. Therefore we modify dominance rules in such a way that a label l is dominated by a label l' only if the partial path ρ has visited all vertices in ρ'. When $x_{\hat{e}} = 0$ instead, the pricing problem is solved removing edge $\hat{e}$ from the graph.

3.3 Primal Heuristic

To speed up the overall branch-and-bound procedure we design a fast primal heuristic that works as follows: (a) we find a shortest path from s to t and we add such a path to a set of shortest paths; (b) if the paths contained in such a set satisfy the target availability, we stop with a feasible solution; (c) otherwise all the edges belonging to shortest paths in the set are removed from the graph, and procedure restarts from step (a). If no path is found, the heuristic stops without any feasible solution. We remark that any solution found by our heuristic is feasible also for RSPP, because no edge is shared between paths, paths are independent, and the probability of their success is greater than $\mathscr{A}$ also for RSPP.

4 Computational Results and Conclusions

We implemented our algorithm in C++ using SCIP framework and CPLEX 12.6 to solve LPs. We generated a random dataset to evaluate the quality of our approach: each vertex is a point, whose coordinates are randomly drawn in the range $(-100, 100)$. For each pair of vertices, we include an edge connecting them with probability γ. For each edge, we fix the cost c_e as the euclidean distance between its endpoints, and the availability probability p_e by choosing at random one value in the set $\{1-5\times10^{-3}, 1-10^{-3}, 1-10^{-4}, 1-10^{-5}\}$. Each instance differs in size ($|V| \in \{10, 15, 20\}$) and density ($\gamma \in \{0.25, 0.5, 0.75\}$), and for each of their combinations we generated 5 instances. Also, we set different availability $\mathscr{A} \in \{0.9, 0.95, 0.99, 0.995, 0.999, 0.9995, 0.9999, 0.99995, 0.99999\}$, for a total of 405 instances.

All our tests have been conducted on a PC equipped with an Intel i7-6700K CPU and 32GB of memory and setting a time limit of 1 h of computing time to evaluate the effort required to solve instances to proven optimality. In Table 1, for each pair of graph size and density, we report the number of instances for which IRSSP is infeasible, the number of instances solved to optimality, and the average number of explored branching nodes, column generation iterations, LP solving time, variable pricing time, overall computing time, and the absolute gap between the availability of an IRSPP solution mapped to an RSPP one and the target $\mathscr{A}$.

Our results show that our algorithm is able to solve up to 97% of the instances, most of them within a few minutes. The convergence of our column generation procedure is also obtained within few iterations and most of the computing time is spent solving the LP subproblems, while the computing time spent pricing new reduced cost variables is on average smaller by an order of magnitude. We can also observe that when a graph is more dense, our algorithm requires more iterations of

Table 1 Computational effort required to solve instances with up to 20 vertices

$\|V\|$	γ	# inf.	# sol.	Nodes	CG iter.	LP t(s)	Price t(s)	Algo. t(s)	$\mathscr{A}$ gap
10	0.25	12	45	6	8	0.0	0.0	0.0	2.57e−06
10	0.5	0	45	85	102	0.1	0.0	0.2	1.14e−04
10	0.75	0	45	1489	1757	41.0	2.7	48.4	7.17e−07
15	0.25	0	45	27	34	0.0	0.0	0.0	1.38e−07
15	0.5	0	45	200	250	0.7	0.1	1.0	4.69e−08
15	0.75	0	45	682	926	12.3	1.6	15.5	2.30e−05
20	0.25	0	45	810	950	10.2	1.3	13.6	1.63e−06
20	0.5	0	42	1957	2551	276.2	27.4	314.4	8.21e−07
20	0.75	0	39	466	1403	625.4	42.9	680.5	2.29e−06

column generation to converge. Furthermore, the violation of the target availability of an IRSPP solution when mapped to an RSPP one is almost negligible.

Our approach proves therefore to be promising. Our plan is to include it in a perspective framework to exactly solve the full RSPP.

References

1. Canale, E., Cancela, H., Robledo, F., Romero, P., Sartor, P.: Diameter constrained reliability: complexity, distinguished topologies and asymptotic behavior. Networks **66**(4) (2015)
2. Gonzalez-Montoro, N., Cherini, R., Finochietto, J.M.: A multiple-link failures enumeration approach for availability analysis on partially disjoint paths. In: 13th International Conference on the Design of Reliable Communication Networks (DRCN) (2017)
3. Pioro, M., Medhi, D.: Routing, Flow, and Capacity Design in Communication and Computer Networks. The Morgan Kaufmann Series in Networking. Elsevier Science (2004)
4. Robledo, F., Romero, P., Saravia, M.: On the interplay between topological network design and diameter constrained reliability. In: 12th International Conference on the Design of Reliable Communication Networks (DRCN) (2016)

A Two-Stage Stochastic Model for Distribution Logistics with Transshipment and Backordering: Stochastic Versus Deterministic Solutions

Rossana Cavagnini, Luca Bertazzi and Francesca Maggioni

Abstract We present a two-stage stochastic program for a distribution logistic system with transshipment and backordering under stochastic demand and we first argue that it is NP-hard. Then, we perform a computational analysis based on a distribution network. In the case with two retailers, we show that modeling uncertainty with a stochastic program leads to better solutions with respect to the ones provided by the deterministic program, especially if limited recourse actions are admitted. Although there are special cases in which the deterministic and the stochastic solutions select the same retailers towards which sending items, in general, the deterministic solution cannot be upgraded in order to find the optimal solution of the stochastic program. Finally, in the case with four retailers, transshipment can provide more flexibility and better results.

Keywords Optimization under uncertainty · Transshipment · Backordering
Stochastic solution analysis

1 Introduction

In recent years, competition pressure has increased and logistics has become more and more crucial for the success of companies due to its impact on costs and service levels. An efficient distribution system is fundamental to satisfy customers' requests

R. Cavagnini (✉) · F. Maggioni
University of Bergamo, Via dei Caniana, 2, Bergamo, Italy
e-mail: r.cavagnini@studenti.unibg.it

F. Maggioni
e-mail: francesca.maggioni@unibg.it

L. Bertazzi
University of Brescia, Contrada Santa Chiara, 50, Brescia, Italy
e-mail: luca.bertazzi@unibs.it

© Springer Nature Switzerland AG 2018

P. Daniele and L. Scrimali (eds.), *New Trends in Emerging Complex Real Life Problems*, AIRO Springer Series 1,
https://doi.org/10.1007/978-3-030-00473-6_15

with reduced lead times and with a good service level. Traditionally, the distribution network is organized as a hierarchical process in which the flow of goods is shipped from the uppermost level of the distribution chain to the lowest. One of the purposes of this paper is to study a more flexible distribution network, where the shipment of products between locations at the same level of the distribution system is admitted. This strategy is called *transshipment* and it allows companies to reduce stock out risks, to share surplus stocks and to improve warehouses management, coping with demand uncertainty.

Based on the inventory system, ordering and transshipment characteristics, [12] present a complete review of the transshipment literature. Examples of stochastic transshipment problems are [5], where fixed replenishment costs are taken into account, while [11] considers the unidirectional transshipment problem, where locations have different backordering and stockout costs. Backordering is not considered in [15], while [16] studies the multi-location transshipment problem including lead times. Finally, [14] proposes a stochastic transshipment model for humanitarian emergencies.

Our contribution is to provide insights about the importance of considering uncertainty in a distribution system with transshipment and backordering.

The remainder of the paper is organized as follows. Section 2 presents the problem description and formulation. Section 3 shows our computational results and, finally, in Sect. 4, conclusions and research perspectives are outlined.

2 Problem Description and Formulation

The analyzed problem deals with a *single echelon* distribution system composed of a single supplier and a set $\mathscr{I}$ of M retailers with a *centralized decision making*. Transshipment is admitted and, in order to keep track of the origin and destination of product flows, we represent retailers performing transshipment by index i and retailers receiving transshipped quantities by index j ($i \in \mathscr{I}$, $j \in \mathscr{I}$). In this problem transshipment is *intra-level* (since it involves only retailers), *bi-directional* (each retailer can both tranship products to other retailers and receive products from them) and *reactive* (it is performed in emergency situations, after demand realization). We deal with a *single product complete pooling* transshipment (retailer i can not keep any inventory quantity if retailer j has a shortage of product), where the *priority principle* is respected (each retailer satisfies its demand at first and then transshipment is performed if necessary), backordering to supplier is allowed and, consequently, the demand can potentially be covered with supplied quantities, with transshipment quantities and with backordered quantities. The unsatisfied demand represents a lost sale. Since retailers are supposed to be close to each other, lead times are considered negligible. Our problem is described on two time intervals: t_0, which represents the time at which we have to take the decision about the quantities to ship from the supplier to retailers and t_1, in which, after demand realization, we decide the quantities to tranship and the quantities to backorder.

Moreover, the problem is characterized by risk presence: the demand is a phenomenon which can not be exactly forecast, but it is stochastic. We denote by d all possible values for the demand, that is a random variable having discrete (mutually indipendent) probability distributions $\mathscr{D}_i$, defined over the support $\mathscr{U}_1 = \{\underline{d}, \ldots, \overline{d}\}$, where $0 < \underline{d} \leq \overline{d}$. Furthermore, we represent by $\mathscr{S}$ the set of scenarios s, $s = 1, \ldots, S$ and by pr^s the probability of each scenario $s \in \mathscr{S}$, so that d_i^s denotes the demand realization for retailer i in scenario s. The measure adopted to evaluate the system performance is the total expected cost.

At time t_0, the decision variables of this model are x_i, which represent the decisions to take at the first stage, i.e. the quantity to ship from the supplier to each retailer i, taking into account the supplier's total inventory availability q and the associated unit inventory cost h_0. We introduce a capacity C_i for each vehicle employed in the shipment of units from the supplier to retailer i and an integer variable v_i, standing for the number of total vehicles used to serve retailer i by direct shipping. The transportation cost between the supplier and each retailer is represented by a variable cost f_i, proportional to the number of shipped units and by a fixed component F_i, paid for each vehicle used.

If retailer j has to face a demand d_j^s greater than the initial inventory level $\overline{I}_{i0}$ plus the quantity x_i received from the supplier, transshipment and/or backordering can be used to avoid stock-out. Thus, at t_1 the decision variables are represented by y_{ij}^s which stand for the quantity to transship from retailer i to retailer j, for each possible scenario s, after the demand realization d_i^s and by b_i^s which represent the quantity to backorder from the supplier for each retailer and for each possible scenario s, after demand realization d_i^s. On one hand, we introduce a capacity C^T for vehicles used to transship units (note that the capacity of vehicles used to ship units from supplier to retailers is typically bigger than the capacity of vehicles used for transshipment) and integer variables V_{ij}^s representing the number of vehicles employed for transshipment from retailer i to retailer j for each scenario s. The total transshipment cost is composed of a unit cost t_{ij} for each transshipped unit and a fixed cost T_{ij} for each vehicle used. On the other hand, backordering is done by using vehicles with the same capacity C_i of vehicles used for the shipment from the supplier to retailer i and we represent the number of vehicles used for backordering with the variables r_i^s. The total backordering cost is composed of a unit backordering cost g_i for each backordered unit and a fixed cost G_i for each vehicle used. Finally, the variables I_i^s represent the balance quantity at each retailer i for each scenario s and they are given by the sum of the initial inventory level $\overline{I}_{i0}$ plus the quantity received from the supplier, the quantity received through transshipment and through backordering minus the sum of the customers' demand and of the transshipped units. If this quantity is positive, it stands for the inventory level and the associated unit cost is represented by h_i. If the quantity is negative, then the balance quantity stands for the stock-out quantity and retailer j has to pay a unit penality cost p_j. In particular, if the product surplus at retailer i is transshipped to retailer j, but it is not sufficient to fully cover the shortage of product of retailer j, and no quantities are backordered, retailer i has neither inventory nor stock-out costs, while retailer j has to face stock-out costs for the unsatisfied demand. We also consider the warehouse capacity Q_i for each retailer i.

Consequently, we formulate the following integer non linear two stage stochastic programming model.

Model $\mathscr{T}$

$$\min h_0(q - \sum_{i \in \mathscr{I}} x_i) + \sum_{i \in \mathscr{I}} (f_i x_i + F_i v_i) + \\ + \sum_{s \in \mathscr{S}} pr^s [h_0(q - \sum_{i \in \mathscr{I}} x_i - \sum_{i \in \mathscr{I}} b_i^s) + \sum_{i \in \mathscr{I}} (g_i b_i^s + G_i r_i^s) + \\ + \sum_{i \in \mathscr{I}} h_i \max\{I_i^s, 0\} + \sum_{i \in \mathscr{I}} \sum_{j \in \mathscr{I} : i \neq j} (t_{ij} y_{ij}^s + T_{ij} V_{ij}^s) - \sum_{j \in \mathscr{I}} p_j \min\{I_j^s, 0\}] \tag{1}$$

s.t.

$$\sum_{i \in \mathscr{I}} (x_i + b_i^s) \leq q \qquad s \in \mathscr{S} \tag{2}$$

$$I_i^s = \overline{I}_{i0} + x_i + b_i^s - d_i^s + \sum_{j \in \mathscr{I} : i \neq j} (y_{ji}^s - y_{ij}^s) \qquad i \in \mathscr{I}, s \in \mathscr{S} \tag{3}$$

$$I_i^s \leq Q_i \qquad i \in \mathscr{I}, s \in \mathscr{S} \tag{4}$$

$$x_i \leq C_i v_i \qquad i \in \mathscr{I} \tag{5}$$

$$b_i^s \leq C_i r_i^s \qquad i \in \mathscr{I}, s \in \mathscr{S} \tag{6}$$

$$y_{ij}^s \leq C^T V_{ij}^s \qquad i \in \mathscr{I}, j \in \mathscr{I} : j \neq i, s \in \mathscr{S} \tag{7}$$

$$x_i \geq 0 \, integer \qquad i \in \mathscr{I} \tag{8}$$

$$y_{ij}^s \geq 0 \, integer \qquad i \in \mathscr{I}, j \in \mathscr{I} : j \neq i, s \in \mathscr{S} \tag{9}$$

$$b_i^s \geq 0 \; integer \qquad i \in \mathscr{I}, s \in \mathscr{S} \tag{10}$$

$$v_i \geq 0 \; integer \qquad i \in \mathscr{I} \tag{11}$$

$$r_i^s \geq 0 \; integer \qquad i \in \mathscr{I}, s \in \mathscr{S} \tag{12}$$

$$V_{ij}^s \geq 0 \, integer \qquad i \in \mathscr{I}, j \in \mathscr{I} : j \neq i, s \in \mathscr{S} \tag{13}$$

$$I_i^s \; free \qquad i \in \mathscr{I}, s \in \mathscr{S} \tag{14}$$

where the objective function (1) represents the minimization of the total expected cost, obtained through the sum of the supplier's inventory cost, the total shipment costs from supplier to retailers, the expected supplier's inventory costs, the total expected backordering cost, the total expected retailers' inventory cost, the total expected transshipment costs and the expected stock-out costs. Constraints (2) implies that the total quantity shipped from the supplier to all retailers (through usual shipment and backordering) cannot be greater than the supplier's initial inventory. Constraints (3) are the balance constraints. Constraints (4) imply that the balance quantity (computed as in (3)) cannot exceed the warehouse capacity Q_i for each retailer i. Constraints (5)–(7) link together the decision variables x_i, b_i^s and y_{ij}^s with the respective integer variables v_i, r_i^s and V_{ij}^s so that if the first ones are positive, these quantities are splitted in a certain number of vehicles represented by the latter ones, considering the respective vehicles capacities C_i and C^T and, consequently, the associated fixed costs F_i, G_i and T_{ij} are charged in the objective function. Finally, constraints from (8) to (14) are variables definition constraints. Due to the non-linearity of Model $\mathscr{T}$, we linearize it following the approach described in [4] and we call the linearized problem "Model $\mathscr{T}^{\mathscr{L}}$". Finally, we notice that Model $\mathscr{T}^{\mathscr{L}}$ can be reduced to the *Fixed Charge Transportation Problem* (see [6, 13]) and hence, it is NP-hard.

3 Computational Results

Model $\mathscr{T}^{\mathscr{L}}$ was implemented in Python 3.6.1 using the Gurobi 7.5.1 solver, and run on an Intel Core i7-7500U 2.70 GHz and 8GB RAM personal computer. Due to the complexity of Model $\mathscr{T}^{\mathscr{L}}$, the running is stopped when a 1% relative gap to the optimal solution or a time limit of 1 h is reached.

We first consider the case with two retailers (i.e. $\mid \mathscr{I} \mid = 2$). Our instances are inspired by a real case presented in [1], in which the uncertain demand of pallets should be satisfied by using trucks with limited capacity. The support of the demand probability distribution is in the set of integer numbers in the interval [30, 130], while the probability distribution is given by a Beta distribution (α, β), where α=20 and $\beta = 16$, having average demand $\mathbb{E}(d) = 85.55556$ pallets. The supplier's inventory level q is equal to 200 pallets, the capacity C_i of the vehicles used for shipment and backordering to all retailers is equal to 34 pallets, the capacity C^T of the vehicle used for transshipment is 17 pallets, while the retailers' warehouse capacity Q_i is equal to 170 pallets. Furthermore, we define the value P of a pallet to be equal to 1053 Euros, and since the unit inventory costs approximatively correspond to 5% of the value of a pallet of 100 kg, we set the supplier's inventory cost equal to 5% P, and the retailers' inventory costs equal to 6% P. Moreover, since the penalty cost corresponds to a lost sale and to a reputation damage, we let p_j equal to 1.5 P. As in [1], we consider a unit shipment cost of a pallet with 100–200 kg weight on a distance up to 500 km equal to 93.60 Euros and a fixed shipment cost equal to $\frac{f_i C_i}{\theta}$, where $\theta = 0.5$. Finally, considering that the fixed transshipment and backordering costs are computed as a function of the unit transshipment and backordering costs, 25 different instances are generated by combining all possible values, as displayed

Table 1 Transshipment and backordering fixed and unit costs

Cost	Extremely low case (EL)	Low (L)	Medium (M)	High (H)	Extremely high case (EH)
t_{ij}	0	$\frac{0.75 f_i}{2} = 35.1$	$\frac{f_i}{2} = 46.8$	$\frac{1.25 f_i}{2} = 58.5$	$+\infty$
T_{ij}	0	$\frac{t_{ij} C^T}{0.5} = 1193.4$	$\frac{t_{ij} C^T}{0.5} = 1591.2$	$\frac{t_{ij} C^T}{0.5} = 1989$	$+\infty$
g_i	0	$0.75 f_i = 70.2$	$f_i = 93.6$	$1.25 f_i = 117$	$+\infty$
G_i	0	$\frac{g_i C}{0.5} = 4773.6$	$F_i = 6364.8$	$\frac{g_i C}{0.5} = 7956$	$+\infty$

in Table 1. We notice that Model $\mathscr{T}^{\mathscr{L}}$ can be reduced into different special cases, which facilitate a trade-off analysis. In particular, in the "Extremely High case", obtained by assigning to transshipment and backordering costs a very high value (for example, equal to infinity), we get one instance in which both transshipment and backordering are not allowed, four instances in which only backordering is allowed and four instances in which only transshipment is allowed. The same parameters are considered also in the case with four retailers, (i.e. $|\mathscr{I}| = 4$), apart from q which is equal to 350 pallet.

In order to determine the right number of scenarios which have to be considered for the stochastic setting, we perform the in-sample stability analysis identifying as benchmark scenario tree, the one with 500 scenarios. The out-of-sample stability analysis in the benchmark tree is obtained with 300 scenarios.

3.1 Stochastic Solution Analysis

In this section, we perform the stochastic solution analysis considering the benchmark scenario tree with 500 scenarios and computing the indicators presented in [3] and in [10]. Table 2 displays the average results for the two retailers case, where with "Other" we refer to instances not belonging to any special case (i.e. the ones in which both transshipment and backordering are allowed). First, the availability of a perfect information about the future is more important if recourse decisions (i.e. backordering and transshipment) are not allowed or just transshipment is admitted with an $EVPI$ of 12.07% in the first case and approx. 10% in the second. The case in which only backordering is allowed is the most flexible with an $EVPI$ of 1.72%, as new quantities can be introduced in the system through the recourse decision, while when only transshipment is allowed, there can be a flow of goods between retailers, but further quantities are not available. Concerning the Value of Stochastic Solution, *VSS*, results show there are more advantages in including stochasticity in the cases where no recourse actions are admitted or only less flexible recourse actions are allowed (i.e. transshipment). In order to understand why the deterministic

Table 2 Average values for the stochastic solution analysis indicators for every special case with two retailers

Cases	RP	WS	EVPI (%)	EEV	VSS (%)	ESSV	LUSS (%)	EIV	LUDS (%)
No transshipment No backordering	56941.68	50066.85	12.07	57688.54	1.31	56941.68	0.00	56941.68	0.00
Only backordering	40856.24	40153.84	1.72	40956.26	0.25	40856.24	0.00	40876.56	0.05
Only transshipment	53557.80	48337.35	9.75	54600.35	1.95	53557.80	0.00	53567.35	0.02
Other	39487.43	38979.73	1.29	39723.29	0.60	39512.12	0.06	39504.06	0.04

Table 3 Values for the stochastic solution analysis indicators for every special case with four retailers and "Medium" cost level

Cases	RP	WS	EVPI (%)	EEV	VSS (%)
No transshipment No backordering	120144.82	109191.65	10.03	131098.04	9.12
Only backordering	112319.31	109191.65	2.86	122822.55	9.35
Only transshipment	107613.10	103077.98	4.40	111960.39	4.04
Other	105432.62 (2.27%)	103077.98	2.28	106535.91	1.05

solution is worse compared to the stochastic one, we compute the *LUSS* and the *LUDS* indicators. Through the *LUSS*, we see that in the cases where no recourse decisions or just one of them are admitted, the deterministic solution identifies the same retailers selected by the stochastic solution, but with wrong delivered quantities. In the other cases, the retailers receiving zero quantities are different in the stochastic and in the deterministic solution and, as a consequence, the poor performance is due both to the selection of retailers and to the selection of the quantities. Through the *LUDS*, we notice that the solution is perfectly upgreadable only if both backordering and transshipment are not allowed, meaning that these quantities are always lower or equal to the ones suggested by the stochastic program. For all other cases, the *LUDS* is not null, meaning that the deterministic solution is only partially upgreadable (at least in one case, the stochastic solution delivers a lower number of pallets than the one suggested by the deterministic solution).

Finally, we focus on the case with four retailers. Due to the computational complexity of the problem, with the exception of the case "No transshipment, No backordering", we analyze only the instances whose costs of the allowed strategy are set at a "Medium" level (i.e. only one instance for each case is considered). Results are displayed in Table 3. We specify that after 549,090 s, the gap to the optimal solution of the *RP* for the "Other" case was not closed and we calculate only the *EVPI* and the *VSS*, since the other indicators require further constraints which make the model even more difficult to get solved to optimality. Differently from Table 2, now, if only backordering is allowed the cost is higher than the case in which only transshipment is admitted, while for the *EVPI*, the previous results are confirmed. Concerning the VSS, the results are now different, as there are more advantages in including stochasticity in the case where only backordering is allowed. Even if with only backordering, the quantities delivered in the first stage are fewer, transshipment is cheaper if only few quantity adjustments are needed and the presence of more retailers provides more flexibility to the distribution system.

4 Conclusions

We presented a real problem arising in logistics and after modeling it with an integer stochastic program, we stated that this is NP-hard. Furthermore, we show that with two retailers, a decision-maker has a greater advantage by including uncertainty, especially if no recourse actions or only transshipment is admitted. We also show that in some cases, the selection of retailers to which quantities should be delivered is the same both in the deterministic and in the stochastic solution. Nevertheless, the deterministic solution can be upgrated only in the special case where no recourse actions are allowed. Conversely, with four retailers, transshipment provides more flexibility. Future research could be devoted to analyze the multistage version of this problem by exploiting lower bounds (see [7, 8]) and, as in [1], to compare the stochastic solution to the one obtained through a rolling-horizon heuristic. Another stream of research could be analyzing robust optimization approaches (see [9]) or adapting approaches presented in [2].

References

1. Bertazzi, L., Maggioni, F.: A stochastic multi-stage fixed charge transportation problem: worst-case analysis of the rolling horizon approach. Eur. J. Operation. Res. (2017)
2. Bertazzi, L., Maggioni, F.: Solution approaches for the stochastic capacitated traveling salesmen location problem with recourse. J. Optim. Theor. Appl. **166**, 321–342 (2015)
3. Birge, J., Louveaux, R.: Introduction to Stochastic Programming. Springer Science & Business Media, Francois (2011)
4. Cavagnini, R., Bertazzi, L., Maggioni, F., Hewitt, M.: A Two-stage Stochastic Optimization Model for the Bike Sharing Allocation and Rebalancing Problem. (submitted) (2018)
5. Herer, Y.T., Rashit, A.: Lateral stock transshipments in a two-location inventory system with fixed and joint replenishment costs. Naval Res. Logist. (NRL). **46**, 525–547 (1999)
6. Klose, A.: Single-sink fixed-charge transportation: applications and exact solution algorithms. Working Papers, Department of Mathematical Sciences, University of Aarhus, vol. 5 (2006)
7. Maggioni, F., Allevi, E., Bertocchi, M.: Monotonic bounds in multistage mixed-integer stochastic programming. Comput. Manag. Sci. **13**, 423–457 (2016)
8. Maggioni, F., Pflug, G.: Bounds and approximations for multistage stochastic programs. Siam J. Optim. **26**(1), 831–855 (2016)
9. Maggioni, F., Potra, F.A., Bertocchi, M.: A scenario-based framework for supply planning under uncertainty: stochastic programming versus robust optimization approaches. Computat. Manag. Sci. **14**, 5-44 (2017)
10. Maggioni, F., Wallace, S.W.: Analyzing the quality of the expected value solution in stochastic programming. Ann. Operat. Res. **200**, 37–54 (2012)
11. Olsson, F.: An inventory model with unidirectional lateral transshipments. Eur. J. Operat. Res. **200**, 725–732 (2010)
12. Paterson, C., Kiesmüller, G., Teunter, R., Glazebrook, K.: Inventory models with lateral transshipments: a review. Eur. J. Operat. Res. **210**, 125–136 (2011)
13. Roberti, R., Bartolini, E., Mingozzi, A.: The fixed charge transportation problem: an exact algorithm based on a new integer programming formulation. Manag. Sci. **61**, 1275–1291 (2014)
14. Rottkemper, B., Fischer, K., Blecken, A.: A transshipment model for distribution and inventory relocation under uncertainty in humanitarian operations. Socio-Econom. Plann. Sci. **46**, 98–109 (2012)

15. Wee, K.E., Dada, M.: Optimal policies for transshipping inventory in a retail network. Manag. Sci. **51**, 1519–1533 (2005)
16. Yücesan, E., et al.: Stochastic optimization for transshipment problems with positive replenishment lead times. Int. J. Prod. Econom. **135**, 61–72 (2012)

An Optimization Model to Rationalize Public Service Facilities

M. Cavola, A. Diglio and C. Piccolo

Abstract Facility Location Models (FLMs) have been widely applied in the context of both private and public sector, to decide the best configuration of new facilities to be located in a given area. In the last years, due to the general interest to reduce costs and improve efficiency, several works focused on problems aimed at modifying the territorial configuration of existing facilities, in terms of number, position and/or capacities, etc. In this work, we propose a new mathematical model to support territorial re-organization decisions in non-competitive contexts. The model assumes the presence of a set of facilities providing different types of services to users (multi-type facilities) and explores the possibility to improve the efficiency of the system by implementing different rationalization actions; i.e., facility closure, service closure, capacity reallocation among services at a given facility. The model aims at finding a trade-off solution between the service efficiency and the need of ensuring a given accessibility level to users. It has been tested on a set of randomly generated instances, to show that a good range of problems can be solved to optimality through the use of a commercial solver (CPLEX).

Keywords Facility location models · Territorial re-organization · Public sector

1 Introduction

Making optimal decisions in the management of public facilities poses critical challenges for strategic, tactical and operational planning, as such decisions can have a strong and lasting impact on service level and performances. In recent years, due

M. Cavola · A. Diglio (✉) · C. Piccolo
Department of Industrial Engineering, University of Naples Federico II,
P.le Tecchio 80, 80125 Naples, Italy
e-mail: antonio.diglio@unina.it

M. Cavola
e-mail: manuel.cavola@unina.it

C. Piccolo
e-mail: carmela.piccolo@unina.it

© Springer Nature Switzerland AG 2018

P. Daniele and L. Scrimali (eds.), *New Trends in Emerging Complex Real Life Problems*, AIRO Springer Series 1,
https://doi.org/10.1007/978-3-030-00473-6_16

to austerity measures resulting in cuts to local authorities and public sector organizations, a strong interest has been devoted to the definition of more efficient management practices. Within this context, public authorities had to deal with budget restrictions while preserving as much as possible service levels and quality of provision. Therefore, institutions have been often interested in the rationalization of the existing network of active public service facilities by closing down, downsizing, upsizing or merging some of them, but also by cutting down operation hours and implementing virtuous demand management policies to pursue the provision of more balanced service levels. Such policies have been having an impact on several essential services, such as education, healthcare, environmental services.

In this context, facility location decisions play a crucial role at a strategic level. The class of models traditionally used in the OR literature to support such decisions are *Facility Location Models* (FLM). They aim at identifying the best position to assign to a set of structures in a given location space, to satisfy a demand (actual or potential) [1, 2]. Depending on the context, on the objectives to be achieved and on the constraints to be satisfied (geographical, environmental, economic and technological), various versions of the models have been defined and a wide range of application areas have been identified [3–7], also in the context of public sector [8, 9], in which the main goal is to combine efficiency goals with the need of guaranteeing a good and equitable access of users to the provided services [10, 11]. In the last years, due to the above described economic context, the territorial reorganization problem of existing facilities systems started to attract attention also from the scientific community. Such problems, usually motivated by occurred or potential changes in the operating conditions of the system under consideration (demand distribution, budget restrictions, requirements imposed by local and/or central authorities, etc.), consists in modifying the current organization of facilities in the location space (number, position, capacities, services, etc.). Also in this case, a multi-stakeholder perspective is required as the need for achieving economic efficiencies has to be combined with the need to preserve the quality of provided service to users.

Bruno et al. [12] proposed a classification framework, with the attempt of systematizing the literature focused on this emerging class of problems. They distinguish between *single-period* and *multi-period* models, on the basis of the planning horizon in which decisions are taken (see for instance [13] and [14]), and between *ex-ante* and *ex-post* models, if decisions are taken before or after changes motivating re-organization occur (see for instance [15] and [12]). In this work, we refer to the class of *ex-post single period models*. Within this class, Wang et al. [16] explored decisions about the re-positioning of existing facilities in different points of the location space; ReVelle et al. [17] analyzed decisions concerning the closure of whole facilities, both in a competitive and non- competitive context. Bruno et al. [18] addressed a real problem arising in the education sector, concerning the merging of existing facilities in clusters with common functions managed in a centralized way. Guerriero et al. [19] analyzed a real problem in the healthcare sector, related to the re-organization of the hospital network of an Italian Region.

In this work, we propose a novel mathematical model that, with reference to a system of facilities providing different types of services (*multi-type facilities*), decides among different rationalization actions to be implemented with the aim of improving the efficiency of the system. The proposed model extends the contribution by Bruno et al. [12] as, besides the closure of whole facilities and/or of single services, it considers further actions, as the service shrinkage and the capacity reallocation among services at a given facility. The remainder of the paper is organized as follows. In Sect. 2, the description of the problem and its formulation in terms of FL model is provided. In Sect. 3, results obtained by testing the model on a set of randomly generated instances are presented. Finally, some conclusions and directions for future research are drawn.

2 A Mathematical Model for the Territorial Re-organization of Public Service Facilities

We consider, in a given location space, a set of facilities $j \in J$ providing multiple services $k \in K$, with limited capacities C_{kj}. For each service k, the demand from any node $i \in I\ (a_{ik})$ is assigned to the closest facility. As each facility does not necessarily provide any type of service, a demand node may also patronize different facilities for different services. With reference to such a system, the problem consists in evaluating possible rationalization actions aimed at improving the efficiency of the system; in particular, the actions that can be undertaken are the closure of single services, the closure of whole facilities and/or the internal capacity reconversion, i.e. the reallocation of capacities among services at the same facility. Each action certainly produces a saving, in terms of operative costs that can be avoided, but it has some negative effects. On one side, part of the demand should be most likely reallocated to farther facilities, with a discomfort in terms of accessibility. On the other side, in the final configuration, active services should require capacity expansion to satisfy the re-allocated demand. In this context, the model aims at identifying the set of rationalization actions to be implemented, in order to achieve a minimum benefit in terms of operating cost reduction, preserving the accessibility of users to the provided services and optimizing the cost incurred to expand the capacities of active services and cover all the demand.

In order to formulate such model, we denote with:

- I The set of demand nodes, indexed by i ($|I| = n$)
- J The set of existing facilities, indexed by j ($|J| = m$)
- K The set of different types of services, indexed by k ($|K| = q$)
- l_{kj} The binary label equal to 1 if facility j currently provides service k
- C_{kj} The capacity of service k at facility j
- a_{ik} The total demand coming from node i for service k
- α_j^{ik} The fraction of demand a_{ik} initially assigned to facility j $\left(0 \leq \alpha_j^{ik} \leq 1\right)$
- d_{ij} Distance between the demand node i and the facility j

c_{kj} The unit cost to expand the capacity of service k at facility j
g_{tk}^{j} The cost to transfer a unit of capacity from service t to service k at facility j
f_{kj} The benefit deriving from the closure of service k at facility j
f_j The additional benefit deriving from the closure of the whole facility j
B The minimum benefit to be obtained
F A very large arbitrary number $\left(F = \max_{i\in I, j\in J}\{d_{ij}\}\right)$

and we introduce the following groups of decision variables:

s_{kj} The binary decision variable equal to 1 if and only if service k provided at facility j is closed
y_j The binary decision variable equal to 1 if and only facility j is closed
r_{kj}^{-} The binary decision variable equal to 1 if and only if the capacity of service k at facility j is reduced and/or shrunk
r_{kj}^{+} The binary decision variable equal to 1 if and only if the capacity of service k at facility j is expanded
x_{j}^{ik} The binary decision variable representing the assignment of demand a_{ik} to facility j in the final configuration
Δ_{kj} The non-negative decision variable denoting the extra-capacity needed for service k at facility j to satisfy the re-allocated demand
q_{tk}^{j} The non-negative decision variable representing the amount of capacity re-allocated from service t to service k of facility j

With this notation, the model can be formulated as follows:

$$\text{minimize} \sum_{j\in J}\sum_{k\in K} c_{kj}\Delta_{kj} + \sum_{t\in K:t\neq k} g_{tk}^{j} q_{tk}^{j} \tag{1}$$

subject to

$$s_{kj} + r_{kj}^{+} + r_{kj}^{-} \leq l_{kj} \quad \forall j \in J, \forall k \in K \tag{2}$$

$$q_{kk}^{j} \leq C_{kj}s_{kj} \quad \forall j \in J, \forall k \in K \tag{3}$$

$$C_{jk}s_{jk} + r_{kj}^{-} \leq q_{kk}^{j} + \sum_{t\in K:t\neq k} q_{kt}^{j} \leq C_{kj}\left(r_{kj}^{-} + s_{kj}\right) \quad \forall j \in J, \forall k \in K \tag{4}$$

$$r_{kj}^{+} \leq \Delta_{kj} + \sum_{t\in K:t\neq k} q_{tk}^{j} \leq r_{kj}^{+}M \quad \forall j \in J, \forall k \in K \tag{5}$$

$$y_j - \frac{\sum_{k\in K} s_{kj}}{\sum_{k\in K} l_{kj}} \leq 0 \quad \forall j \in J \tag{6}$$

$$y_j + \sum_{k\in K}\left(l_{kj} - s_{kj}\right) \geq 1 \quad \forall j \in J \tag{7}$$

$$\sum_{j\in J}\sum_{k\in L_j} f_{kj}s_{kj} + \sum_{j\in J} f_j y_j \geq B \tag{8}$$

$$\sum_{j \in J} x_j^{ik} = 1 \quad \forall i \in I,\ \forall k \in K \tag{9}$$

$$x_j^{ik} + s_{kj} \le l_{kj} \quad \forall i \in I, \forall j \in J, \forall k \in K \tag{10}$$

$$\sum_{t \in J} d_{it} x_t^{ik} + \left(F - d_{ij}\right)\left(l_{kj} - s_{kj}\right) \le F \quad \forall i \in I,\ \forall j \in J,\ \forall k \in K \tag{11}$$

$$\sum_{i \in I} a_{ik}\, x_j^{ik} \le C_{kj} + \Delta_{kj} + \sum_{t \in K} (q_{tk}^j - q_{kt}^j) \quad \forall j \in J,\ \forall k \in K \tag{12}$$

$$s_{kj}, r_{kj+}, r_{kj-}, y_j, x_j^{ik} \in \{0, 1\} \quad \forall k \in K,\ \forall j \in J \tag{13}$$

$$q_{kt}^j \ge 0,\ \Delta_{kj} \ge 0 \quad \forall i \in I,\ \forall j \in J,\ \forall k, t \in K \tag{14}$$

The objective function (1) is the total cost incurred to cover the reallocated demand, given by the cost for capacities' expansion $\left(\sum_{j \in J} \sum_{k \in K} c_{kj} \Delta_{kj}\right)$ and internal capacities' reconversion $\left(\sum_{t \in K: t \ne k} g_{tk}^j\, q_{tk}^j\right)$.

Constraints (2) assure that each service k at any facility j may be closed $\left(s_{kj} = 1\right)$, expanded $\left(r_{kj}^{+} = 1\right)$ or shrunk $\left(r_{kj}^{-} = 1\right)$. Constraints (3–5) define the mechanism for internal capacities' reallocation. In particular, constraints (3) impose that the artificial variables q_{kk}^j (self-reallocations) may assume positive values only for closed services $\left(s_{kj} = 1\right)$; while, for shrunk services, only re-allocations toward different services are feasible. Constraints (4) regulate the internal capacity reconversion and self-reallocation mechanisms of both closed and shrunk services. In particular, they impose that the capacity of a shrinking service k may be used only for internal reconversions and that, at least, one unity of capacity is ceded; while the capacity of closed services k has to be fully reallocated, either toward other services t $\left(\sum_{t \in K: t \ne k} q_{kt}^j\right)$ or toward itself $\left(q_{kk}^j\right)$. The self-re-allocation does not require any cost; hence, when the model closes a facility, it will tend to re-allocate all the related capacity toward itself, unless the transfer toward different services is needed.

Constraints (5) assure that each expanding service k receives at least one unit of additional capacity, either ex novo $\left(\Delta_{kj}\right)$ or from other services $\left(\sum_{t \in K: t \ne k} q_{tk}^j\right)$, and no limit on the maximum incoming capacity is imposed, being M a very large number. Conditions (6–7) are logical constraints useful to define the relation between variables s_{kj} and y_j. In particular, each facility j is constrained to be open as long as the number of closed services is lower than the total number of provided services (6); while, it is constrained to be closed when all its provided services are closed (7). Constraint (8) expresses the need for the planner to obtain a minimum benefit B. Conditions (9–12) rule the demand reallocation mechanism. In particular, conditions (9) guarantee the total coverage of demand for any service k Constraints

(10–11) impose that the demand from any node i for each service k is assigned to the closest facility (11), among those still providing it (10). Constraints (12) indicate that the total demand allocated to a service k at a facility j has not to exceed its total capacity C_{kj}, including any additional capacity, either introduced ex novo $\left(\Delta_{kj}\right)$ or deriving from the other services $\left(\sum_{t \in K: t \neq k} q_{tk}^{j}\right)$, and excluding any amount of capacity ceded toward other services of the same facility $\left(\sum_{t \in K} q_{kt}^{j}\right)$. Note that the total additional $\left(\Delta_{kj} + \sum_{t \in K: t \neq k} q_{tk}^{j}\right)$ and ceded capacity $\left(\sum_{t \in K} q_{kt}^{j}\right)$ cannot be simultaneously positive quantities.

Finally, Constraints (13–14) define the domain of the decision variables.

3 Results

In order to show the capability of the proposed model to provide interesting solutions to the above introduced rationalization problem, in the following an illustrative example is provided, in which the solution obtained with reference to a very limited size instance is represented and commented. In Fig. 1, a system with 3 facilities ($J = \{A, B, C\}$), providing three services ($K = \{1, 2, 3\}$), is represented. Each node j is subdivided in sectors, associated to the provided services; the size of single sectors is proportional to the related capacities $\left(C_{kj}\right)$, while the size of the whole node is proportional to the total capacity of the facility j $\left(C_j = \sum_k C_{kj}\right)$. According to the allocation of demand nodes ($I = \{a, b, c, d, e, f, g, h, i, l\}$) to available facilities, each service k at each facility j is characterized by a total captured demand $\left(A_{ik} = \sum_i a_{ik} \alpha_{ij}^{k}\right)$, represented, in the figure, as saturation level of the related sector. In Fig. 2, the solution obtained by imposing the closure of two services is represented. It can be noticed that the model closes service 2 at facility B and service 3 at facility C ($s_{2B} = s_{3C} = 1$). Accordingly, users are re-allocated toward facilities A e B for service 3 and toward C for service 2. As consequence, such services need to be expanded $\left(r_{3A}^{+} = r_{3B}^{+} = r_{2C}^{+} = 1\right)$, as they are not able to cover the reallocated demand with their own residual capacities.

From Fig. 2, it is possible to notice that service 2 at facility C and service 3 at facility B are expanded through internal capacities coming from closed services; while, in the case of service 3 at facility A, the needed capacity partially comes from service 1 $\left(r_{1A}^{-} = 1\right)$, that is shrunk, and partially added ex-novo (green arrow in the figure).

In order to test the model, a set of random instances has been generated. The number of users has been fixed equal to 100, while the number of facilities and services have varied in the set {8, 10, 12}. The other parameters of the instances have been randomly generated according with the procedure introduced in [12]. In particular, it is worth to underline that f_{kj} values have been fixed equal to 1 for each pair (k, j), while f_j values equal to 0 for each facility j; this way B represents the

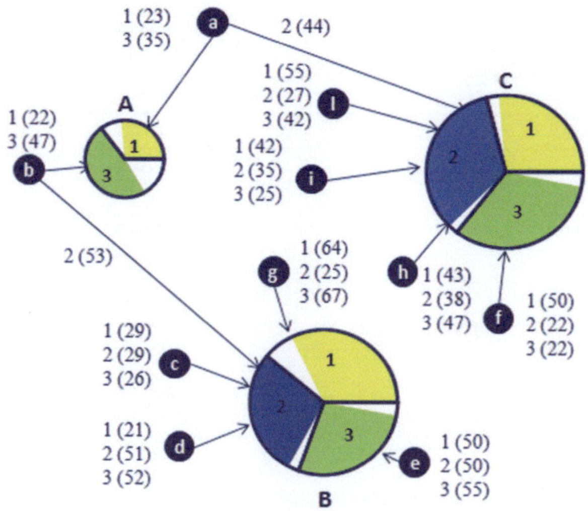

Fig. 1 Instance 1: initial configuration

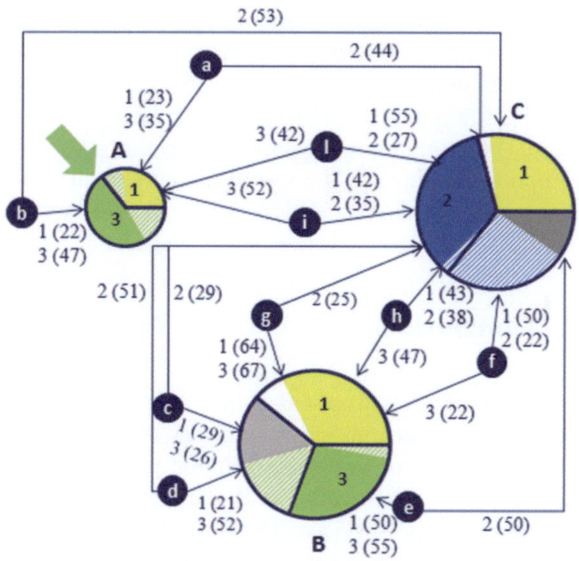

Fig. 2 Instance 1: final solution

minimum number of services to be closed and, from now on, it will be indicated by p. For each pair $(|J|, |K|)$, 5 different instances have been generated and solved. The test instances have been solved for two different values of parameter p, equal to 10% and 20% of the total number of available services. In Table 1 the computational times to optimally solve the model are reported.

Results show that the CPU time generally depends on the combination of the cardinality of the sets J and K. In particular, the number of offered services seems to affect the computational times more than the number of active facilities. This is

Table 1 Computational times (CPU times)

			$p = 0, 10 \sum_{k,j} l_{kj}$			$p = 0, 20 \sum_{k,j} l_{kj}$		
$\|J\|$	$\|I\|$	$\|K\|$	Min	Max	Average	Min	Max	Average
8	100	8	8.56	19.43	12.40	22.01	166.17	66.22
		10	9.09	87.69	32.90	27.36	147.36	78.93
		12	5.65	122.41	39.18	11.52	251.59	136.176
10	100	8	6.36	37.19	20.46	24.53	205.68	122.00
		10	6.83	55.47	21.98	23.78	457.43	337.58
		12	35.25	179.38	92.15	240.20	3766.17	1470.39
12	100	8	6.07	25.13	11.54	23.33	282.45	126.936
		10	10.82	74.05	39.586	12.91	286.80	166.532
		12	173.83	397.07	260.78	508.49	2762.67	1854.00

due to the fact that the capacity re-allocation options do not depend on the number of existing facilities but on the number of offered services at single facilities. Another parameter affecting the number of re-allocation options is the number p of closed services; indeed, as it increases, more demand is re-allocated and more extra-capacity at the remaining services is needed. Note that by fixing the cardinalities of sets J and K, CPU times increase a lot by varying the percentage of closed services from 10% to 20%. These preliminary results suggest that a good range of problems can be solved to optimality through the use of a commercial solver, but appropriate algorithmic approaches should be developed to cope with larger instances.

4 Conclusions

In this work, we proposed a new mathematical model to support territorial re-organization decisions in non-competitive contexts. The model assumes the presence of a set of facilities offering different types of services to users (multi-type facilities) and explores, together with the closure of whole facilities, also the possibility of closing single services. The objective function is represented by the extra-cost to be paid in order to satisfy the reallocated demand, while constraints expressing the need of obtaining a target benefit from the shrinkage process are included. The presented preliminary results suggest that a good range of problems can be solved to optimality through the use of a commercial solver, but appropriate algorithmic approaches should be developed to cope with larger instances. Finally, as concerns the model, new versions may be formulated by allowing the shrinking of the services without re-allocate any demand, the introduction of constraints about the maximum expansion allowed for existing facilities, equity constraints for the final distribution of the demand, and real-world applications should be addressed.

References

1. Eiselt, H.A., Marianov, V. (eds.): Foundations of Location Analysis. Springer, New York (2011)
2. Laporte, G., Nickel, S., Saldanha-Da-Gama, F. (eds.): Location Science. Springer, Berlin (2014)
3. Erkut, E., Karagiannidis, A., Perkoulidis, G., Tjandra, S.A.: A multicriteria facility location model for municipal solid waste management in North Greece. Eur. J. Oper. Res. **187**(3), 1402–1421 (2008)
4. Melo, M.T., Nickel, S., Saldanha-Da-Gama, F.: Facility location and supply chain management—a review. Eur. J. Oper. Res. **196**(2), 401–412 (2009)
5. Daskin, M.S.: Network and Discrete Location: Models, Algorithms, and Applications. Wiley (2011)
6. Farahani, R.Z., Hekmatfar, M., Arabani, A.B., Nikbakhsh, E.: Hub location problems: a review of models, classification, solution techniques, and applications. Comput. Ind. Eng. **64**(4), 1096–1109 (2013)
7. Sterle, C., Sforza, A., Amideo, A.E.: Multi-period location of flow intercepting portable facilities of an intelligent transportation system. Socio-Econ. Plan. Sci. **53**, 4–13 (2016)
8. Marianov, V., Serra, D.: Location problems in the public sector. Facil. Locat. Appl. Theor. **1**, 119–150 (2002)
9. Araya, F., Dell, R., Donoso, P., Marianov, V., Martínez, F., Weintraub, A.: Optimizing location and size of rural schools in Chile. Int. Trans. Oper. Res. **19**(5), 695–710 (2012)
10. Marsh, M.T., Schilling, D.A.: Equity measurement in facility location analysis: a review and framework. Eur. J. Oper. Res. **74**(1), 1–17 (1994)
11. Barbati, M., Piccolo, C.: Equality measures properties for location problems. Optim. Lett. **10**(5), 903–920 (2016)
12. Bruno, G., Genovese, A., Piccolo, C.: Capacity management in public service facility networks: a model, computational tests and a case study. Optim. Lett. **10**(5), 975–995 (2016)
13. Bruno, G., Genovese, A., Piccolo, C.: Territorial amalgamation decisions in local government: models and a case study from Italy. Socio-Econ. Plan. Sci. **57**, 61–72 (2017)
14. Wilhelm, W., Han, X., Lee, C.: Computational comparison of two formulations for dynamic supply chain reconfiguration with capacity expansion and contraction. Comput. Oper. Res. **40**(10), 2340–2356 (2013)
15. Sonmez, A.D., Lim, G.J.: A decomposition approach for facility location and relocation problem with uncertain number of future facilities. Eur. J. Oper. Res. **218**(2), 327–338 (2012)
16. Wang, Q., Batta, R., Bhadury, J., Rump, C.M.: Budget constrained location problem with opening and closing of facilities. Comput. Oper. Res. **30**(13), 2047–2069 (2003)
17. ReVelle, C., Murray, A.T., Serra, D.: Location models for ceding market share and shrinking services. Omega **35**(5), 533–540 (2007)
18. Bruno, G., Esposito, E., Genovese, A., Piccolo, C.: Institutions and facility mergers in the Italian education system: Models and case studies. Socio-Econ. Plan. Sci. **53**, 23–32 (2016)
19. Guerriero, F., Miglionico, G., Olivito, F.: Location and reorganization problems: the Calabrian health care system case. Eur. J. Oper. Res. **250**(3), 939–954 (2016)

An Efficient and Simple Approach to Solve a Distribution Problem

C. Cerrone, M. Gentili, C. D'Ambrosio and R. Cerulli

Abstract We consider a distribution problem in a supply chain consisting of multiple plants, multiple regional warehouses, and multiple customers. We focus on the problem of selecting a given number of warehouses among a set of candidate ones, assigning each customer to one or more of the selected warehouses while minimizing costs. We present a mixed integer formulation of the problem of minimizing the sum of the total transportation costs and of the fixed cost associated with the opening of the selected warehouses. We develop a heuristic and a metaheuristic algorithm to solve it. The problem was motivated by the request of a company in the US which was interested both in determining the optimal solution of the problem using available optimization solvers, and in the design and implementation of a simple heuristic able to find good solutions (not farther than 1% from the optimum) in a short time. A series of computational experiments on randomly generated test problems is carried out. Our results show that the proposed solution approaches are a valuable tool to meet the needs of the company.

Keywords Supply chain · Greedy · Carousel Greedy

C. Cerrone
Department of Biosciences and Territory, University of Molise,
Via Francesco De Sanctis, 1, 86100 Campobasso, CB, Italy
e-mail: carmine.cerrone@unimol.it

M. Gentili (✉)
Industrial Engineering Department, University of Louisville,
132 Eastern Parkway, Louisville, KY 40292, USA
e-mail: monica.gentili@louisville.edu; mgentili@unisa.it

M. Gentili · C. D'Ambrosio · R. Cerulli
Department of Mathematics, University of Salerno, Giovanni Paolo II, 132,
84084 Fisciano, SA, Italy
e-mail: cdambrosio@unisa.it

R. Cerulli
e-mail: raffaele@unisa.it

© Springer Nature Switzerland AG 2018

P. Daniele and L. Scrimali (eds.), *New Trends in Emerging Complex Real Life Problems*, AIRO Springer Series 1,
https://doi.org/10.1007/978-3-030-00473-6_17

1 Introduction

We consider a distribution problem in a multi-commodity supply chain consisting of multiple plants, multiple regional warehouses, and multiple customers. The problem addressed consists of selecting a given number of warehouses among a set of candidates, and minimizing the sum of the total transportation costs and of the fixed cost associated with the opening of the selected warehouses such that all customers are served and capacity limits at the plants are respected. The problem was motivated by a consulting project for a company located in the US which was interested in offering a web service by solving the problem on-line for its own customers. The specific request of the company was twofold. On the one hand, the company was interested in comparing the computational times required by different optimization solvers (both commercial and free ones) to determine the optimal solution of the problem, and, on the other hand, it was interested in the design of a simple heuristic approach with a good trade-off between quality of the solution and required computational time (the goal was a solution no more than 1% worse than the optimum).

The problem addressed belongs to the general class of location problems [5, 6] and can be seen as an extension of the well known fixed charge facility location problem which includes shipments from plants to distribution centers and multiple commodities [4]. The fixed charge facility location problem and its extension in supply chain design have been widely studied in the literature, and several exact and heuristic approaches have been proposed [3].

In this paper we propose to use a greedy algorithm and a novel metaheuristic approach, namely a Carousel Greedy approach, which responds to the company requirement of simplicity and which proved to be efficient in solving a wide class of problems [1, 2]. An extensive experimental analysis is carried out to compare computational times required by the solvers and to study the trade-off between quality and computational time required by our heuristic approaches.

The paper is organized as follows. A formal description of the problem along with the mathematical formulation is given in the next section. Our heuristic approaches are described in Sect. 3. The experimental analysis is presented in Sect. 4. Conclusions are discussed in Sect. 5.

2 Problem Description

Let S be a set of plants, C a set of final customers which need to be served, W a set of warehouses connected to the plants and customers, and P the set of products which can be produced at each plant. Given a set D, we denote by $|D|$ its cardinality. Each plant $s \in S$ can produce a maximum given quantity u_{sp} of product $p \in P$ and can ship it to a warehouse $w \in W$ for a unit transportation cost equal to σ_{swp}. From each warehouse $w \in W$, a product $p \in P$ is shipped to a customer $c \in C$ for a unit transportation cost equal to σ_{wcp}. Each plant has also a total production capacity which is denoted by u_s. Each customer $c \in C$ has an associated demand d_{cp} for product $p \in P$. Finally, each warehouse $w \in W$ can be activated with a fixed activation cost equal to σ_w. We want to activate at most k warehouses among the candidate ones so that the total cost, given by the sum of the activation costs and the sum of the transportation costs, is minimized, and all the customer demands are satisfied without exceeding capacity at the plants. Let us define the following set of variables: x_{swp} is the quantity of product $p \in P$ which is produced at plant $s \in S$ and shipped to warehouse $w \in W$; x_{wcp} is the quantity of product $p \in P$ which is shipped from warehouse $w \in W$ to customer $c \in C$; $y_w \in \{0, 1\}$ which is equal to one if warehouse $w \in W$ is activated and equal to zero otherwise. The mathematical formulation of our problem is as follows.

$$z = \min \sum_{w \in W} \sigma_w y_w + \sum_{p \in P} \sum_{s \in S} \sum_{w \in W} \sigma_{swp} x_{swp} + \sum_{p \in P} \sum_{w \in W} \sum_{c \in C} \sigma_{wcp} x_{wcp} \tag{1}$$

$$\sum_{w \in W} y_w \leq k \tag{2}$$

$$\sum_{w \in W} x_{swp} \leq u_{sp} \qquad \forall\, p \in P, s \in S \tag{3}$$

$$\sum_{w \in W} \sum_{p \in P} x_{swp} \leq u_s \qquad \forall\, s \in S \tag{4}$$

$$\sum_{w \in W} x_{wcp} \geq d_{cp} \qquad \forall\, p \in P, c \in C \tag{5}$$

$$x_{swp} \leq u_{sp} y_w \qquad \forall\, p \in P, s \in S, w \in W \tag{6}$$

$$x_{wcp} \leq d_{cp} y_w \qquad \forall\, p \in P, c \in C, w \in W \tag{7}$$

$$\sum_{s \in S} x_{swp} = \sum_{c \in C} x_{wcp} \qquad \forall\, p \in P, w \in W \tag{8}$$

$$y_w \in \{0, 1\} \qquad \forall\, w \in W \tag{9}$$

$$x_{swp} \geq 0 \qquad \forall\, p \in P, s \in S, w \in W \tag{10}$$

$$x_{wcp} \geq 0 \qquad \forall\, p \in P, c \in C, w \in W \tag{11}$$

Objective function (1) minimizes the sum of the total activation costs and of the total transportation costs. Constraint (2) ensures the activation of at most k warehouses. Constraints (3) state that the total amount of product $p \in P$ shipped from plant s cannot exceed the capacity of the plant to produce that product. Constraints (4) are additional capacity constraints which require that the total production capacity is not exceeded in the plant. Customer demands are satisfied thanks to constraints (5). Constraints (6) and (7) are logical constraints which ensure that flow cannot transit through a warehouse if it is not activated. Finally, constraints (8) are flow conservation constraints for each warehouse.

3 Our Solution Approaches

We use two solution approaches to solve the problem described in the previous section: a greedy approach and a carousel greedy approach [2].

Greedy Algorithm
The main idea of the greedy algorithm is to choose at each step, according to a greedy criterion, one of the warehouses to be chosen and to stop when k warehouses are selected. The final solution is then obtained by solving the mathematical model (1)–(11) by setting $y_w = 1$ for each warehouse w which was greedily chosen. The greedy criterion, at each step, evaluates the overall cost of including a warehouse in the solution and chooses to select the warehouse corresponding to the minimum cost. Specifically, during the first iteration, in order to choose the first warehouse to be included in the solution, the algorithm solves $|W|$ linear programming problems, one for each warehouse that needs to be evaluated. In particular, the linear problem associated with any given warehouse $w \in W$ is obtained by model (1)–(11) by fixing to one the variable y_w of the warehouse being evaluated, and it is as follows:

$$z_w^{(1)} = \min \sigma_w + \sum_{p \in P} \sum_{c \in C} \sigma_{scp} x_{scp} \tag{12}$$

$$\sum_{c \in C} x_{scp} \leq u_{sp} \qquad \forall\, p \in P, s \in S \tag{13}$$

$$\sum_{c \in C} \sum_{p \in P} x_{scp} \leq u_s \qquad \forall\, s \in S \tag{14}$$

$$\sum_{s \in S} x_{scp} \geq d_{cp} \qquad \forall\, p \in P, c \in C \tag{15}$$

$$x_{scp} \geq 0 \qquad \forall\, p \in P, c \in C, s \in S \tag{16}$$

In the above formulation, we have only one set of variables, namely variables x_{scp}, which denote the quantity of product p to be shipped from plant s to customer c. The unit transportation cost σ_{scp} in the objective function (12) is obtained as the sum of the unit transportation cost for shipping product p from plant s to warehouse w plus the unit transportation cost of shipping product p from warehouse w to customer c, i.e. $\sigma_{scp} = \sigma_{swp} + \sigma_{wcp}$.

The selected warehouse, after the solution of $|W|$ linear programming problems, is the warehouse w^*, such that the associated value z_{w^*} is the minimum, i.e., $w^* = \text{argmin}_{w \in W} z_w$. In the successive step, $|W| - 1$ linear programming problems will be solved to choose the next warehouse to be activated. Specifically, with any warehouse w to be evaluated, the linear programming problem to be solved is:

$$z_w^{(2)} = \min \sigma_{w^*} + \sigma_w + \sum_{p \in P} \sum_{c \in C} \sigma_{scp} x_{scp}$$
$$\text{subject to } (13)-(16) \tag{17}$$

where the unit transportation costs σ_{scp} in the objective function (17) are defined as follows:

$$\sigma_{scp} = \min\{\sigma_{sw^*p} + \sigma_{w^*cp}, \sigma_{swp} + \sigma_{wcp}\}$$

The algorithm proceeds in this way until k warehouses are selected.

In our experiments, we solved the linear programming problems associated with each warehouse by means of an optimization solver.

Carousel Greedy Algorithm
The carousel greedy algorithm is a generalized method to enhance greedy algorithms, originally proposed in [2]. The aim is to obtain a procedure which is almost as fast and simple as the greedy procedure (on which it is based), while achieving accuracy levels similar to those of a metaheuristic. The main idea underlying a carousel greedy algorithm is that during the execution of a constructive heuristic, the later decisions are likely to be more informed and valid than the earlier ones. Given this observation, the carousel greedy procedure for our specific problem starts from a feasible solution F (that is, a set of selected warehouses which satisfies constraints (2)–(8)) obtained through a greedy procedure, and operates in three main phases:

- In the first phase a partial (unfeasible) solution R is built from F by removing a number of warehouses equal to a given percentage β of the size of F.
- In the second phase, the partial solution R is modified by iteratively removing from it the oldest elements (that is, removing warehouses which were the first to be selected) and adding new ones. This step is repeated for a pre-defined number α of iterations. The choice for adding new warehouses to R are made according to the same greedy criterion used by the greedy algorithm. The output of this step is a new partial solution R'.
- In the final phase, the partial solution R' is completed, using the greedy algorithm, to produce a feasible solution.

4 Experimental Analysis

We ran a wide set of experiments on randomly generated instances. We implemented our algorithms in Java (version 9) and use a PC with Core i7 2.6 Ghz and 16 Gb RAM. We obtained the optimal solution of the problem by solving the mathematical model using both Cplex (version 12.7) and LPSolve (version 5.5). We ran the greedy and the carousel greedy using both Cplex and LPSolve to solve the mathematical models required by the greedy criterion. Our analysis had the aim of (i) comparing the computational time of the commercial solver Cplex and of the free solver LPSolve, and (ii) performing a sensitivity analysis of the quality of the solution and of the computational time required by our greedy and carousel greedy algorithms.

We generated several instances by varying the following: the total number of products ($|P| = 5, 10, 20$), the total number of plants ($|S| = 5, 10, 15$), the total number of customers ($|C| = 100, 200, 400$), the total number of candidate warehouses ($|W| = 50, 100, 200$), and the total number of warehouses to be activated ($k = 10, 20$). Customers, plants and warehouses were randomly distributed on a grid of size 5,000 × 5,000. We assumed transportation costs not to depend on the product type and to be proportional to the distances between either supply and warehouses or warehouses and customers. Distances σ_{sw}, between a plant s and a warehouse w, are perturbed euclidean distances, that is $\sigma_{sw} = \delta_{sw} \pm 0.05\delta_{sw}$, where δ_{sw} is the euclidean distance and the perturbation is either added or subtracted at random. Distances σ_{wc}, between a warehouse w and a customer c, are computed analogously. Demand d_{cp} of product p at customer c is uniformly distributed within [0, 100]. Supply at the plants for each product was randomly generated so that the sum of all the supply for that product is less than a predefined percentage of the total demand, that is $\sum_s u_{sp} \leq \gamma \sum_c d_{cp}$. We set $\gamma = 0.27$. Finally, total supply u_s at each plant is randomly generated so that it is strictly less than the sum of all the supply of each product at the plant, i.e., $u_s \leq \varepsilon \sum_p u_{sp}$. We set $\varepsilon = 0.22$.

For each combination of input parameters, we generated three different instances and reported the average of the obtained results.

Results are shown in Tables 1 and in Fig. 1.

Table 1 shows the results of the mathematical model (Cplex column), the greedy algorithm (Greedy column), and the carousel greedy algorithm (CG column) on the set of instances when $k = 10$, which were the most difficult to solve. The results of the carousel greedy algorithm are reported for different values of the parameters α and β. The table shows the computational time (in seconds) required by Cplex to solve the instance to optimality as well as the computational time and the percentage gap from the optimum for the other approaches. The gap is computed as $\frac{Opt-UB}{Opt}$ where Opt is the solution value returned by Cplex, and UB is the solution value returned either by the greedy algorithm or by the carousel greedy algorithm. We set a time limit of 1,800 s for the solver. The symbol * in the table indicates Cplex reached the time limit on at least one of the instances of the scenario. This happened for scenarios 16 and 17; hence, the corresponding gaps for the greedy and of the carousel greedy algorithms are not computed with respect to the optimal solution for these scenarios but with respect to the best solution found within the time limit. The mathematical models to be solved in the greedy criterion were solved using Cplex. As expected, the computational time for the three approaches increases with larger instances. On average (considering only scenarios 1–15, for which Cplex did not reach the time limit), the greedy algorithm returns a solution 10 times faster than Cplex; the carousel greedy 4 times faster than Cplex when $\alpha = 3$ and $\beta = 0.1$, otherwise, with the other settings, the carousel greedy is 6 times faster than Cplex. The percentage gap from the optimum is always less than 6% for the greedy (with a maximum gap equal to 5.4%). Using the settings $\alpha = 3$ and $\beta = 0.1$, the threshold of 1% is met by the carousel greedy algorithm for all the scenarios but one (scenario 11, for which the gap is equal to 1.2%).

Table 1 Results of the mathematical model, the greedy algorithm, and the carousel greedy on the set of instances when $k = 10$

Scenario					Cplex	Greedy		CG: $\alpha = 1, \beta = 0.1$		CG: $\alpha = 1, \beta = 0.2$		CG: $\alpha = 3, \beta = 0.1$	
ID	$\|P\|$	$\|S\|$	$\|C\|$	$\|W\|$	Time (s)	Time (s)	Gap (%)	Time (s)	Gap (%)	Time (s)	Gap (%)	Time (s)	Gap (%)
1	5	5	100	50	3.91	1.64	1.6	3.31	0.4	3.31	0.4	4.45	0.3
2	5	5	100	100	15.56	2.54	2.3	6.29	0.5	6.29	0.5	8.44	0.3
3	5	5	100	200	62.12	5.28	2.0	11.38	1.1	11.38	1.1	15.92	0.8
4	5	5	200	50	7.84	3.76	1.9	7.02	0.1	7.02	0.1	9.18	0.1
5	5	5	200	100	99.28	5.81	2.6	13.98	0.8	13.98	0.8	18.27	0.4
6	5	5	200	200	229.10	11.71	2.1	27.25	0.9	27.25	0.9	35.46	0.9
7	5	5	400	50	63.33	7.66	1.0	15.59	0.3	15.59	0.3	20.30	0.2
8	5	5	400	100	128.35	16.48	2.0	33.71	0.9	33.71	0.9	43.24	0.6
9	5	5	400	200	718.56	33.07	2.6	61.59	1.3	61.59	1.3	82.27	0.6
10	5	10	200	100	211.30	17.90	5.4	31.76	0.5	31.76	0.5	40.17	0.4
11	5	15	200	100	238.92	30.07	4.1	51.70	1.4	51.70	1.4	68.37	1.2
12	10	5	200	100	102.72	16.97	1.4	33.29	0.9	33.29	0.9	43.82	0.4
13	10	10	200	100	193.52	56.78	2.0	79.05	0.8	79.05	0.8	103.01	0.5
14	10	15	200	100	772.73	109.06	4.4	145.82	1.4	145.82	1.4	190.42	0.8
15	20	5	200	100	891.05	47.10	2.1	76.71	0.8	76.71	0.8	107.23	1.0
16	20	10	200	100	1093.85*	117.92	5.3	196.71	1.2	196.71	1.2	257.07	0.3
17	20	15	200	100	1803.27*	222.53	−11.6	366.06	−14.1	366.06	−14.1	509.01	−14.4
Average (scenarios 1–15)					*249.22*	*24.39*	*2.5*	*39.90*	*0.8*	*39.90*	*0.8*	*52.70*	*0.6*
St. Dev. (scenarios 1–15)					*284.31*	*27.78*	*1.2*	*37.10*	*0.4*	*37.10*	*0.4*	*48.98*	*0.3*
Max (scenarios 1–15)					*891.05*	*109.06*	*5.4*	*145.82*	*1.4*	*145.82*	*1.4*	*190.42*	*1.2*
Min (scenarios 1–15)					*3.91*	*1.64*	*1.0*	*3.31*	*0.1*	*3.31*	*0.1*	*4.45*	*0.1*

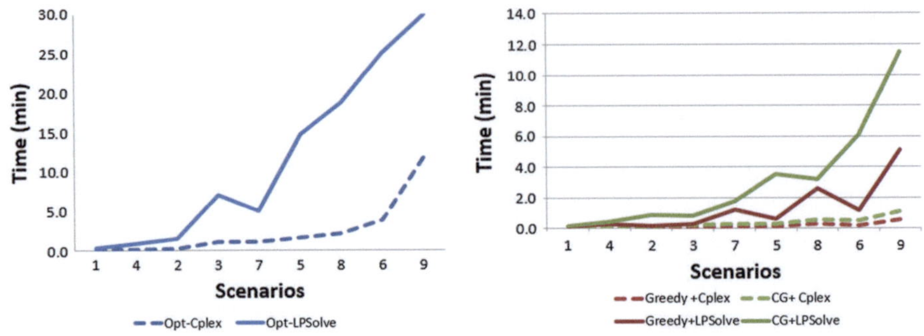

Fig. 1 Running times of the proposed approaches when using two different solvers, namely Cplex and LPSolve

Finally, Fig. 1 shows the analysis of the running times when two different solvers, namely Cplex and LPSolve, are used to solve the mathematical models. Within the time limit of 1,800 s, the free solver LPSolve was able to solve to optimality for only the first nine scenarios; consequently, the results are reported only for this subset of scenarios. The graph on the left compares the running time for the two solvers, while the graph on the right compares the running time for the greedy and for the carousel greedy when the two different solvers are used to solve the mathematical models required by the greedy criterion. On average, Cplex is six times faster than LPSolve, which results in the greedy and carousel greedy being six and seven times faster on average, respectively, when using Cplex than when using LPSolve.

5 Conclusions

Motivated by the request from a company in the US, we address a distribution problem in a multi-commodity supply chain consisting of multiple plants, multiple regional warehouses, and multiple customers. Our focus was on the problem of selecting a given number of warehouses among a set of candidate ones, assigning each customer to one or more of the selected warehouses while minimizing total costs. The company was interested in the design and implementation of a simple heuristic able to find a solution not farther than 1% from the optimum in a short time. We designed a greedy and a carousel greedy approach and tested the trade-off between quality of the solution and running time on a set of randomly generated instances. Our results show that the proposed heuristic approaches are a valuable tool to meet the needs of the company.

References

1. Carrabs, F., Cerrone, C., D'Ambrosio, C. Raiconi, A.: Column generation embedding carousel greedy for the maximum network lifetime problem with interference constraints. Springer Proc. Math. Stat. **217**, 151–159 (2017)
2. Cerrone, C., Cerulli, R., Golden, B.: Carousel greedy: a generalized greedy algorithm with applications in optimization. Comput. Operat. Res. **85**, 97–112 (2017)
3. Daskin, M.S., Snyder, L.V. Berger, R.T.: Facility location in supply chain design. In Logist. Syst. Design Optim. 39–65. Springer (2005)
4. Geoffrion, A.M., Graves, G.W.: Multicommodity distribution system design by benders decomposition. Manag. Sci. **20**(5), 822–844 (1974)
5. Melo, M.T., Nickel, S., Saldanha-Da-Gama, F.: Facility location and supply chain management-a review. Eur. J. Operat. Res. **196**(2), 401–412 (2009)
6. Mirchandani, P.B., Francis, R.L.: Discrete Location Theory (1990)

First-Time Interaction Under Revenue-Sharing Contract and Asymmetric Beliefs of Supply-Chain Members

Tatyana Chernonog

Abstract The paper provides a thorough investigation of a first-time interaction between a retailer and a manufacturer who are unreliable in a cost function of the manufacturer. We consider a two-echelon supply chain of a single customized product, where parties interact via a revenue-sharing contract. The general model is formulated as a Retailer-Stackelberg game with two-sided information asymmetry. We derive the equilibrium strategy and parties' profits when: (i) information is complete, (ii) hidden information asymmetry is present, and (iii) known information asymmetry is present. For a third scenario, we propose two different contracts to induce a Pareto-optimal information-sharing equilibrium.

Keywords Revenue sharing contract · Asymmetric information
Supply chain management

1 Introduction

The globalization and customization of the supply chain have rapidly accelerated through the last decade, since the growth of the internet has presented supply chains with many significant opportunities for cost reduction and service improvements [10]. On the one hand, to increase profits, retailers continuously look for new low-cost or brand-new manufacturers, and the internet offers an efficient way to connect many potential retailers and manufacturers. This trend, in turn, increases the likelihood that a lack of prior interactions will mean the retailers do not know how efficient the manufacturers are [8]. On the other hand, with the proliferation of social media and online publishing, styles and trends now change faster than ever, and customization helps manufacturers gain insights from customized designs and fine-tune products to stay one step ahead of the competition.

T. Chernonog (✉)
Department of Management, Bar-Ilan University, 5290002 Ramat-Gan, Israel
e-mail: Tatyana.Chernonog@biu.ac.il

© Springer Nature Switzerland AG 2018

P. Daniele and L. Scrimali (eds.), *New Trends in Emerging Complex Real Life Problems*, AIRO Springer Series 1,
https://doi.org/10.1007/978-3-030-00473-6_18

The paper provides a thorough investigation of a first-time interaction between a retailer and a manufacturer in a supply chain of a single customized product. We suppose both the retailer and the manufacturer acknowledge the presence of inherent uncertainty in the supplier's production process. In line with Gurnani and Shi [8], we assume the supply-chain parties have different estimates of the manufacturer's cost function. Following the last trend in a market, the retailer is a leader of a supply chain and parties interact under a widely used revenue-sharing contract [1–3, 5].

To our knowledge, only two papers have analyzed revenue-sharing contracts with asymmetric information under consignment contracts. Kong et al. [9] consider a revenue-sharing policy in which the supplier sells the product to two competing retailers under one-sided information asymmetry. Zhang and Chen [17] study information sharing in a two-echelon supply chain, in which both the supplier and the retailer possess partial information on the demand. Studies dealing with two-sided information asymmetry tend to focus on wholesale-price contracts [4, 8, 12, 13, 18].

The Importance and the innovative contribution of this paper lie in its generality. For general revenue and cost functions of the manufacturer, and for a general set of decisions, we have investigated the effect of information asymmetry in the manufacturer's cost function. To the best of our knowledge, no studies in the operations management literature have investigated the first-time interaction between a retailer and a manufacturer of a customized product, who operate under a revenue-sharing contract under two-sided information asymmetry.

2 Model Formulation

Consider a manufacturer who distributes a customized product to customers via a dominant retailer. Let $s \in S$ be a vector of actions in a feasible set S of positive real numbers, such as selling price, product quality, shipping fee, or investment in research and development or in brand promotion undertaken by the manufacturer. The total revenue is denoted by $R(s)$. Let $\{C_x(s) : x \in \Omega\}$, where Ω is an index set, be a collection of the manufacturer's possible cost function, which would all make sense. According to the normative approach (e.g., [7]), both the retailer and the manufacturer assess a prior density function $f_r(\cdot)$ and $f_m(\cdot)$ over the set $\{x \in \Omega_i\}$, $i \in \{r, m\}$. These density functions reflect the retailer's and manufacturer's beliefs regarding the actual cost function. Note $f_r(\cdot)$ may be the industry-wide estimated measures of performance, and the difference in the parties' beliefs may be attributed to the manufacturer's proximity to the production process. Some of the actions that affect R may not affect C_x, and vice versa. To avoid trivial cases, we assume $s \in S$ exist for which $R(s) - E_i[u(C_x(s))] > 0$, $i \in \{r, m\}$, where $E_i[\cdot]$ denotes the expectation calculated based on party i's beliefs and $u(\cdot)$ denotes the utility function of the manufacturer. Because we consider a first-time interaction involving a customized product, both estimates could be inaccurate. However, for their planning purposes, the retailer and the manufacturer would use their own best beliefs [8].

The manufacturer and the retailer interact via a revenue-sharing contract under which the retailer demands a fraction η of the revenue from selling the product. Therefore, the manufacturer, via the vector of actions s, dictates the total profit of the supply chain, whereas the retailer dictates the revenue-sharing rule. The profit of the retailer and the expected utility function of the manufacturer's profit are, respectively,

$$\Pi_r(\eta|s) = \eta R(s) \tag{1}$$

and

$$E_i[u(\Pi_m(s|\eta,\ f_i))] = (1-\eta)R(s) - E_i[u(C_x(s))], \quad i \in \{r,\ m\}. \tag{2}$$

Lemma 1 *$E[\Pi_m(s|\eta,\ f_i)]$ decreases in η for any s.*

Proof Straightforward from (2).

We model the manufacturer-retailer relationship as a sequential non-cooperative game in which the retailer is the leader of the supply chain and the manufacturer is the follower (Retailer-Stackelberg). In what follows, we derive the equilibrium strategy and parties' profits under the following conditions: (i) information is complete—the retailer's and manufacturer's beliefs regarding the actual cost function are common knowledge; (ii) hidden information asymmetry is present—the retailer does not know that the manufacturer's beliefs regarding the cost function are different from her estimation; and (iii) known information asymmetry is present—the retailer knows that the manufacturer's belief regarding the cost function is different from her estimation; however, this belief is the private information of the manufacturer.

2.1 *The Case of Complete Information*

We first assume $f_r(\cdot)$ and $f_m(\cdot)$ are common knowledge; that is, the retailer and the manufacturer know each other's estimates. Under this assumption, the retailer sets her equilibrium revenue share η^* by solving

$$\begin{aligned} &\max_{0\le\eta\le1} \{\Pi_r(\eta|s(\eta|f_m)) = \eta R(s(\eta|f_m))\} \\ &s.t. \quad s(\eta|f_m) \equiv \arg\max_{s\in S}\{E_m[u(\Pi_m(s|\eta,\ f_m))]\} \end{aligned}. \tag{3}$$

Clearly, it is an optimization problem in which the decision maker is the retailer and the manufacturer follows by determining his set of actions, denoted by $s^* \equiv s(\eta^*|f_m)$, so that the pair (η^*, s^*) is a Nash equilibrium. We assume both maxima exist. This assumption holds, for example, when both $R(s)$ and $C(s)$ are continuous and either S

is a compact set or $\lim_{s\to\infty}[R(s) - E_m[u(C_x(s))]] \leq 0$.[1] Note η may take the value of 0 or 1, in which case, no contract will be signed.

Proposition 1 *Under complete information, the equilibrium strategy is not affected by the retailer's belief about the manufacturer's cost function.*

Proof Straightforward from (3).

By Proposition 1, the retailer has to assess the manufacturer's beliefs regarding the cost function, but neither party has to know the retailer's belief.

Representative example

Consider a revenue-sharing contract in the mobile app supply chain [1, 2], where the specific vector of the manufacturer's actions are retail price, p, and q, whereby an increase in q causes an increase in demand but requires some level of investment. Examples of q might be the quality level or the degree of promotion or innovation invested in the mobile application. Hereafter, we refer to q as the quality level of the product. The vector of actions is $s = (p, q)$, $0 \leq p, q < \infty$ and the revenue of the supply chain is $R(p, q) = pD(p, q)$, where $D(p, q)$ is a demand function. We adopt the linear demand function used by Xie et al. [16] and El Ouardighi and Kim [14], according to which

$$D(p, q) = a - \alpha p + \beta q,$$

where a is the base customer demand, α is the customer's sensitivity to selling price, and β is the customer's sensitivity to product quality. Thus, total revenue is

$$R(p, q) = pD(p, q).$$

As in the case of virtual-product models [6], the distribution of mobile apps is characterized by a negligible unit distribution cost and ample capacity to fulfill demand. Therefore, our model does not include either holding or shortage costs. Let the cost function be

$$C(q) = \gamma q^2/2,$$

where the scale coefficient γ is a random variable and $1/\gamma$ reflects the manufacturer's investment efficiency [15]. The random variable γ can take the value γ_1 and γ_2, where $\gamma_2 > \gamma_1 > 0$. Thus, γ_1 represents a scenario in which the manufacturer's investment efficiency is high, whereas γ_2 represents a low manufacturer's investment efficiency. The retailer believes the probability that the manufacturer is highly efficient is μ_r, whereas the manufacturer believes this probability is μ_m. For simplicity, we denote each party's expected scale coefficients by $\bar{\gamma}_i \equiv \mu_i\gamma_1 - (1 - \mu_i)\gamma_2$, $i \in \{r, m\}$. For this example, and assuming the manufacturer is risk neutral, the optimization problem (3) can be rewritten as

[1] $s \to \infty$ means that at least one component of S tends to infinity.

$$\max_{0\le\eta\le1}\{\Pi_r(\eta|p(\eta|\mu_m),\, q(\eta|\mu_m)) = \eta p(\eta|\mu_m)D(p(\eta|\mu_m),\, q(\eta|\mu_m))\}$$
$$s.t.\quad (p(\eta|\mu_m),\, q(\eta|\mu_m)) \equiv \arg\max_{p,q}\left\{E[\Pi_m(p,\, q|\eta,\, \mu_m) \equiv (1-\eta)pD(p,\, q) - \bar{\gamma}_m q^2/2]\right\}.$$

Solving the above optimization problem, we find that if $\frac{\beta^2}{2\alpha} < \bar{\gamma}_m < \frac{\beta^2}{\alpha}$, then at equilibrium, $\eta^* = \frac{2\alpha\bar{\gamma}_m}{\beta^2} - 1$, $p^* = \frac{a\bar{\gamma}_m}{2(2\alpha\bar{\gamma}_m-\beta^2)}$, $q^* = \frac{a(\beta^2-\alpha\bar{\gamma}_m)}{\beta(2\alpha\bar{\gamma}_m-\beta^2)}$, $\Pi_r(\eta^*|p^*,\, q^*) = \frac{a^2\bar{\gamma}_m^2\alpha}{4\beta^2(2\alpha\bar{\gamma}_m-\beta^2)}$, and $E_m[\Pi_m(p^*,\, q^*|\eta^*)] = \frac{a^2\bar{\gamma}_m^2(\beta^2-\alpha\bar{\gamma}_m^2)}{2\beta^2(2\alpha\bar{\gamma}_m^2-\beta^2)}$. Note condition $\frac{\beta^2}{2\alpha} < \bar{\gamma}_m < \frac{\beta^2}{\alpha}$ ensures an interior equilibrium revenue share and finite value of equilibrium price and quality. For these reasons, hereafter we assume this condition holds.

2.2 *The Case of Hidden Information Asymmetry*

We assume the retailer does not know that the manufacturer's beliefs regarding the cost function are different from the industry-wide estimated $f_r(\cdot)$. Under this assumption, the retailer sets her equilibrium revenue share $\hat{\eta}^*$ by solving

$$\max_{0\le\eta\le1}\{\Pi_r(\eta|s(\eta|f_r)) = \eta R(s(\eta|f_r))\}$$
$$s.t.\quad s(\eta|f_r) = \arg\max_{s\in S}\{E_r[u(\Pi_m(s|\eta,\, f_r))]\}. \tag{4}$$

As in previous subsection, we assume both maxima exist. This assumption holds, for example, when both $R(s)$ and $C(s)$ are continuous and either S is a compact set or $\lim_{s\to\infty}[R(s) - E_r[u(C_x(s))]] \le 0$. Then, given $\hat{\eta}^*$, the manufacturer, based on his personal beliefs about the actual cost function, $f_m(\cdot)$, determines the optimal set of actions, denoted $\hat{s}^* \equiv s(\hat{\eta}^*|f_m)$, by solving

$$\max_{s\in S}\left\{E_m\left[u\left(\Pi_m(s|\hat{\eta}^*,\, f_m)\right)\right]\right\}, \tag{5}$$

so that the pair $(\hat{\eta}^*,\, \hat{s}^*)$ is a Nash equilibrium.

Proposition 2 *Under hidden information asymmetry, the equilibrium strategy is affected by the retailer's beliefs about the manufacturer's cost function as well as the manufacturer's belief.*

Proof Straightforward from (4) and (5).

Representative example (continued)
Following (4) and (5), we find that under hidden information asymmetry at equilibrium, $\hat{\eta}^* = \frac{2\alpha\bar{\gamma}_r-\beta^2}{\beta^2}$, $\hat{p}^* = \frac{a\bar{\gamma}_m}{2(\alpha(\bar{\gamma}_m+\bar{\gamma}_r)-\beta^2)}$, $\hat{q}^* = \frac{a(\beta^2-\alpha\bar{\gamma}_r)}{\beta(\alpha(\bar{\gamma}_m+\bar{\gamma}_r)-\beta^2)}$, $\Pi_r(\hat{\eta}^*|\hat{p}^*,\, \hat{q}^*,\, \mu_m) = \frac{a^2\bar{\gamma}_r^2\alpha(2\alpha\bar{\gamma}_r-\beta^2)}{4\beta^2(\alpha(\bar{\gamma}_m+\bar{\gamma}_r)-\beta^2)^2}$ and $E_m\left[\Pi_m(\hat{p}^*,\, \hat{q}^*|\hat{\eta}^*,\, \mu_m)\right] = \frac{a^2\bar{\gamma}_m(\beta^2-\alpha\bar{\gamma}_r)}{2\beta^2(\alpha(\bar{\gamma}_m+\bar{\gamma}_r)-\beta^2)}$.

2.3 Comparison of Complete Information and Hidden Information Asymmetry

Unsurprisingly, the hidden information asymmetry concerning the cost function may cost the retailer in terms of reduced profits, as the following proposition claims.

Proposition 3 $\Pi_r(\eta^*|s^*) \geq \Pi_r(\hat{\eta}^*|\hat{s}^*)$.

Proof By comparing the constraints of (3) and (5), we found the optimal manufacturer's set of actions s is obtained by substituting η^* and $\hat{\eta}^*$ into $s(\eta|f_m)$, for the complete-information and for the hidden-information-asymmetry cases, respectively. This result is due to the manufacturer's personal belief about his cost function, and his decision which is based on it. Thus, $\Pi_r(\eta^*|s^*) = \Pi_r(\eta^*|s(\eta^*)) \geq \Pi_r(\eta|s(\eta))\ \forall \eta$, and hence, $\Pi_r(\eta^*|s^*) \geq \Pi_r(\hat{\eta}^*|s(\hat{\eta}^*)) = \Pi_r(\hat{\eta}^*|\hat{s}^*)$.

By Proposition 3, the retailer may benefit, but never loses, from knowing the manufacturer's belief about the cost function. The next proposition corresponds to the profit of the manufacturer in this situation.

Proposition 4 $E_m[u(\Pi_m(s^*|\eta^*, f_m))] \geq E_m\left[u\left(\Pi_m(\hat{s}^*|\hat{\eta}^*, f_m)\right)\right]$ *if and only if* $\eta^* < \hat{\eta}^*$.

Proof Let $\eta^* < \hat{\eta}^*$. Then $E_m\left[u\left(\Pi_m(\hat{s}^*|\hat{\eta}^*, f_m)\right)\right] \leq E_m\left[u\left(\Pi_m(\hat{s}^*|\eta^*, f_m)\right)\right] \leq E_m[u(\Pi_m(s^*|\eta^*, f_m))]$. The first inequality arises because, by Lemma 1, $E[\Pi_m(s|\eta, f_m)]$ decreases in η for any s, and the second arises because $s^* = \arg\max_{s \in S} E_m[\Pi_m(s|\eta^*, f_m)]$. For the converse, let $\eta^* > \hat{\eta}^*$. Then $E_m[u(\Pi_m(s^*|\eta^*, f_m))] \leq E_m\left[u\left(\Pi_m(s^*|\hat{\eta}^*, f_m)\right)\right] \leq E_m\left[u\left(\Pi_m(\hat{s}^*|\hat{\eta}^*, f_m)\right)\right]$. The first inequality is valid by Lemma 1, and the second follows from $\hat{s}^* = \arg\max_{s \in S} E_m\left[u\left(\Pi_m(s|\hat{\eta}^*, f_m)\right)\right]$. □

By Proposition 4, depending on the relative values of η^* and $\hat{\eta}^*$, the manufacturer expects to gain or lose in a hidden-information-asymmetry case. Combining the results of Propositions 3 and 4, we conclude that information sharing between parties benefits the retailer and may hurt the manufacturer.

For simplicity of presentation, let $E_m[\Pi^*] \equiv \Pi_r(\eta^*|s^*) + E_m[\Pi_m(s^*|\eta^*, f_m)]o$ and $E_m\left[\hat{\Pi}^*\right] \equiv \Pi_r(\hat{\eta}^*|\hat{s}^*) + E_m\left[\Pi_m(\hat{s}^*|\hat{\eta}^*, f_m)\right]$, $i \in \{r, m\}$. Consequently, the manufacturer will not share information voluntarily.

Corollary 1 *If* $\eta^* < \hat{\eta}^*$, *then* $E_m[\Pi^*] \geq E_m\left[\hat{\Pi}^*\right]$.

Proof Straightforward from Propositions 3 and 4. □

Representative example (continued)
The result of Proposition 4 can be simplified to $E_m[\Pi_m(p^*, q^*|\eta^*, \mu_m)] \leq E_m\left[\Pi_m(\hat{p}^*, \hat{q}^*|\hat{\eta}^*, \mu_m)\right]$ iff $\bar{\gamma}_m < \bar{\gamma}_r$. In addition, we can extend the result of Corollary 1 as follows: $E_m[\Pi^*] \leq E_m\left[\hat{\Pi}^*\right]$ iff $\bar{\gamma}_m < \bar{\gamma}_r$.

2.4 The Case of Known Information Asymmetry

We assume the retailer knows that the manufacturer's belief regarding the cost function is different from the industry-wide estimated $f_r(\cdot)$; however, $f_m(\cdot)$ is the private information of the manufacturer. Propositions 3 and 4 show that although the retailer benefits when the manufacturer reveals his private information, the manufacturer might lose. Consequently, the manufacturer will not share his private information voluntarily. If $E_m[\Pi^*] \geq E_m\left[\hat{\Pi}^*\right]$ when $\eta^* > \hat{\eta}^*$, the retailer can induce the manufacturer to reveal his belief by paying a fee. The latter is called a side payment, defined by Leng and Zhu [11] as a monetary transfer between two members of a supply chain made to improve the chain-wide performance (it is also known as a transfer payment, compensation, reimbursement, etc.). For additional discussion on the setting optimal side payment, see Avinadav et al. [3]. Otherwise, if $E_m[\Pi^*] < E_m\left[\hat{\Pi}^*\right]$ when $\eta^* > \hat{\eta}^*$, the information sharing cannot be achieved through a simple side payment. In the last case, a natural question to ask is what type of contract will result in a Pareto-optimal information-sharing equilibrium.

Following Mishra et al. [13] and using the results of Lemma 1, we propose a contract to motivate the manufacturer to reveal his personal belief. This contract can be stated as follows: $\eta^{**} = \min(\eta^*, \hat{\eta}^*)$. This contract states that the revenue share the retailer charges when the manufacturer reveals his personal belief will not exceed that under no information sharing. In effect, the contract prevents the retailer from raising her revenue share after the manufacturer reveals his belief. The contract can be easily implemented because, as we stated earlier, the retailer's belief is the industry-wide estimated measure of performance, which is common knowledge. Consequently, the manufacturer can ensure that the retailer's revenue share is not higher than under no information sharing.

Representative example (continued)

Because $E_m[\Pi^*] \leq E_m\left[\hat{\Pi}^*\right]$ iff $\bar{\gamma}_m < \bar{\gamma}_r$, in this case, the side-payment contract is not appropriate for the retailer. Thus, the retailer may increase her profit by using contract $\eta^{**} = \min(\eta^*, \hat{\eta}^*)$ to incentivize the manufacturer to reveal his belief.

3 Conclusions

The importance and the innovative contribution of this paper lie in its generality. For general revenue and cost functions of the manufacturer, and for a general set of decisions, we have investigated the effect of information asymmetry in the manufacturer's cost function. We find that under complete information, the equilibrium strategy is not affected by the retailer's belief about the manufacturer's cost function, whereas under hidden information asymmetry, the equilibrium strategy is affected by the retailer's belief about the manufacturer's cost function as well as the manufacturer's belief. We compare results in these two cases to identify conditions under which information sharing will occur in a case of known information asymmetry. For the last case, we propose two different contracts to induce a Pareto-optimal information-sharing equilibrium.

References

1. Avinadav, T., Chernonog, T., Perlman, Y.: Consignment contract for mobile apps between a single retailer and competitive developers with different risk attitudes. Eur. J. Oper. Res. **246**(3), 949–957 (2015)
2. Avinadav, T., Chernonog, T., Perlman, Y.: The effect of risk sensitivity on a supply chain of mobile applications under a consignment contract with revenue sharing and quality investment. Int. J. Prod. Econ. **168**, 31–40 (2015)
3. Avinadav, T., Chernonog, T., Perlman, Y.: Mergers and acquisitions between risk-averse parties. Eur. J. Oper. Res. **259**(3), 926–934 (2017)
4. Bian, W., Shang, J., Zhang, J.: Two-way information sharing under supply chain competition. Int. J. Prod. Econ. **178**, 82–94 (2016)
5. Chernonog, T.: Consignment contract with revenue sharing in online retailing under strategic information sharing. Working paper, Bar-Ilan University, 2018.
6. Chernonog, T., Avinadav, T.: Profit criteria involving risk in price setting of virtual products. Eur. J. Oper. Res. **236**(1), 351–360 (2014).
7. Fishburn, G.: Tax evasion and inflation. Aust. Econ. Pap. **20**(37), 325–332 (1981).
8. Gurnani, H., Shi, M.: A bargaining model for a first-time interaction under asymmetric beliefs of supply reliability. Manag. Sci. **52**(6), 865–880 (2006)
9. Kong, G., Rajagopalan, S., Zhang, H.: Revenue sharing and information leakage in a supply chain. Manag. Sci. **59**(3), 556–572 (2013)
10. Lancioni, R.A., Smith, M.F., Oliva, T.A.: The role of the Internet in supply chain management. Ind. Mark. Manag. **29**(1), 45–56 (2000)
11. Leng, M., Zhu, A.: Side-payment contracts in two-person nonzero-sum supply chain games: review, discussion and applications. Eur. J. Oper. Res. **196**(2), 600–618 (2009)
12. Mishra, B.K., Raghunathan, S., Yue, X.: Information sharing in supply chains: incentives for information distortion. IIE Trans. **39**(9), 863–877 (2007)
13. Mishra, B.K., Raghunathan, S., Yue, X.: Demand forecast sharing in supply chains. Prod. Oper. Manag. **18**(2), 152–166 (2009)
14. El Ouardighi, F., Kim, B.: Supply quality management with wholesale price and revenue-sharing contracts under horizontal competition. Eur. J. Oper. Res. **206**, 329–340 (2010)
15. Xiao, T., Yang, D.: Risk sharing and information revelation mechanism of a one manufacturer and one-retailer supply chain facing an integrated competitor. Eur. J. Oper. Res. **196**(3), 1076–1085 (2009).
16. Xie, G., Yue, W., Wang, S., Lai, K.K.: Quality investment and price decision in a risk-averse supply chain. Eur. J. Oper. Res. **214**, 403–410 (2011)
17. Zhang, J., Chen, J.: Coordination of information sharing in a supply chain. Int. J. Prod. Econ. **143**(1), 178–187 (2013)
18. Zhu, X., Mukhopadhyay, S.K., Yue, X.: Role of forecast effort on supply chain profitability under various information sharing scenarios. Int. J. Prod. Econ. **129**(2), 284–291 (2011)

Monte Carlo Sampling for the Probabilistic Orienteering Problem

Xiaochen Chou, Luca Maria Gambardella and Roberto Montemanni

Abstract The Probabilistic Orienteering Problem is a variant of the orienteering problem where customers are available with a certain probability. Given a solution, the calculation of the objective function value is complex since there is no linear expression for the expected total cost. In this work we approximate the objective function value with a Monte Carlo Sampling technique and present a computational study about precision and speed of such a method. We show that the evaluation based on Monte Carlo Sampling is fast and suitable to be used inside heuristic solvers. Monte Carlo Sampling is also used as a decisional tool to heuristically understand how many of the customers of a tour can be effectively visited before the given deadline is incurred.

Keywords Probabilistic Orienteering Problem · Monte Carlo Sampling
Heuristic algorithms

1 Introduction

The Probabilistic Orienteering Problem (POP) is a variant of the orienteering problem where the presence of customers are stochastic. The objective is to serve a selected subset of the given customers, in such a way that the expected profit is maximized within a given time budget (the deadline). Each customer will be available for visit

X. Chou · L. M. Gambardella · R. Montemanni (✉)
IDSIA - Dalle Molle Institute for Artificial Intelligence (USI-SUPSI), Manno, Switzerland
e-mail: roberto@idsia.ch

X. Chou
e-mail: xiaochen@idsia.ch

L. M. Gambardella
e-mail: luca@idsia.ch

© Springer Nature Switzerland AG 2018

P. Daniele and L. Scrimali (eds.), *New Trends in Emerging Complex Real Life Problems*, AIRO Springer Series 1,
https://doi.org/10.1007/978-3-030-00473-6_19

with a certain probability, and a certain prize is collected when the customer is served before deadline. We aim at simultaneously maximizing the total prize collected and minimizing the total travel time. Therefore, the expected profit is the difference between the expected total prize and the travel time.

Analytical approximations for the expected total travel time have been proposed in [1, 2], however, the expected total prize is hard to compute, for the reason that it depends not only on presence probability of the customers, but also on probability of the customers being served before the deadline. One of the ways to deal with such an objective function is to approximate using Monte Carlo sampling, which has become a state-of-the-art approach for several stochastic/probabilistic vehicle routing problems such as the Probabilistic Traveling Salesman Problem with Deadlines [3] and the Orienteering Problem with Stochastic Travel and Service Times [4]. Here we will investigate the use of a Monte Carlo Sampling heuristics for evaluating the objective function of the POP. Such an approach is designed with the aim of using it within fast heuristic solvers for the POP itself.

2 Problem Definition

We denote with $V = \{0, 1 \ldots n, n + 1\}$ a set of n customers with the depot being node 0 and the destination being $n + 1$. Let t_{ij} be the deterministic travelling time from customer i to customer j. There is a global deadline D which is a given time budget. When a customer i is served before the deadline, a prize p_i is collected. The availability of a customer i is modeled by a Bernoulli variable $b_i = \{0, 1\}$ which takes value 1 with presence probability π_i. The probability of each customer is independent from the others. Depot and destination must be visited and get no prize, therefore $\pi_o = \pi_{n+1} = 1$ and $p_0 = p_{n+1} = 0$.

A tour $\tau : i_0 = 0, i_1, i_2, \ldots, i_q, i_{q+1} = n + 1$ is defined as a sequence of q customers selected to be served, plus the depot and the destination node. The prize collected in a tour is $P(\tau)$, and the travel time is $T(\tau)$. With a given coefficient C, the objective function is the difference between the expected total prize and the expected total travel time:

$$u(\tau) = \mathbb{E}[P(\tau)] - C\mathbb{E}[T(\tau)] \quad (1)$$

3 Objective Function Evaluator Based on Monte Carlo Sampling

The objective function can be computed by using Monte Carlo Sampling. First, for an instance with n customers, we generate a set of scenarios, i.e. different fully connected graphs with different available customers, generated by sampling accordingly to the presence probability π_i of each customer. Then we use Monte Carlo Sampling to estimate the objective function by random sampling s of these scenarios, computing the (deterministic) objective function values for each of the s scenarios and finally average these values. The approximation can be computed in time $O(ns)$.

Due to the heuristic nature of the approach, different sets of samples can lead to different results when evaluating a single solution, which means solution A is better than solution B for one set of samples, while for another set of samples it could be the opposite.

In Sects. 4.3 and 4.4 we will provide some experiments to understand which could be promising values for the parameter s, regulating the number of samples.

It is important to observe that in the POP, given a solution $\tau : i_0 = 0, i_1, i_2, \ldots, i_q, i_{q+1}, \ldots, i_n, i_{n+1} = 0$, the deadline is incurred after a certain node i_q, from where only travel time is increasing but no more prize is collected. In Sect. 4.2 we will use the Monte Carlo evaluator for finding this node. The decision maker is intended to serve only customers 0 to i_q, therefore identifying such a customer is an important decision.

4 Experimental Results

4.1 Data Sets and Evaluation Environment

The experiments we present were carried out on a Quad-Core Intel Core i7 processor running at 2.0 GHz with 8 GB of RAM. The code was written in C++. We tested the Monte Carlo evaluator on the POP benchmark instances generated by Angelelli et al. in [1]. The test instances are available at: http://or-brescia.unibs.it/instances.

Basing on the TSP benchmark instances from the TSPLIB95 library, the characteristics of the POP benchmark instances are set as follows:

- The location values of the customers $V = \{0, 1, \ldots, n\}$ are from the corresponding TSP instances. The destination $V = n + 1$ has the same location value as the depot $V = 0$.
- The global Deadline D takes value of $\omega \cdot T_{max}$, with T_{max} being the optimal value of the TSP over all nodes and $\omega = \{\frac{1}{4}, \frac{1}{2}, \frac{3}{4}\}$, representing three different types of deadlines.
- The prizes are generated according to two rules. Either $p_i = 1$ for all customers, or p_i takes value of a pseudo-randomly generated integer in {1, 2, ..., 100} for each customer i.
- The probabilities are set according to two rules. Either $\pi_i = 0.5$ for all customers, or π_i takes value of a random number in the interval $[0.25, 0.75]$ for each customer i.

For the coefficient C between the travel time and the prize in the objective function, we take $C = 0.001$ as in [1].

4.2 Customers Selection

As mentioned in Sect. 3, when sampling we want to find the last node visited before the deadline. Instead of using simple strategy such as calculating the average of maximum number of customers served before the deadline, we calculate the value of objective function for each node with all samples, supposing that it is the last customer visited, then the maximum value will show the last node visited before the deadline, and we choose the most promising last node consequently. The approach is heuristic but it seems to provide good results in practice.

First, we test on 2 instances with the same location values and prize values, but different presence probability types. In this 2 instances, the prize $p_i = 1, \forall i \in V$, and the presence probability π_i is either fixed with value 0.5 or random (Figs. 1 and 2). The x-axis shows all nodes $\{0, 1, \ldots, n\}$ in the order of a given solution, and the y-axis shows the approximated value for the objective function at each node, as described above.

From the results we can observe that the slopes are similar for both instances with random and fixed presence probability, and in both cases there exists a significant maximum at node 14 showing the best node to stop serving customers.

Then we try on 2 instances with the same location values, but with random prize p_i for each customer. The presence probability π_i is still either fixed with value 0.5 or random (Figs. 3 and 4). This time total prize collected at local maximum has much higher value than in Figs. 1, 2, but travel time after local maximum is increasing in the same way and the coefficient $C = 0.001$ keeps the same, therefore the value is almost flat after the local maximum. A significant maximum can be observed, showing again the best stopping point.

By using this method, the evaluation for the cost of a given tour τ can be performed only on the subset of customers more likely to be served before the deadline, which

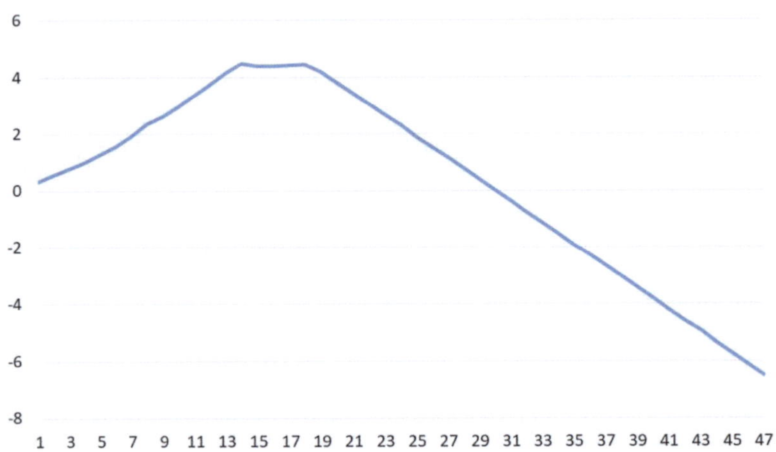

Fig. 1 $u(\tau)$ with $p_i = 1, \pi_i = 0.5$

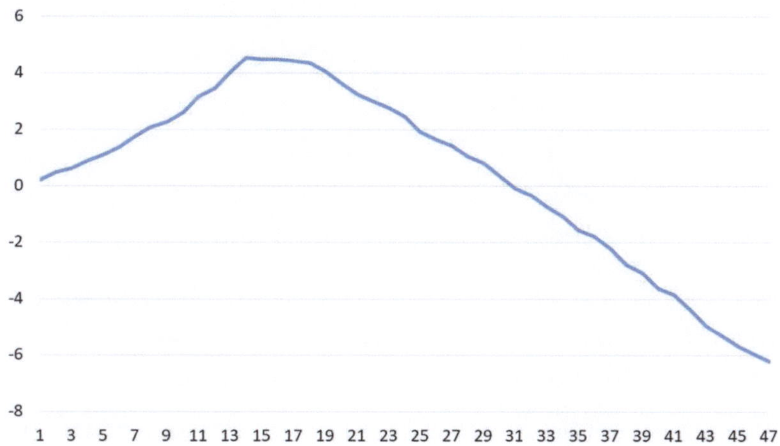

Fig. 2 $u(\tau)$ with $p_i = 1, \pi_i$ random

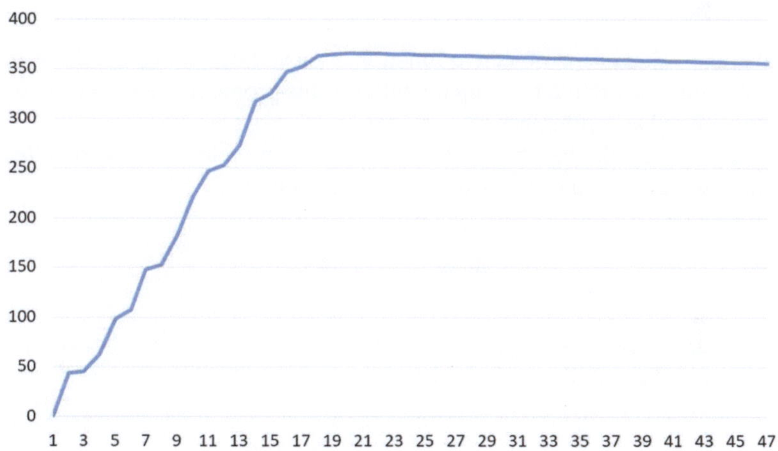

Fig. 3 $u(\tau)$ with p_i random, $\pi_i = 0.5$

will speed up the evaluation and lead to more accurate results within appropriate heuristic solvers.

4.3 Number of Samples

In this part, we study the influence of parameter s when approximating the objective functions using Monte Carlo Sampling.

We run 50 times the Monte Carlo Sampling for 8 different POP instances, with different number of samples $s \in \{10, 20, \ldots, 90, 100, 200, \ldots, 1000\}$, and compute

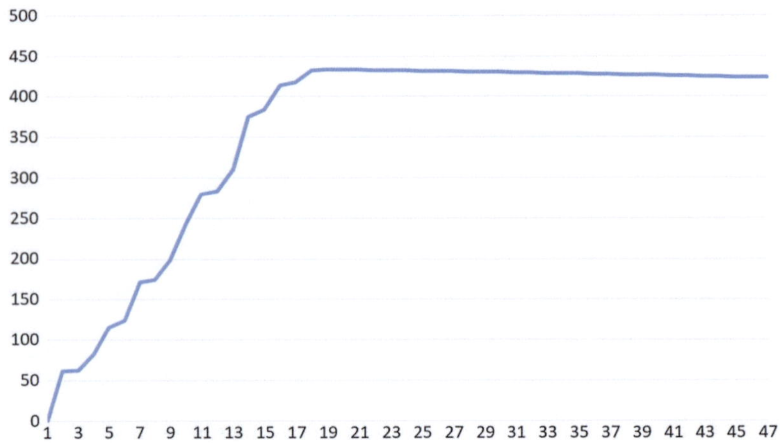

Fig. 4 $u(\tau)$ with p_i random, π_i random

$\overline{x}_s$ the average value of the objective function achieved by s samples. Since there is no exact evaluator available to compare with for this problem, the results reported in [1] are considered as a reference x_{ref}.

For analysis we calculate two indicators. One is the relative difference between $\overline{x}_s$ and x_{ref}, which indicates the accuracy of the results:

$$\delta_1 = \left| \frac{\overline{x}_s - x_{ref}}{\overline{x}_s} \right| \tag{2}$$

The second indicator is relative standard deviation for each number of sample s, which indicates the stability and consistency:

$$\delta_2 = \sqrt{\frac{\sum_{i=1}^{50}(x_i - \overline{x})}{N-1}} \cdot \frac{100}{\overline{x}} \tag{3}$$

Notice: For both indicators we calculate the relative values instead of absolute values, this allows us to compare the results obtained on different instances.

We test on 8 representative instances, Figs. 5 and 6 show the values of indicator δ_1 and δ_2 in percentage on y-axis, with different number of samples s on x-axis. Each curve presents a test instance.

In Fig. 5 we can see that generally δ_1 reduce to 1% when $s \geq 50$, but δ_2 in Fig. 6 is still relatively high. Considering δ_1 and δ_2 together, two interesting points can be observed. The first one is the point where values decrease fast, it happens around $s = 50$, another is the first converging point after fluctuation, which takes place around $s = 400$. Therefore, if a small number of samples is wanted, 50 could be the choice, while considering stability 400 is a better choice. If we make a more

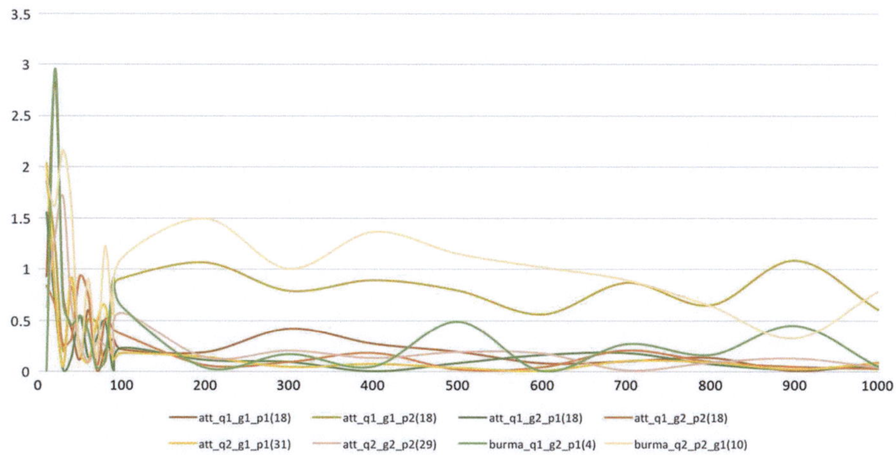

Fig. 5 Relative difference between $\overline{x}_s$ and x_{ref} for 8 test instances

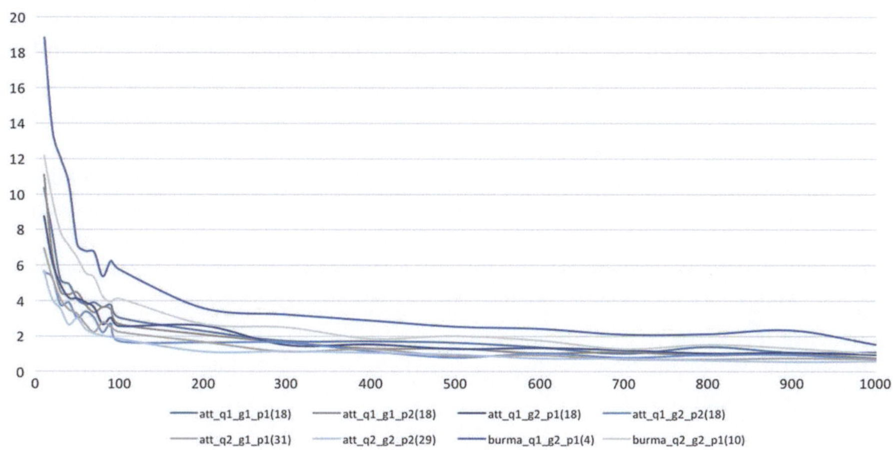

Fig. 6 Relative standard deviation for 8 test instances

instance-dependent analysis, for some instances values of s lower than 50 works well, for others higher values are needed.

4.4 Computation Times

We are also interested in experimental computational times: the theoretical computational time for the evaluation of a solution is $O(ns)$, where n is the dimension of an instance and s the number of samples. But how does this translate in practice?

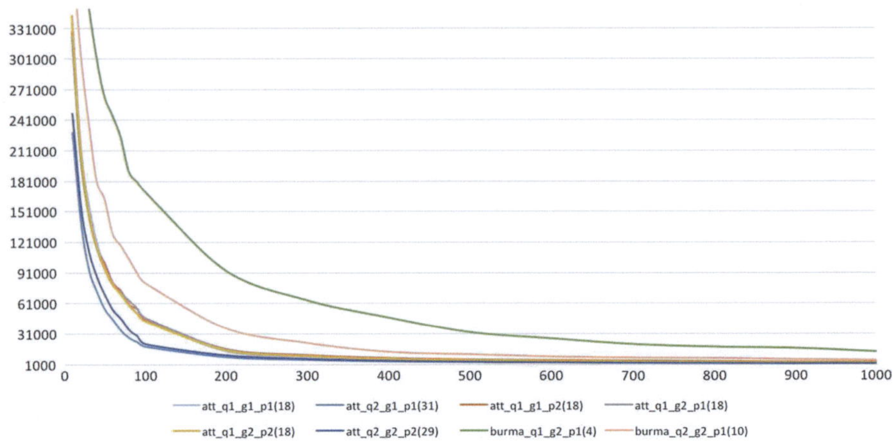

Fig. 7 Computational time for 8 test instances

In this test, we count number of evaluations per second for different number of samples. The test instances are the same 8 representative POP instances as in Sect. 4.3. The number of samples s is on x-axis and the number of evaluations per second with different s is on y-axis. Each curve presents a test instance.

In Fig. 7 we can see that, for $s \leq 100$, the number of evaluations per second drops dramatically, and for $s > 100$ there is a trend of gradual decrease. This shows that the number of samples s has a significant influence on the computational time.

If we compare the results vertically, we will see that evaluations for smaller instances always have higher speed. This shows that the size of instances n is another influencing factor for computational time.

In general, the evaluation process appears to be extremely fast and suitable to be used inside heuristic solvers.

5 Conclusion

In this work we approximated the objective function of the Probabilistic Orienteering Problem based on Monte Carlo Sampling. We also used the same Monte Carlo Sampling procedure to decide how many customers of a given tour should be visited in order to maximize the profit. Such a characteristic is going to be very useful once the evaluator is embedded into a metaheuristic algorithm, which is the next step of our research. A computational study on the performance of the evaluator we propose, both in terms of precision and speed, was also presented.

Acknowledgements Xiaochen Chou was supported by the Swiss National Science Foundation through grant 200020_156259: "Hybrid Sampling-based metaheuristics for Stochastic Optimization Problems with Deadlines".

References

1. Angelelli, E., Archetti, C., Filippi, C., Vindigni, M.: The probabilistic orienteering problem. Comput. Oper. Res. **81**, 269–281 (2017)
2. Campbell, A.M., Thomas, B.W.: Probabilistic traveling salesman problem with deadlines. Transp. Sci. **42**(1), 1–21 (2008)
3. Weyland, D., Montemanni, R., Gambardella, L.M.: Heuristics for the probabilistic traveling salesman problem with deadlines based on quasi-parallel monte carlo sampling. Comput. Oper. Res. **40**(7), 1661–1670 (2013)
4. Papapanagiotou, V., Montemanni, R., Gambardella, L.M.: Hybrid sampling-based evaluators for the orienteering problem with stochastic travel and service times. J. Traffic Logist. Eng. **3**(2), 108–114 (2015)

A Genetic Algorithm Framework for the Orienteering Problem with Time Windows

Claudio Ciancio, Annarita De Maio, Demetrio Laganà, Francesco Santoro and Antonio Violi

Abstract The Orienteering Problem (OP) is a routing problem which has many applications in logistics, tourism and defense. Given a set of nodes, where each node represents a Point of Interest (POI), the orienteering problem aims to design a tour leaving from a starting POI, visiting a subset of POIs and finally arriving at the ending POI. The objective of the problem is to maximize the total score of the visited POIs while the total travel time and the total cost of the route do not exceed two predefined thresholds. Each POI is characterized by a score, a position, a visit time, and a time window in which the POI can be visited. This problem is often investigated to develop tourism trip planning mobile applications. Usually these apps must be able to generate good solutions in few seconds. Therefore, the use of efficient heuristic approaches to find good quality solutions is needed. In this paper we present a genetic algorithm framework combined with some local search operators to deal with the analyzed problem.

Keywords Orienteering problem · Tourist trip design problem · Local search Genetic algorithm

---

C. Ciancio (✉) · D. Laganà · A. Violi
DIMEG, University of Calabria, Rende, Italy
e-mail: claudio.ciancio@unical.it

D. Laganà
e-mail: demetrio.lagana@unical.it

A. Violi
e-mail: antonio.violi@unical.it

A. De Maio
DISC, University of Milano Bicocca, Milan, Italy
e-mail: annarita.demaio@unimib.it

F. Santoro
ITACA srl., via P. Bucci 41/C, 87036 Rende, Italy
e-mail: santoro@itacatech.it

© Springer Nature Switzerland AG 2018

P. Daniele and L. Scrimali (eds.), *New Trends in Emerging Complex Real Life Problems*, AIRO Springer Series 1,
https://doi.org/10.1007/978-3-030-00473-6_20

1 Introduction

The Orienteering Problem (OP), also denoted as the Selective Travelling Salesperson Problem or the Maximum Collection Problem, is a well known NP-hard combinatorial optimisation problem [1]. The problem is characterized by a starting node, an ending node, and a set of nodes related to point of interest (POI) that can be visited, each with a positive score and a positive service time. The objective of the problem is to find a path from the starting node to the end node that maximizes the total score collected from the visited nodes. The orienteering problem with time windows (OPTW) is a natural extension of OP in which the path from the start and the end node have to start and finish predetermined time and each POI have to be visited in a certain time window [2]. The OPTW is often investigated to develop tourism trip planning mobile applications and in general to solve the tourist trip design problem (TTDP). The objective of TTDP is to help tourists to determine the visiting places among numerous possible options in a customized trip [3]. Typically, most tourists try to visit as many locations as possible in the available time during a trip. However, with the limitations of time and budget, they might not be able to visit every place. To select the best route the user typically states her needs, interests and constraints based upon selected parameters [4]. The OPTW will then be solved to select those places that have an high scores taking into account the user constraints. Since the first publications appeared, many exact algorithms have been developed to solve the basic OP and the OPTW. A classification of the exact algorithms is presented in Feillet et al. [5]. However, the use of heuristic approach in even more interesting since the TTDP applications need to find a good quality solution in few seconds. A genetic algorithm approach is presented in Sect. 3. Preliminary computational results are then reported in Sect. 4.

2 Mathematical Formulation

Parameters

N	set of nodes including the set of POI and the starting and ending point of the route;
N^+	set of POI;
H	set of activity types;
s	starting node;
e	ending node;
t_s	time in which the user wants to start the journey;
t_e	time in which the user wants to end the journey;
p_i	score assigned by the customer to the POI i;
a_i	starting time in which the POI i can be visited;
b_i	ending time in which the POI i can be visited;
t_{ij}	travel time to reach the node j from the node i;

c_{ij} travel cost to reach the node j from the node i;
l_{ij} distance from the node i to the node j;
t_i time necessary to visit the POI i;
c_i cost to visit the POI i;
v_i activity type of the POI i;
L maximum length of the route;
C maximum cost of the route;

Decision Variables

x_{ij} binary variable equal to 1 if the node j is visited immediately after the node i;
u_i time in which the node i is visited by the customer;

Based on these definitions, the mathematical formulation of the problem is the following:

$$\max \sum_{i \in N^+} \sum_{j \in N \setminus \{i\}} p_i x_{ij} \tag{1}$$

$$\sum_{i \in N \setminus \{s\}} x_{si} = \sum_{i \in N \setminus \{e\}} x_{ie} = 1 \tag{2}$$

$$\sum_{i \in N^+ \setminus \{k\}} x_{ik} = \sum_{i \in N^+ \setminus \{k\}} x_{ki} \leq 1 \qquad \forall k \in N \tag{3}$$

$$\sum_{i \in N} \sum_{j \in N \setminus \{i\}} l_{ij} x_{ij} \leq L \tag{4}$$

$$\sum_{i \in N} \sum_{j \in N \setminus \{i\}} (c_{ij} + c_i) x_{ij} \leq C \tag{5}$$

$$u_i + t_i + t_{ij} \leq u_j + (1 - x_{ij}) M \qquad \forall i \in N \quad \forall j \in N \setminus \{i\} \tag{6}$$

$$a_i \leq u_i \leq b_i \qquad \forall i \in N \tag{7}$$

$$x_{ij} \in \{0, 1\} \qquad \forall i \in N \quad \forall j \in N \setminus \{i\} \tag{8}$$

Constraint (2) guarantees that the tour starts and ends at the predefined starting and ending nodes. Constraints (3) are the flow conservation constraints and ensure that a node is visited at most once. Constraints (4) and (5) impose a maximum length and cost of the route. Finally, constraints (6) prevent the construction of subtours while constraints (7) are used to satisfy the time windows of the visited POIs.

2.1 Additional Constraints

The formulation (1)–(8) is the basic version of the orienteering problem with time windows. However to make the model more usefull for real life applications related to tourism we took into account the following additional constraints:

- some POI can be visited only on some moments of the day. However, one time window is not enough to model this constraint. For example, a museum could be opened in the morning from 8:00 am to 12:00 and in the afternoon from 3:00 to 6:00 pm.
- the user can specify a time interval in which to have a break of at least a given threshold. For example, a break of at least 30 min must be assigned in the time interval 12:00 am–2:00 pm.
- some users may be interested in different types of activities and they wish to split their time among them. To guarantee a certain degree of diversification the user can specify a maximum time allowed to each type of activities.
- tourists typically enjoy relaxing and having breaks as much as they enjoy visits to POIs [4]. A maximum activity time is imposed to give the user enough time for breaks and resting.

3 Genetic Algorithm

Genetic algorithm (GA) is an optimization technique that tries to replicate the evolution process, in which individuals with the best features have more possibilities of surviving and reproducing [6]. GA undertakes to evolve the solution, during its execution, according to the following basic pattern:

1. Random generation of the first population of solutions;
2. Application of a fitness function to the solutions belonging to the current population;
3. Selection of the best solutions based on the value of the fitness function;
4. Generation of new solutions using crossover and mutation;
5. Repetition of steps 2–3–4 for n iterations;
6. Selection of the best found solution.

3.1 Chromosome Encoding/Decoding

The first step of this method is to encode the features of the route into specific chromosomes. A chromosome is a sequence of id assigned to each POI Fig. 1.

The following definitions are needed:

- let v be the current node. Initially $v = s$;
- let $\hat{t}$ be the current time. Initially $\hat{t} = t_s$;
- let RC be the residual budget. Initially $RC = C$;
- let RL be the residual travel distance that the user is willing to perform. Initially $RL = L$;
- let W_i the set of time windows in which it is possible to visit the POI i. Each element $w_{ki} \in W_i$ is characterized by a starting and an ending time $[a_{ki}, b_{ki}]$;

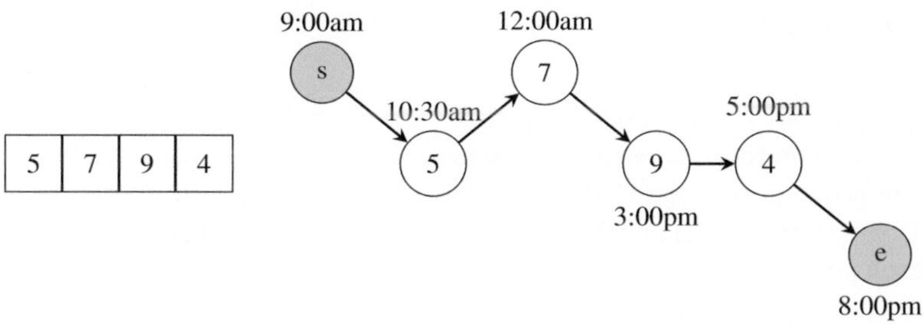

Fig. 1 A chromosome and its corresponding route

When a chromosome is decoded the lowest possible value u_i that allow to respect the set of constraints is assigned to each POI that composes the chromosome.

$$u_i = \max\{\hat{t} + t_{vi}, \min_{w_{ki} \in W_i : b_{ki} <= \hat{t} + t_{vi} + t_i} a_{ki}\} \tag{9}$$

Once a time u_i is assigned to the POI i the value of the current node v is updated to i and the next POI is examined. The current time and residual budget and travel distance are also updated.

$$\hat{t} = \hat{t} + t_{vi} + t_i \tag{10}$$

$$RC = RC - c_{vi} - c_i \tag{11}$$

$$RL = RL - l_{vi} \tag{12}$$

If it is not possible to assign a feasible value of u_i of or the final value of RC or RL is negative it means that is not possible to build a feasible route with that sequence of POIs and the chromosome is discarded from the population.

3.2 Selection

The selection of the chromosomes to produce a new generation is an extremely important step of the algorithm. The most promising chromosomes will be included in the next generation and will be used as "parents" in the crossover operations. A chromosome is more likely to be selected if the value of its correspondent fitness function is high. A probability p_k is assigned to each chromosome $k \in K$ through the roulette wheel selection according to its fitness value f_k:

$$p_k = \frac{f_k}{\sum_{k' \in K} f_{k'}} \tag{13}$$

a random number in the interval [0, 1] is then generated to select one chromosome.

3.3 Crossover

A new population is iteratively generated to explore new areas of the feasible region. In the genetic algorithm, crossover is a genetic operator used to create new chromosomes from one generation to the next. Two different crossover operations named 1-point crossover (Fig. 2) and 2-point crossover (Fig. 3) have been used.

At each generation a $pop_1 - pop_2$ chromosomes are created by using these two operators.

3.4 Mutation

Mutation is another genetic algorithm used to maintain genetic diversity from one generation of a population to the next. This operation consists of randomly altering the value of one element of the chromosome according to a mutation probability Fig. 4.

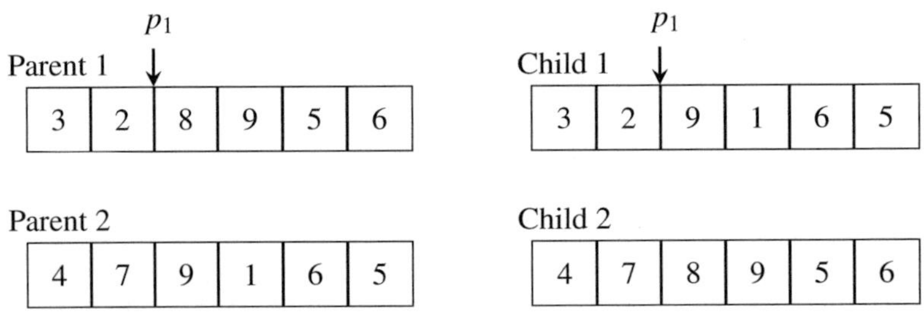

Fig. 2 1-point crossover

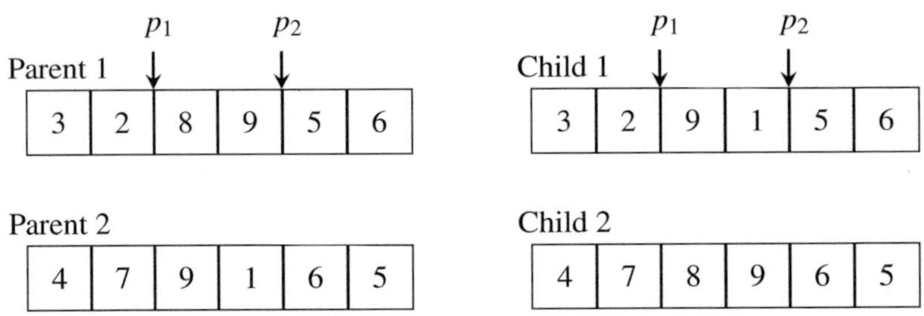

Fig. 3 2-point crossover

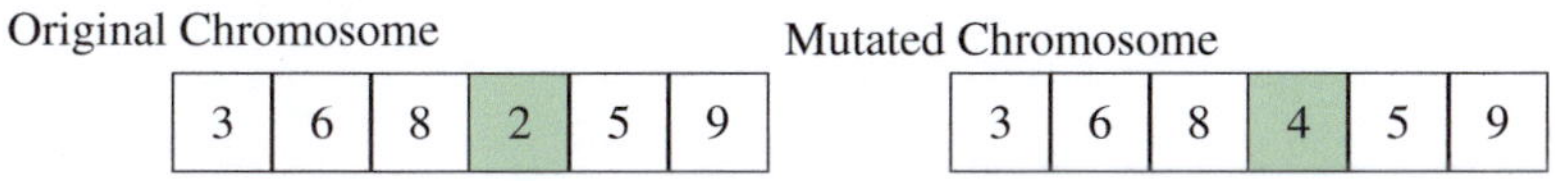

Fig. 4 Mutation operator

3.5 Local Search

When a new chromosome is created two local search operators are used with a certain probability to build better routes:

- addOperator: with this operator a new chromosome is built by trying to add a POI j in the route. The position in which the POI j is added is selected by trying to minimize the travel cost variation.
- swapOperator: with this operator a new chromosome is built by trying to add a POI j in a place of another POI j' included in the current route. The POI j' is the one with the lowest score that allow to create a feasible route.

3.6 Final Solution

Selection, crossover and mutation are repeated iteratively until one of these conditions is satisfied:

1. the processing time exceeds a maximum time;
2. the number of iterations performed exceeds a maximum number.

4 Computational Tests

This section illustrates the results of some preliminary computational analysis. The goal is to prove the effectiveness of the proposed algorithm. The algorithm was implemented as single thread code in Java and all tests are performed on a desktop computer equipped with an Intel Core i7 processor with 2.6 GHz, 8 GB RAM, and running Windows 10. A set of instances was created by changing the number of available POI and the maximum budget. These instances have been solved through the Cplex MIP solver to calculate the optimal solution and then using the proposed heuristic algorithm for both the problem described in the formulation (1)–(8) and by adding the set of constraints described in Sect. 2.1. In particular we imposed a break of at least 30 min in the time interval [12:00–14:30], a maximum time T_h of 4 h for

Table 1 Computational results

Input data				Optimal solution					Heuristic solution			
ID	POI	C (€)	L	z^*	Time (s)	C^* (€)	L^*	z	z_c	Gap (%)	$\hat{C}$ (€)	$\hat{L}$
1	30	50	200	45	1	49.5	62.5	45	45	0.0	49.5	62.5
2	30	75	200	69	89	73.3	86.5	69	67	0.0	73.8	89.2
3	30	100	200	72	357	98.4	132.4	72	69	0.0	97.4	107.2
4	35	50	200	41	1	48.6	58.2	41	38	0.0	48.6	58.2
5	35	75	200	62	149	73.2	106.0	62	62	0.0	73.8	109.1
6	35	100	200	71(94)	7200	90.3	71.7	72	68	30.5	97.2	81.2
7	40	50	200	38	3	49.6	98.2	38	38	0.0	48.3	96.5
8	40	75	200	80	984	74.9	104.4	79	74	1.2	74.6	103.2
9	40	100	200	78(104)	7200	84.7	113.6	78	74	33.3	84.7	113.6
10	45	50	200	70	2	49.6	73.0	70	70	0.0	49.6	73.0
11	45	75	200	69	2025	73.9	069.5	69	69	0.0	74.3	71.7
12	45	100	200	64(106)	7200	83.4	87.0	68	64	55.5	79.4	77.4
13	50	50	200	48	1	49.4	86.8	48	48	0.0	47.4	87.0
14	59	75	200	60(63)	7200	73.4	97.3	59	59	6.7	73.0	95.4
15	50	100	200	72(85)	7200	99.7	68.45	71	68	19.7	98.7	108.8

each type of activities and a minimum rest time of 2 h. The results are reported in Table 1 where the columns z^*, z and z_c report the value of the optimal solution and the solution achieved through the genetic algorithm for the two analyzed formulations. A maximum time of two hours was set for the Cplex MIP solver while 20 s were used for the heuristic approach. It is possible to see from the table that the MIP solver is able to find in few seconds the optimal solution for all the instances with a budget of 50€. However the time to solve the problem drastically rises by increasing the maximum budget and the number of POI. In fact, the time of 2 h is not enough to find the optimal solution of 5 instances. The heuristic algorithm was able to obtain a better or equal solutions for 12 of the 15 analyzed instances with only 20 s. The results obtained are very encouraging and assess the effectiveness of the proposed approach.

5 Conclusions

The Orienteering Problem with time windows is an NP-hard problem derived from several practical situations. We presented a genetic algorithms approach to tackle efficiently with this problem. The computational results show that the algorithm is able to find good quality solutions in a small computational time. The additional constraints added in the second run allow the algorithm to find solutions that better fits with the user needs. Different improvements to make the algorithm more general and efficient can still be added to the framework proposed in this paper. Further studies will be developed to take into account multimodal transportation and hotel selection. Moreover, in order to obtain better quality solutions or decrease time consumption, it would be interesting to develop new operators or adapt the ones developed for other routing problems. Other approaches based on alternative local search moves, tabu search, simulated annealing will also be tried to maximize the performance. Further studies will also be developed to analyze the potential application of a similar algorithm on other classical variants of OP such as the team OP, the time dependent OP as well as recent ones such as the stochastic OP, the generalized OP or the clustered OP.

References

1. Golden, B.L., Levy, L., Vohra, R.: The orienteering problem. Naval Res. Logist. **34**(3), 307–318 (1987)
2. Kantor, M.G., Rosenwein, M.B.: The orienteering problem with time windows. J. Oper. Res. Soc. **43**(6), 629–635 (1992)
3. Vincent, F.Y., Jewpanya, P., Ting, C.J., Redi, A.P.: Two-level particle swarm optimization for the multi-modal team orienteering problem with time windows. Appl. Soft Comput. **61**, 1022–1040 (2017)

4. Gavalas, D., Konstantopoulos, C., Mastakas, K., Pantziou, G.: Mobile recommender systems in tourism. J. Netw. Comput. Appl. **39**, 319–333 (2014)
5. Feillet, D., Dejax, P., Gendreau, M.: Traveling salesman problems with profits. Transp. Sci. **39**(2), 188–205 (2005)
6. Ciancio, C., Ambrogio, G., Gagliardi, F., Musmanno, R.: Heuristic techniques to optimize neural network architecture in manufacturing applications. Neural Comput. Appl. **27**(7), 2001–2015 (2016)

A Financial Optimization Model with Short Selling and Transfer of Securities

Gabriella Colajanni and Patrizia Daniele

Abstract In this paper we present a financial mathematical model, based on networks, aiming at maximizing the profits while simultaneously minimizing the risk. In addition, our model is characterized by short selling, which consists in the sale of non-owned financial instruments with subsequent repurchase, and transfer of securities. We propose an Integer Nonlinear Programming (INLP) Problem, whose solution provides us with the optimal distribution of securities to be purchased and sold.

Keywords Financial problems · Risk management · Multicriteria decision-making · Multi-period portfolio selection problems · Short Selling Transfer of securities

1 Introduction

In financial literature, a portfolio is assumed as a set of financial assets or investments which are owned by an individual (an investor) or a financial institution and consist of various financial instruments such as shares of a company (often referred as equities), government bonds, and so on. A multi-period portfolio selection problem as a Markowitz mean-variance optimization problem (see [8, 9]) with intermediaries and the addition of transaction costs and taxes (on the capital gain) has already been studied in [5]. One of the newest models is the one for calculating optimal portfolio weights developed by Black and Litterman (see [3, 4]). In 1969, Samuelson (see [13]) and Merton (see [10]), taking inspiration from Mossin's work (see [11]), formulate and solve a many-period generalization.

G. Colajanni (✉) · P. Daniele
Department of Mathematics and Computer Science, University of Catania,
Viale Andrea Doria, 6, Catania, Italy
e-mail: colajanni@dmi.unict.it

P. Daniele
e-mail: daniele@dmi.unict.it

© Springer Nature Switzerland AG 2018
P. Daniele and L. Scrimali (eds.), *New Trends in Emerging Complex Real Life Problems*, AIRO Springer Series 1,
https://doi.org/10.1007/978-3-030-00473-6_21

Other extensions of the Markowitz model are studied (for example see [1] in which the variance has been replaced by the Value-at-Risk, [6] in which a multiobjective optimization problem is solved by using a Multiple Criteria Decision Aiding method, [7] where a Mixed Integer Linear Programming problem with transaction costs is studied, [12, 14] where multiperiod mean-variance models are analyzed). In [2] the authors used Mixed Integer Programming methods to construct portfolios, reducing the number of different stocks.

Here, we make the previous multi-period model ([5]) more realistic by adding not only transaction costs, taxes (on the capital gain) and time-length for some financial securities, but also the *short selling* and the *transfer of financial assets*. Short selling is the sale of a security that is not owned by the seller or that the seller has borrowed. Short selling is motivated by the belief that a security's price will decline, enabling it to be bought back at a lower price to make a profit. Short selling may be prompted by speculation, or by the desire to hedge the downside risk of a long position in the same security or a related one. Since the risk of loss on a short sale is theoretically infinite, short selling should only be used by experienced traders, who are familiar with the risks. While short selling is frequently vilified and short sellers viewed as ruthless operators out to destroy companies, the reality is that short selling provides liquidity to markets and prevents stocks from being bid up to ridiculously high levels. Although abusive short-selling practices, such as rumor-mongering to drive a stock lower, are illegal, short selling, when done properly, can be a good tool for portfolio risk management. The transfer of financial securities consists in relocating one or more assets from a bank, or a financial intermediary, to another one; in such a way, the investor can seize the opportunities offered by the commercial initiatives of the various financial institutions. The plan of the rest of the paper is as follows. In Sect. 2 we introduce the financial model consisting of financial securities, issuers, investors, and intermediaries. We derive the optimization problem of each investor based on the maximization of his expected gain and the minimization of his risk portfolio. In Sect. 3 we apply the model to a numerical example consisting of a financial network with two issuers, two financial securities and an investor. In Sect. 4 we summarize the obtained results.

2 The Mathematical Model

As in [5], we consider a financial network consisting of: n financial securities, and the typical one is denoted by i; S issuers of financial securities, such as companies, banks, countries, etc., and the typical one is denoted by s; K investors (security purchasers) and the typical one is denoted by k; B financial intermediaries, and the typical one is denoted by b. In addition, we consider a partition of the financial securities by means of the sets $\mathscr{A}_1, \ldots, \mathscr{A}_s, \ldots, \mathscr{A}_S$, where $\mathscr{A}_s$ represents the set of financial securities made available by issuer s. A representation of the financial network is depicted in Fig. 1. We analyze the model in a discrete time horizon: $1, \ldots, j, \ldots, t$.

We introduce the following binary variables:

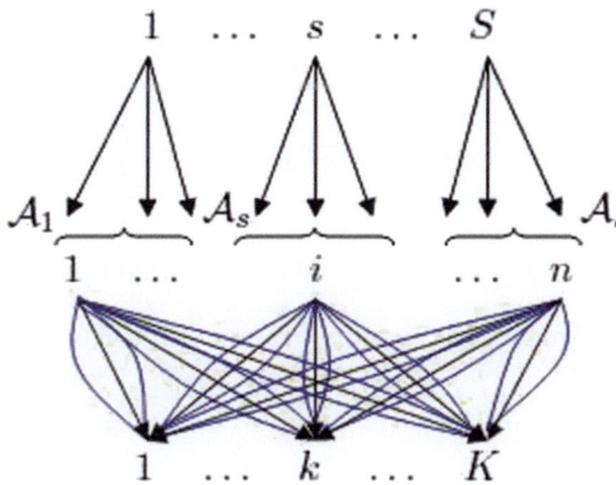

Fig. 1 Financial network

$$x_{i,j}^k = \begin{cases} 1 \text{ if security } i \text{ is purchased by } k \text{ at time } j \\ 0 \text{ otherwise} \end{cases} \quad \forall i = 1, \ldots, n, \quad \forall j = 1, \ldots, t, \quad \forall k = 1, \ldots, K;$$

$$y_{i,j}^k = \begin{cases} 1 \text{ if security } i \text{ is sold by } k \text{ at time } j \\ 0 \text{ otherwise} \end{cases} \quad \forall i = 1, \ldots, n, \quad \forall j = 1, \ldots, t, \quad \forall k = 1, \ldots, K;$$

$$z_{1i}^b = \begin{cases} 1 \text{ if security } i \text{ is purchased by } b \\ 0 \text{ otherwise} \end{cases} \quad \forall i = 1, \ldots, n, \quad \forall b = 1, \ldots, B;$$

$$z_{2i}^b = \begin{cases} 1 \text{ if security } i \text{ is sold by } b \\ 0 \text{ otherwise} \end{cases} \quad \forall i = 1, \ldots, n, \quad \forall b = 1, \ldots, B.$$

Further, in order to take into account the short selling, we introduce the following binary variables:

$$w_{i,j}^k = \begin{cases} 1 \text{ if security } i \text{ is purchased by } k \text{ at time } j \\ 0 \text{ otherwise} \end{cases} \quad \forall i = 1, \ldots, n, \quad \forall j = 2, \ldots, t, \quad \forall k = 1, \ldots, K;$$

$$h_{i,j}^k = \begin{cases} 1 \text{ if security } i \text{ is sold by } k \text{ at time } j \\ 0 \text{ otherwise} \end{cases} \quad \forall i = 1, \ldots, n, \quad \forall j = 1, \ldots, t-1, \quad \forall k = 1, \ldots, K.$$

Let $C_{i,j}$ be the purchase cost; $\gamma_k^b \cdot C_{i,j} + C_k^b$ the commission to the chosen financial intermediary, given by a percentage of the purchase cost, γ_k^b and a flat fee; $\mathbb{E}[D_{i,j}]$ and $\mathbb{E}[P_{i,j}]$ the expected values of dividends in the case of shares or interests in the case of bonds and the expected values of amounts of money to pay (for example in the case of an increase in the corporate capital), respectively; $\mathbb{E}[R_{i,j}]$ the expected selling price; $\beta_k^b \cdot R_{i,j} + F_k^b$ the charge to the chosen financial intermediary; $\alpha_i^k \left(\frac{\left| \mathbb{E}[u_{i,j}] \right| + \mathbb{E}[u_{i,j}]}{2} \right)$ the taxation on the capital gain, if it is positive and where

$$\mathbb{E}[u_{i,j}] = \mathbb{E}[R_{i,j}] - C_{i,\bar{j}} + \sum_{\hat{j}=\bar{j}+1}^{j} \Big(\mathbb{E}[D_{i,\hat{j}}] - \mathbb{E}[P_{i,\hat{j}}] \Big),$$

where $\bar{j}$ and j indicate the purchase and selling time respectively, with $1 \leq \bar{j} < j \leq t$.

Some financial securities have a time length which we denote by τ_i, therefore $S = \bar{j} + \tau_i$ represents its expiration time.

Let $\mathbb{E}[N_{i,\bar{j}+\tau_i}]$ be the expected nominal value and $\alpha_i^k \left(\frac{\left| \mathbb{E}[g_{i,\bar{j}+\tau_i}] \right| + \mathbb{E}[g_{i,\bar{j}+\tau_i}]}{2} \right)$ the capital gains tax where, in this case, $\mathbb{E}[g_{i,\bar{j}+\tau_i}] = \mathbb{E}[N_{i,\bar{j}+\tau_i}] - C_{i,\bar{j}} + \sum_{\hat{j}=\bar{j}+1}^{\bar{j}+\tau_i} \Big(\mathbb{E}[D_{i,\hat{j}}] - \mathbb{E}[P_{i,\hat{j}}] \Big)$.

We refer the reader to [5] for a detailed description of the model. Here, let also T_i^b denote the financial title transfer fee, M_b the maximum time limit, fixed by financial intermediary b, within which investor k is obliged to short covering (namely, he has to buy the not owned securities), $\mathbb{E}[p_{i,\bar{j}}] = \mathbb{E}[R_{i,j}] - \mathbb{E}[C_{i,\bar{j}}]$ the new capital gain, $I_i^b(\bar{j} - j)$ the interest, which is a function of time, to be paid to the broker who lends the security which has to be sold in the short selling.

In agreement with Nagurney and Ke (see [12]) we assume that the decision-makers seek not only to increase their net revenues but also to minimize risk with the risk being considered as the possibility of suffering losses compared to the expected profit. Therefore, we introduce $(\sigma_p^k)^2$ the risk on the portfolio. Then, the objective function to maximize is as follows:

$$\mathbb{E}[e_p^k] - \eta_k (\sigma_p^k)^2 = \sum_{i=1}^{n} \Bigg\{ \sum_{\bar{j}=1}^{t-1} x_{i,\bar{j}}^k \Bigg[- C_{i,\bar{j}} - \sum_{b=1}^{B} z_{1i}^b \cdot (\gamma_k^b C_{i,\bar{j}} + C_k^b)$$

$$+ \sum_{j=\bar{j}+1}^{\min\{\bar{j}+\tau_i,t\}} \Bigg(\mathbb{E}[-P_{i,j} + D_{i,j}] + y_{i,j}^k \Bigg(\mathbb{E}[R_{i,j}] - \alpha_i^k \left(\frac{\left| \mathbb{E}[u_{i,j}] \right| + \mathbb{E}[u_{i,j}]}{2} \right)$$

$$- \sum_{b=1}^{B} z_{2i}^b \cdot (\beta_k^b \mathbb{E}[R_{i,j}] + F_k^b) - \sum_{\hat{j}=j+1}^{\min\{\bar{j}+\tau_i,t\}} \Big(\mathbb{E}[-P_{i,\hat{j}} + D_{i,\hat{j}}] \Big) \Bigg) \Bigg) \Bigg]$$

$$+\sum_{\bar{j}=1}^{t-\tau_i-1} x_{i,\bar{j}}^k\Bigg[(1-\sum_{j=\bar{j}+1}^{\bar{j}+\tau_i} y_{i,j}^k)\left(\mathbb{E}[N_{i,\bar{j}+\tau_i}]-\alpha_i^k\left(\frac{\left|\mathbb{E}[g_{i,\bar{j}+\tau_i}]\right|+\mathbb{E}[g_{i,\bar{j}+\tau_i}]}{2}\right)\right)\Bigg]$$

$$+\sum_{\bar{j}=t-\tau_i}^{t-1} x_{i,\bar{j}}^k\Bigg[(1-\sum_{j=\bar{j}+1}^{t} y_{i,j}^k)\mathbb{E}[N_{i,t}]\Bigg]-\sum_{b=1}^{B}\Big(z_{1i}^b\cdot\sum_{\bar{b}\neq b} z_{2i}^{\bar{b}}\cdot T_i^b\Big)$$

$$+\sum_{j=1}^{t-1} h_{i,j}^k\Bigg[\mathbb{E}[R_{i,j}]+\sum_{b=1}^{B} z_{1i}^b\Bigg[-(\beta_k^b\mathbb{E}[R_{i,j}]+F_k^b)-\sum_{\bar{j}=j+1}^{\min\{j+M_b,t\}} w_{i,\bar{j}}^k\Big(\mathbb{E}[C_{i,\bar{j}}]$$

$$+(\gamma_k^b\mathbb{E}[C_{i,\bar{j}}]+C_k^b)+I_b(\bar{j}-j)+\alpha_i^k\left(\frac{\left|\mathbb{E}[p_{i,\bar{j}}]\right|+\mathbb{E}[p_{i,\bar{j}}]}{2}\right)\Big)\Bigg]\Bigg]\Bigg\}-\eta_k(\sigma_p^k)^2.$$

The problem formulation is as follows:

$$\max\left\{\mathbb{E}[e_p^k]-\eta_k(\sigma_p^k)^2\right\} \tag{1}$$

$$\sum_{k=1}^{n}\sum_{j=1}^{t-1} x_{i,j}^k \leq 1 \quad \forall i=1,\ldots,n \tag{2}$$

$$y_{i,j}^k \leq \sum_{\bar{j}=j-\tau_i+1}^{j-1} x_{i,\bar{j}}^k \quad \forall i=1,\ldots,n, \quad \forall j=2,\ldots,t \tag{3}$$

$$y_{i,j}^k \leq \frac{\sum_{\bar{j}=2}^{j-1}(1-y_{i,\bar{j}}^k)}{j-2} \quad \forall i=1,\ldots,n, \quad \forall j=3,\ldots,t \tag{4}$$

$$(\sigma_{p_k})^2 \leq \bar{R}_k \tag{5}$$

$$\sum_{i=1}^{n}\sum_{j=1}^{t-1} x_{i,j}^k C_{i,j} \leq \bar{B}_k \tag{6}$$

$$\sum_{b=1}^{B} z_{1i}^b = \sum_{j=1}^{t-1} x_{i,j}^k \quad \forall i=1,\ldots,n \tag{7}$$

$$\sum_{b=1}^{B} z_{2i}^b = \sum_{\bar{j}=2}^{t} y_{i,\bar{j}}^k \quad \forall i=1,\ldots,n \tag{8}$$

$$\sum_{\substack{\max\{j<\bar{j}:\\ D_{i,j}>0\}}}^{\bar{j}}\ \sum_{z\in\mathscr{A}_s}\sum_{k=1}^{K}(x_{z,j}^k-y_{z,j}^k)\geq 1 \quad \forall s\in\mathscr{S},\ D_{i,\bar{j}}>0 \tag{9}$$

$$\sum_{\bar{j}=j+1}^{\min\{j+M_b,t\}} w_{i,\bar{j}}^k = h_{i,j}^k \quad \forall i = 1, \ldots, n \quad \forall j = 1, \ldots, t-1 \tag{10}$$

$$\sum_{k=1}^{K} \sum_{j=1}^{t-1} h_{i,j}^k \leq 1 \quad \forall i = 1, \ldots, n \tag{11}$$

$$\sum_{\bar{j}=1}^{j} x_{i,\bar{j}}^k \Big(\sum_{\hat{j}=\bar{j}+1}^{j} \big(1 - y_{i,\hat{j}}^k\big)\Big) \leq 1 - h_{i,j}^k \quad \forall i = 1, \ldots, n \quad \forall j = 1, \ldots, t-1 \tag{12}$$

$$x_{i,j}^k, y_{i,j}^k, z_{1i}^b, z_{2i}^b, h_{i,j}^k, w_{i,j}^k \in \{0, 1\}$$
$$\forall i = 1, \ldots, n, \quad \forall j = 1, \ldots, t, \quad \forall b = 1, \ldots, B. \tag{13}$$

The meaning of constraint (2) is that it is possible to buy the same security only once and it can be purchased by a single investor (but there are numerous coincident securities). Constraint (3) means that it is possible to sell a security only if it has been purchased previously and has not yet expired. Constraint (4) means that you can sell a stock only if it has not yet been sold. Constraint (5) means that there is a risk limit, $\bar{R}_k$, which represents the maximum risk limit that the investor is willing to accept. Constraint (6) means that there is a budget limit, $\bar{B}_k$, which represents the maximum available budget for an investor. Constraints (7) and (8) mean that for each security, only one financial intermediary can be chosen for purchasing and selling activities. Constraint (9) means that each issuer must sell at least one security during the dividend distribution periods, where the dividend $D_{i,\bar{j}}$ at time $\bar{j}$ of security $i \in \mathscr{A}_s$ is given by:

$$D_{i,\bar{j}} = \frac{U_{\bar{j}}^s - R_{\bar{j}}^s}{\sum_{\substack{\max\{j<\bar{j}:\\ D_{i,j}>0\}}}^{\bar{j}} \sum_{z \in \mathscr{A}_s} \sum_{k=1}^{K} (x_{z,j}^k - y_{z,j}^k)}.$$

Constraint (10) means that if investor k sells the title i in the short selling market, he is obliged to buy back the same security within the time established by the financial intermediary. Constraint (11) states that it is possible to sell in the short selling market the same security more than once. Finally, constraint (12) affirms that investor k cannot sell security i in the short selling market, if he owns security i.

3 A Numerical Example

In this section we apply the model to a numerical example that consists of a financial network with two issuers, two financial securities and an investor, as depicted in Fig. 2.

We consider also two financial intermediaries and we analyze the model in the following time horizon: 1, ..., 5.

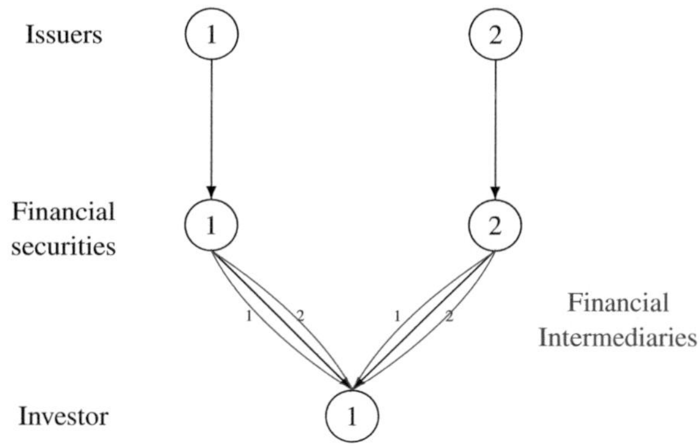

Fig. 2 Network Topology for the Numerical Example

To solve the example we used Matlab on a laptop with an Intel Core2 Duo processor and 4 GB RAM.

We assume the following data are given:

C_{ij}	$j=1$	$j=2$	$j=3$	$j=4$
$i=1$	5	6	6	7
$i=2$	9	9	8	9

$\mathbb{E}[R_{ij}]$	$j=2$	$j=3$	$j=4$	$j=5$
$i=1$	7	7	8	1
$i=2$	20	15	10	1

$\mathbb{E}[-P_{ij}]$	$j=2$	$j=3$	$j=4$	$j=5$
$i=1$	0	0	-2	0
$i=2$	0	0	0	0

$\mathbb{E}[U_{ij}-R_{ij}]$	$j=2$	$j=3$	$j=4$	$j=5$
$i=1$	0	3	0	15
$i=2$	0	0	0	0

We also assume that $\tau_1 = 2$ and that the second financial security does not expire, so we require $\tau_2 = 4$. Further, we assume that the nominal value of each security at maturity or at time $j = 5$ coincides with its current value (cost) $\mathbb{E}[N_{i,\bar{j}+\tau_i}] = C_{i,\bar{j}+\tau_i}$, $\mathbb{E}[N_{i,5}] = C_{i,4}$, that the maximum budget and risk values are $\bar{B} = 25$ and $\bar{R} = 15$ respectively, that the percentages of taxation are $\alpha_1 = 15\%$ and $\alpha_2 = 10\%$, that commission costs are given by $\beta^1 = \gamma^1 = 5\%$, $\beta^2 = \gamma^2 = 15\%$, $C^1 = F^1 = 0.5$ and $C^2 = F^2 = 2$, that $\eta = 0.2$ is the risk aversion index, $(\sigma_{1j}) = (2, 2, 2, 2, 2)$, $(\sigma_{2j}) = (1, 1, 1, 1, 1)$ the variances of the titles and $\rho_{12j} = 0\ \forall j = 1, \ldots, 5$ meaning that the two titles are completely unrelated.

We get the following optimal solutions:

$$x^*_{14} = 1; \qquad x^*_{1j} = 0\ \forall j = 1, 2, 3;\ x^*_{2j} = 0\ \forall j = 1, 2, 3, 4;$$

$$y^*_{1j} = y^*_{2j} = 0 \quad \forall j = 2, 3, 4, 5;$$

$$h^*_{1j} = 0\ \forall j = 1, 2, 3, 4; \quad h^*_{22} = 1, \quad h^*_{2j} = 0\ \forall j = 1, 3, 4;$$

$$w^*_{1j} = 0\ \forall j = 2, 3, 4, 5; \quad w^*_{24} = 1, \quad w^*_{2j} = 0\ \forall j = 2, 3, 5;$$

$$z_{11}^{1*} = z_{12}^{1*} = z_{22}^{1*} = 1, \quad z_{11}^{2*} = z_{12}^{2*} = z_{21}^{1*} = z_{21}^{2*} = z_{22}^{2*} = 0.$$

These optimal solutions clearly show that the most convenient choice for the investor is to buy security 1 at time 4 and never sell it; using the short selling, to sell security 2 at time 2 and to buy it at time 4, through the financial intermediary 1.

In this case the total gain is: 21.20.

If, now, we assume that the commission costs are given by $\beta^1 = \gamma^2 = 5\%$, $\beta^2 = \gamma^1 = 15\%$, $C^1 = F^2 = 0.5$ and $C^2 = F^1 = 2$, we see that it is more convenient to choose financial intermediary 1 for the selling and, after a transfer of security 2, financial intermediary 2 for the buying. In this case the total gain is: 19.7.

4 Conclusions

The presented financial model determines which securities every investor has to buy and sell, which financial intermediary he has to choose and at what time it is more convenient to buy and sell a security in order to maximize his own profit and minimize his own risk, taking into account not only the presence of the financial intermediaries (therefore the transaction costs or commissions), the capital gains taxes and that some financial securities have a time length, but also the short selling and the transfer of financial assets.

Acknowledgements The research of the authors was partially supported by the research project "Modelli Matematici nell'Insegnamento-Apprendimento della Matematica" DMI, University of Catania. This support is gratefully acknowledged.

References

1. Benati, S., Rizzi, R.: A mixed-integer linear programming formulation of the optimal mean/Value-at-Risk portfolio problem. Eur. J. Oper. Res. **176**, 423–434 (2007)
2. Bertsimas, D., Darnell, C., Soucy, R.: Portfolio construction through mixed-integer programming at Grantham, Mayo, Van Otterloo and Company. Interfaces **29**(1), 49–66 (1999)
3. Black, F., Litterman, R.: Asset allocation: combining investor views with market Equilibrium. J. Fixed Income 7–18 (1991)
4. Black, F., Litterman, R.: Portfolio optimization. Financ. Anal. J. **48**(5), 28–43 (1992)
5. Colajanni, G., Daniele, P.: A Financial Model for a Multi-period Portfolio Optimization Problem with a variational formulation. In: Khan, A., Kbis, E., Tammer, C. (eds.) Variational Analysis and Set Optimization: Developments and Applications in Decision Making (2018) (in press)
6. Greco, S., Matarazzo, B., Slowinski, R.: Beyond Markowitz with multiple criteria decision aiding. J. Bus. Econ. **83**, 29–60 (2013)
7. Kellerer, H., Mansini, R., Speranza, M.G.: Selecting portfolios with fixed costs and minimum transaction lots. Ann. Oper. Res. **99**, 287–304 (2000)
8. Markowitz, H.M.: Portfolio selection. J. Financ. **7**, 77–91 (1952)
9. Markowitz, H.M.: Portfolio Selection: Efficient Diversification of Investments. Wiley, New York (1959)

10. Merton, R.C.: Lifetime portfolio selection under uncertainty: the continuous-time case. Rev. Econ. Stat. **51**(3), 247–257 (1969)
11. Mossin, J.: Optimal multiperiod portfolio policies. J. Bus. **41**(2), 215–229 (1968)
12. Nagurney, A., Ke, K.: Financial networks with intermediation: risk management with variable weights. Eur. J. Oper. Res. **172**(1), 40–63 (2006)
13. Samuelson, P.A.: Lifetime portfolio selection by dynamic stochastic programming. Rev. Econ. Stat. **51**(3), 239–246 (Aug)
14. Steinbach, M.C.: Markowitz revisited: mean-variance models in financial portfolio analysis. Soc. Ind. Appl. Math. **43**(1), 31–85 (2001)

Strong Nash Equilibria for Cybersecurity Investments with Nonlinear Budget Constraints

Patrizia Daniele and Laura Scrimali

Abstract This paper investigates the existence of strong Nash equilibria in a cybersecurity investment supply chain game theory model. We consider a supply chain network consisting of retailers and consumers at demand markets with each retailer being faced with nonlinear budget constraints on his security investments. We also assume that the demand for the product at each demand market is known and fixed and, hence, the conservation law of each demand market must be fulfilled. The model is a Generalized Nash equilibrium model for which we define a variational equilibrium, that allows us to give a variational inequality formulation. Our aim is to give a necessary condition to be a strong Nash equilibrium of the model in terms of a system of variational inequalities.

Keywords Cybersecurity · Investments · Supply chains · Game theory
Nash equilibrium · Strong Nash equilibrium

1 Introduction

Supply chains highly depend on information technology to enhance effectiveness as well as efficiency and to support communications and coordination among the network of suppliers, manufacturers, distributors, and even freight service providers. However, information technology, if not properly secured, can increase the vulnerability of supply chains to cyberattacks. Many examples exist of cyber attacks

P. Daniele (✉) · L. Scrimali
Department of Mathematics and Computer Science, University of Catania,
Viale Andrea Doria, 6, Catania, Italy
e-mail: daniele@dmi.unict.it

L. Scrimali
e-mail: scrimali@dmi.unict.it

© Springer Nature Switzerland AG 2018

P. Daniele and L. Scrimali (eds.), *New Trends in Emerging Complex Real Life Problems*, AIRO Springer Series 1,
https://doi.org/10.1007/978-3-030-00473-6_22

infiltrating supply chains. Only the first six months of 2017 have seen an inordinate number of cybersecurity meltdowns, which weren't just standard corporate breaches. In August 2016 the mysterious hacking group known as the Shadow Brokers claimed to have breached the spy tools of the elite NSA-linked operation known as the Equation Group. In May 2017 a strain of ransomware called WannaCry spread around the world, damaging hundreds of thousands of targets, including public utilities and large corporations. One month later, another wave of ransomware infections, called Petya, NotPetya and a few other names, hit targets all over the world. It infected networks in multiple countries like the US pharmaceutical company Merck, Danish shipping company Maersk, and Russian oil giant Rosnoft, but researchers suspected that the ransomware actually masked a targeted cyberattack against Ukraine. In March 2017, WikiLeaks published a data collection containing 8,761 documents that were probably stolen by the CIA which contained an extensive documentation of alleged spying operations. In February 2017, the internet infrastructure company Cloudflare announced that a bug in its platform caused random leakage of potentially sensitive customer data. Two days before France's presidential runoff in May 2017, hackers downloaded a collection of 9 GB of emails leaked from the party of the current French president.

Numerous companies and organizations have now realized that investing in cybersecurity is necessary.

In this paper, following [2], we present a cybersecurity investment supply chain game theory model consisting of retailers and consumers at demand markets with each retailer being faced with a nonlinear budget constraint on his security investments. Since the strategy of a given retailer is affected by the strategies of the other retailers, the governing concept is a Generalized Nash equilibrium (GNE). We also assume that the demand for the product at each demand market is known and fixed and, hence, the conservation law of each demand market must be fulfilled. We make use of a *variational equilibrium*, which is a special kind of GNE. The variational equilibrium allows for a variational inequality formulation of the GNE model.

It is well-known that in a GNE each retailer has no unilateral incentive to deviate, however a retailer may benefit from forming coalitions with other retailers in order to resist efficiently to cyberattacks. Therefore, we focus on strong Nash equilibrium (SNE), see [1], which establishes that there is no coalition of retailers that can improve their payoffs by collective deviation. Thus, we provide a necessary condition to be SNE in terms of a system of variational inequalities.

The paper is organized as follows. In Sect. 2 we recall some basic definitions and properties of Nash, strong Nash and generalized Nash equilibria. In Sect. 3 we present the supply chain game theory model of cybersecurity investments. In Sect. 4 we provide a necessary condition for the existence of a strong Nash equilibirum for our model.

2 Basic Definitions and Properties

We consider a normal form game $G = (N, (X_i)_{i\in N}, (u_i)_{i\in N})$, where N is the finite set of players, X_i is the set of strategies of player $i \in N$, and $u_i : X \to \mathbb{R}$, with $X = \prod_{i\in N} X_i$, is the player i's payoff function.

Let $\mathscr{P}(N)$ be the power set of N, containing all possible player coalitions and let C be a nonempty set of $\mathscr{P}(N)$. For each coalition $C \subseteq N$, let $-C = \{i \in N : i \notin C\}$ be the set of the rest of players. If C is a singleton $\{i\}$, instead of $-\{i\}$ we will simply write $-i$. Using these notations, given two strategy profiles $x, y \in X$, (x_C, y_{-C}) denotes the strategy in which players from C play the strategy profile x and the remaining of players, namely $-C$, play the strategy y.

The Nash equilibrium [9] is a strategy profile such that no player can unilaterally change her/his strategy to increase her/his payoff.

Definition 1 A strategy profile $x^* \in X$ is a Nash equilibrium of the game G if for all $i \in N$

$$u_i(x_i^*, x_{-i}^*) \geq u_i(x_i, x_{-i}^*), \quad \forall x_i \in X_i.$$

A Pareto efficient (or optimal) strategy is a situation in which no player can improve his/her payoff without decreasing the payoff of someone else.

Definition 2 A strategy profile $x^* \in X$ is Pareto efficient if there does not exist a strategy for all $x \in X$ such that

$$u_i(x) \geq u_i(x^*), \quad \forall i \in N.$$

The strong Nash equilibrium [1] is a strategy for which no coalition of players has a profitable deviation that improves the payoff of each member of the coalition.

Definition 3 A strategy profile $x^* \in X$ is a strong Nash equilibrium if for all $C \subseteq N$ there does not exist a strategy for all x_C such that

$$u_i(x_C, x_{-C}^*) > u_i(x^*), \quad \forall i \in C.$$

Definition 3 implies that any strong Nash equilibrium is Pareto efficient [10]. Moreover, a Nash equilibrium that is also Pareto efficient is a strong Nash equilibrium [3]. If we consider deviating coalitions with a unique player, the strong Nash equilibrium reduces to the Nash equilibrium.

The Generalized Nash equilibrium extends the classical Nash equilibrium concept, by assuming that each player's strategy set may depend on the opponents' strategies x_{-i}. Thus, let $X_i(x_{-i})$ be the set of strategies of player i when the other players choose x_{-i}.

Definition 4 A strategy profile $x^* \in X(x^*) = \prod_{i\in N} X_i(x_{-i}^*)$ is a Generalized Nash equilibrium if for all $i \in N$

$$u_i(x_i^*, x_{-i}^*) \geq u_i(x_i, x_{-i}^*), \quad \forall x_i \in X_i(x_{-i}^*).$$

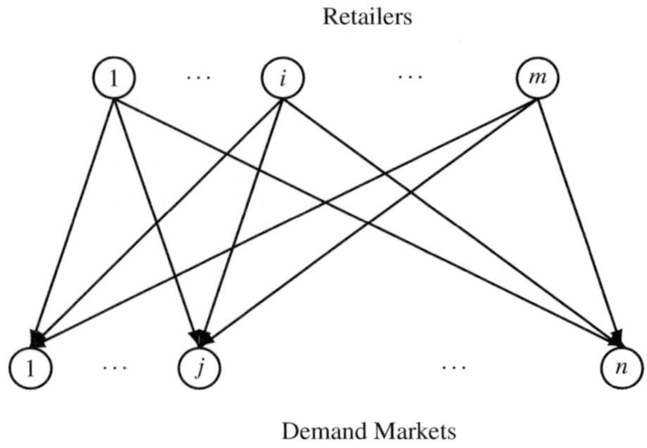

Fig. 1 The bipartite structure of the supply chain network game theory model

3 The Model

We now recall the supply chain game theory model of cybersecurity investments with nonlinear budget constraints introduced in [2].

It consists of m retailers and consumers at n demand markets (Fig. 1). Each retailer $i; i = 1, \ldots, m$, can transact with demand market $j; j = 1, \ldots, n$, with Q_{ij} denoting the product transaction from i to j. For each retailer $i; i = 1, \ldots, m$, we denote by $s_i; i = 1, \ldots, m$ his cybersecurity or, simply, security, level. We group the product transactions for retailer $i; i = 1, \ldots, m$, into the n-dimensional vector Q_i and then we group all such retailer transaction vectors into the mn-dimensional vector Q. The security levels of the retailers are grouped into the m-dimensional vector s.

Then, the cybersecurity level in the supply chain network is the average security and is denoted by $\bar{s}$, where $\bar{s} = \sum_{i=1}^{m} \frac{s_i}{m}$.

The retailers seek to maximize their individual expected utilities, consisting of expected profits, and compete in a noncooperative game in terms of strategies consisting of their respective product transactions and security levels.

In contrast to the models in [2, 7, 8], the demand at each demand market j, d_j, is assumed to be fixed and known and must satisfy the following conservation law:

$$d_j = \sum_{i=1}^{m} Q_{ij}, \quad j = 1, \ldots, n. \tag{1}$$

Also, the product transactions have to satisfy capacity constraints and must be nonnegative, so that we have the following conditions:

$$0 \leq Q_{ij} \leq \bar{Q}_{ij}, \text{ with } \sum_{i=1}^{m} \overline{Q}_{ij} > d_j \quad i = 1, \ldots, m; j = 1, \ldots, n. \tag{2}$$

The cybersecurity level of each retailer i must satisfy the following constraint:

$$0 \leq s_i \leq u_{s_i}, \quad i = 1, \ldots, m, \tag{3}$$

where $u_{s_i} < 1$ for all i; $i = 1, \ldots, m$, and $s_i = 0$ means that retailer i has no security.

The demand price of the product at demand market j, $\rho_j(d, s)$; $j = 1, \ldots, n$, depends both on the vector of demands and the network security. In view of the conservation of flow equations above, we can define $\hat{\rho}_j(Q, s) \equiv \rho_j(d, s)$; $j = 1, \ldots, n$. We assume that the demand price functions are continuously differentiable and concave.

There is an investment cost function h_i; $i = 1, \ldots, m$, associated with achieving a security level s_i with the function assumed to be increasing, continuously differentiable and convex. An example of an $h_i(s_i)$ function that satisfies these properties is

$$h_i(s_i) = \alpha_i \left(\frac{1}{\sqrt{(1 - s_i)}} - 1 \right) \quad \text{with } \alpha_i > 0.$$

The term α_i enables distinct retailers to have different investment cost functions based on their size and needs. Such functions have been introduced by [11] and also utilized by [7, 8]. Each retailer is faced with a limited budget for cybersecurity investment. Hence, the following nonlinear budget constraints must be satisfied:

$$\alpha_i \left(\frac{1}{\sqrt{(1 - s_i)}} - 1 \right) \leq B_i; \quad i = 1, \ldots, m, \tag{4}$$

that is, each retailer can't exceed his allocated cybersecurity budget.

The profit f_i of retailer i; $i = 1, \ldots, m$ is the difference between his revenue and his costs associated, respectively, with production and transportation, that is:

$$f_i(Q, s) = \sum_{j=1}^{n} \hat{\rho}_j(Q, s) Q_{ij} - c_i \sum_{j=1}^{n} Q_{ij} - \sum_{j=1}^{n} c_{ij}(Q_{ij}), \tag{5}$$

where $c_{ij}(Q_{ij})$ are convex functions.

If there is a successful cyberattack on a retailer i; $i = 1, \ldots, m$, retailer i incurs an expected financial damage given by

$$D_i p_i,$$

where D_i, the damage incurred by retailer i, takes on a positive value, and p_i is the probability of a successful cyberattack on retailer i, where:

$$p_i = (1 - s_i)(1 - \bar{s}), \quad i = 1, \ldots, m, \tag{6}$$

with the term $(1 - \bar{s})$ denoting the probability of a cyberattack on the supply chain network and the term $(1 - s_i)$ denoting the probability of success of such an attack on retailer i. We assume that such a probability is a given data on the basis of statistical observations.

Each retailer i; $i = 1, \ldots, m$, hence, seeks to maximize his expected utility, $E(U_i)$, corresponding to his expected profit given by:

$$E(U_i) = (1 - p_i)f_i(Q, s) + p_i(f_i(Q, s) - D_i) - h_i(s_i) = f_i(Q, s) - p_i D_i - h_i(s_i). \tag{7}$$

Let us remark that, because of the assumptions, $-E(U_i)$ is a convex function. Let $\mathbb{K}^i$ denote the feasible set corresponding to retailer i, where

$$\mathbb{K}^i \equiv \{(Q_i, s_i)|0 \le Q_{ij} \le \bar{Q}_{ij}, \forall j,\ 0 \le s_i \le u_{s_i},$$

$$\text{and the budget constraint } h_i(s_i) - B_i \le 0, \text{ holds for } i\}.$$

We also define

$$\mathbb{K} \equiv \Big\{(Q, s) \in \mathbb{R}^{mn+m} : -Q_{ij} \le 0,\ Q_{ij} - \overline{Q}_{ij} \le 0,\ -s_i \le 0,$$

$$s_i - u_{s_i} \le 0,\ h(s_i) - B_i \le 0,\ i = 1, \ldots, m,\ j = 1, \ldots, n\Big\}.$$

In addition, we define the set of shared constraints $\mathscr{S}$ as follows:

$$\mathscr{S} \equiv \{Q|(1) \text{ holds}\}.$$

We now state the following definition.

Definition 5 (*A Supply Chain Generalized Nash Equilibrium in Product Transactions and Security Levels*) A product transaction and security level pattern $(Q^*, s^*) \in \mathbb{K}$, $Q^* \in \mathscr{S}$, is said to constitute a supply chain Generalized Nash equilibrium if for each retailer i; $i = 1, \ldots, m$,

$$E(U_i(Q_i^*, s_i^*, Q_{-i}^*, s_{-i}^*)) \ge E(U_i(Q_i, s_i, Q_{-i}^*, s_{-i}^*)), \quad \forall (Q_i, s_i) \in \mathbb{K}^i, \forall Q \in \mathscr{S}. \tag{8}$$

Hence, according to the above definition, a supply chain Generalized Nash equilibrium is established if no retailer can unilaterally improve upon his expected utility (expected profit) by choosing an alternative vector of product transactions and security level, given the product flow and security level decisions of the other retailers and the demand constraints.

We now provide the linkage that allows us to analyze and determine the equilibrium solution via a variational inequality through a variational equilibrium [4, 5].

Definition 6 (*Variational Equilibrium*) A product transaction and security level pattern (Q^*, s^*) is said to be a variational equilibrium of the above Generalized Nash equilibrium if $(Q^*, s^*) \in \mathbb{K}$, $Q^* \in \mathscr{S}$, is a solution of the variational inequality

$$-\sum_{i=1}^{m}\sum_{j=1}^{n} \frac{\partial E(U_i(Q^*, s^*))}{\partial Q_{ij}} \times \left(Q_{ij} - Q^*_{ij}\right) - \sum_{i=1}^{m} \frac{\partial E(U_i(Q^*, s^*))}{\partial s_i} \times \left(s_i - s^*_i\right) \geq 0,$$
$$\forall (Q, s) \in \mathbb{K}, \forall Q \in \mathscr{S}; \tag{9}$$

namely, $(Q^*, s^*) \in \mathbb{K}$, $Q^* \in \mathscr{S}$, is a supply chain Generalized Nash equilibrium product transaction and security level pattern if and only if it satisfies the variational inequality

$$\sum_{i=1}^{m}\sum_{j=1}^{n}\left[c_i + \frac{\partial c_{ij}(Q^*_{ij})}{\partial Q_{ij}} - \hat{\rho}_j(Q^*, s^*) - \sum_{k=1}^{n} \frac{\partial \hat{\rho}_k(Q^*, s^*)}{\partial Q_{ij}} \times Q^*_{ik}\right] \times (Q_{ij} - Q^*_{ij})$$
$$+\sum_{i=1}^{m}\left[\frac{\partial h_i(s^*_i)}{\partial s_i} - \left(1 - \sum_{k=1}^{m}\frac{s^*_k}{m} + \frac{1 - s^*_i}{m}\right) D_i - \sum_{k=1}^{n} \frac{\partial \hat{\rho}_k(Q^*, s^*)}{\partial s_i} \times Q^*_{ik}\right]$$
$$\times (s_i - s^*_i) \geq 0, \quad \forall (Q, s) \in \mathbb{K}, \forall Q \in \mathscr{S}. \tag{10}$$

For convenience, we define now the feasible set $\mathscr{K}$ where $\mathscr{K} \equiv \mathbb{K} \cap \mathscr{S}$.

Problem (10) admits a solution since the classical existence theorem, which requires that the set $\mathscr{K}$ is closed, convex, and bounded and the function entering the variational inequality is continuous, is satisfied (see also [6]).

4 Strong Nash Equilibria

Today, sophisticated cyberattacks are coming down hard on companies and consumers also because of the increasing international collaborations between cybercrime groups. For this reason, a retailer has to make big efforts to contrast cyberattacks, but, often, fighting alone cannot be efficient. However, joining forces and forming coalitions with other retailers may result in substantial benefits. Therefore, we are interested in analyzing SNE and provide a necessary condition for the existence of a SNE for our model.

Theorem 1 *A necessary condition for a strategy profile* $(Q^*, s^*) \in \mathscr{K}$ *to be a strong Nash equilibrium is to be a solution to the following system of variational inequalities:*

$$\sum_{i=1}^{m}\sum_{j=1}^{n}\frac{\partial E(U_i(Q^*,s^*))}{\partial Q_{ij}}\times(Q_{ij}-Q^*_{ij})-\sum_{i\in N}\frac{\partial E(U_i(Q^*,s^*))}{\partial s_i}\times(s_i-s_i^*)\geq 0$$

$$\sum_{i\in C}\sum_{j=i}^{n}w_i\frac{\partial E(U_i(Q^*,s^*))}{\partial Q_{ij}}\times(Q_{ij}-Q^*_{ij})-\sum_{i\in C}w_i\frac{\partial E(U_i(Q^*,s^*))}{\partial s_i}\times(s_i-s_i^*)\geq 0,\forall C\subseteq N,$$

$$\forall(Q,s)\in\mathcal{K}, \tag{11}$$

for some $w_i \in \mathbb{R}^m_+$ *and* $\sum_{i=1}^{m} w_i = 1.$

Proof Let $(Q^*, s^*) \in \mathcal{K}$ be a strong Nash equilibrium of the cybersecurity investment supply chain game, then it has to be a Nash equilibrium and Pareto efficient for all possible coalitions. Thus, as a Nash equilibrium, SNE solves variational inequality (9), which is nothing but the first variational inequality in (11).

Moreover, a Pareto efficient product transaction and security level pattern, $(Q^*, s^*) \in \mathcal{K}$, can be expressed as a solution to the multiobjective optimization problem

$$\min(-E(U_1(Q,s)),\ldots,-E(U_m(Q,s)))$$
$$\text{subject to } (Q,s)\in\mathcal{K}.$$

Due to the assumptions on $-E(U_i)$ and the structure of set $\mathcal{K}$, the classical scalarization approach allows us to consider an equivalent minimum scalar problem. In particular, we focus on Pareto efficient solutions with respect to all coalitions, namely, we have to solve the problem

$$\min(-\sum_{i\in C}w_iE(U_i(Q,s)),\quad \forall C\subseteq N$$
$$\text{subject to } (Q,s)\in\mathbb{K},\ Q\in\mathcal{S}, \tag{12}$$

where $w_i \in \mathbb{R}^m_+$ and $\sum_{i=1}^m w_i = 1$. Finally, applying the minimum principle, we find that a Pareto efficient solution verifies the other inequalities of the system (11). Therefore, we conclude that if $(Q^*, s^*) \in \mathcal{K}$ solves system (11), then it is a strong Nash equilibrium of the cybersecurity investment supply chain game. □

Acknowledgements The research of the authors was partially supported by the research project PON SCN 00451 CLARA—CLoud plAtform and smart underground imaging for natural Risk Assessment, Smart Cities and Communities and Social Innovation.

References

1. Aumann, R.: Acceptable points in general cooperative *n* person games. Ann. Math. Studies **40**, 287–324 (1959)
2. Daniele, P., Maugeri, A., Nagurney, A.: Cybersecurity investments with nonlinear budget constraints: analysis of the marginal expected utilities. In: Daras NJ, Rassias MT (Eds.) Operations

Research, Engineering, and Cyber Security: Trends in Applied Mathematics and Technology. Springer Optimum Applications, vol. 113, pp. 117–134 (2017)

3. Gatti, N., Rocco, M., Sandholm, T.: On the verification and computation of strong Nash equilibrium. In: 13, International Foundation for Autonomous Agents and Multiagent Systems on Proceedings of the 2013 International Conference on Autonomous Agents and Multi-agent Systems, AAMAS? Richland, SC, ISBN 978-1-4503-1993-5, pp. 723–730 (2013)
4. Kulkarni, A.A., Shanbhag, U.V.: On the variational equilibrium as a refinement of the generalized Nash equilibrium. Automatica **48**, 45–55 (2012)
5. Luna, J.P.: Decomposition and Approximation Methods for Variational Inequalities, with Applications to Deterministic and Stochastic Energy Markets. PhD Thesis, Instituto Nacional de Matematica Pura e Aplicada, Rio de Janeiro, Brazil (2013)
6. Maugeri, A., Raciti, F.: On existence theorems for monotone and nonmonotone variational inequalities. J. Convex Anal. **16**(3–4), 899–911 (2009)
7. Nagurney, A., Daniele, P., Shukla, S.: A supply chain network game theory model of cybersecurity investments with nonlinear budget constraints. Ann. Oper. Res. **248**(1), 405–427 (2017)
8. Nagurney, A., Nagurney. L.S., Shukla, S.: A Supply chain game theory framework for cybersecurity investments under network vulnerability. In: Daras, N.J., Rassias, M.T. (eds.) Computation, Cryptography, and Network Security, Springer International Publishing Switzerland, pp. 381–398 (2015)
9. Nash, J.F.: Non-cooperative games. Ann. Math. **54**, 286–295 (1951)
10. Nessah, R., Tazdait, T.: Absolute optimal solution for a compact and convex game. Eur. J. Op. Res. **224**(2), 353–361 (2013)
11. Shetty, N., Schwartz, G., Felegehazy, M., Walrand, J.: Competitive Cyber-Insurance and Internet Security. In: Proceedings of the Eighth Workshop on the Economics of Information Security (WEIS 2009), June 24–25 University College London, England (2009)

Situation Awareness and Environmental Factors: The EVO Oil Production

Massimo de Falco, Nicola Mastrandrea, Wathiq Mansoor and Luigi Rarità

Abstract The paper considers simulation results for a supply network, that deals with Extra Virgin Olive (EVO) oil production, an activity that is typical of Southern Italy. The phenomenon is studied by differential equations, that focus on goods on arcs and queues for the exceeding goods. Different numerical schemes are used for simulations. A strategy of Situation Awareness allows defining a possible choice of the input flow to the supply network. The achieved results indicate that Situation Awareness permits to find good compromises for the modulation of production queues and the optimization of the overall system features.

Keywords Situation Awareness · Production systems · Simulations

---

M. de Falco · N. Mastrandrea
Dipartimento di Scienze Aziendali - Management & Innovation Systems, University of Salerno, Via Giovanni Paolo II, 132, 84084 Fisciano (SA), Italy
e-mail: mdefalco@unisa.it

N. Mastrandrea
e-mail: nmastrandrea@unisa.it

W. Mansoor
University of Dubai, Academic City, United Arab Emirates
e-mail: wmansoor@ud.ac.ae

L. Rarità (✉)
Dipartimento di Ingegneria Industriale, University of Salerno, Via Giovanni Paolo II, 132, 84084 Fisciano (SA), Italy
e-mail: lrarita@unisa.it

© Springer Nature Switzerland AG 2018

P. Daniele and L. Scrimali (eds.), *New Trends in Emerging Complex Real Life Problems*, AIRO Springer Series 1,
https://doi.org/10.1007/978-3-030-00473-6_23

1 Introduction

The wealth of Italian regions is connected to different features that often deal with phenomena of distribution and production of goods. In order to increase the Italian prestige at national and international levels, a particular attention is devoted to marketing strategies, that focus on the export of cultivated goods. Such a situation implies efforts to guarantee a relevant treatment of products by supply networks, and this is studied in Campania, an Italian region where production and distribution of Extra Virgin Olive (EVO) oil, obtained by olives, represent fundamental economic activities. Indeed, in this case the primary factor is the production, as its possible delays generate negative consequences on the delivery of the final product to the final users.

The aim of this work is to analyze supply networks for the EVO oil production, also focusing on environment factors that always determine a constraint in terms of input flows for production systems. In this case, supply networks are modelled by a fluid-dynamic model, that deals with Partial and Ordinary Differential Equations (PDEs, ODEs). Input flows to such networks are determined via an approach of Situation Awareness, applied to the EVO oil production. The advantages are evident: on one hand, the chosen model foresees space-time dynamics of goods; on the other, Situation Awareness indicates, considering environment parameters, possible inputs for the production systems, in order to avoid situations of queues of goods to process.

Various mathematical models [1, 2] have been used to study supply networks. Some approaches are discrete and based on individual parts; others are continuous and deal with differential equations (see [3, 4] for applications to road networks). The first work, that considers a continuous approach, is by Armbruster et al. [5], who obtained a conservation law [6, 7], that describes the parts densities. Other works have been introduced to define further phenomena within supply systems [8]. In this case, we refer to: A model proposed in [9, 10], that uses conservation laws for densities of parts and queues for each supplier; A discretization scheme for PDEs and ODEs [11, 12], that analyzes an Upwind method for PDEs and and Euler scheme for ODEs with different space meshes and fixed time grid mesh (details are in [13]). For other numerical schemes, useful for the computational efficiency, see [14], while important applications in science and engineering are in [15].

As there is a strong necessity of controlling the production processes and hence the input flows to the supply networks, Situation Awareness is useful. According to Endsley's opinion [16], Situation Awareness deals with "the perception of the elements in the environment within a volume of time and space, the comprehension of their meaning, and the projection of their status in the near future". Such a model consists of three different levels, that provide indications to plan decisions for input flows of supply networks: perception, by which all elements of the surrounding environment are perceived; comprehension, that studies which data from the environment are suitable for the goals to reach; projection, namely the capability of projecting the recognized elements in future times. The expected advantages of

Situation Awareness are simply achieved by the features of the three levels, that allow taking decisions for specific domains.

Finally, simulations are made: a real example of supply system for the EVO oil production is considered. In this case, input flows refer to two different cases: decisions planned by the leadership of a farm in Campania region (Italy); decisions that, considering real environment data, are determined by a model of Situation Awareness. It is shown that Situation Awareness is able to accelerate the system dynamics, in terms of emptying of queues in some parts of the system.

The paper is organized as follows. Section 2 presents Situation Awareness for the context of the EVO oil production. Section 3 describes a mathematical model for supply networks and the proposed numerical method. Section 4 presents the simulation results for a real case study. Conclusions end the paper in Sect. 5.

2 Situation Awareness for EVO Oil Production

We focus on an application of Situation Awareness by considering the Endsley's model [16] within the scenario of the EVO oil production for a real farm in Campania region (Italy). Figure 1 shows the approach.

In particular, the *environment* considers conditions by which a good EVO oil depend, namely: weather, rain and/or wind, humidity. A *situation* describes states for good growths of olives and consists of three different steps:

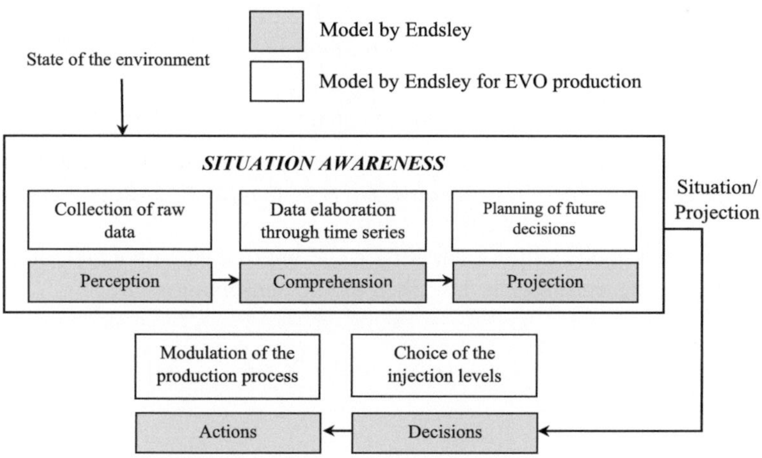

Fig. 1 Abstract vision of Situation Awareness for the described context

- *Perception*: environment data are acquired.
- *Comprehension*: the kept data are elaborated. Notice that combinations of parameters for olives growth could provide a safe forecast on the quality of the obtained EVO oil. In this paper, time series are used for the comprehension step.
- *Projection*: suitable future decisions are used to make work plans, with consequent definition of a Decision Support System (DSS).

The DSS, that is not reported in detail here, is a support for the leadership of the farm under consideration. The rules of DSS are based on Fuzzy Logic and this allows defining correct levels of olives injection to the production networks.

3 Supply Networks: Model and Numerics

An ODE-PDE approach for supply systems (see [10]), that recalls [5, 9], is described. In particular, we consider: conservation laws for density of goods over the suppliers; temporal variation of queues for the transition of parts among suppliers. Finally, we analyze some numerical schemes for the proposed mathematical model.

Now, we focus on the model. A supply network is a graph with set of arcs and vertices indicated, respectively, by $\mathcal{J}$ and $\mathcal{V}$. A generic arc $j \in \mathcal{J}$ is represented by a real interval $\left[a_j, b_j\right]$, that models a supplier, with possible infinite endpoints. We have arcs of type: incoming if $b_j < +\infty$, or outgoing if $a_j > -\infty$. In case of outgoing arcs, there exist queues. Each vertex $v \in \mathcal{V}$ connects a set of incoming arcs, $Inc(v) \subset \mathcal{J}$, and a set of outgoing arcs $Out(v) \subset \mathcal{J}$. Finally, distributions coefficients $(\alpha_{v,j})_{j\in Out(v)}$ such that $\alpha_{v,j} \in]0, 1[$ and $\sum_{j\in Out(v)} \alpha_{j,v} = 1 \ \forall\ v \in \mathcal{V}$, are useful to indicate the percentage of flux that, flowing from v, distributes to the supplier j.

For each arc $j \in \mathcal{J}$, define the following quantities: $\rho_j\,(t,x) \in \left[0, \rho_j^{\max}\right]$, density of parts at point x and time t; $f_j\left(\rho_j\,(t,x)\right) := \min\left\{\mu_j, v_j\rho_j(t,x)\right\}$, flux function; $\mu_j > 0$, maximum processing capacity; $L_j > 0$, length; $T_j > 0$, processing time; $v_j := L_j/T_j$, processing velocity.

Assuming that goods within each arc $j \in \mathcal{J}$ are treated with a maximal flux μ_j and velocity v_j, the dynamics is described by the conservation law:

$$\frac{\partial}{\partial t}\rho_j\,(t,x) + \frac{\partial}{\partial x} f_j\left(\rho_j\,(t,x)\right) = 0, \quad \forall\, x \in \left[a_j, b_j\right], \quad t > 0, \tag{1}$$

$$\rho_j\,(0,x) = \rho_{j,0}\,(x) \geq 0, \quad \rho_j\left(t, a_j\right) = \frac{f_{j,inc}\,(t)}{v_j}, \tag{2}$$

where the initial condition, $\rho_{j,0}$, and the inflow, $f_{j,inc}\,(t)$, have to be provided.

If arc $j \in Out(v)$ with $v \in \mathcal{V}$, there exists a time dependent queue, $q_j\,(t)$, that follows the equation:

$$\frac{d}{dt}q_j(t) = \alpha_{v,j} \sum_{m \in Inc(v)} f_m(\rho_m(b_m, t)) - f_{j,inc}(t). \tag{3}$$

We assume that, for each arc $j \in \mathcal{J}$:

$$f_{j,inc}(t) := \begin{cases} \gamma_j(t), \text{if } a_j = -\infty, \\ \min\left\{\alpha_{v,j} \sum_{m \in Inc(v)} f_m(\rho_m(b_m, t)), \mu_j\right\}, \text{if } q_j(t) = 0, \\ \mu_j, \text{if } q_j(t) > 0, \end{cases} \tag{4}$$

where $\gamma_j(t)$ is an assigned input function on the left boundary $\{(a_j, t) : t \geq 0\}$.

Now, we focus on some numerical schemes. For the just described approach, we deal with a scheme, that focuses on the upwind method for PDEs (1) and an explicit Euler scheme for ODEs (3) assuming different space meshes, see deeper details in [13].

For the generic arc $j \in \mathcal{J}$, define a numerical grid in $[0, L_j] \times [0, T]$ with points $(x_i, t^n)_j$, $i = 0, \ldots, N_j$, $n = 0, \ldots, \eta_j$, where: N_j is the number of segments into which the j-th supplier is divided; η_j is the number of segments into which $[0, T]$ is divided. Then, indicate the approximated density at point (x_i, t^n) and the value of the approximated queue buffer occupancy at time t^n, respectively, by ${}^j\rho_i^n$ and q_j^n .

For the generic arc $j \in \mathcal{J}$, set a fixed time grid mesh Δt and different space grid meshes $\Delta x_j = v_j \Delta t$. In this case, grid points are of type $(x_i, t^n)_j = (i\Delta x_j, n\Delta t)$, $i = 0, \ldots, N_j$, $n = 0, \ldots, \eta_j$.

Finally, the Upwind scheme for the parts density of arc j is:

$$\frac{{}^j\rho_i^n - {}^j\rho_i^{n+1}}{{}^j\rho_i^n - {}^j\rho_{i-1}^n} = \frac{\Delta t}{\Delta x_j} v_j, \; j \in \mathcal{J}, \; i = 0, \ldots, N_j, n = 0, \ldots, \eta_j, \tag{5}$$

and the CFL condition (see [12]) holds as $\Delta t = \min\left\{\frac{\Delta x_j}{v_j} : j \in \mathcal{J}\right\}$.

For queues, if $a_j < -\infty$, the explicit Euler method reads as:

$$q_j^{n+1} - q_j^n + \Delta t \; f_{j,inc}^n = \Delta t \; \alpha_{v,j} \sum_{k \in Inc(v)} f_k({}^k\rho_{N_k}^n), \quad n = 0, \ldots, \eta_j, \tag{6}$$

where:

$$f_{j,inc}^n = \begin{cases} \min\left\{\alpha_{v,j} \sum_{k \in Inc(v)} f_k({}^k\rho_{N_k}^n), \mu_j\right\}, & \text{if } q_j^n(t) = 0, \\ \mu_j, & \text{if } q_j^n(t) > 0. \end{cases}$$

For suitable numerical corrections for $f_{j,inc}^n$, see [13].

Finally, if $a_j = -\infty$, boundary data are used by referring to ghost cells and the inflows defined by $\gamma_j(t)$, see Eq. (4).

4 Simulations

We analyze some results of a real supply network, that deals with the EVO oil production, see Fig. 2. The network is used inside a little farm in Campania region (Italy) and, considering the interpretation given in [10], each arc is a machine or a conveyor belt.

Each arc has a role described as follows. Arc 1 is a conveyor belt that transports olives. According to a distribution coefficient 0.5, olives are equally distributed to arcs 2 and 3, that are machines useful for peeling and skins separations of olives. Arcs 4–7 work for olive pressing. Arcs 8 and 9 consider the oil collection. Finally, arc 10 is useful for bottling operations.

Considering the described numerical scheme, the network is simulated by $\Delta t = 0.0125$ and: $L_i = T_i = 1, i = 1, \ldots, 10$; $\mu_1 = 600$; $\mu_{10} = 16$; $\mu_j = 35 - 2j$, $j = 2, \ldots, 9$; $\rho_i(0, x) = 0$, $i = 1, \ldots, 10$; $q_i(0) = 0$, $i = 2, \ldots, 10$; total simulation time $\overline{T} = 500$; input profile $\gamma(t)$ for arc 1 chosen as:

$$\gamma(t) = \begin{cases} t, & 0 \leq t < 50, \\ 50, & 50 < t \leq 80, \\ 130 - t, & 80 < t \leq 130. \end{cases} \tag{7}$$

Function (7) obeys a behaviour that is really considered inside the little farm in discussion. In fact, olives are first injected according to a linear increasing profile (phase of hard injection); Then, a constant one; Finally, a decreasing one (phase of light injection).

Figures 3, 4 present the various queues, from which we get that: Queue $q_2(t)$ has a smoother profile as it is directly connected to arc 2; Because of μ_j, $j = 1, \ldots, 10$, slopes for queues $q_j(t)$, $j = 3, \ldots, 9$, are different from the ones of $q_{10}(t)$; Queues dynamics is slow although the behaviour of $\gamma(t)$ is zero $\forall\, t > 130$. This last aspect is clearly shown by $q_{10}(t)$ that becomes zero at $t \simeq 380 \gg 130$.

Queues occur due to $\gamma(t)$, v_j and μ_j, $j = 1, \ldots, 10$. As the EVO oil production system cannot be redesigned in terms of processing times and maximal fluxes, the behaviour of $\gamma(t)$ implies the dynamics of the supply network. Possible optimization techniques for the supply systems modelled via fluid-dynamic approaches is still

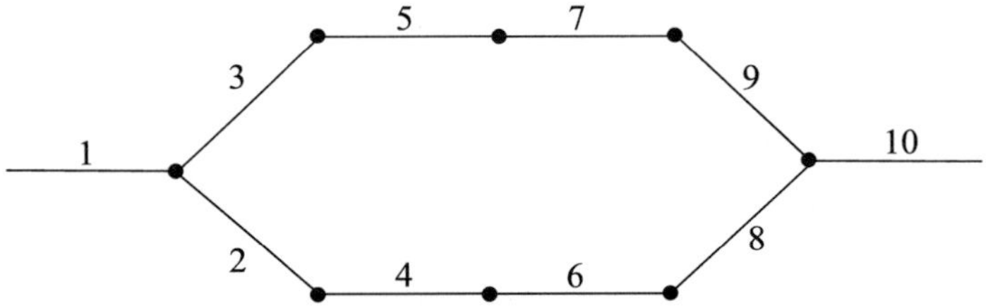

Fig. 2 Supply network for the EVO oil production

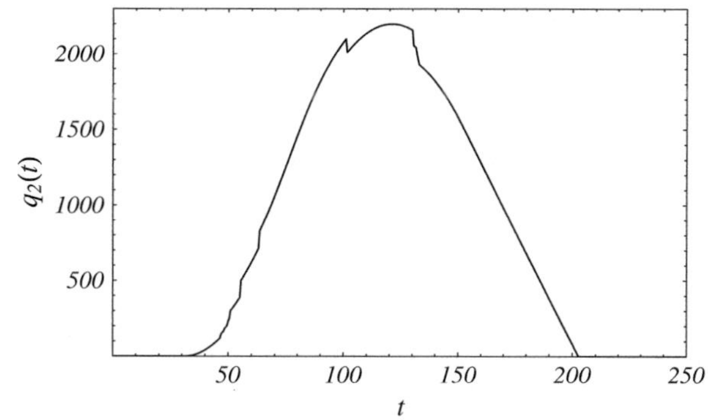

Fig. 3 q_2 versus time t

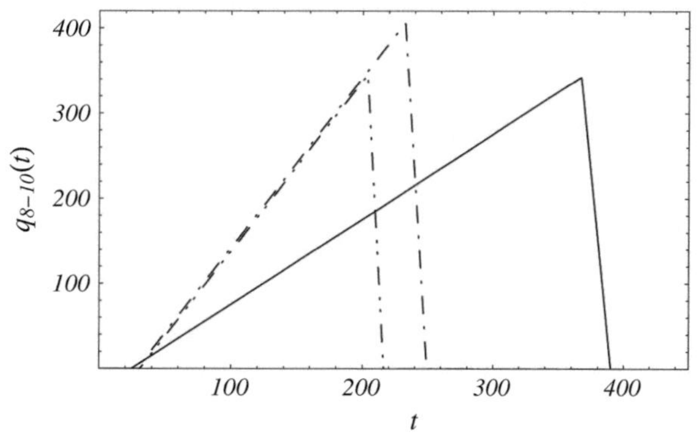

Fig. 4 Queues $q_j(t)$, $j = 8, 9, 10$; $q_8(t)$: dot dot dashed line; $q_9(t)$: dot dashed line; $q_{10}(t)$ continuous line

under investigation now, especially in terms of DSSs that indicate correct choices of $\gamma\ (t)$ to erase queues.

In this case, an approach, based on Situation Awareness, allows analyzing the environment data in order to get that $\overline{\gamma}\ (t)$, a new possible choice for $\gamma\ (t)$, focuses on only constant levels, namely:

$$\widetilde{\gamma}\ (t) = \begin{cases} 40, \ 0 \le t < 45, \\ 90, \ 45 \le t \le 90, \\ 60, \ 90 < t \le 150, \\ 20, \ t > 150. \end{cases} \tag{8}$$

The new profile (8) is obtained due to the typical Italian weather conditions in months useful for olives, from August to December. Considering a time t expressed in days, a Situation Awareness criterion suggests constant levels of injections, namely: Light injection from August 1st to September 15th (about 45 days); High profile from September 15th to October 30th; Medium injection from November 1st to December 30th; Low profile from December 30th.

The system performances due to $\gamma(t)$ and $\widetilde{\gamma}(t)$ are defined by the cost functional:

$$\Gamma := \sum_{j=1}^{10} \left(\int_0^{\overline{T}} \frac{q_j(t)}{10} dt \right),$$

that describes the average area due to queues, dependent on the input profiles. We have that:

$$\Gamma(\gamma(t)) \simeq 128988, \quad \Gamma(\widetilde{\gamma}(t)) \simeq 97559.$$

Notice that the adoption of a new profile does not erase queues but allows only to decrease them. This last aspect is still in investigation.

5 Conclusions

Focusing on the model for supply systems proposed in [9, 10], a real case of production network for the EVO oil has been studied.

In particular, using a procedure based on Situation Awareness, the simulations have showed that input profiles are able to modulate production queues, but not to erase them completely.

Further studies, based on Situation Awareness and Fuzzy Logic for the comprehension and the projection phases, are going to be developed in order to obtain robust optimization criteria for the performances of supply networks.

References

1. Daganzo, C.: A Theory of Supply Chains. Springer, New York, Berlin, Heidelberg (2003)
2. Helbing, D., Lämmer, S.: Supply and production networks: from the bullwhip effect to business cycles. In: Armbruster, D., Mikhailov, A.S., Kaneko, K. (eds.) Networks of Interacting Machines: Production Organization in Complex Industrial Systems and Biological Cells, pp. 33–66. World Scientific, Singapore (2005)
3. Cascone, A., D'Apice, C., Piccoli, B., Rarità, L.: Circulation of car traffic in congested urban areas. Commun. Math. Sci. **6**(3), 765–784 (2008)
4. Manzo, R., Piccoli, B., Rarità, L.: Optimal distribution of traffic flows at junctions in emergency cases. Eur. J. Appl. Math. **23**(4), 515–535 (2012)
5. Armbruster, D., Degond, P., Ringhofer, C.: A model for the dynamics of large queueing networks and supply chains. SIAM J. Appl. Math. **66**, 896–920 (2006)
6. Bressan, A.: Hyperbolic Systems of Conservation Laws—The One—Dimensional Cauchy Problem. Oxford University Press, Oxford (2000)
7. Dafermos, C.: Hyperbolic Conservation Laws in Continuum Physics. Springer (1999)
8. Armbruster, D., Degond, P., Ringhofer, C.: Kinetic and fluid models for supply chains supporting policy attributes. Bull. Inst. Math. Acad. Sin. (N.S.), **2**, 433–460 (2007)
9. Göttlich, S., Herty, M., Klar, A.: Network models for supply chains. Commun. Math. Sci. **3**, 545–559 (2005)

10. Göttlich, S., Herty, M., Klar, A.: Modelling and optimization of supply chains on complex networks. Commun. Math. Sci. **4**, 315–330 (2006)
11. de Falco, M., Gaeta, M., Loia, V., Rarità, L., Tomasiello, S.: Differential quadrature-based numerical solutions of a fluid dynamic model for supply chains. Commun. Math. Sci. **14**(5), 1467–1476 (2016)
12. Leveque, R.J.: Finite Volume Methods for Hyperbolic Problems. Cambridge University Press, Cambridge, 2002
13. Cutolo, A., Piccoli, B., Rarità, L.: An upwind-Euler scheme for an ODE-PDE model of supply chains. SIAM J. Sci. Comput. **33**(4), 1669–1688 (2011)
14. Tomasiello, S.: DQ based methods: theory and application to engineering. In: Leng, J., Sharrock, W. (eds.) Handbook of Research on Computational Science and Engineering: Theory and Practice. IGI Global, pp. 316–346 (2013)
15. Kamarian, S., Salim, M., Dimitri, R., Tornabene, F.: Free Vibration Analysis of Conical Shells Reinforced with Agglomerated Carbon Nanotubes. Int. J. Mech. Sci. **108–109**(1), 157–165 (2016)
16. Endsley, M.R.: Toward a theory of situation awareness in dynamic systems. Hum. Factors J. Hum. Factors Erg. Soc. **37**(1), 32–64 (1995)

A Freight Adviser for a Delivery Logistics Service e-Marketplace

Annarita De Maio, Antonio Violi, Demetrio Laganà and Patrizia Beraldi

Abstract We introduce the study of an application for the Macingo Technologies, a company that manages an e-marketplace for logistics services. A carrier has to pick-up and delivery freight from different points and is able to accept further delivery requests through the web company platform. In order to investigate the convenience in accepting extra-deliveries, the company wants to develop a decisional support system that suggests a list of convenient deliveries, and as a consequence, the best itinerary for the pick-up and delivery points. We study a *Vehicle Routing Problem* in which a subset of mandatory customers has to be visited by the vehicle, while the goal consists of maximizing the net revenue with respect to the routing cost to serve the set of mandatory customers. Some preliminary computational results are presented, showing the validity of the proposed approach.

Keywords Vehicle routing · Pick-up and delivery · Time windows

1 Introduction

https://www.Macingo.com is one of the largest Italian community for sharing bulky good transport: from cars to motorbikes, from boats to industrial products, the website connects who has a request for delivery (private individuals and companies) with

---

A. De Maio (✉) · A. Violi · D. Laganà · P. Beraldi
DIMEG, Università della Calabria, Rende, Italy
e-mail: annarita.demaio@unical.it

A. Violi
e-mail: antonio.violi@unical.it

D. Laganà
e-mail: demetrio.lagana@unical.it

P. Beraldi
e-mail: patrizia.beraldi@unical.it

A. Violi
INNOVA s.r.l., Rome (RM), Italy

© Springer Nature Switzerland AG 2018

P. Daniele and L. Scrimali (eds.), *New Trends in Emerging Complex Real Life Problems*, AIRO Springer Series 1,
https://doi.org/10.1007/978-3-030-00473-6_24

carriers that are compatible from a geographical, merchandise and capacity point of view. The e-marketplace was created by *Macingo Technologies srl*, that developed a platform for matching supply and demand of transportation services. The scientific literature deeply investigated the logistics e-marketplace: [1] describes the best features and classification of this type of markets; [2–4] introduce a range of interesting mathematical problems related to the way these negotiations take place; some case studies about the collaboration between carriers and its benefits are introduced by [5, 6]; finally a general overview about collaboration in freight transportation is presented by [7].

The main idea of the Macingo's business was initially to offer a web service in which carriers are able to saturate their vehicle capacity in surplus, taking in charge customers' requests submitted on the web platform. In general, carriers offer directly on the site their availability of cargo to thousands of transport requests that arrive on the platform every month. The customer that needs to transport a good places the request on the site and receives offers from interested carriers. If he finds an interesting offer, he can proceed to book the transport directly on the site paying a minimum commission. The company offers a very basilar service actually. No auction or optimization for prices, routes, loads is provided to the users. On loads and geographical evaluation base, each carrier receive a list of possible service requests and analyses them by himself. In this way he can decides to send a private offer for transportation to a particular customer.

Macingo wants to improve the service, introducing a decisional support system for the logistics operator that use the platform. The new service will suggest to each carrier a list of request to accept that are considered convenient from a profit point of view. In particular, the new service should work as follow: the carrier inserts on the platform a list of nodes for pick-up and delivery freight that correspond to the request he has already in charge, in terms of geographical positions, loads and transportation costs. Starting from the actual carrier status, the platform evaluates if there are extra convenient requests that it can suggest to the carrier, considering the incremental profit he can obtain accepting them. At the end, the platform describes the best route to follow. If no request is convenient, the platform suggest only the best route for the pre-inserted deliveries.

In this work, we study the mathematical formulation that drives the engine of the platform for modelling the problem related to the new service that Macingo wants to introduce. The problem is formulated as a variant of Vehicle Routing Problem with pick-up and delivery and time windows, with a set of mandatory requests and the maximization of the incremental profit; where the expression "incremental profit" refers to the net revenue (total profit minus total routing cost, resulting from the service of not mandatory requests compared to the fixed routing cost to serve all the mandatory request).

Vehicle Routing Problems (*VRPs*) spread out in the optimization of goods distribution management in supply chains. This approach design an optimal route for a fleet of vehicles that deliver a set of customers, with a set of different constraints. The *VRP* is deeply studied because of its wide applicability and its importance in determining efficient strategies for reducing operational costs in distribution networks. In practice, several variants of the problem exist because of the diversity in real-life applications. A short literature overview is presented in the following: *VRP* was introduced for the first time by [8]; while a great description of the classical version and its variants is presented in different contributions, like [9–11]. The *VRP* generalizes the well-known NP-Hard Travelling Salesman Problem but is much more difficult to solve in practice. Lots of solution methods were presented in the past: good exact solution methods were introduced by [12, 13], while heuristics approaches are described in [14, 15]. The literature is also rich of great contribution on the pick-up and delivery variant: two different classes are described in [16]. The first refers to the problem in which each vehicle can both deliver goods and collect waste but it must complete all deliveries before starting to collect (*Vehicle Routing Problems with Backhauls (VRPB)*). The second class is composed by problems where goods are moved between pick-up and delivery points (*Vehicle Routing Problems with Pick-ups and Deliveries (VRPPD)*), in which different sub-classes could be recognized (customers are either delivery or pick-up customers but cannot be both, customers can be both, simultaneously pick-up and delivery, etc ...). For a detailed overview in the field the reader can refer to [11, 16, 17].

In this work, we introduce a variant of the pick-up and delivery case with time windows. In this case, a carrier has to process a set of deliveries that he has in charge, visiting a set of pick-up and delivery points, that are fixed, starting his route from the depot. The carrier has to decide if it is convenient to saturate the capacity of his vehicle accepting further deliveries, maximizing his additional revenue. He also needs to know which is the best route that satisfy all the constraints, including time windows restrictions. The paper is organized as follows: Sect. 2 introduces the problem formulation and its features, Sect. 3 presents some preliminary computational results and Sect. 4 reports some concluding remarks.

2 Problem Formulation

In this section, the mathematical formulation based on binary arc-variables is presented. We consider a complete directed graph $G = (V, A)$, where V is the set of vertices and A is the set of arcs. We suppose that G is an Euclidean graph, so the triangular inequality holds. Let m be the number of requests to be evaluate, each one made up of a pair of vertices (pick-up and delivery). A set of pick-up vertex $P = \{1, 2, \dots, m\} \subset V$ and delivery vertex $D = \{m+1, m+2, \dots, n\} \subset V$ can be visited by the vehicle starting from the depot (with index zero and $n+1$). The subset $F = P_f \times D_f$ represents the set of the mandatory requests, with $P_f \subset P$ and $D_f \subset D$. The set with all vertices is denoted by $V = P \cup D \cup \{0\} \cup \{n+1\}$. The parameter c_{ij} is the non-negative cost of the arc $(i, j) \in A$, that satisfies the triangular inequality, while the parameter t_{ij} is the time for traversing the correspondent arc.

We consider only one vehicle that has capacity C. Different parameters are associate with each node. A quantity q_i represent the demand/supply at vertex (pick-up vertices are associated with a positive value, delivery vertices with a negative value, while it is equal to zero at depot). Each node defines the earliest and the latest time to start the service, e_i and l_i respectively; the duration of the service d_i, $i \in P \cup D$ and the revenue R_i, $\forall i \in D_f$ for visiting the node. We consider also a maximum time for routing TM and the cost of the routing CVR that the logistics operator initially estimated for serving the mandatory customers with the original route.

The continuous variables B_i indicate the beginning of the service at node i (with B_{0s} and B_{0r} respectively starting and ending time service at the depot), while Q_i is the load of the vehicle when leaving the vertex i. The binary variable z_i is equal to 1 if the node i is visited; the binary variable x_{ij} is equal to 1 if vehicle travels directly from vertex i to vertex j. This problem can be formulated as follows:

$$Max \sum_{i \in D_f} R_i z_i - \Delta C \tag{1}$$

s.to.

$$\sum_{j:(0,j) \in A} x_{0j} = 1 \tag{2}$$

$$\sum_{i:(i,0) \in A} x_{i0} = 1 \tag{3}$$

$$\sum_{i:(i,j) \in A} x_{ij} - \sum_{i:(j,i) \in A} x_{ji} = 0 \qquad \forall j \in P \cup D \tag{4}$$

$$\sum_{j \in V} x_{ij} = z_i \qquad \forall i \in V \tag{5}$$

$$z_i = z_{i+|P|} \quad \forall i \in P \tag{6}$$

$$B_{i+|P|} \geq B_i \quad \forall i \in P \tag{7}$$

$$z_i = 1 \qquad \forall i \in F \tag{8}$$

$$B_i \leq l_i z_i \qquad \forall i \in V \tag{9}$$

$$B_i \geq e_i z_i \qquad \forall i \in V \tag{10}$$

$$B_{0r} - B_{0s} \leq TM \tag{11}$$

$$B_j \geq B_i + d_i + t_{ij} - M_{ij}(1 - x_{ij}) \qquad \forall (i,j) \in A \tag{12}$$

$$Q_j \geq (Q_i + q_j) * x_{ij} \qquad \forall (i,j) \in A \tag{13}$$

$$max\{0, q_i\} \leq Q_i \leq min\{C, C + q_i\} \qquad \forall i \in V \tag{14}$$

$$B_i, B_{0s}, B_{0r} \geq 0 \qquad \forall i \in V \tag{15}$$

$$x_{ij} \in \{0, 1\} \qquad \forall (i,j) \in A, \tag{16}$$

$$Q_i \geq 0 \qquad \forall i \in V. \tag{17}$$

The objective function (1) maximizes the additional profit that the operator can reach if he accepts extra-delivery; the profit is computed considering the revenue in accepting new deliveries minus the additional cost of the new route compared with the original one, $\Delta C = \sum_{(i,j)\in A} c_{ij}x_{ij} - CVR$. Constraints (2)–(3) force the tour to start and finish at the depot, constraints (4) are the typical path constraints, constraints (5) impose the visit of each selected node. Constraints (6)–(7) impose the necessity to pick-up the freight before deliver. Constraints (8) force a group of nodes to be visited (in order to satisfy the demand of the set of scheduled customers), constraints (9)–(10) impose the time windows for visiting each node, constraint (11) defines the maximum duration of the route, constraints (12) set a minimum time for the service beginning of customer j in a determined route and also guarantee that no sub-tours occur. The constant M_{ij} is a large enough number (for instance, $M_{ij} = l_i + t_{ij} - e_j$). Constraints (13) impose that each node of delivery has to be visited after its correspondent pick-up node in terms of loads, while constraints (14) are the vehicle capacity constraints. Constraints (15)–(17) define the variables nature.

Constrains (13) are non linear in this shape. In order to linearise them we introduce the new variable $m_{ij} = Q_i * x_{ij}$ and we substitute it into (13) in the following way:

$$Q_j \geq m_{ij} + q_j x_{ij} \;\; \forall (i,j) \in A \tag{18}$$

We consider an upper bound for the Q_i that is equal to the vehicle capacity C; the additional constrains needed for the linearisation are described below:

$$m_{ij} \leq Cx_{ij} \;\; \forall (i,j) \in A \tag{19}$$

$$m_{ij} \leq Q_i \;\; \forall (i,j) \in A \tag{20}$$

$$m_{ij} \geq Q_i - (1 - x_{ij})C \;\; \forall (i,j) \in A \tag{21}$$

$$m_{ij} \geq 0 \;\; \forall (i,j) \in A \tag{22}$$

3 Computational Results

The problem described in Sect. 2 was implemented in C++ by using IBM Concert Technology and CPLEX 12.6,[1] and run on an Intel Core i7-6500U 2.50 GHz and 8 GB RAM personal computer. Some instances are generated ad hoc in order to test the

[1]https://www.ibm.com/products/ilog-cplex-optimization-studio.

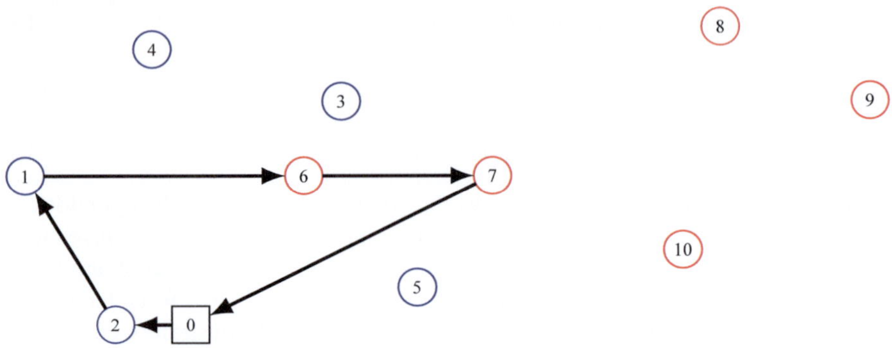

Fig. 1 Examples of the initial routing for the logistic operator without considering new offers from the platform. Total routing cost equal to 674

model. Instances are labelled as *pPdDfF*, where P is the number of pick-up points, D is the number of delivery points and F is the number of scheduled deliveries selected. Because of the absence of real data for now, in the tests we consider the CVR equal to the TSP solution value for the set made up of the fixed pick-up and delivery points and the depot, respecting the time window constraints. We estimate also the R_i values on the base of the carrier experience.

Our decisional support model is able to select the convenient delivery in order to produce an additional earning for the logistics operator. In order to clarify how the model works, a simple example is showed in the following pictures.

In Figs. 1 and 2 an instance with $P = \{1, \ldots, 5\}$, $D = \{6, \ldots, 10\}$ and $F = \{1,2\text{–}6,7\}$ is described. The solution scheme for the initial situation is represented below, the total routing cost computed with the TSP is equal to 674, the logistics operator has to serve the pickup-delivery couple 1–6 and 2–7.

In Fig. 2, the revenue for serving each extra delivery point is considered. The model has to advice the logistics operator about which extra-delivery is convenient to take in charge. In the new solution, the built route serves the couples 1–6, 2–7, 4–9; with a total routing cost equal to 1052 and a total earning equal to 500 for serving the couple 4–9. Considering the objective function, the extra earning obtaining serving the couple 4–9 is equal to $500 - 1052 + 674 = 122$. The situation after solving the model is described below:

In the example showed, the logistics operator can accept a new delivery in order to improve his final earning.

A set of instances with different cardinalities was tested. The results are described in the table, that is organized as follow: column *Instance* introduces the instance name, column *Initial TSP* reports the TSP cost for the scheduled deliveries, column *New R.C.* reports the new total routing cost obtained by solving the model, column *T. Earning* reports the objective function value, column *N. Extra Del.* introduces the extra delivery number that the model assigns to the logistics operator, the last two columns describe the execution time and the duality gap provided by Cplex, respectively (Table 1).

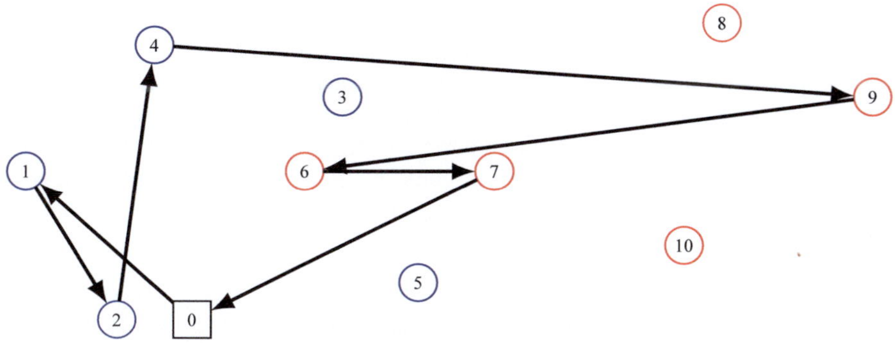

Fig. 2 Examples of the final routing for the logistic operator considering new offers from the platform. Total earning equal to 122

Table 1 Computational results

Instance	Initial TSP	New R.C.	T. Earning.	N. Extra Del.	Time (s)	Gap %
5P5D2F	674	1052	122	1	0.23	0.00
5P5D2F	674	924.17	612.83	2	0.72	0.00
6P6D2F	674	1047.94	829.05	2	3.98	0.00
6P6D3F	795.23	1395.67	602.32	2	5.72	0.00
7P7D3F	795.23	1296.67	622.33	2	66.38	0.00
8P8D3F	795.23	1395.67	600.32	3	229.36	0.00

Starting from two different configurations of initial scheduled deliveries, we test some instances of growing size, in order to show that the model is able to modify the previous scheduling and suggesting new convenient solutions for the logistics operator. The duality gap provided by Cplex and showed in the last column proves that all the instances are solved to optimality. The results show that the model is able to assign extra-requests in all the analysed instances, maximizing the additional profit. It is also possible to underline that the model could be effective also in the case in which it is not convenient to suggest extra-request: the model gives in output the TSP route of the mandatory nodes. In this way, if the carrier plans in a worst way his route, he is able to modify his travel.

4 Conclusion

In this work, we introduce a variant of the classical *VRP* with pick-up and delivery and time windows. The model is formulated for matching the need of a company that manages an e-marketplace for logistics services and wants to introduce a new function for suggesting its customers about the best and convenient requests to accept.

The company wants also to provide a support about the best route to follow in that case. As described in the previous section, the mathematical model is able to give a good support for selecting the best options and rescheduling the routing for the pick-up and delivery points, taking into account different constraints. Starting from this evidence, it is possible to affirm that the model faces the company needs perfectly, and in the future an heuristic algorithm could be developed for solving instances with a big number of pick-up and delivery points, in order to meet real-life market request.

Acknowledgements This work has been partially supported by POR Calabria FESR-FSE 2014–2020, with the grant for research project "Smart Macingo", CUP J18C17000580006.

References

1. Nandiraju, S., Regan, A.: Freight transportation electronic marketplaces: a survey of the industry and exploration of important research issues. Transp. Res. Rec. (2008)
2. Figliozzi, M.A., Mahmassani, H.S., Jaillet, P.: A framework for the study of Carrier strategies in an auction based transportation marketplace. Transp. Res. Board J. Transp. Res. Board **1854**, 162–170 (2003)
3. Ihde, T.: Dynamic Alliance Auctions: a Mechanism for Internet Based Transportation Markets. Physica Verlag, Hedelberg (2004)
4. Triki, C., Oprea, S., Beraldi, P., Crainic, T.G.: The stochastic bid generation problem in combinatorial transportation auctions. Eur. J. Oper. Res. **236**(3), 991–999 (2014)
5. Audy, J.F., DÁmours, S., Rousseau, L.M.: Cost allocation in the establishment of a collaborative transportation agreement: an application in the furniture industry. J. Oper. Res. Soc. **62**(6), 960–970 (2010)
6. Frisk, M., Gothe-Lundgren, M., Jornsten, K., Ronnqvist, M.: Cost allocation in collaborative forest transportation. Eur. J. Oper. Res. **205**, 448–458 (2010)
7. Agarwal, R., Ergun, O.: Network design and allocation mechanisms for carrier alliances in liner shipping. Oper. Res. **58**, 1726–1742 (2010)
8. Dantzig, G.B., Ramser, J.H.: The truck dispatching problem. Manag. Sci. **6**, 80–91 (1959)
9. Golden, B.L., Raghavan, S., Wasil, E.A.: The Vehicle Routing Problem: latest Advances and New Challenges. Springer, New York, (2008)
10. Laporte, G.: Fifty years of vehicle routing. Transp. Sci. 408–416, (2009)
11. Toth, P., Vigo, D.: The Vehicle Routing Problem. SIAM Monographs on Discrete Mathematics and Applications. Society for Industrial and Applied Mathematics, Philadelphia (2002a)
12. Baldacci, R., Christofides, N., Mingozzi, A.: An exact algorithm for the vehicle routing problem based on the set partitioning formulation with additional cuts. Math. Program. Ser. A **115**, 351–385 (2008)
13. Fukasawa, R., Longo, H., Lysgaard, M., Poggi de Aragao, M., Reis, E., Uchoa, R., Werneck, F.: Robust branch-and-cut-andprice for the capacitated vehicle routing problem. Math. Program. Ser. A **106**, 491–511 (2006)
14. Cordeau, J.-F., Gendreau, M., Laporte, G., Potvin, J.-Y., Semet, F.: A guide to vehicle routing heuristics. J. Oper. Res. Soc. **53**(5), 512–522 (2002)
15. Vidal, T., Crainic, T.G., Gendreau, M., Prins, C.: Heuristics for multi-attribute vehicle routing problems: a survey and synthesis. Eur. J. Oper. Res. **231**(1), 1–21 (2013)
16. Parragh, S.N., Doerner, K.F., Hartl, R.F.: A survey on pickup and delivery problems. J. fur Betriebswirtschaft **21**(51), 58 (2008)
17. Berbeglia, G., Cordeau, J.-F., Gribkovskaia, I., Laporte, G.: Static pickup and delivery problems: a classification scheme and survey. TOP **15**(1), 1–31 (2007)

Specification and Aggregate Calibration of a Quantum Route Choice Model from Traffic Counts

Massimo Di Gangi and Antonino Vitetta

Abstract This paper analyses certain aspects related to the route choice model in transport systems. The effects of an interference term have been taken into consideration in addition to the effect of a traditional covariance term. Both the specification and calibration of an interference term in a quantum route choice model are shown in the context of an assignment model. An application to a real system is reported where the calibration of QUMs (Quantum Utility Models) was performed using traffic counts. Results are compared with traditional and consolidated models belonging to the Logit family. Based on the theoretical and numeric results, it is highlighted how the interference term and quantum model can consider other aspects (such as information) with respect to traditional RUMs (Random Utility Models).

Keywords Assignment · Path choice · Quantum

1 Introduction

In this paper, the effects of other sources of information within random utility route choice models are considered. Such effects are included in an interference term. Route choice is a component of assignment models; it simulates user behaviour in a transport network, having defined the origin and the destination of the journey, the departure time and the mode.

At the route choice level, some user behaviours are simulated by a utility function. (i) With regard to the kind of uncertainty, Random Utility Models (RUMs) [1] or Fuzzy Utility Models (FUMs) [2] can be considered. (ii) Concerning the perception of overlapping alternatives, RUMs simulate the overlapping perception between

---

M. Di Gangi (✉)
Dipartimento di Ingegneria, Università degli Studi di Messina, 98166 Messina, Italy
e-mail: mdigangi@unime.it

A. Vitetta
DIIES, Università degli Studi Mediterranea di Reggio Calabria, 89124 Reggio Calabria, Italy
e-mail: vitetta@unirc.it

© Springer Nature Switzerland AG 2018
P. Daniele and L. Scrimali (eds.), *New Trends in Emerging Complex Real Life Problems*, AIRO Springer Series 1,
https://doi.org/10.1007/978-3-030-00473-6_25

alternatives considering an additive term (i.e. Path size [3], C-Logit [4]) or a covariance (i.e. Nested Logit [5], Probit [6], Gammit [7]). (iii) In relation to experience on the past and information, within-day [8] and/or day-to-day models [9] simulate these aspects.

RUMs or FUMs assume that each user assumes a choice set of clearly perceived alternatives and chooses one of the alternatives belonging to the choice set. The analyst associates a probability choice to each of the alternatives, and probabilities for the alternatives are estimated (a) without or (b) with a covariance in perception.

In real-life situations, the user may perceive that certain alternatives interfere with one another, so the analyst cannot consider interfering alternatives as clearly separate and interference should be explicitly considered. The interference simulates the lack of information on one or more alternatives. To take account of these aspects, a new class of models has been proposed [10], called Quantum Utility Models (QUMs).

QUM assumes that each user perceives a choice set of alternatives in which all or some of the alternatives interfere. This effect can also take account of the availability of information within the trip. The analyst associates a quantum probability choice to each perceived alternative with interference, (c) not considering or (d) considering covariance.

Thus, the above-defined cases are: (a) no covariance, no interference; (b) covariance, no interference; (c) no covariance, interference; and (d) covariance, interference. In the decision-making process, if the absence of the interference effect is assumed, QUM gives the same specification of RUM [11].

In addition to the paper reported in the literature related to RUMs, QUM has been specified in this paper considering interference at the generation level of alternatives, route choice and assignment model and procedure with RUMs and QUMs, calibration of a QUM from counted flow, and application to a real case. The results were compared with traditional models. The main advancement proposed in this paper is the aggregated calibration from traffic counts of the values of the QUM parameters and comparison with RUMs.

Following a short summary of assignment models in Sect. 2, QUMs are specified in Sect. 3 and integrated in the context of assignment models. The results of a test performed in a real case are reported in Sect. 4; while Sect. 5 contains a summary of achieved results.

2 Assignment Model

The assignment model considers user behaviour within the route choice model. The (equilibrium) assignment model can be formulated as the combination of a demand model and a supply model leading to a fixed-point model [12].

In a congested transport network, the flow vector $\boldsymbol{f}$ is the solution of the fixed-point problem expressed by:

$$\boldsymbol{f} = \boldsymbol{v}(\boldsymbol{\gamma}(\boldsymbol{f}), \boldsymbol{I}, \Delta, \boldsymbol{d}; \boldsymbol{\beta})$$

where:

$\boldsymbol{c} = \boldsymbol{\gamma}\ (\boldsymbol{f})$ expresses the dependence of the link cost vector **c** and the link flow vector $\boldsymbol{f}$ by means of a cost function $\boldsymbol{\gamma}$;
$\boldsymbol{I}$ is the set of perceived alternatives;
$\boldsymbol{\Delta}$ is the link-route incidence matrix;
$\boldsymbol{d}$ is the demand vector;
$\boldsymbol{\beta}$ is the vector of control variable to be estimated

To solve the fixed-point problem, Robbins and Monro [13] propose the Method of Successive Averages (MSA) applied to the Stochastic User Equilibrium (SUE) problem as proposed by Sheffy and Powell [6]. Two variants of MSA can be considered: the Flow Averaging (MSA-FA) algorithm that updates link flows [14], and the Cost Averaging (MSA-CA) algorithm that updates link costs [12]. In this paper, the MSA-FA algorithm is considered.

The link flow depends on the control variables included in vector $\boldsymbol{\beta}$. They could be estimated starting from counted flows and an initial value $\boldsymbol{\beta}^{in}$. The counted flows are reported in the vector $\boldsymbol{f}^{ob}$.

Considering the case of Minimum Least Square method, the optimization problem can thus be formulated as:

Control variables $\boldsymbol{\beta}$

Optimization problem $min\ z_1(\boldsymbol{\beta}, \boldsymbol{\beta}^{in}, \mathbf{Z}_{\beta}) + z_2\left(\boldsymbol{f}, \boldsymbol{f}^{ob}, \mathbf{Z}_f\right)$

Constraints $\boldsymbol{f} = \boldsymbol{\nu}(\boldsymbol{c}, \boldsymbol{I}, \mathbf{1}, \boldsymbol{d}; \boldsymbol{\beta})$

$$c = \gamma(f)$$

$$d \geq 0$$

$$z_1(\boldsymbol{\beta}, \boldsymbol{\beta}^{in}, \mathbf{Z}_{\beta}) = (\boldsymbol{\beta} - \boldsymbol{\beta}^{in})^T \mathbf{Z}_{\beta}^{-1} (\boldsymbol{\beta} - \boldsymbol{\beta}^{in})$$

$$z_2\left(\boldsymbol{f}, \boldsymbol{f}^{ob}, \mathbf{Z}_f\right) = (\boldsymbol{f} - \boldsymbol{f}^{ob})^T \mathbf{Z}_f^{-1} (\boldsymbol{f} - \boldsymbol{f}^{ob})$$

where:

$\mathbf{Z}_{\beta}$ and $\mathbf{Z}_f$ are the variance-covariance matrices of $\boldsymbol{\beta}$ and $\boldsymbol{f}$ respectively
z_1 is a distance function between simulated and initial control variables
z_2 is a distance function between simulated and counted flow values.

3 QUM and RUM Route Choice for Assignment

Considering a specific origin/destination (o/d) pair j, the route choice model is generally defined in terms of (i) generation of the set of the perceived alternatives, and (ii) choice from among those alternatives. Aa new approach combining RUM and QUM is shown below. Very often, these two levels are simulated using the same type of model (i.e. introducing a new term in the utility specification of RUM or FUM [15]).

In this paper, a new model introducing the interference term and quantum models for the first level (generation) is proposed.

Route choice behaviour modelling through RUM. In relation to the route choice level, given o/d pair j, RUMs evaluate the probability $p_{k,j}$ of choosing route k within the choice set of perceived alternatives. The choice of perceived alternatives is the same for both levels. Assuming the utilities of the alternatives distributed as an independent Gumbel probabilistic function, some specifications for $p_{k,j}$ can be obtained (Multinomial Logit, Path Size-Logit, C-Logit).

The main advantages of these models consist in their simple structure and choice probabilities obtained in a closed form. Due to hypotheses on utility distribution, the main drawback of Multinomial Logit (MNL) is that routes belonging to the choice set I_j are considered independent. To overcome this drawback, some modifications to the MNL have been introduced. Such modifications maintain the simple Logit structure and introduce a correction term within the expected value of the utility function in place of the correlation among alternatives.

Ben-Akiva and Bierlaire [3] introduced the Path-Size Logit (PSL) model for an application of discrete choice theory for aggregate alternatives. Cascetta et al. [4] proposed a modification of the MNL model (named C-Logit), introducing a commonality factor to measure the degree of similarity of each route with the other routes of the choice set I_j. As stated by Prato [16], the main disadvantage of C-Logit is that this specification simulates only a part of commonality effects.

Assuming the utility of the alternatives distributed as a Normal probabilistic function, the Multinomial Probit (MNP) is obtained. One formulation applicable in a real-size system was proposed by Daganzo and Sheffy [17] where the covariance is explicitly specified. MNP cannot be expressed in a closed form.

Extension of route choice behaviour modelling through RUM and QUM. Users consider certain criteria in generating alternative routes (i.e. minimum time, maximum reliability, etc.). QUMs introduce an interference term to take into account the fact that some criteria can interfere with one another and cannot be perceived as clearly separated. The interference simulates the lack of information. This could also simulate the effect of information obtained during the trip. In the RUM approach, the analyst estimates a choice probability for each alternative belonging to the choice set. The overlapping effects are considered by means of a covariance term. It implies that the user perceives a unique utility for each alternative, possibly with attributes shared with other alternatives. Within the proposed model, it is possible to simulate the two levels in different ways:

- (i) generation of the perceived criteria as a choice set (generated with N criteria) with elementary criterion c; (ii) the choice is modelled with QUM
- (i) generation of a choice set with elementary routes k; (ii) the choice is modelled with RUM

Introducing the interference term $r_{k,j}$, the probability $q_{k,j}$ of choosing an alternative k of o/d pair j can be expressed as [11]:

$$q_{k,j} = p_{k,j} + r_{k,j} \,\forall_j, k$$

where:

$p_{k,j}$ is the RUM component

In the case that N criteria are considered for the generation of the choice set, the interference term is:

$$\mathrm{r_{k,j}} = 2 \sum_{\mathrm{c=1...N-1}} \sum_{\mathrm{c'=N+1...N}} (p_{k|c} p_c p_{k|c'} p_{c'} cos(\theta_{kc,kc'}))^{0,5}$$

where:

$\mathrm{p_c}$ is the RUM for choosing criterion c;
$p_{k|c}$ is the RUM for choosing route k conditioned to choose criterion c;
$\theta_{kc,kc'}$ is the interference angle between criterion c and c′ for alternative k

Values of parameters of the interference terms assume values that respect the following constraints: $r_{k,j} \in [-p_{k,j}, 1 - p_{k,j}]; \sum_{k \in Ij} r_{k,j} = 0$.

This model can take into account both overlapping and interference effects. As can be noted, the interference effects are additive.

4 Experimentation

The main objective of this section is to apply RUMs and QUMs in a real case in order to calibrate QUMs from traffic counts and to evaluate the performances of QUMs in comparison with traditional RUMs.

Two criteria are considered for the generation level: minimum travel time in congestion condition T and maximum reliability R. Routes are generated using a procedure proposed by De La Barra [18]. For the generation level, QUM is adopted with the interference specified as:

$$\theta_{T,R} = (\beta_0 b_{T,R} - 1)\pi/2$$

where:

$b_{T,R}$ is the number of times that a route is selected as the best in terms of T with the De La Barra procedure (normalized in the range [−1, 1]);
β_0 is a calibrated parameter;

Note that $b_{T,R}$ represents a measure of interference between R and T.

For the choice level, Multinomial Logit, Path Size-Logit or C-Logit RUMs are considered.

The interference angle, type of criteria and RUM specification considered here are simply an example to validate the model. Other specifications for the interference angle, criteria and RUM can be adopted without losing the generality of this approach.

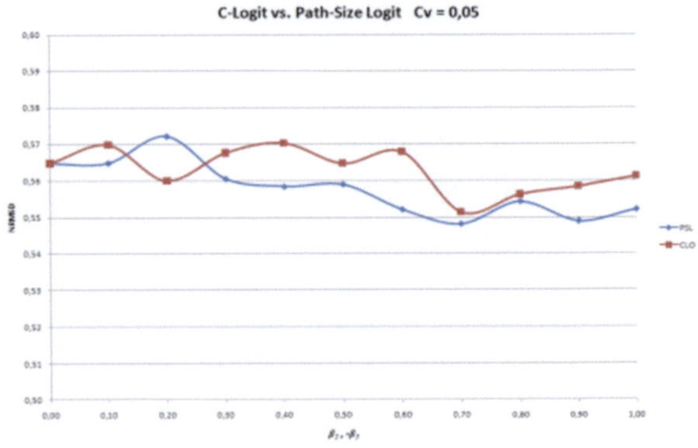

Fig. 1 Normalized root-mean-square deviation for RUMs

The test network used in this context is a real network in Benevento, a city in southern Italy with a population of about 62 000. The main characteristics of the graph of the road network are: 66 internal zones and 14 external zones, 759 nodes and 1 579 arcs. Observed flows are derived from surveys. The link (and flows) considered in the reported experiments belongs to 77 road trunks yielding 139 counting sections.

In the MSA-FA procedure, stochastic network loading was done using Logit-based techniques where routes were explicitly generated. For each o/d pair, the route search procedure is iterated 1 000 times and a maximum of 10 routes are generated.

Variance is computed assuming a non-negative relative standard deviation (coefficient of variation *Cv*) and less than 1 common for all O/D pairs. Parameter α of the Logit model was computed congruently with variance for each O/D pair.

To compare the values of simulated flows with counted ones, the value of the normalized root-mean-square deviation (NRMSD) was considered, where lower values indicate less residual variance. Considering the range of the values assumed for parameters, a linear grid-search procedure was considered, avoiding a more sophisticated approach [19].

In the following applications, a value for *Cv* equal to 0.05 was considered.

Values assumed for the other parameters of the models are: (i) Path-size Logit (PSL in the following) β_2 from 0.00 to 1.00 with a step of 0.10; (ii) C-Logit (CLO in the following) β_3 from 0.00 to -1.00 with a step of -0.10. Note that values of 0.00 coincide with the Logit model.

The results, in terms of NRMSD obtained for each MSA-FA experiment are reported in Fig. 1, where values in the axis of CLO parameters β_3 have changed sign to enhance readability.

Values assumed for the QUM parameters considered in the choice level are the best (in terms of NRMSD) obtained in the previous set of experiments. These values are: $\beta_2 = 0.70$ for Path-size Logit and $\beta_3 = -0.70$ for C-Logit. The values adopted for Quantum parameter β_0 vary from 0.00 to 2.25. Note that the value of $\beta_0 = 0.00$ coincides with no interference effect.

The results, in terms of NRMSD, obtained for each MSA-FA experiment are reported in Fig. 2, where QUM, QPS and QCL indicate results (in terms of quan-

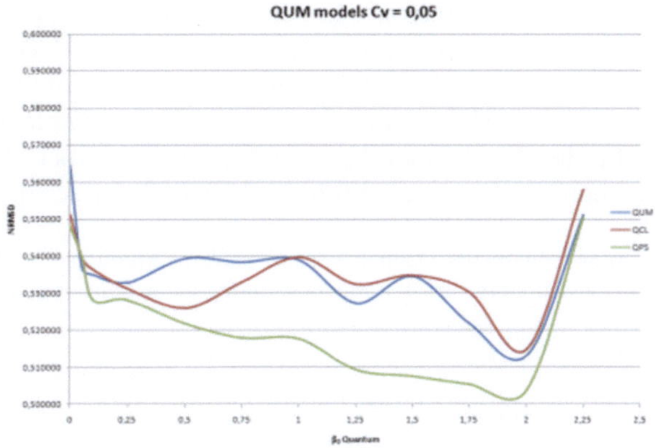

Fig. 2 Normalized root-mean-square deviation for quantum models

tum probabilities) obtained using Multinomial Logit, Path Size-Logit or C-Logit, respectively. As shown in Fig. 2, the presence of the interference term improves the reproduction of observed flows. Considering models that introduce covariances among routes, the interference term plays a significative role.

The presence of the interference term in the PSL model improves the reproduction of observed flows. Since path size indicates the fraction of the route that constitutes a "full" alternative, this can be explained by the fact that other behavioural aspects play a considerable role if the covariance term is strongly considered.

Considering C-Logit, the presence of the interference term in this model generally improves the reproduction of observed flows. This can be explained by the fact that the reduction of the utility of a route because of its similarity with respect to others does not take into consideration other behavioural effects that can be considered by means of QUMs.

In general, it can be said that, in a large number of experiments, the interference term reduces the *NRMSD* indicating that the covariance term does not simulate all the elements considered in the user's choice process.

5 Conclusions

In this paper, the effects of the interference term, in addition to the effects of traditional covariance effect term, are considered within route choice models in a transport system. Traffic flow counts obtained from a real network were used to calibrate RUMs and QUMs. From calibration results, in the Logit family, QUMs better reproduce observed flow than RUMs.

Based on the theoretical and numeric results, this confirms that, in application, the interference term simulates a different aspect with respect to the covariance effect. The interference term and quantum model simulate other aspects and have to be considered for simulation of the decision process.

These results must also be confirmed in other real-size systems and other contexts. QUMs consider overlapping and interference effects. The quantum probability is evaluated as the sum of two terms and, without interference, probability is obtained from RUMs. RUMs can be considered as a particular case of QUMs.

Further developments could relate to more experiments considering the effects of the interference term in Probit and Gammit-based route choice models.

Acknowledgements Authors wish to thank the Municipality of Benevento for having made available the data of the urban traffic plan during the national research project PRIN 2009.

References

1. Ben-Akiva, M.E., Lerman, S.R.: Discrete Choice Analysis: Theory and Application to Travel Demand. MIT Press, Cambridge, MA (1985)
2. De Maio, M.L., Vitetta, A.: Route choice on road transport system: a fuzzy approach. J. Intell. Fuzzy Syst. **28**(5), 2015–2027 (2015)
3. Ben-Akiva, M.E., Bierlaire, M.: Discrete choice methods and their applications to short term travel decisions. In: Hall, R.W. (ed.) Handbook of Transportation Science, pp. 5–34. Kluwer, Dordrecht, The Netherlands (1999)
4. Cascetta, E., Nuzzolo, A., Russo, F., Vitetta, A.: A modified logit route choice model overcoming path overlapping problems: specification and some calibration results for interurban networks. In: Lesort, J.B. (ed.) Proceedings of the Thirteenth International Symposium on Transportation and Traffic Theory, pp. 697–711. Pergamon, Lyon, France (1996)
5. Ben-Akiva, M.E.: Structure of passenger travel demand models. Transp. Res. Rec. 526 (1974)
6. Sheffi, Y., Powell, W.B.: An algorithm for the equilibrium assignment problem with random link times. Networks **12**, 191–207 (1982)
7. Cantarella, G.E., Binetti, M.G.: Stochastic assignment with Gammit path choice models. In: Patriksson, M. (ed.) Transportation Planning, pp. 53–67. Kluwer Academic Publisher (2002)
8. Cascetta, E., Cantarella, G.E.: A day-to-day and within-day dynamic stochastic assignment model. Transp. Res. Part A **25**(5), 227–291 (1991)
9. Watling, D., Hazelton, M.L.: The dynamics and equilibria of day-to-day assignment models. Netw. Spat. Econ. **3**(3), 349–370 (2003)
10. Busemeyer, J.R., Bruza, P.D.: Quantum Models of Cognition and Decision. Cambridge University Press (2012)
11. Vitetta, A.: A quantum utility model for route choice in transport systems. Travel Behav. Soc. **3**, 29–37 (2016). https://doi.org/10.1016/j.tbs.2015.07.003
12. Cantarella, G.E.: A general fixed-point approach to multi-mode multi-user equilibrium assignment with elastic demand. Transp. Sci. **31**, 107–128 (1997)
13. Robbins, H., Monro, S.: A stochastic approximation method. Ann. Math. Stat. **22**(3), 400–407 (1951)
14. Daganzo, C.: Stochastic network equilibrium with multiple vehicle types and asymmetric, indefinite arc cost jacobians. Transp. Sci. **17**, 282–300 (1983)
15. Cascetta, E., Russo, F., Viola, F.A., Vitetta, A.: A model of route perception in urban road networks. Transp. Res. Part B **36**(7), 577–592 (2002)
16. Prato, C.G.: Route choice modeling: past, present and future research directions. J. Choice Model. **2**(1), 65–100 (2009)
17. Daganzo, C., Sheffi, Y.: On stochastic models of traffic assignment. Transp. Sci. **11**, 253–274 (1977)

18. De La Barra, T., Perez, B., Anez, J.: Multidimensional path search and assignment. In: 21st PTRC Summer Annual Meeting, pp. 307–320 (1993)
19. Jiménez, A.B., Lazaro, J.L., Dorronsoro, J.R.: Finding optimal model parameters by discrete grid search. In: Corchado, E., et al. (eds.) Innovations in Hybrid Intelligent Systems, ASC, vol. 44, pp. 120–127. Springer, Berlin, Heidelberg (2007)

Modeling and Solving the Packet Routing Problem in Industrial IoT Networks

Luigi Di Puglia Pugliese, Dimitrios Zorbas and Francesca Guerriero

Abstract The IEEE802.15.4-TSCH (Time Slotted Channel Hopping) is a recent Medium Accesss Control (MAC) protocol designed for Industrial Internet of Things (IIoT) applications. The data transmissions in TSCH networks are performed according to a tight schedule computed by either a centralized entity or by the network nodes. The higher the schedule length, the higher the energy consumption of the network nodes and the end-to-end delay. In this paper, we address the problem of finding optimal routing topologies that minimize the schedule length. The problem can be viewed as a particular instance of the spanning tree problem with cost associated with each arc and a proper defined function that accounts for the schedule length. We propose a formulation for the problem along with optimal solution approaches. The computational results are carried out by considering realistic instances. The aim of the experimental phase is to evaluate the influence of the problem's characteristics on the optimal solution and to assess the behavior of the proposed solution approaches.

Keywords Routing problem · Tree · Mixed integer linear program · IoT

1 Introduction

The advent of the fourth industrial revolution brings forward some new challenges in manufacturing and automation. Internet of Things (IoT) applications use data collected by physical things to minimize manufacturing mistakes and reduce the cost. However, IIoT applications require efficient communications among the devices with

L. Di Puglia Pugliese (✉) · F. Guerriero
DIMEG, University of Calabria, Rende, Italy
e-mail: luigi.dipugliapugliese@unical.it

F. Guerriero
e-mail: francesca.guerriero@unical.it

D. Zorbas
Department of Informatics, University of Piraeus, Piraeus, Greece
e-mail: dzorbas@unipi.gr

© Springer Nature Switzerland AG 2018

P. Daniele and L. Scrimali (eds.), *New Trends in Emerging Complex Real Life Problems*, AIRO Springer Series 1,
https://doi.org/10.1007/978-3-030-00473-6_26

low energy cost and high end-to-end reliability [12]. Time-slotted communication combined with a channel hopping MAC such as the IEEE802.15.4-TSCH [9] has been proposed to provide the required high reliability at the link-layer.

In IEEE802.15.4-TSCH networks, the time is divided in slotframes where each slotframe consists of equal size timeslots. At each timeslot, a node can retain one of the following states: transmit a data packet, receive a data packet, or remain in sleep mode and conserve energy. Typically, depending on the application, each node transmits a number of packets per slotframe, which are forwarded towards a data collector (*root*) either in a single-hop or multi-hop manner.

The timeslots that will be allocated to each node, are defined by the network scheduler. In the literature, there are several scheduling algorithms which compute a schedule so that all the packets are forwarded to the *root* within each slotframe. The total number of timeslots that are allocated to the nodes by the scheduler constitutes the schedule length. The schedule length plays a critical role in the network operation. The longer the schedule, the higher the consumption of the nodes and the longer the end-to-end delay of the network. Besides, the schedule length heavily depends on the applied routing protocol and the characteristics of the network topology.

The IEEE802.15.4-TSCH standard adopts an IPv6 Routing Protocol for Low-Power and Lossy Network, called RPL [11], in order to build and maintain the routes from the nodes to the *root*. RPL focuses on a low-cost construction and maintenance of the network topology and does not guarantee the optimality of the schedule. Moreover, since an optimal schedule does not imply the optimality of the link qualities (e.g. packet delivery ratio), in this paper, we also take into account the cost of the constructed routing tree. Since link reliability generally worsens with the distance [7], we measure the routing cost as the sum of the relative distance between the nodes that constitute the edges of the data collection tree. As a consequence, the problem we need to examine is related to the choice of best routing scheme, in terms of schedule length and routing cost.

We formulate the problem as a mixed integer linear program. The model is solved minimizing either the cost or the schedule length with the aim of analyzing the characteristic of the problem. In addition, ad-hoc solution strategies are designed in order to overcome the difficulties in handling min-max function, used to evaluate the schedule length.

The proposed solution can be used either as an optimal static routing solution when the link qualities are known and do not considerably change through time, or as a benchmark when designing low-power distributed protocols.

This is the first work that deals with the problem of optimal routing in IEEE802.15.4-TSCH networks. Even though many other studies in the literature examine the problem of optimal routing, most of them focus on the minimization of the node energy consumption by proposing a joint optimization transmission power control and routing scheme [1, 3–5, 8]. Note that the transmission power control can be achieved in our work once the routing topology with optimal schedule length has been computed. Karnik et al. [2] solve a similar routing problem maximizing the network throughput. This work assumes data transmissions using a different MAC layer, thus, the schedule length is not taken into account.

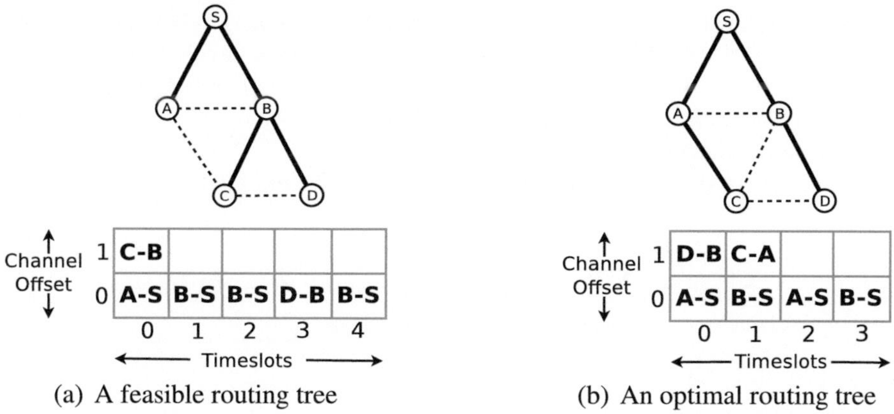

Fig. 1 A feasible and an optimal routing solution in terms of schedule length for a network with 4 nodes (A, B, C, D) and a *root* (S)

The rest of the paper is organized as follows. Section 2 explains how the routing process affects the schedule length of a TSCH network. Section 3 reports the formulation of the problem. Sections 4 and 5 are devoted to the solution approaches. Finally, Sect. 6 shows the numerical results and Sect. 7 concludes the paper.

2 Connection Between Routing and Schedule Length

To better understand the connection between routing and schedule length, we introduce the example of Fig. 1. Two routing solutions and their corresponding data transmission schedules for a network with 4 nodes and a *root* (node S) are displayed. We assume that each node generates 1 packet per slotframe and all the links have the same cost. The bold lines correspond to the selected routing links and the dashed lines to the radio links. Channel offsets are assigned by the scheduler to allow parallel transmissions. The left solution exhibits a schedule length equal to 5. We can observe that the *root* is inactive at timeslot 3. On the other hand, the right routing tree presents the optimal performance with 4 timeslots reserved in total. No *root* inactive timeslots appear.

3 Problem Formulation

We formulate the problem as a multi-commodity flow problem on a directed graph $D(N, A)$, where N is the set of nodes and A is the set of arcs, that identify the links among the nodes. Each node $i \in N \setminus \{root\}$ is associated with a commodity. A cost

l_{ij} is associated with each arc $(i, j) \in A$ and a number of packets q_i is associated with each node $i \in N$. We define binary variables y_{ij}^k, $\forall (i, j) \in A$, $k \in N \setminus \{root\}$ stating whether the flow for commodity k to the *root* passes through arc (i, j). Let x_{ij} be a binary variable indicating whether arc (i, j) is in the solution, that is, $y_{ij}^k = 1$ for at least one commodity k.

The set of constraints, defining the routing tree T are reported in what follows.

$$\sum_{j:(i,j)\in A} y_{ij}^k - \sum_{j:(j,i)\in A} y_{ji}^k = \begin{cases} 1 & \text{if } i = k \\ -1 & \text{if } i = root \quad \forall k \in N \setminus \{root\}, \\ 0 & \text{otherwise} \end{cases} \tag{1}$$

$$x_{ij} \geq y_{ij}^k, \quad \forall (i, j) \in A, k \in N \setminus \{root\}, \tag{2}$$

$$y_{ij}^k \geq 0, \forall (i, j) \in A, k \in N \setminus \{root\}, x_{ij} \in \{0, 1\} \forall (i, j) \in A. \tag{3}$$

Equation (1) represent the flow conservation constraints for each commodity k. Equation (2) define the value of variables x. Equation (3) state the domain of the decision variables. Constraints (1)–(3) define a tree T, that is, a connected and acyclic graph. Thus, there exists a unique path from each node $k \in N$ to the root node in T.

The schedule length LT can be modeled as follows [6].

$$LT = \max\{2Q_M - q_M, Q_{root}\}, \tag{4}$$

where $M = \arg\max_{i \in N}\{Q_i : x_{iroot} = 1\}$ and Q_i $\forall i \in N$ are non-negative variables indicating the number of packets that reaches node i and defined as $Q_i = \sum_{(u,v)\in T_i} x_{uv} q_u + q_i$, where T_i is the subtree of T rooted at node i. Let z_i, $\forall i \in N : (i, root) \in A$ be binary variables indicating whether M, in Eq. (4), is obtained for node $i \in N$.

In order to mathematically represent the function LT (to be minimized), we have to consider the following constraints:

$$LT \geq Q_{root}, \tag{5}$$

$$LT \geq 2Q_M - \sum_{i\in N:(i,root)\in A} z_i q_i, \tag{6}$$

$$Q_i \geq \sum_{(v,i)\in A} Q_v x_{vi} + q_i \quad \forall i \in N, \tag{7}$$

$$Q_M \geq Q_i - H(1 - x_{i,root}), \quad \forall i \in N, \tag{8}$$

$$z_i Q_M \leq Q_i, \quad \forall i \in N : (i, root) \in A, \tag{9}$$

$$\sum_{i\in N:(i,root)\in A} z_i = 1. \tag{10}$$

Equations (5) and (6) are used to linearize LT. Equation (7) state the value of Q_i. Equation (8) set Q_M to the maximum Q_i among all nodes i directly connected to the root. Equations (9) and (10) define the value of z_i. In particular, if $Q_i \neq Q_M$, then $z_i = 0$. Additionally, constraints (9) and (10) are satisfied only for $z_i = 1$ with $Q_i = Q_M$. The parameter H is a big number that can be set equal to $\sum_{i \in N} q_i$.

Constraints (7) and (9) are non linear due to the presence of the terms $Q_v x_{vj}$ and $z_i Q_M$, respectively. In general, a term of the form $\alpha\delta$, where α is a continuous variable and δ is an integer variable can be linearized by defining a new continuous variable $\phi = \alpha\delta$ and adding the following constraints

$$\phi \geq \alpha - H(1-\delta), \tag{11}$$

$$\phi \leq H\delta, \tag{12}$$

$$\phi \leq \alpha. \tag{13}$$

Thus, the linearized constraints (7) take the following form

$$Q_i \geq \sum_{(v,i)\in A} \hat{Q}_{vi} + q_i \quad \forall i \in N, \tag{14}$$

where $\hat{Q}_{vi} = Q_v x_{vi}$, and constraints (8) can be simplified in the following way

$$Q_M \geq \hat{Q}_{iroot}, \quad \forall (i, root) \in A. \tag{15}$$

To linearize the terms $z_i Q_M$ in constraints (9), we introduce non-negative variables z_i^M, $\forall i \in N : (i, root) \in A$. Constraints (9) are formulated as follows:

$$z_i^M \leq Q_i, \quad \forall i \in N : (i, root) \in A. \tag{16}$$

The set of constraints (1)–(3), (5), (6), (10)–(16) defines our problem as a mixed integer linear formulation.

The problem with objective function the minimization of the tree cost as a function of the quality of each link $l_{ij} = \frac{distance(i,j)}{max\_distance}$ between nodes i and j, i.e.

$$\sum_{(i,j)\in A} l_{ij} x_{ij}, \tag{17}$$

reduces to the spanning tree problem which is polynomially solvable. On the other hand, model (1)–(3), (5), (6), (10)–(16) with objective function the minimization of the schedule length LT is more difficult to solve due to the min-max function (4). In addition, the minimization of the cost does not guarantee the minimization of LT. Let P^{LT} be the problem where the schedule length is minimized and P^c be the problem with cost minimization. We propose two iterative approaches for solving P^{LT} which make use of problem P^c amended with further constraints at each iteration.

4 Logarithmic Search Approach

We argue to solve P^{LT} by making use of P^c, where we impose a limitation γ on the value of function (4). Thus, model (1)–(3), (5), (6), (10)–(16) with minimization of (17) is solved for a given value of γ with the constraint

$$LT \leq \gamma. \tag{18}$$

The main idea is to explore the criteria-space by solving problem $P(\gamma) = \{\min(17), \text{ s.t. } (1)\text{–}(3), (5), (6), (10)\text{–}(16), (18)\}$ considering different value of γ, providing a value β for function (4). We perform a logarithmic search given a lower bound lb and an upper bound ub on the value of β. Given $\gamma = \lceil \frac{ub+lb}{2} \rceil$, if problem $P(\gamma)$ is infeasible, then $lb = \gamma$, otherwise $ub = \beta$. It is worth observing that a valid lower bound on function (4) is given by $\sum_{i \in N} q_i$. In addition, a valid ub is given by the value of (4) calculated for the optimal solution of P^c. Given the iteration count k, the proposed strategy, referred in the sequel as LSA, is reported in Algorithm 1.

Algorithm 1: LSA

initialization: $k = 0$, $lb = \sum_{i \in N} q_i$, $\gamma^{(k)} = lb$, $\beta^{(k)} = \infty$, $\beta^* = \infty$

Solve $P(\gamma^{(0)})$;
if $P(\gamma^{(0)})$ *is feasible* **then**
  STOP, $\beta^* = LT[P(\gamma^{(0)})]$;
else
  Define $ub = LT[P^c]$;
  if $P(ub - 1)$ *is unfeasible* **then**
    STOP, $\beta^* = ub$;
  else
    $k+ = 1$;
    $\gamma^{(k)} = \lceil \frac{ub+lb}{2} \rceil$;
    while $ub - lb > 1$ **do**
      Solve $P(\gamma^{(k)})$;
      if $P(\gamma^{(k)})$ *is feasible* **then**
        $\beta^{(k)} = LT[P(\gamma^{(k)})]$;
        $ub = \beta^{(k)}$;
      else
        $lb = \gamma^{(k)}$;
      $k+ = 1$;
      $\gamma^{(k)} = \lceil \frac{ub+lb}{2} \rceil$;
$\beta^* = \beta^{(k)}$;

5 Logic Cuts Approach

P^c is solved by excluding the optimal trees determined in the previous iterations. In particular, given $X^k = \{(i,j) : x_{ij} = 1\}$ be the set of arcs in the optimal solution to P^c at iteration k, the following constraint is considered in the next iteration

$$\sum_{(i,j)\in X^k} x_{ij} \leq |X^k| - 1. \tag{19}$$

$P^c(k) = \{\min(17), \text{s.t. } (1)–(3), (5), (6), (10)–(16), (19)\}$ looks for the optimal $(k+1)$-spanning tree. In order to limit the number of iterations, we add to $P^c(k)$ the constraint $LT \leq LT[P^c(k-1)] - 1$ at each iteration. The steps of the proposed logic cuts approach, referred in the sequel as LCA, are depicted in Algorithm 2.

Algorithm 2: LCA

```
initialization: k = 0, X^0 = ∅, lb = Σ_{i∈N} q_i, β^(k) = ∞, β* = ∞
Solve P(lb);
if P(lb) is feasible then
    STOP, β* = lb;
else
    Define ub = LT[P^c];
    if P(ub − 1) is unfeasible then
        STOP, β* = ub;
    else
        Solve P^c(0)
        while P^c(k) is feasible do
            β^(k) = LT[P^c(k)];
            Construct set X^{k+1};
            k+ = 1;
            Solve P^c(k);
β* = β^(k);
```

6 Computational Results

In this section, we provide evidence on the bi-objective nature of the problem addressed and we give computational analysis on the behavior of the proposed solution approaches, that is, LSA and LCA. We compare our findings with the results of the RPL protocol using the—commonly used—expected transmission count (ETX) as the link quality metric. In ETX, we count the number of successive transmissions needed to deliver a number of packets without error at their destination.

We consider three scenarios with number of nodes equal to 20, 30, and 40. We randomly generate 25 instances for each scenario where the nodes are randomly

Table 1 Average computational results for problem P^c and P^{LT} and comparison with the RPL

	P^c			P^{LT}				RPL	
Nodes	Cost	LT	Time	Cost	LT	Time	#timeLimit	Cost	LT
20	11.29	59.52	0.09	14.14	47.60	461.18	6	14.59	53.48
30	15.08	99.20	0.23	21.01	63.44	1800.00	25	21.20	78.40
40	17.41	133.68	0.66	26.83	87.08	1800.00	25	26.83	104.60
AVG	14.59	97.47	0.33	20.66	66.04	1353.73	56	20.87	78.83

scattered on a terrain with $100 \times 100\,\text{m}^2$ size. The position of the *root* is also chosen randomly. We assume that there is a link between any two nodes if the signal power at the receiver is higher than the required sensitivity power [10]. We exclude links that achieve a failure rate higher than 1/3 over 10 measurements. Finally, we allow each node to send from 1 to 3 packets per slotframe. The model was solved by using CPLEX 12.51 and the tests were carried out on an Intel(R) core(TM) i7-4720HQ CPU 2.60 GHz 8 GB RAM machine.

6.1 *Model Evaluation*

Table 1 shows results obtained when solving model P^c and P^{LT} and those obtained with RLP. We report results for the three scenario with 20, 30, and 40 nodes averaged over the 25 instances. We report the cost given by expression (17) under column "cost", the value of expression (4) under column "*LT*", the execution time (in seconds) under column "time" for both P^c and P^{LT}, and the number of instances for which the model runs out of time limit for P^{LT}. We impose a time limit of 1800 s.

The results highlight the difficulty in solving P^{LT}. Using model P^{LT}, the solver CPLEX is able to provide the optimal solution within the time limit imposed for 19 instances over the 25 instances with 20 nodes (see Table 1). No optimal solution is provided for the instances with 30 and 40 nodes. The results, collected in Table 1, clearly underline the conflicting nature of cost and *LT*. Indeed, the average minimum value of *LT* is 66.04 and the corresponding cost is 20.66. Whereas, when the cost is minimized, the average value of *LT* grows to 97.47 with a cost of 14.59. The proposed model outperform the RPL protocol. The schedule length can be improved by up to 20% while achieving a similar routing tree cost.

6.2 *Comparison of the Proposed Algorithms*

Table 2 shows the average numerical results on the instances solved by both approaches. In particular, we report the number of instances not solved to optimality under

Table 2 Numerical results averaged over the instances solved to optimality by both LSA and LCA

	LSA					LCA				
Nodes	LT	Time	Iter	#timeLimit	#UB	LT	Time	Cuts	#timeLimit	#UB
20	45.86	62.60	0.77	3	25	45.86	53.49	1.05	2	25
30	59.35	89.62	0.00	8	20	59.35	111.22	0.00	8	21
40	81.67	179.24	0.00	15	18	81.67	182.56	0.00	16	16
AVG	62.29	110.49	0.26	26	63	62.29	115.76	0.35	26	62

column #timeLimit and the number of instances for which a solution is available under column #UB.

LCA outperforms LSA for the instances with 20 nodes. Indeed, the former is 1.17 times faster than the latter. LSA behaves the best for the instance with 30 nodes. Indeed, it is 1.24 times faster than LCA. The two approaches behave quite similar for the instances with 40 nodes. Overall, LCA is 1.05 times slower than LSA. Both approaches are able to solve to optimality the 65% of the instances despite the 25% of the available optimal solutions obtained by solving P^{LT} (see last row of column #timeLimit of Table 1).

7 Conclusions

In this paper, we address the problem of routing packets in IEEE802.15.4-TSCH networks. The problem has many applications in the field of Industrial Internet of Things. We formulate the problem as a mixed integer linear program with the aim of minimizing the schedule length. Two solution approaches are designed which iteratively solve restricted problems. The computational results highlight the difficulty of the problem, the conflicting nature of the minimization of the schedule length and the minimization of the cost, and the promising behavior of the proposed solution approaches for solving to optimality the problem.

References

1. Cruz, R.L., Santhanam, A.V.: Optimal routing, link scheduling and power control in multihop wireless networks. In: Twenty-Second Annual Joint Conference of the IEEE Computer and Communications Societies, INFOCOM 2003, vol. 1, pp. 702–711. IEEE (2003)
2. Karnik, A., Iyer, A., Rosenberg, C.: Throughput-optimal configuration of fixed wireless networks. IEEE/ACM Trans. Netw. (TON) **16**(5), 1161–1174 (2008)
3. Liu, F., Tsui, C.-Y., Zhang, Y.J.: Joint routing and sleep scheduling for lifetime maximization of wireless sensor networks. IEEE Trans. Wirel. Commun. **9**(7), 2258–2267 (2010)
4. Llorca, J., Sterle, C., Tulino, A.M., Choi, N., Sforza, A., Amideo, A.E.: Joint content-resource allocation in software defined virtual CDNs. In: IEEE International Conference on Communication Workshop, ICCW 2015, pp. 1839–1844 (2015)

5. Llorca, J., Tulino, A.M., Sforza, A., Sterle, C.: Optimal content distribution and multi-resource allocation in software defined virtual CDNs. In: Springer Proceedings in Mathematics and Statistics, ODS 2017, pp. 275–286 (2017)
6. Palattella, M.R., Accettura, N., Grieco, L.A., Boggia, G., Dohler, M., Engel, T.: On optimal scheduling in duty-cycled industrial iot applications using IEEE802.15.4e TSCH. IEEE Sens. J. **13**(10), 3655–3666 (2013)
7. Petrova, M., Riihijarvi, J., Mahonen, P., Labella, S.: Performance study of IEEE 802.15.4 using measurements and simulations. In: 2006 Wireless Communications and Networking Conference, WCNC 2006, vol. 1, pp. 487–492. IEEE (2006)
8. Saad, M.: Joint optimal routing and power allocation for spectral efficiency in multihop wireless networks. IEEE Trans. Wirel. Commun. **13**(5), 2530–2539 (2014)
9. Watteyne, T., Palattella, M., Grieco, L.: Using IEEE 802.15.4e time-slotted channel hopping (TSCH) in the internet of things (IoT): problem statement (2015)
10. Win, M.Z., Pinto, P.C., Shepp, L.A.: A mathematical theory of network interference and its applications. Proc. IEEE **97**(2), 205–230 (2009)
11. Winter, T., Thubert, P., Brandt, A., Hui, J., Kelsey, R., Levis, P., Pister, K., Struik, R., Vasseur, J.P., Alexander, R.: RPL: IPv6 routing protocol for low-power and lossy networks (2012)
12. Xu, L.D., He, W., Li, S.: Internet of things in industries: a survey. IEEE Trans. Ind. Inf. **10**(4), 2233–2243 (2014)

An Origin-Destination Based Parking Pricing Policy for Improving Equity in Urban Transportation

Mariano Gallo and Luca D'Acierno

Abstract In this paper we propose to optimise parking pricing fares in urban areas with the aim to improve transportation equity; the optimisation approach is applied to an origin-destination parking pricing policy that can differentiate the tariffs for each origin-destination pair, considering the difference in accessibility, in particular with public transport services. An optimisation model is implemented, and a solution algorithm is proposed. Model and algorithm are tested on the case study of Naples (Italy), where the quality of transit services is very different between zones and OD pairs; therefore, differentiating parking fares as a function of origin and destination of the trip may be very useful for rebalancing accessibilities among zones, aiming to improve transportation equity.

Keywords Transportation · Parking pricing · Equity · Optimisation Accessibility

1 Introduction

Parking pricing is widely used in almost all middle-large and large European cities. In some cases, it is adopted only for municipality cash reasons since this policy can collect money rapidly without significant monetary investments. A more important reason for its application is the promotion of transit systems use since the parking pricing increases the car use cost and, therefore, tends to move users towards public transport, even without further investments on the transit system.

M. Gallo (✉)
Dipartimento di Ingegneria, Università del Sannio, piazza Roma 21, 82100 Benevento, Italy
e-mail: gallo@unisannio.it

L. D'Acierno
Dipartimento di Ingegneria Civile, Edile e Ambientale, Università di Napoli Federico II, via Claudio 21, 80125 Naples, Italy
e-mail: luca.dacierno@unina.it

© Springer Nature Switzerland AG 2018

P. Daniele and L. Scrimali (eds.), *New Trends in Emerging Complex Real Life Problems*, AIRO Springer Series 1,
https://doi.org/10.1007/978-3-030-00473-6_27

The parking pricing, which belongs to the transport pricing policies that also include road pricing, is more diffused in urban areas than the road pricing since it is simpler to be adopted. However, this policy represents only a 'second best' approach to road user charging [1] since not all facilities may be priced.

A review of parking measures and policies can be found in [2, 3]; some case studies of road pricing and parking policies are reported in [4].

Usually, parking pricing policies are destination-based: the fare depends on the parking zone and, often, the more central the zone, the higher the fare. Moreover, in some cities, residents can park freely in their zone or paying a low-cost annual subscription. The destination-based approach can be not equitable if the transit system has significantly different levels of service among city zones; indeed, people who have the availability of good public transport services for their trips and, then, have an actual alternative to personal car, pay the same parking fare of people destined in the same zone who do not have the service or have a low-quality transit service. A more equitable policy should provide different fares on different origin-destination (OD) pairs, reducing parking costs for these disadvantaged people; parking fares, instead, should be higher for OD pairs that are well served by the public transport system. A policy based on OD parking pricing could improve equity of the whole system, acting on the accessibilities among OD pairs, since they depend both on transit service supply and parking pricing fares.

In this paper, we propose an OD parking pricing policy where the fares are optimised so to maximise equity in terms of origin-destination accessibility. An OD parking pricing policy was previously proposed in [5] with a different objective (the minimisation of society's global costs, regardless of equity), while an origin-destination taxi faring was proposed in [6], with the aim to improve equity of the transit systems. Equity in transportation is an important topic that, recently, has been studied by several researchers, such as, for instance, [7, 8].

The paper is organised as follows: Sect. 2 describes the problem and formulates an optimisation model; a solution algorithm is proposed in Sect. 3; numerical results on a real-scale case are summarised in Sect. 4; finally, Sect. 5 concludes.

2 Problem Description and Model Formulation

Our goal is the fare design of on-street public parking areas; moreover, we assume that the transportation demand is known and the multimode transportation supply (mass transit system and road system) is modelled.

Usually, parking fares are defined as a function only of the destination area (destination-based), regardless of the origin of the trip. In this paper, we propose to design parking fares considering both the origin (identified with the residence of the car owner) and the destination. In this case, all cars should be provided with an identification card indicating the residence zone of car owner (the same card usually used as the license for free parking in the residence zone), and the parking signs should show the different fares as a function of the origin zone.

To solve this problem, we propose an optimisation model where the decision variables are the OD parking fares (one for each OD pair), and the objective function should measure the transportation equity. In particular, we propose to use the same objective function proposed in [6], that is the variance of the logsum variables of the mode choice model divided by the car distance, for each OD pair. More in detail, we introduce the following terms:

$$s''_{OD}(\boldsymbol{V}_{OD}) = s'_{OD}(\boldsymbol{V}_{OD})/D^{car}{}_{OD} \qquad \forall\, OD \tag{1}$$

$$s'_{OD}(\boldsymbol{V}_{OD}) = s_{OD}(\boldsymbol{V}_{OD})/\theta \qquad \forall\, OD \tag{2}$$

$$s_{OD}(\boldsymbol{V}_{OD}) = \theta \ln \sum_{m} \exp(V^{m}{}_{OD}/\theta) \qquad \forall\, OD \tag{3}$$

where $\boldsymbol{V}_{OD}$ is the vector of systematic utilities referring to different modes m, $V^{m}{}_{OD}$, $D^{car}{}_{OD}$ is distance by car from zone O to zone D on the minimum path, $s'_{OD}(\boldsymbol{V}_{OD})$ is the logsum variable, θ is the parameter of the Logit model, $s_{OD}(\boldsymbol{V}_{OD})$ is the EMPU (*Expected Maximum Perceived Utility*) variable and Eq. (3) calculates the EMPU variable using a Logit model. Further details on EMPU variable, logsum variable, Logit model and systematic utilities can be found in [9]. The logsum variable is, then, directly proportional to the EMPU variable that is considered a measure of the accessibility [10] between O and D. In our problem, we have to divide it by θ for practical reasons (the parameter θ is not explicitly known since it is included in the coefficients of the Logit model and calibrated with them) and by the distance so to underline the effects of the quality of the connections between zones: two OD pairs with the same quality of connecting services will have the same value of $s''_{OD}(\boldsymbol{V}_{OD})$ independently on the distance.

The objective function is, then, given by $var(s''(\boldsymbol{V}_{OD}(\boldsymbol{y})))$; it is able to represent the transportation (in)equity: the lower the value of the objective function, the higher the equity (theoretically, it is equal to 0 if there is perfect equity). The constraints of the problem refer to minimum and maximum values of parking fares and to the discrete feature of them (indeed, even if theoretically these fares can be continuous variables, the actual applicability of the proposed policy requires a limited number of feasible fares). We consider as decision variable for an OD pair an integer number, y_{OD}, multiplied by a fixed fare value, *ffv*, that can be, for instance, assumed equal to 0.5 €/h or other fractions of the currency. Hence, the optimisation model may be formulated as follows:

$$\boldsymbol{y}^{opt} = \underset{\boldsymbol{y}}{\operatorname{argmin}}\ var(s''(\boldsymbol{V}_{OD}(\boldsymbol{y}))) \tag{4}$$

subject to

$$y_{OD} \text{ integer} \qquad \forall\, OD \tag{5}$$

$$0 \le y_{OD} \le y_{max} \qquad \forall\, OD \tag{6}$$

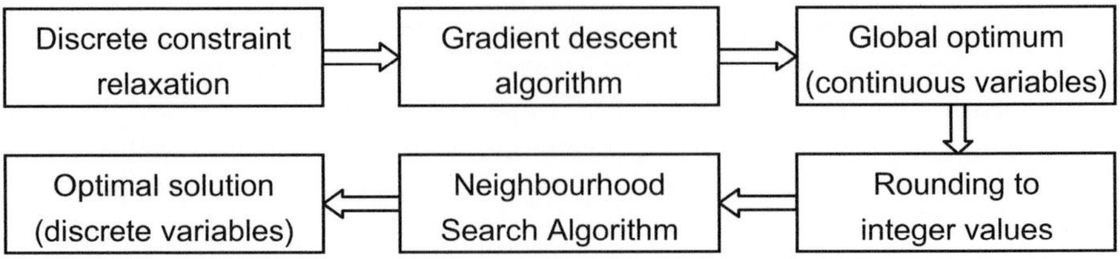

Fig. 1 Solution method

where $\boldsymbol{y}$ [$\boldsymbol{y}^{opt}$] is the [optimal] vector of decision variables, y_{OD}, one for each OD pair, y_{max} is the maximum value of the decision variable.

Analogously to [6], it is possible to state that objective function (4) is convex. Indeed, we can write:

$$var(s''(\boldsymbol{V}_{OD}(\boldsymbol{y}))) = \Sigma_{OD}(s''_{OD}(\boldsymbol{V}_{OD}(\boldsymbol{y})) - s''_M(\boldsymbol{V}_{OD}(\boldsymbol{y})))^2/(n_{OD} - 1)$$

which is convex, since the terms $(s''_{OD}(\boldsymbol{V}_{OD}(\boldsymbol{y})) - s''_M(\boldsymbol{V}_{OD}(\boldsymbol{y})))$ are strictly decreasing and, therefore, convex with the increase in a value of y_{OD}; the quadratic function of a convex function is convex and a linear combination of convex functions is a convex function as well. Therefore, the optimisation problem has only one local optimum that corresponds to the global one, if we assume continuous variables. Naturally, assuming variables as discrete, theoretically more than one solution may correspond to the optimum.

3 Solution Method

For solving the optimisation model (4)–(6), which is a non-linear discrete model, numerous methods and algorithms can be used; most of them are based on constraint relaxation (such as branch-and-bound, or Lagrangian relaxation) and heuristic rounding. Considering that the problem could be theoretically formulated with continuous variables (the discrete assumption is necessary only for implementing the policy in real-world cases) and that the objective function is convex, continuous and derivable under this assumption, we propose to solve first the continuous problem with a standard gradient algorithm. Successively, we round the solution to the nearest integer one and, finally, use a neighbourhood search (NS) method to identify the local optimal discrete solution that is nearest to the global optimal continuous one. In Fig. 1 the proposed solution method is reported.

The used NS method examines at each iteration the neighbourhood of the current solution and generates the next solution as the best one of all solutions belonging to the neighbourhood. Each neighbourhood contains all solutions obtained by changing the value of a variable, y_{OD}, increasing or decreasing its value (satisfying the constraints), and maintaining unaltered the other values. Therefore, in our case, a neighbourhood may contain at most $2 \cdot n_{OD}$ solutions. The algorithm ends when all solutions in the neighbourhood are worse than the current solution, which can be considered a local optimum.

4 Numerical Results

We tested the proposed approach on a real-scale case, the city of Naples (Italy) that has about one million inhabitants. The choice of this case study was driven by a feature that is necessary for developing the proposed policy: transit system is inequitable; indeed, some OD pairs are well connected by high-frequency metro or funicular lines, other ones are connected only with low-frequency services (rail or bus lines) and other ones are only marginally served or are not served at all by public transport services.

An essential point of the procedure is to identify the parking fare zones, which are zones of the city that will have the same fares; since the objective of the policy is to improve the equity, the fare zones are identified considering their similarity concerning transit supply. We partitioned the study area into 20 zones, as reported in Fig. 2a; these zones were obtained from the union of the 54 traffic zones used for simulating the transportation demand on the territory. Note that, the zone 20 is a suburban zone where currently there are not parking fares. Figure 2b–d report, respectively, the road network, the urban rail network and the bus lines; these elements were implemented in a multimodal transportation supply model with the software Omnitrans 6.0. This software allowed to obtain all data on performances of all transportation modes for each OD pair, that is pedestrian times, transit times (onboard, access/egress, waiting), transit transfers, car travel time and distance. Moreover, for each OD pair also the monetary transit costs were added, according to the current transit fare framework, as well as the car monetary costs as the sum of a travel cost (0.25 €/km) and a parking cost, depending on the parking fares, assuming an average parking duration. Main features of the case study are summarised in Table 1.

The all-mode demand matrix used in the test was the same used in [6] that has given acceptable results for a real-scale application. The mode choice model was adapted from [11] and calibrated for the city of Naples. Since this is a real-scale test and not a real test, we assumed this model to be valid, without performing further specifications and calibrations required in the case of a real application. The specification of the model, whose coefficient values are reported Table 2, is the following:

$$V^{car}{}_{OD} = \beta_b{}^{car} \cdot T^{car}{}_{OD} + \beta_{mc}{}^{car} \cdot C^{car}{}_{OD} + \beta_{park}{}^{car} \cdot C^{park}{}_{OD} + \beta_{ca} \cdot CarAv \quad \forall OD$$

$$\begin{aligned} V^{tran}{}_{OD} = {} & \beta_b{}^{tran} \cdot T^{tran\_b}{}_{OD} + \beta_{ped}{}^{tran} \cdot T^{tran\_ae}{}_{OD} + \beta_w \cdot T^{tran\_w}{}_{OD} + \\ & + \beta_{tr} \cdot N^{tran\_t}{}_{OD} + \beta_{mc}{}^{tran} \cdot C^{tran}{}_{OD} \end{aligned} \quad \forall OD$$

$$V^{ped}{}_{OD} = \beta_{ped} \cdot T^{ped}{}_{OD} \quad \forall OD$$

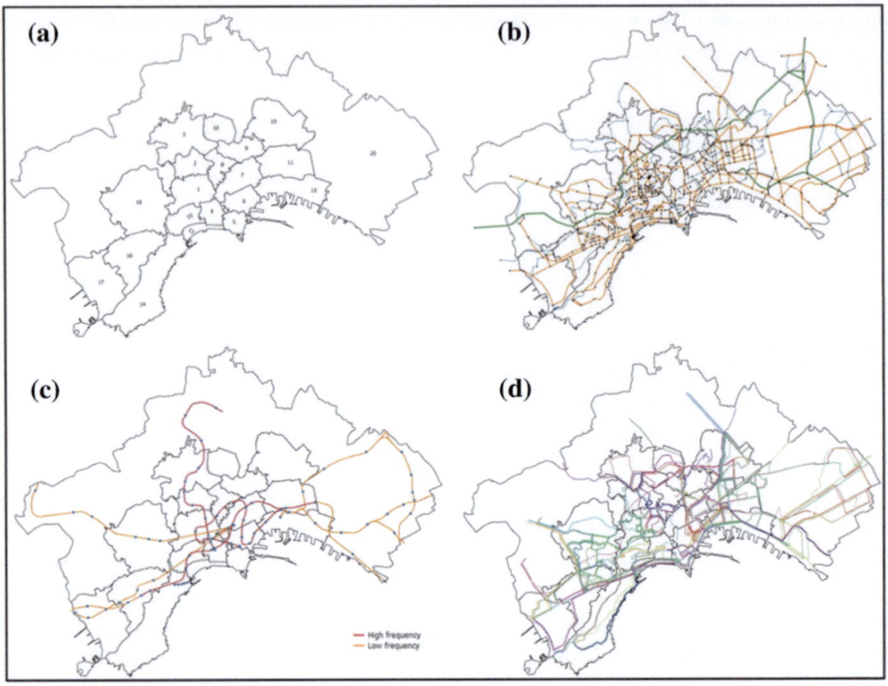

Fig. 2 Case study: **a** fare zones; **b** road network; **c** urban rail network; **d** bus lines

Table 1 Features of the case study

Inhabitants	962,003	Directed rail links	108
Traffic zones (internal centroids)	54	Rail segments	54
External centroids	16	Rail network length (km)	55.2
Fared zones	20	Rail lines (including funiculars)	12
Directed road links	1,774	Rail station (including funiculars)	60
Road segments	887	Bus lines	100
Road nodes	622	Main bus stops	370
Road network length (km)	335.7		

where ${V^{car}}_{OD}\left[{V^{tran}}_{OD}; {V^{ped}}_{OD}\right]$ is the systematic utility of car [(mass) transit; pedestrian] mode from zone O to zone D on the minimum path, ${T^{car}}_{OD}[{T^{tran\_b}}_{OD}; {T^{tran\_ae}}_{OD}; {T^{tran\_w}}_{OD}; {N^{tran\_t}}_{OD}]$ is the expected travel time by car [transit onboard travel time; access/egress transit time; waiting transit time; number

Table 2 Coefficients of the mode choice model

Attribute	Coefficient
${T^{car}}_{OD}, {T^{tran\_b}}_{OD}$	−1.02
${C^{car}}_{OD}, {C^{tran}}_{OD}$	−0.20
${C^{park}}_{OD}$	−0.40
CarAv	2.29
${T^{tran\_ae}}_{OD}, {T^{ped}}_{OD}$	−1.72
${T^{tran\_w}}_{OD}$	−2.57
${N^{tran\_t}}_{OD}$	−0.29

Table 3 Modal split

Mode	Model results (%)	ISTAT data (%)
Private motorised modes	55.6	48.33
Pedestrian	18.4	22.43
Transit modes	26.1	28.53
Other modes	–	0.72

of transfers] from zone O to zone D on the minimum transit path, ${C^{car}}_{OD}\left[{C^{tran}}_{OD}\right]$ is the expected monetary cost by car (only travel costs) [(mass) transit system] from zone O to zone D on the minimum path, ${C^{park}}_{OD}$ is the parking cost on the OD pair, *CarAv* is the average number of cars available per family.

Applying this model, considering a car travel cost of 0.25 €/km, an average parking duration of 4 h and a parking cost of 1.5 €/h on all zones except for zone 20 where the parking is not fared, the model generates the modal split reported in Table 3; the same table reports the modal split in Naples obtained by the ISTAT data [12] for systematic (i.e. home-work and home-school) trips. These results can be accepted for a real-scale test case.

In the application, we have compared results corresponding to three different scenarios: (a) the starting scenario, described before, (b) a scenario without parking fares and (c) the optimised scenario. Scenario (c) is obtained by applying the model (1)–(3) solved with the algorithm described in Sect. 3. The objective function value was equal to 0.375628 in the starting scenario, 0.135585 in the non-fared scenario and 0.096145 in the optimised scenario. Figure 3 reports the values of $s''_{OD}(V_{OD})$ for each internal OD pair for these three scenarios.

5 Conclusions

This paper proposes an origin-destination based parking pricing policy aiming to increase equity in transportation. The principle of the policy is that the OD pairs that are not served with a good quality public transport system should pay less

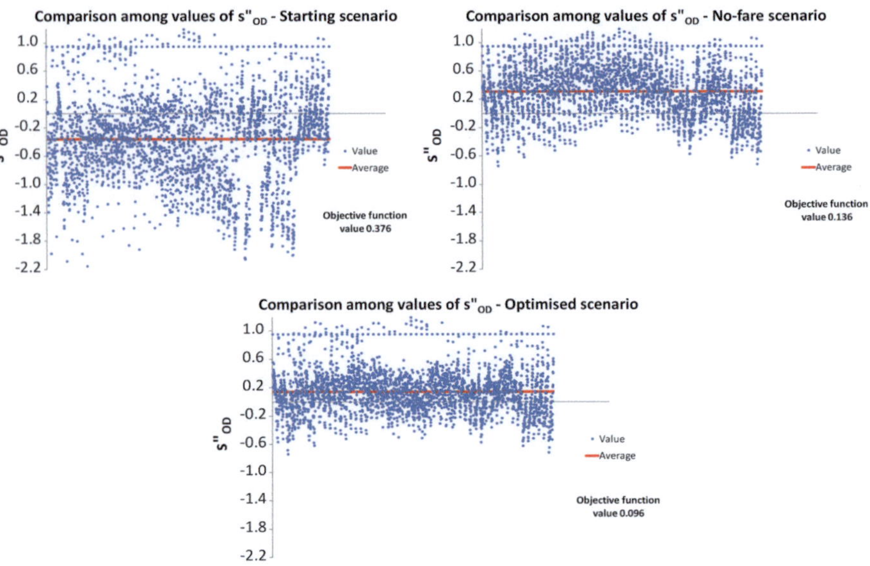

Fig. 3 Values of $s''_{OD}(\boldsymbol{V}_{OD})$ for each internal OD pair

for parking, considering that the whole society finances (mass-)transit systems. We propose an optimisation model and a solution algorithm for implementing the policy; the proposed approach is tested on a real-scale case study giving promising results.

Future research will be addressed to test the proposal considering different mode choice models, focusing on the elasticities of the modal split with car parking costs, to evaluate other impacts of the policy, such as consumption and emissions, and to optimise also the zoning phase besides the fares.

References

1. Verhoef, E.T.: Second-best congestion pricing in general networks. Heuristic algorithms for finding second-best optimal toll levels and toll points. Transp. Res. B **36**, 707–729 (2002)
2. Palmer, D., Ferris, C.: Parking measures and policies research review. Transport Research Laboratory—TRL. Project Record PPRO 4/45/4 (2010)
3. Inci, E.: A review of the economics of parking. Econ. Transp. **4**, 501–563 (2015)
4. Jansson, J.O.: Road pricing and parking policy. Res. Transp. Econ. **29**, 346–353 (2010)
5. D'Acierno, L., Gallo, M., Montella, B.: Optimisation models for the urban parking pricing problem. Transp. Policy **13**, 34–48 (2006)
6. Gallo, M.: Improving equity of urban transit systems with the adoption of origin-destination based taxi fares. Socio-Econ. Plann. Sci. (2018) (in press)
7. Jones, P., Lucas, K.: The social consequence of transport decision-making: clarifying concepts, synthesising knowledge and assessing implications. J. Transp. Geogr. **21**, 4–16 (2012)
8. Caggiani, L., Camporeale, R., Ottomanelli, M.: Facing equity in transportation network design problem: a flexible constraints based model. Transp. Policy **55**, 9–17 (2017)

9. Cascetta, E.: Transportation systems analysis: models and applications. Springer, New York, USA (2009)
10. Ben-Akiva, M., Lerman, S.: Discrete Choice Analysis. MIT Press, Cambridge, USA (1985)
11. Cascetta, E.: Modelli per i sistemi di trasporto. Teoria e applicazioni, UTET, Torino, Italy (2006)
12. ISTAT I dati del censimento (2011). http://dati-censimentopopolazione.istat.it

Bayesian Optimization for Full Waveform Inversion

Bruno G. Galuzzi, Riccardo Perego, Antonio Candelieri and Francesco Archetti

Abstract Full Waveform Inversion (FWI) is a computational method to estimate the physical features of Earth subsurface from seismic data, leading to the minimization of a misfit function between the observed data and the predicted ones, computed by solving the wave equation numerically. This function is usually multimodal, and any gradient-based method would likely get trapped in a local minimum, without a suitable starting point in the basin of attraction of the global minimum. The starting point of the gradient procedure can be provided by an exploratory stage performed by an algorithm incorporating random elements. In this paper, we show that Bayesian Optimization (BO) can offer an effective way to structure this exploration phase. The computational results on a 2D acoustic FWI benchmark problem show that BO can provide a starting point in the parameter space from which the gradient-based method converges to the global optimum.

Keywords Bayesian optimization · Full waveform inversion · Inversion problems

B. G. Galuzzi (✉) · R. Perego · A. Candelieri · F. Archetti
Department of Computer Science, Systems and Communications,
University of Milano-Bicocca, viale Sarca 336, 20125 Milan, Italy
e-mail: brunogaluzzi@gmail.com; bruno.galuzzi@unimib.it

R. Perego
e-mail: r.perego12@campus.unimib.it

A. Candelieri
e-mail: antonio.candelieri@unimib.it

F. Archetti
Consorzio Milano-Ricerche, via Roberto Cozzi, 53, 20126 Milan, Italy
e-mail: francesco.archetti@unimib.it

© Springer Nature Switzerland AG 2018

P. Daniele and L. Scrimali (eds.), *New Trends in Emerging Complex Real Life Problems*, AIRO Springer Series 1,
https://doi.org/10.1007/978-3-030-00473-6_28

1 Bayesian Optimization

Bayesian optimization (BO) [11] is a suitable global optimization algorithm to find a global minimum x^* of a black-box, usually expensive-to-evaluate, objective function $f(x)$, where $x \in X \subset R^d$ is a point in a d-dimensional bounded-box space X.

In BO the objective function is modelled as a realization of a stochastic process, typically a Gaussian Process (GP) on a probability space (Ω, Σ, P). A GP, which defines a prior distribution over the function f, is completely specified by its mean $\mu(x) : X \rightarrow R$ and a definite positive covariance function $k(x, x') : X^2 \rightarrow R$,

$$f(x) \approx GP(\mu(x); k(x, x')) \tag{1}$$

The BO algorithm starts with an initial set of k points $\{x_i\}_{i=1}^k \in X$ and the associated observations $\{y_i\}_{i=1}^k$, with $y_i = f(x_i)$. At each iteration $t \in \{k+1, N\}$, the GP prior is updated using the Bayes rule, to obtain posterior distribution conditioned on the current training set $S_t = \{(x_i, y_i)\}_{i=1}^t$ containing the past evaluated points and observations. For any point $x \in X$, the posterior mean $\overline{\mu}_t(x)$ and the posterior variance $\overline{\sigma}_t^2(x)$ of the GP, conditioned on S_t, are known in closed-form:

$$\mu_t(x) = K(X_t, x)^t \left[K(X_t, X_t) + \lambda I\right]^{-1} Y_t \tag{2}$$

$$\sigma_t^2(x) = k(x, x) - K(X_t, x)^t \left[K(X_t, X_t) + \lambda I\right]^{-1} K(X_t, x) \tag{3}$$

where $K(X_t, X_t)$ is the $t \times t$ matrix whose (ij)th entry is $k(x_i, x_j)$, $K(X_t, x)$ (respectively Y_t) is the $t \times 1$ vector whose ith entry is $k(x_i, x)$ (respectively y_i) and λ is the noise variance. A new point x_{t+1} is then selected and evaluated to provide an observation $y_{t+1} = f(x_{t+1})$. This new pair $\{(x_{t+1}, y_{t+1})\}$ is added to the current training set S_t, to define the training set for the next iteration $S_{t+1} = S_t \cup \{(x_{t+1}, y_{t+1})\}$.

The new point to evaluate is selected by solving an auxiliary optimization problem, typically of the form:

$$x_{t+1} = \arg\max U_t(x; S_t) \tag{4}$$

where U_t is an acquisition function to maximize. The rationale is that, because the optimization run-time or cost is dominated by the evaluation of the expensive objective function f, time and effort should be dedicated to choosing a promising point to evaluate, by solving the auxiliary problem. Solving this auxiliary problem does not involve the evaluation of the expensive objective function f, but only the posterior quantities of the GP and, thus, is considered cheap.

In this paper we focus on two of the most used acquisition function: the expected improvement (EI) [12], and the confidence bound (CB) [1] (lower/upper confidence bound, LCB/UCB, for minimization and maximization problems, respectively). EI is defined as follows:

$$EI_t(x; S_t) = \left(f^+ - \mu_t(x)\right) \Phi\left(\frac{f^+ - \mu_t(x)}{\sigma_t(x)}\right) + \sigma_t(x) \cdot N\left(\frac{f^+ - \mu_t(x)}{\sigma_t(x)}\right) \tag{5}$$

where, considering a minimization problem, $f^+ = \min_{x_i \in X_{1:t}} f(x_i)$ is the best value found after t evaluations (aka best seen), and $\Phi(\cdot)$ and $N(\cdot)$ are the normal cumulative distribution and the density probability function, respectively. The LCB formula is:

$$LCB_t(x; S_t) = \mu_t(x) - k\sigma_t(x) \tag{6}$$

where $k \geq 0$ is a parameter to manage the exploration/exploitation trade-off: a larger k drives exploration. In this paper we investigate the application of BO to seismic inversion problems comparing the EI and CB acquisition functions. An example of its application in the model inversion optimization problems, in the field of haemodynamic, is in [15].

2 Introduction to Seismic Inversion Problems

The estimation of the geological properties of the subsurface can be obtained by means of seismic acquisition (Fig. 1a), in which artificially induced seismic waves, created by an impulse source, propagate through the subsurface. The receivers, distributed on the surface along a line ($2D$ seismic) detect the returning waves and measure the arrival times and the amplitudes of the waves at different distances, or offset. The seismic data are organized in a seismograms $d_{obs}(t, x_r)$, where $t \in [0, T]$ is the recording time and $\{x_r\}_{r=1}^{nr} \subset R^2$ is the set of receivers locations. Figure 1b shows a seismogram acquired during a marine seismic acquisition.

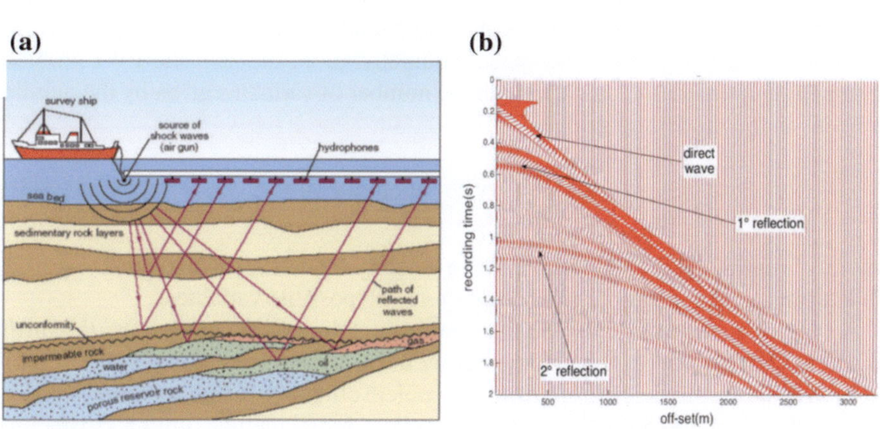

Fig. 1 **a** Example of a marine seismic acquisition, taken from [18]. **b** Example of a marine seismogram

A seismic inversion problem [7], in case of acoustic approximation, consists of the estimation of an acoustic velocity subsurface model $v_p(\mathbf{x})$, that explains the events observed in a seismogram (reflections, refractions), and can be formulated as a minimization problem

$$v^* = \arg\min_{v \in V} F(v) \tag{7}$$

with V the set of possible acoustic velocity models, and $F(v) \geq 0$ a misfit function

$$F(v) = ||d_{obs} - d_{pred}(v)||, \tag{8}$$

measuring the difference between the observed and the predicted seismograms $d_{pred}(v)$, computed by means of some seismic modelling algorithm.

In case of $2D$ acoustic approximation, the generation and propagation of seismic waves is modelled by the $2D$ acoustic wave equation [13]:

$$\ddot{p}(\mathbf{x}, t) - v(\mathbf{x})^2 \Delta p(\mathbf{x}, t) = \delta(\mathbf{x} - \mathbf{x}_0)s(t) \tag{9}$$

where $t \in [0, T]$ is the recording time, $\mathbf{x} = (x, z) \in D \subset R^2$ is a bi-dimensional space domain, p is the acoustic pressure of the wave, x_0 is the location of the source, and $s(t)$ is the seismic source. The predicted seismograms d_{pred} correspond to the restriction of the solution of the acoustic wave equation to the receivers locations.

The solution of the wave equation can be obtained using an explicit finite-difference (FD) scheme, where the space domain D is sampled through a uniform grid spacing dx, along the horizontal and vertical direction, obtaining a regular grid D based on $nx \cdot nz$ grid nodes, with $i = 1, \ldots, nx \cdot nz$, and with x_i, obtained scrolling the grid nodes along the rows.

It is important to remark that each evaluation of the misfit function is computationally expensive. Indeed, it requires the solution of the wave equation, which mandates for relatively small values of space sampling dx and time sampling dt [5]. According to a FD grid for the space parametrization, the acoustic FWI problem becomes an optimization problem with the number of variables given by the number of nodes in the modelling grid

$$\arg\min_{v \in V} F(v) \approx \arg\min_{v \in V^i} F(v^i) \tag{10}$$

where V^i represents the set of the P-wave velocity models discretized on the grid D. This means an optimization problem with $nx \cdot nz$ possible variables.

An important aspect of FWI is that the computation of the gradient of the misfit function, $\nabla_v f(v)$, can be done efficiently by means of the adjoint method [16]. This fact allows us to solve such minimization problem efficiently using an iterative procedure, updating an initial model v_0 with a gradient-based method, until a satisfactory match between the observed and the predicted data is obtained. However, as already noted in [2], the misfit functions are characterized by the presence of multiple local minima, and such local optimization approach will converge to a local minimum if the used starting model is not in the basin of attraction of the global minimum.

Global optimization algorithms can estimate such starting model, and meta-heuristic optimization methods have been already proposed for FWI, specifically Genetic Algorithms [10, 17] and Simulated Annealing [6, 9]. In this paper we propose BO as global optimization approach to identify a promising starting model for a synthetic 2D acoustic FWI benchmark problem, in the field of seismic exploration.

3 The Marmousi Benchmark

This benchmark problem consists of the estimation of the acoustic velocity model of Fig. 2a, that represents the upper central part of the Marmousi model [19], from a FWI procedure on a set of synthetic seismograms. This model contains 192 and 48 grid points in the z-and x-direction, respectively, with a grid spacing of 24 m. The first two rows of the modelling grid represent the water layer, whose velocity and depth are considered known and fixed a priori. We considered 16 seismograms, recorded by a spread of 192 receivers, equally spaced 24 m. Both the sources and the receivers are at a depth of 24 m, and the recording time is $T = 4$ s. The synthetic seismograms are obtained by solving the acoustic wave equation (9), using an efficient FD scheme, whose details of the implementation can be found in [8]. As misfit function, we used the sum of the all the L^2-norm difference between the observed and the synthetic seismograms, whereas as local optimization algorithm for FWI, we used the conjugate gradient method [14], one of most common gradient method to solve FWI [7].

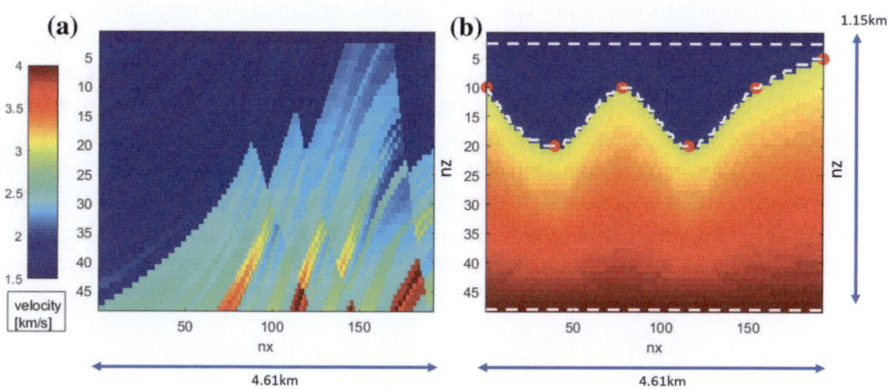

Fig. 2 **a** The portion of the Marmousi velocity model used as true model for the FWI procedure. **b** An example velocity model derived from the proposed sparse parametrization technique

4 Experimental Setting and Results

The starting model is estimated by means of BO using a sparse parameterization technique to reduce the number of parameters of modelling grid, formed by $192 \cdot 48$ grid nodes, to only 10 parameters, using a set of three interfaces and four velocities (Fig. 2b), as described in [6]. The first and the third interfaces are associated with the seabed (between 2nd and the 3rd row of the modelling grid) and the bottom of the model (situated at the 48th row) and are considered flat. The second interface represents a possible velocity contrast between the seabed and the bottom of the model and is represented by a set of six nodes $\{(x_k, z_k)\}_{k=1}^{6}$ (the red point in Fig. 2b) with x_k evenly distributed and fixed along the x-direction and z_k that can range along the z-direction. We use a non-oscillatory spline to interpolate the interface across the nodes (the dot white spline of Fig. 2b). The four velocities prescribed the velocity just below the first interface, just above the second interface, just below the second interface, and just above the third interface, respectively. The velocities of the grid nodes between interfaces are obtained using linear interpolation in the vertical direction.

The overall number of variables needed for the parametrization is 10: a real variable for each one of the 4 velocities and 6 discrete variables for the z_k components of the nodes at the second interface. Velocities are in the range 1.5–4 km/s, while the range for the 6 discrete variables is between 4 and 47. An overall number of 1000 function evaluations was fixed for each experiment, with 10 independent runs for each acquisition function (i.e., EI and LCB).

Figure 3a, b show the best velocity models obtained for EI and LCB, respectively. The first model obtains a value of the misfit function of 1868.1, whereas the second one obtains a value of 1723.5. These two models have been used as different starting points for the local optimization procedure on the modelling grid. Figure 3c, d show the corresponding velocity models obtained by the local optimization, after 1000 iterations of conjugate gradient. The final models obtained are very similar to the actual one, except in some areas near the lateral and the bottom boundaries, where, however, the seismic illumination is poor. The main result is that the starting models estimated by the BO, independently by the acquisition function used (EI vs. LCB), can be considered quite near the basin of attraction of the global minimum of the misfit function, corresponding to the actual model. However, the differences between the two starting models obtained at the end of the BO, respectively for EI (Fig. 3c) and LCB (Fig. 3d), lead to slightly different final models at the end of the entire procedure, with the lower misfit function value for LCB of 9.4 with respect to a value of 57.7 for EI.

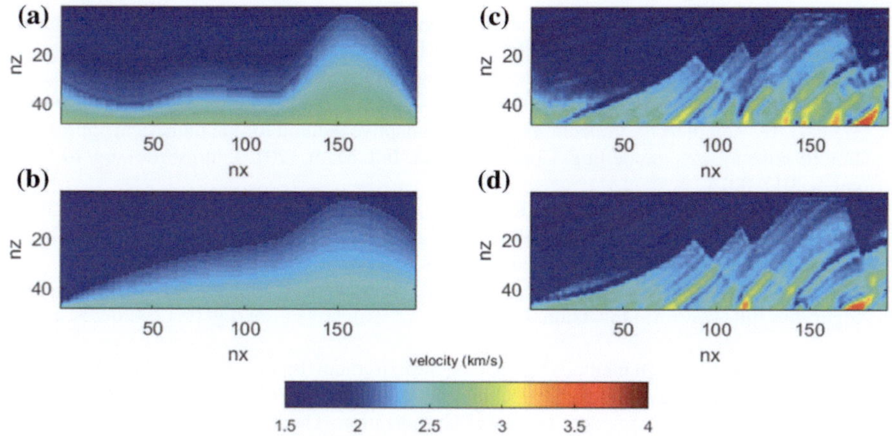

Fig. 3 **a** Best velocity model obtained by EI. **b** Best velocity model obtained by LCB. **c** Final velocity model obtained by the local optimization using (**a**) as starting model. **d** Final velocity model obtained by the local optimization using (**b**) as starting model

5 Conclusions

According to the emerging interest on BO, to solve black-box and expensive optimization processes [3, 4], we have proposed the BO as a global optimization algorithm for the estimation of a promising starting model for FWI. We considered two alternative acquisition functions for BO and test them on a 2D acoustic FWI benchmark problem, namely the Marmousi model. The low error between the actual and the final models obtained, independently on the acquisition function, makes the proposed approach well suited for the seismic inversion problems, such as FWI, usually characterized by highly non-linearity, multiple local minima, and an expensive-to-evaluate misfit function.

References

1. Auer, P.: Using confidence bounds for exploitation-exploration trade-off. J. Mach. Learn. Res. (2002). https://doi.org/10.1162/153244303321897663
2. Beydon, W.B.: First born and rytov approximations: modeling and inversion conditions in a canonical example. J. Acoust. Soc. Am. (1988). https://doi.org/10.1121/1.396537
3. Candelieri, A., Giordani, I., Archetti, F., Barkalov, K., Meyerov, I., Polovinkin, A., Sysoyev, A., Zolotykh, N.: Tuning hyperparameters of a SVM-based water demand forecasting system through parallel global optimization. Comput. Oper. Res. (2018) (Article in Press)
4. Candelieri, A., Perego, R., Archetti, F.: Bayesian optimization of pump operations in water distribution systems. J. Glob. Optim. (Article in Press)
5. Courant, R., Friedrichs, K., Lewy, H.: On the partial difference equations of mathematical physics. IBM J. (1967). https://doi.org/10.1147/rd.112.0215

6. Datta, D., Sen, M.K.: Estimating a starting model for full-waveform inversion. Geophysics (2016). https://doi.org/10.1190/geo2015-0339.1
7. Fichtner, A.: Full Seismic Waveform Modelling and Inversion. Springer Science & Business Media (2010). https://doi.org/10.1007/978-3-642-15807-0
8. Galuzzi, B., Zampieri, E., Stucchi, E.: A Local adaptive method for the numerical approximation in seismic wave modelling. Commun. Appl. Ind. Math. (2017). https://doi.org/10.1515/caim-2017-0014
9. Galuzzi, B., Zampieri, E., Stucchi, E.: Global optimization procedure to estimate a starting velocity model for local Full Waveform Inversion. Optimi. Decis. Sci. Methodol. Appl. (2017). https://doi.org/10.1007/978-3-319-67308-0_18
10. Mazzotti, A., Bienati, N., Stucchi, E., Tognarelli, A., Aleardi, M., Sajeva, A.: Two-grid genetic algorithm full-waveform inversion. Lead. Edge. (2016). https://doi.org/10.1190/tle35121068.1
11. Močkus, J.: On Bayesian methods for seeking the extremum. In: Optimization Techniques IFIP Technical Conference. Springer (1975). https://doi.org/10.1007/978-3-662-38527-2_55
12. Močkus, J.: Bayesian Approach to Global Optimization: Theory and Applications. Springer Netherlands (1989). https://doi.org/10.1007/978-94-009-0909-0
13. Morton, K.W., Mayers, D.F.: Numerical Solution of Partial Differential Equations: An Introduction. Cambridge University Press, Cambridge (2005)
14. Nocedal, J., Wright, S.: Numerical Optimization. Springer (2006). https://doi.org/10.1007/978-0-387-40065-5
15. Perdikaris, P., Karniadakis, G.: Model inversion via multi-fidelity Bayesian optimization: a new paradigm for parameter estimation in haemodynamics, and beyond. J. R. Soc. Interface. (2016). https://doi.org/10.1098/rsif.2015.1107
16. Plessix, R.: A review of the adjoint-state method for computing the gradient of a functional with geophysical applications. Geophys J. Int. (2006). https://doi.org/10.1111/j.1365-246X.2006.02978.x
17. Sajeva, A., Aleardi, M., Stucchi, E., Bienati, N., Mazzotti, A.: Estimation of acoustic macro models using a genetic full-waveform inversion: applications to the Marmousi model. Geophysics (2016). https://doi.org/10.1190/geo2015-0198.1
18. The Open University, Walton Hall, Milton Keynes, MK7 6*AA* [Online]. http://www.open.edu/openlearn/
19. Versteeg, R.: The Marmousi experience: velocity model determination on a synthetic complex data set. Lead. Edge. (1994). https://doi.org/10.1190/1.1437051

A Decomposition-Based Heuristic for the Truck Scheduling Problem in a Cross-Docking Terminal

Manlio Gaudioso, M. Flavia Monaco and Marcello Sammarra

Abstract We consider the truck scheduling problem at a cross docking terminal with many inbound and outbound doors, under the assumption of constant handling time for all the trucks, the objective being to minimize the completion time of the whole process. We propose a mathematical model together with a Lagrangian Relaxation scheme. We discuss the structural properties of the relaxed problem and derive a Lagrangian heuristic able to compute, at the same time, good feasible solutions and increasing lower bounds. The numerical results show that the Lagrangian decomposition is a promising approach to the solution of such problems.

Keywords Lagrangian relaxation · Scheduling · Heuristics

1 Introduction

A cross docking terminal in a distribution network is a transhipment node, where inbound loads have to be unloaded from a set of trucks, then scanned, sorted on the basis of the customer demands, moved across the dock, and finally loaded onto the outbound trucks, for immediate delivery. The unloading and loading phases are performed at the terminal doors, which are assumed to be able to process one truck at a time; suitable equipment and skilled work teams operate at the doors. The main problem to face consists in defining the unloading and loading sequences for trucks, so as to synchronize at best the two phases, and to avoid, or at least to minimize, the storage need. The cross docking problems may differ from each other on the basis of

M. Gaudioso · M. F. Monaco
DIMES, Universitá della Calabria, Rende, Italy
e-mail: gaudioso@dimes.unical.it

M. F. Monaco
e-mail: monaco@dimes.unical.it

M. Sammarra (✉)
ICAR, Consiglio Nazionale delle Ricerche, Rende, Italy
e-mail: sammarra@icar.cnr.it

© Springer Nature Switzerland AG 2018

P. Daniele and L. Scrimali (eds.), *New Trends in Emerging Complex Real Life Problems*, AIRO Springer Series 1,
https://doi.org/10.1007/978-3-030-00473-6_29

organizational features of the terminals (one/many doors, absence/presence of temporary storage areas), information assumed to be available (quantity and typology of goods to be unloaded/loaded, unloading/loading/moving times), specific constraints, and objectives to be considered in the decisional process (see [1, 2, 4] for a comprehensive survey of the literature). Here we consider the truck scheduling problem at a cross docking terminal with many inbound and outbound doors (Multi-Gate Cross Docking problem—MGCD), under the assumption of constant handling time for all the trucks, the objective being to minimize the completion time of the whole process. The outline of the paper is the following: in Sect. 2 we propose a mathematical model; in Sect. 3 we discuss the properties of a Lagrangian Relaxation of the problem. In Sect. 4 we derive a Lagrangian heuristic able to compute, at the same time, good feasible solutions and increasing lower bounds. The proposed algorithm is an extension to the Multi-Gate case of the one presented in [3] for the Single-Gate cross docking problem. The numerical experience is reported in Sect. 5. Conclusions are drawn in Sect. 6.

2 The MGCD Model

To derive the mathematical model for the MGCD we will make use of the notation defined in Table 1. As in [3] we assume that: all trucks are ready at the beginning of the planning horizon, all of them require the same processing time, the transhipment time is negligible, and preemption is not allowed. As a consequence of the above assumptions, we can consider a discretized planning horizon, where each time-slot is the processing time of a truck. Defining the decision variables:

- $x_{ik}^g = 1$ if the truck $i \in I$ is unloaded at gate $g \in G^I$ in the time-slot $k \in K$, 0 otherwise;
- $y_{jh}^g = 1$ if the truck $j \in J$ is loaded at gate $g \in G^O$ in the time-slot $h \in H$, 0 otherwise;
- C_M the makespan;

Table 1 Notation

Name	Definition
G^I	Set of inbound gates ($\|G^I\| = g^I$)
G^O	Set of outbound gates ($\|G^O\| = g^O$)
I	Set of inbound trucks ($\|I\| = n$)
J	Set of outbound trucks ($\|J\| = m$)
$I_j \subseteq I$	Set of inbound trucks supplying the outbound truck $j \in J$ ($\|I_j\| \geq 1$)
$J_i \subseteq J$	Set of outbound trucks supplied by the inbound truck $i \in I$ ($\|J_i\| \geq 1$)
K	Set of time-slots for the inbound truck service ($\|K\| = n$)
H	Set of time-slots for the outbound truck service ($\|H\| = L \geq m + n$)

the MGCD problem we address can be modeled as follows:

$$\min\ C_M \tag{1}$$

$$\sum_{g\in G^I}\sum_{k\in K} x_{ik}^g = 1 \qquad \forall i \in I \tag{2}$$

$$\sum_{i\in I} x_{ik}^g \le 1 \qquad \forall k \in K,\ g \in G^I \tag{3}$$

$$\sum_{g\in G^O}\sum_{h\in H} y_{jh}^g = 1 \qquad \forall j \in J \tag{4}$$

$$\sum_{j\in J} y_{jh}^g \le 1 \qquad \forall h \in H,\ g \in G^O \tag{5}$$

$$\sum_{g\in G^O}\sum_{h\in H} h y_{jh}^g - \sum_{g\in G^I}\sum_{k\in K} k x_{ik}^g \ge 1 \qquad \forall i \in I, j \in J_i \tag{6}$$

$$C_M - \sum_{g\in G^O}\sum_{h\in H} h y_{jh}^g \ge 0 \qquad \forall j \in J \tag{7}$$

$$x_{ik}^g \in \{0, 1\} \qquad \forall i \in I, k \in K, g \in G^I \tag{8}$$

$$y_{jh}^g \in \{0, 1\} \qquad \forall j \in J, h \in H, g \in G^O \tag{9}$$

In the above model, (2)–(3) are the assignment constraints of each inbound truck to an inbound gate and a time-slot; (4)–(5) are the analogous constraints for the outbound trucks; constraints (6) impose the precedence relation between the processing of inbound and outbound trucks exchanging some goods. These constraints are instrumental to define the freight consolidation for the outbound trucks. Finally, constraints (7) define the objective function (1). Relaxing the coupling constraints (6) in a Lagrangian fashion, the resulting problem will decompose in two assignment-like subproblems.

3 The Lagrangian Relaxation of MGCD

Let λ be the vector of Lagrangian multipliers associated to constraints (6), with components $\lambda_{ij} \ge 0,\ \forall i \in I,\ j \in J_i$; defining

$$\rho_i = \rho_i(\lambda) = \sum_{j\in J_i} \lambda_{ij} \quad \forall i \in I, \qquad \rho = (\rho_1, \dots, \rho_n)^\top$$

$$\sigma_j = \sigma_j(\lambda) = \sum_{i\in I_j} \lambda_{ij} \quad \forall j \in J, \qquad \sigma = (\sigma_1, \dots, \sigma_m)^\top$$

$$s = s(\lambda) = \sum_{i \in I} \sum_{j \in J_i} \lambda_{ij} = \sum_{i \in I} \rho_i = \sum_{j \in J} \sigma_j$$

it is easy to verify that the Lagrangian Relaxation of the MGCD problem becomes:

$$Z_{LR}(\lambda) = s + Z^I_{LR}(\rho) + Z^O_{LR}(\sigma) \tag{10}$$

where

$$PI(\rho) \quad \begin{cases} Z^I_{LR}(\rho) = \min \sum_{i \in I} \sum_{g \in G^I} \sum_{k \in K} k\rho_i x^g_{ik} \\ \qquad s.t.(2), (3), (8) \end{cases} \tag{11}$$

$$PO(\sigma) \quad \begin{cases} Z^O_{LR}(\sigma) = \min \left(C_M - \sum_{j \in J} \sum_{g \in G^O} \sum_{h \in H} h\sigma_j y^g_{jh} \right) \\ \qquad s.t.(4), (5), (7), (9) \end{cases} \tag{12}$$

$PI(\rho)$ and $PO(\sigma)$ are assignment like problems whose cost coefficients, $c_{ik} = k\rho_i$ and $d_{jh} = h\sigma_j$ respectively, are independent from the gates and satisfy:

$$c_{i1} \leq c_{i2} \leq \cdots \leq c_{in} \quad \forall i \in I$$

$$d_{j1} \leq d_{j2} \leq \cdots \leq d_{jL} \quad \forall j \in J$$

Thanks to the above observation, it is possible to prove that the optimal solutions of both subproblems enjoy some useful properties from which simple solution algorithms can be derived. The steps of these algorithms are detailed in Table 2. Observe that both of them act as a sort of "wrap-around" algorithm, where the trucks are first ordered and then sequentially assigned to the gates in such a way that the gate workload is as balanced as possible.

4 A Lagrangian Heuristic for MGCD

For a given vector of Lagrangian multipliers $\bar{\lambda}$, the optimal solutions $x^*(\bar{\rho})$ of PI and $(y^*(\bar{\sigma}), C^*_M(\bar{\sigma}))$ of PO are independent schedules of inbound/outbound trucks at the corresponding gates. If such schedules satisfy the relaxed constraints, then $x = x^*(\bar{\rho})$, $y = y^*(\bar{\sigma})$ are also feasible schedules for the MGCD problem with a makespan $C_M = C^*_M(\bar{\sigma})$ as computed by the algorithm for solving $PO(\sigma)$ in Table 2. In case one or more constraints (6) are violated by the Lagrangian solution, feasibility can be achieved by simple forward shifting operations on the outbound schedules, as shown in Table 3.

The shifting heuristic of Algorithm 2 can be embedded within an iterative procedure for solving the Lagrangian Dual problem:

Table 2 Algorithms for solving the Lagrangian relaxation problem

Algorithm for $PI(\rho)$	
1	Sort the trucks $i \in I$ by nonincreasing values of ρ_i
2	Sequentially assign the k-th group of g^I trucks to the time slot k (one truck for gate)
3	If the last group contains less than g^I trucks, assign the trucks to the first available gates
Algorithm for $PO(\sigma)$	
1	Sort the trucks $j \in J$ by nonincreasing values of σ_j
2	If $\sum_{j\in J} \sigma_j \leq 1$ then $C_M = \lceil m/g^O \rceil$, else $C_M = L$
3	Sequentially assign the t-th group of g^O trucks to the time slot $h = C_M - t + 1$ (one truck for gate)
4	If the last group contains less than g^O trucks, assign the trucks to the first available gates

Table 3 Algorithm 2

	The shifting heuristic
1	$x = x^*(\bar{\rho})$
2	Sort the outbound trucks j at the gate $g \in G^O$ with respect to the minimum starting time-slot that makes feasible the corresponding constraint (6), that is $$h^j_{\min} = \max_{k\in K} \left\{ k \sum_{g\in G^I} x^g_{ik} \,\middle|\, i \in I_j \right\} + 1$$
3	For each outbound gate g, schedule the trucks assigned to g with respect to the sorting previously defined

$$\max_{\lambda\geq 0} Z_{LR}(\lambda)$$

We have adopted the Subgradient Algorithm, with the standard updating rule:

$$\lambda^{(r+1)}_{ij} = \max \left\{0, \lambda^{(r)}_{ij} + t^{(r)} g(x^{(r)}, y^{(r)})\right\} \quad \forall i \in I, j \in J_i$$

where $x^{(r)}$ and $y^{(r)}$ are the optimal solutions of (11) and (12) at the iteration r, $g(x^{(r)}, y^{(r)})$ is a subgradient of the Lagrangian function at $\lambda^{(r)}$, and $t^{(r)}$ is the classical stepsize. Therefore, at each iteration, we are able to compute both a feasible solution for the MGCD problem, and an increasing lower bound $Z_{LR}(\lambda^{(r)})$ on the optimal value of the makespan C_M.

5 Numerical Experience

We have tested both the MGCD model and the Lagrangian heuristic on a set of instances with different values of n, m, g^I and g^O. For each fixed pair (n, m), we have randomly generated ten different configurations of the sets J_i, $i = 1, \ldots, n$, such that $1 \leq |J_i| \leq m$. For each configuration, we have considered four different combinations of (g^I, g^O). By this way, our test set consists of 720 instances, divided in two classes: small size and large size instances. The Lagrangian heuristic has been coded in C++ and Cplex 12.6 has been used as benchmark ILP solver. The experiments have been run on a machine equipped with a 3.1 GHz CPU and 16 GB of RAM. We have imposed a time limit of 600 s to Cplex, while we let the Lagrangian heuristic run for 10,000 iterations starting with null multipliers.

The results are reported in Tables 4 and 5, where a single row corresponds to ten instances. As for Cplex, we report the best upper bound U^c and the computation time T^c, as average values, and the number, N^c, of optimal solutions it was able to find within the time limit. For the Lagrangian heuristic we report the best upper bound U^h and the computation time T^h, as average values, and the number N^h of times the objective function value returned by our algorithm is not greater than the one obtained by Cplex. In the last column we report the average relative error: $ARE = 1/10 \cdot \left(\sum_{l=1}^{10} \frac{U_l^h - U_l^c}{U_l^c} \right)$.

A quick glance at the result tables highlights the computational complexity of the MGCD problem: Cplex succeeds in finding the optimal solution in six out of 720 instances. The Lagrangian heuristic performs quite satisfactorily: in a very short computation time, never reaching two seconds, it returns a feasible solution whose objective value is comparable with the best one obtained by Cplex in ten minutes. More in detail, looking at the last column in Table 4, we note that the relative error of our algorithm is almost negligible ($0.0 \leq ARE \leq 0.04$) in 23 out of 36 cases, it is lightly significant ($0.05 \leq ARE \leq 0.08$) in 11 cases, and in the remaining two cases it is $ARE < 0$. Passing to the larger instances in Table 5, we have again $0.0 \leq ARE \leq 0.04$ in the majority of cases (21 out 36), $ARE = 0.05$ in only 3 cases, while the number of cases when $ARE < 0$ grows to 12. Furthermore, taking into account the entire set of instances, both the small and large ones, the maximum absolute value of ARE in favor of the Lagrangian heuristic is 18%, while it is 8% in the opposite case.

Table 4 Small size instances results

				Cplex			Lagrangian heuristic			
n	m	g^I	g^O	U^c	T^c	N^c	U^h	T^h	N^h	ARE
20	20	2	2	15.40	600.00	0	16.10	0.32	3	0.05
		2	3	13.20	493.59	3	13.70	0.31	5	0.04
		3	2	13.40	600.01	0	13.80	0.40	6	0.03
		3	3	10.80	600.01	0	11.30	0.34	5	0.05
20	25	2	2	17.60	600.01	0	18.50	0.47	3	0.05
		2	3	14.70	600.01	0	15.10	0.43	5	0.03
		3	2	15.80	600.01	0	17.10	0.57	0	0.08
		3	3	12.30	600.01	0	12.70	0.68	6	0.03
20	30	2	2	20.20	600.01	0	21.10	0.66	3	0.05
		2	3	16.00	600.01	0	16.80	0.52	2	0.05
		3	2	18.20	600.01	0	19.60	0.63	1	0.08
		3	3	13.70	600.01	0	14.10	0.61	5	0.03
25	20	2	2	18.20	600.01	0	18.40	0.36	8	0.01
		2	3	15.60	554.65	2	15.90	0.47	7	0.02
		3	2	14.60	600.01	0	15.80	0.41	0	0.08
		3	3	12.40	600.01	0	13.00	0.42	4	0.05
25	25	2	2	20.50	600.01	0	20.90	0.50	6	0.02
		2	3	17.00	600.01	0	17.50	0.52	5	0.03
		3	2	17.10	600.01	0	18.10	0.59	3	0.06
		3	3	14.30	600.01	0	14.60	0.56	5	0.02
25	30	2	2	23.80	600.01	0	23.10	0.72	5	−0.01
		2	3	18.60	600.01	0	18.90	0.62	6	0.02
		3	2	19.20	600.01	0	20.70	0.68	0	0.08
		3	3	18.50	600.01	0	16.00	0.63	5	−0.03
30	20	2	2	20.40	600.01	0	21.00	0.49	5	0.03
		2	3	17.90	345.09	1	18.40	0.49	5	0.03
		3	2	16.60	600.01	0	17.20	0.49	3	0.04
		3	3	13.60	600.01	0	14.40	0.48	3	0.06
30	25	2	2	22.60	600.01	0	23.20	0.60	5	0.03
		2	3	19.50	600.01	0	20.10	0.61	4	0.03
		3	2	18.60	600.01	0	19.60	0.69	1	0.05
		3	3	15.50	600.01	0	16.00	0.61	5	0.03
30	30	2	2	25.10	600.01	0	25.50	0.79	7	0.02
		2	3	21.00	600.01	0	21.10	0.78	9	0.00
		3	2	21.00	600.01	0	21.90	0.95	3	0.04
		3	3	17.10	600.00	0	17.60	0.86	4	0.03

Table 5 Large size instances results

				Cplex			Lagrangian heuristic			
n	m	g^I	g^O	U^c	T^c	N^c	U^h	T^h	N^h	ARE
40	40	3	3	23.20	600.01	0	23.90	0.92	3	0.03
		3	4	20.80	600.01	0	20.60	0.91	10	−0.01
		4	3	20.60	600.01	0	21.40	0.99	3	0.04
		4	4	17.70	600.01	0	18.20	0.91	6	0.03
40	45	3	3	24.70	600.01	0	25.60	1.04	2	0.04
		3	4	21.50	600.01	0	22.30	1.05	2	0.04
		4	3	22.30	600.01	0	22.80	1.16	5	0.02
		4	4	19.10	600.01	0	19.80	1.12	3	0.04
40	50	3	3	26.90	600.00	0	27.40	1.26	5	0.02
		3	4	22.90	600.01	0	23.60	1.16	4	0.03
		4	3	23.80	600.01	0	24.90	1.19	1	0.05
		4	4	20.20	600.01	0	21.00	1.25	2	0.04
40	40	3	3	24.50	600.01	0	25.50	0.99	0	0.04
		3	4	22.30	600.03	0	22.60	1.01	7	0.01
		4	3	21.80	600.01	0	22.80	1.11	2	0.05
		4	4	18.70	600.01	0	19.30	1.00	4	0.03
45	45	3	3	31.70	600.01	0	27.30	1.13	5	−0.04
		3	4	31.00	600.01	0	23.40	1.14	9	−0.10
		4	3	23.70	600.01	0	24.20	1.26	5	0.02
		4	4	27.60	600.01	0	20.80	1.16	5	−0.08
45	50	3	3	34.90	600.01	0	29.00	1.32	3	−0.04
		3	4	39.20	600.01	0	24.90	1.25	9	−0.18
		4	3	39.30	600.01	0	26.00	1.42	3	−0.16
		4	4	21.40	600.01	0	22.10	1.34	3	0.03
50	40	3	3	26.50	600.01	0	27.00	1.13	5	0.02
		3	4	23.70	600.01	0	23.80	1.13	7	0.00
		4	3	22.90	600.01	0	24.00	1.17	0	0.05
		4	4	20.20	600.01	0	20.40	1.12	6	0.01
50	45	3	3	27.90	600.01	0	28.40	1.24	6	0.02
		3	4	24.90	600.01	0	25.10	1.24	8	0.01
		4	3	24.50	600.01	0	25.20	1.35	5	0.03
		4	4	27.80	600.01	0	21.90	1.26	4	−0.05
50	50	3	3	31.10	600.01	0	30.30	1.41	5	−0.02
		3	4	38.10	600.01	0	26.50	1.41	4	−0.12
		4	3	33.10	600.01	0	27.10	1.54	4	−0.04
		4	4	35.60	600.01	0	23.10	1.42	6	−0.15

6 Conclusions

In this paper we have proposed a mathematical model and a Lagrangian Relaxation scheme for scheduling trucks at a cross-docking distribution terminal. We have derived some properties of the relaxed problem leading to a Lagrangian heuristic algorithm, that has been tested on a wide set of instances. The numerical experience has shown the efficiency and the effectiveness of our algorithm, if compared to a benchmark ILP solver.

References

1. Boysen, N., Fliedner, M.: Cross dock scheduling: classification, literature review and research agenda. Omega Int. J. Manage. S. **38**(6), 413–422 (2010)
2. Buijs, P., Vis, I.F., Carlo, H.J.: Synchronization in cross-docking networks: a research classification and framework. Eur. J. Oper. Res. **239**(3), 593–608 (2014)
3. Chiarello, A., Gaudioso, M., Sammarra, M.: Truck synchronization at single door cross-docking terminals. OR Spectr. **40**(2), 395–447 (2018)
4. Van Belle, J., Valckenaers, P., Cattrysse, D.: Cross-docking: state of the art. Omega Int. J. Manage. S. **40**(6), 827–846 (2012)

Offline Patient Admission, Room and Surgery Scheduling Problems

Rosita Guido, Vittorio Solina, Giovanni Mirabelli and Domenico Conforti

Abstract Patient admission and surgery scheduling is a complex combinatorial optimization problem. It consists on defining patient admission dates, assigning them to suitable rooms, and schedule surgeries accordingly to an existing master surgical schedule. This problem belongs to the class of NP-hard problems. In this paper, we firstly formulate an integer programming model for offline patient admissions, room assignments, and surgery scheduling; then apply a matheuristic that combines exact methods with rescheduling approaches. The matheuristic is evaluated using benchmark datasets. The experimental results improve those reported in the literature and show that the proposed method outperforms existing techniques of the state-of-the-arts.

Keywords Combinatorial optimization · Patient admission scheduling
Surgery scheduling · Matheuristic

R. Guido (✉) · V. Solina · D. Conforti
de-Health Lab, Department of Mechanical, Energy and Management Engineering,
University of Calabria, Rende, Italy
e-mail: rosita.guido@unical.it

V. Solina
e-mail: vittorio.solina@unical.it

D. Conforti
e-mail: domenico.conforti@unical.it

G. Mirabelli
Department of Mechanical, Energy and Management Engineering,
University of Calabria, Rende, Italy
e-mail: giovanni.mirabelli@unical.it

© Springer Nature Switzerland AG 2018

P. Daniele and L. Scrimali (eds.), *New Trends in Emerging Complex Real Life Problems*, AIRO Springer Series 1,
https://doi.org/10.1007/978-3-030-00473-6_30

1 Introduction

Patient admission scheduling problems (PASPs) concern with deciding which patient to admit and at what time. These problems can be very complex, mainly when different subproblems are tackled at the same time, like patient-to-bed assignment and surgery scheduling problems. The patient bed assignment problem (PBAP) is a sub-task of the PASP and concerns the choose of a suitable room to be assigned to patients by considering medical requirements, patient needs, and hospital resource availability. Despite PBAPs were usually addressed only as bed capacity problems [1, 2] formalized them as an offline and combinatorial optimization problem. Interesting problem extensions were based on it [3, 4]. New realistic situations, like patients with a risk of overstay and postponed admission date were introduced to PASPs in [4]. The goal consists on defining patient admission dates, assigning patients to suitable bed-room-wards, and reduce overcrowded rooms. These problems were named as PASU. The reader is referred to [4] for more details about PASU problems. More complex is the PASU with surgery scheduling problem [5], referred hereafter as PASU-OR problem.

The PBAP of [2] is NP-hard [6], as well as all other similar problems based on it, and heuristic approaches to solve benchmark instances for the aforementioned problems were designed in the literature.

The contribution of this study is twofold. First, it fills the gap in the literature by introducing a mixed integer programming (MIP) model for the PASU-OR problem. It is an extension of that proposed for PASU problems in [7] and an improvement of those proposed in [4, 8] because reduces at minimum the number of decision variables. Second, this study tests an efficient matheuristic procedure and improves the results on a set of benchmark instances available in the literature. The matheuristic was originally designed for PBAPs in [9].

The paper is organized as follows. Section 2 presents the PASU-OR and the MIP model. Section 3 reports computational results on a set of the benchmarks of PASU-OR. The results are compared with those reported in [5] and discussed. Conclusions are drawn in Sect. 4.

2 Problem Statement and an Optimization Model

In the following, we introduce the problem statement and the used notation. The PASU-OR problem is characterized by patient admission date, patient-to-room assignment and surgery scheduling decisions. Patients are characterised by mandatory and preferred medical equipment, gender, admission date (which can be postponed up to a defined date), fixed length of the stay (LOS), medical specialty, room preference. Rooms, located in wards, differ for medical equipment, number of beds, and high/or medium levels of expertise in treating certain pathologies. All patients have to be admitted to the hospital in a defined planning horizon and assigned to suit-

Table 1 Patient and room attributes

Patient	
$AD_p = \left\{a_p, \ldots, a_p^{'}\right\}$	Range of admission dates: a_p and $a_p^{'}$ are the first and the latest possible admission date, respectively
L_p	Length of stay as consecutive nights
$H_p = \left\{a_p, \ldots, z_p^{'} - 1\right\}$	Period during which patient has a stay of L_p, where $z_p^{'} = a_p^{'} + L_p$
$sp_p \in S$	Patient specialty
ME_p	Is the set of mandatory equipment for p
Gender	Male or female
For each patient in PS	
ls_p	Length of surgery (in minutes)
δ_p	Number of days between admission date and surgery date
$SD_p = \{h + \delta_p, h \in AD_p\}$	Range of possible surgery dates
Room	
$\bar{S}_r$	Set of specialties that cannot be treated in room r
C_r	Capacity, i.e., number of beds
E_r	Set of equipment in room $r \in R$
$gp_r \in GP$	Gender policy. $GP = \{1, 2, 3\}$: 1 and 2 denote rooms restricted to male and female patients (RGP), respectively; 3 denotes dependent gender policy (DGP), i.e., patients with the same gender of patients already staying in a room should be assigned
Operating room	
b_{hs}	OR time (in minutes) of surgical specialty $s \in SS$ on day $h \in H$
ov_{hs}	Maximum overtime of surgical specialty $s \in SS$ on day $h \in H$

able hospital rooms in correspondence with their characteristics for a fixed number of consecutive nights.

Let H be a planning horizon, and P be the set of elective patients, indexed by h and p, respectively. L_p denotes the length of stay (LOS), which could be extended by one night for some patients with a risk of overstay. For each patient is known the range of admission dates AD, and H_p that is the range of days between the first possible admission date and the last day of hospitalization. Some patients have to undergo a surgery and exactly δ_p days after their admission. Surgeries are scheduled by considering an already defined master surgical schedule (MSS). In addition, some patients were already assigned to rooms before the current planning phase. Let P_0 be this set of patients. A transfer in a different room is allowed only for them, even if is penalised in order to reduce patient discomfort. Let R, S and SS, be the set of rooms, medical and surgical specialties, indexed by r and s, respectively. The main attributes of patients, rooms and ORs are reported in Table 1.

To simplify the readability of the model formulation, we define $H_p^{ov} = \left\{z_p, \ldots, z_p^{'}\right\}$ as the range of possible overstay nights, and the following subsets of patients and rooms: P_F, P_M, PS, and P^{ov} are the sets of woman, men, patients who have to undergo surgery, and patients with a risk of overstay, respectively. Let $\bar{R}^{dgp} = \{r \in R | gp_r = 3\}$, be the set of rooms with DGP. Let $\bar{R}_p = \left\{r \in R : sp_p \notin \bar{S}_r, ME_p \subseteq E_r\right\}$, and $\bar{R}_p^{dgp} = \left\{r \in \bar{R}_p | gp_r = 3\right\}$, be the subsets of rooms feasible for patient p and those with DGP, respectively.

Hard constraints are on room capacity, mandatory equipment, patient specialty, and patient stay as consecutive nights. A surgery has to be performed in a defined date and scheduled in those OR blocks assigned to the related surgical specialty, as defined by the MSS. Requirements related to preferred equipment, room capacity preference, gender policies, department specialism, delayed admissions, transfers, and overcrowded rooms due to overstay patients, OR underutilization, and OR overtime express desired properties. They are tackled as soft constraints since do not influence the validity of a schedule but impact on its quality. Violated soft constraints are penalised in objective function.

2.1 An Optimization Model for Offline Patient Admission, Rooms and Surgery Scheduling Problems

Before to introduce our optimization model, we define the decision variables and their meaning as follows. $ad_{prh} = 1$ and $x_{prh} = 1$, if patient $p \in P$ is admitted on day $h \in AD_p$ and then assigned to room $r \in \bar{R}_p$ over H_p; $os_{prh} = 1$ if $p \in P^{ov}$ is in room $r \in R$ on day $d \in H_p^{ov}$; $t_p = 1$, if $p \in P_0$ is not assigned to the already occupied room $\bar{r}_p$; $m_{rh} = 1$ if male patients are in room $r \in \bar{R}^{dgp}$ on day $d \in H$, and $bg_{rh} = 1$ if there are both male and female patients. The above binary decision variables take value 0 otherwise. Delayed admission (in days) is denoted by $del_p \geq 0$. A room r is overcrowded on day $h \in H$ if $oc_{rh} > 0$. Finally, variables related to surgeries scheduling are: $sd_{ph} = 1$ if patient p undergoes surgery on day $h \in SD_p$; auxiliary variables are sov_{hs}, tov_h, and u_h denoting daily OR overtime per surgical specialty, overall OR overtime and OR underutilization, respectively. The overall OR time underutilization is computed as deviation of utilised OR time from a constant k_u, defined as $k_u = min\,\{reqOR, avOR\}$, where $reqOR$ and $avOR$ are the overall requested OR time and the overall available OR time. The objective function is a weighted sum of eight terms. The costs are listed in Table 2.

$$\min \sum_{p\in P}\sum_{r\in\bar{R}_p}\sum_{h\in H_p} w_{pr}x_{prh} + \sum_{r\in\bar{R}^{dgp}}\sum_{h\in H} w_g bg_{rh} + w_t\sum_{p\in P_0} t_p + \sum_{p\in P}(w_{del}del_p)+$$

$$+\sum_{r\in R}\sum_{h\in H} w_{oc}oc_{rh} + \sum_{h\in H}\sum_{s\in SS} w_{sov}sov_{hs} + \sum_{h\in H} w_{tov}tov_h + w_u(k_u - \sum_{h\in H} u_h) \quad (1)$$

$$\sum_{r\in\bar{R}_p}\sum_{h\in AD_p} ad_{prh} = 1 \qquad \forall p \in P \quad (2)$$

Table 2 Violations of the soft constraints and related penalty costs

Violation	Soft constraint	Penalty cost	Violation	Soft constraint	Penalty cost
v_1	Preferred equipment	w_{pe}	v_7	Delay	w_{del}
v_2	Room capacity preference	w_{cr}	v_8	Overcrowded room	w_{oc}
v_3	Department specialism	w_{sp}	v_9	Surgical specialty overtime	w_{sov}
v_4	RGP	w_g	v_{10}	Overall OR overtime	w_{tov}
v_5	DGP	w_g	v_{11}	OR underutilization	w_u
v_6	Transfers	w_t			

$$del_p = \sum_{r \in \bar{R}_p} \sum_{h \in AD_p} ad_{prh}(h - a_p) \qquad \forall p \in P \qquad (3)$$

$$\sum_{k=h}^{h+L_p-1} x_{prk} \geq ad_{prh} L_p \qquad \forall p \in P, r \in \bar{R}_p, h \in AD_p \qquad (4)$$

$$\sum_{h \in H_p} x_{prh} = L_p ad_{prh} \qquad \forall p \in P, r \in \bar{R}_p \qquad (5)$$

$$C_r \geq \sum_{\substack{p \in P \\ h \in H_p, r \in \bar{R}_p}} x_{prh} \qquad \forall r \in R, h \in H \qquad (6)$$

$$m_{rh} \geq x_{prh} \qquad \forall r \in \bar{R}_p^{dgp}, p \in P_M, h \in H \qquad (7)$$

$$bg_{rh} + 1 \geq m_{rh} + x_{prd} \qquad \forall r \in \bar{R}^{dgp}, p \in P_F, h \in H_p \qquad (8)$$

$$t_p = \sum_{r \in R \setminus \bar{r}_p} x_{pr1} \qquad \forall p \in P_0 \qquad (9)$$

$$os_{prk} = ad_{pr,k-L_p} \qquad \forall p \in P^{ov}, r \in \bar{R}_p, k \in H_p^{ov} \qquad (10)$$

$$C_r + oc_{rh} \geq \sum_{\substack{p \in P| \\ h \in H_p, r \in \bar{R}_p}} x_{prh} + \sum_{\substack{p \in P^{ov}| \\ h \in H_p^{ov}, r \in \bar{R}_p}} os_{prh} \qquad \forall r \in R, h \in H \qquad (11)$$

$$sd_{p,h+\delta_p} = \sum_{r \in \bar{R}_p} ad_{prh} \qquad \forall p \in PS, h \in AD_p \qquad (12)$$

$$sd_{ph} \leq b_{h,sp_p} \qquad \forall p \in PS, h \in SD_p \qquad (13)$$

$$b_{hs} + sov_{hs} \geq \sum_{p \in PS_{|sp_p = s, h \in SD_p}} ls_p sd_{ph} \qquad \forall h \in H, s \in SS \tag{14}$$

$$\sum_{s \in SS} b_{hs} + tov_h \geq \sum_{p \in PS|h \in SD_p} ls_p sd_{ph} \qquad \forall h \in H \tag{15}$$

$$u_h = \sum_{p \in PS|h \in SD_p} ls_p sd_{ph} - \sum_{s \in SS} sov_{hs} \qquad \forall h \in H \tag{16}$$

$$sov_{hs} \leq uv_s^{sov} \qquad \forall h \in H, s \in S \tag{17}$$

$$tov_h \leq uv^{tov} \qquad \forall h \in H \tag{18}$$

Objective function (1) plans patient admissions, assigns patients to rooms according to quality of care and patient preferences, and schedules surgeries. The first term considers violations $v_1 - v_4$: patient-to-room assignments are penalised per night by w_{pr}, which is the sum of the first four costs. The subsequent four terms penalise $v_5 - v_8$, respectively; the last three terms penalise overall OR overtime computed per all specialties, OR under-utilization and overall OR under-utilization, respectively. Constraints (2) ensure that each patient is admitted only once in AD_p and has to be assigned only one room among those feasible. Constraints (3) evaluate delayed admissions and Constraints (4)–(5) ensure patient stay as consecutive L_p nights. Constraints (6) state that the number of patients assigned to a room cannot be greater than the number of beds. Constraints (7)–(8) capture the presence of male patients and DGP violation if there both male and female patients on the same day, respectively. Constraints (9) evaluate transfers, and Constraints (10)–(11) consider overstays and overcrowded rooms. Constraints (12) and (13) ensure that each patient $p \in PS$ undergoes surgery δ_p days after admission, only once, and on day in which his/her specialty has allotted OR time in the MSS. Constraints (14)–(16) are on surgical specialty overtime, overall OR overtime per day, and OR under-utilization, respectively. Finally, Constraints (17)–(18) impose an upper value to OR overtime. The above defined decision variables complete the MIP model formulation.

3 Computational Results

In this section we present computational results carried out on the small short family [5] for assessing the quality of the matheuristic solution and found good schedules in reasonable times. This family consists of three sets and 15 instances. Main features are summarised in Table 3.

The PASU-OR benchmark are infeasible instances owing a greater demand than resource. To overcome this, the planning horizon is doubled with respect to the original, and room underutilization related to the H is minimized. This term is defined as deviation of the overall room utilization from a constant. The constant is $k_{bed} = min\,\{reqB, avB\}$, where $reqB$ and avB are the overall bed requests and the overall

Table 3 Main features of the short family instances

Family	$\|Dep\|$	$\|R\|$	$\|S\|$	$\|P\|$	$\|OR\|$	$\|H\|$
Short 1	2	25	9	391–439	2	14
Short 2	4	50	18	574–644	4	14
Short 3	6	75	23	821–925	5	14

number of beds, respectively. The term $(k_{bed} - \sum_{h \in H} bu_h)$, denoted by $\bar{v}_{12}$ in Table 4, is penalised and added to objective function (1).

As already stated in Sect. 1, PASU-OR problems are NP-hard and heuristic approaches were devised to find good solutions in a reasonable time because exact solvers are not effective to explore the solution space mainly of large instances. The solution approach here implemented is based on the matheuristic FiNeMath, developed to solve PBAPs in [9]. Metaheuristic algorithms are generic solution procedures based on exploring the solution space by considering an incumbent solution and iteratively changing it in favour of a new solution. FiNeMath exploits complementarity among fix-relax methods, neighbourhood-based searches, and exact solvers. Let s_0 be an initial feasible schedule, and F_0 its objective function value. Some patient-to-room assignments are selected from s_0 randomly and added to the MIP model as constraints. They are thus fixed components, while the remaining ones are repaired by an exact solver in the defined neighbourhood. The procedure is iterated until a stopping criterion is reached. The objective function value is thus improved iteratively by destroying a current schedule and repairing it by an exact solver. A high level pseudocode is provided below, named as Algorithm 1. For more details, the reader is referred to [9].

Algorithm 1 FiNeMath

Require: MIP model, $MaxIter$, s_0 (an initial feasible schedule)
  $i \leftarrow 0$
  while $i \neq MaxIter$ **do**
    Select randomly some patient-to-room assignments from the incumbent solution
    Add these assignments as constraints to MIP model
    Solve the current MIP model
    $i \leftarrow i + 1$
  end while

The number of patient-to-room assignments added as constraints to the MIP model influences both improvements in the objective function value and computational times [9]. The percentage range of fixed assignments was set in the range 10–40%; each MIP problem was solved with a gap of 3%, that decreases up to 1% in the latest iterations; the number of iterations was set to 32. The computational experiments were performed using IBM ILOG CPLEX V12.7.1, Academic license. We used the cost values reported in [5]. We evaluated the improvement/worsening of our results with respect to the best-known mean values R_{CS}, found by a simulated annealing

Table 4 Results on the small family instances

Family	Result	F	Weighted soft constraints								ΔF (%)	Time (s)
			$\sum_{i=1}^{4} \bar{v}_i$	$\bar{v}_5$	$\bar{v}_6$	$\bar{v}_7$	$\bar{v}_8$	$\bar{v}_9 + \bar{v}_{10}$	$\bar{v}_{11}$	$\bar{v}_{12}$		
Short 1	Our	78163.0	13074	1020	280	5228	42	735	48260	9524	4.7	965.4
	R_{CS}	82020.0	13906	80	220	3262	71	1251	46890	16340		155.8
Short 2	Our	119299.6	27408	1350	80	6496	44.6	927	63330	19664	10.5	2084
	R_{CS}	133380.2	30822	70	580	4406	92.2	1392	61750	34268		276.7
Short 3	Our	193331.8	42418	2960	140	11182	76.8	1863	104000	30692	8.1	4445.0
	R_{CS}	210257.8	46280	130	620	7811	148.8	2742	101390	51136		408.7

approach in [5]. All the best-known values R_{CS} were improved. Table 4 lists the results in terms of mean values per set of instances, and reports the single components of the objective function. In the last column there are the percentage improvement values, which were evaluated by $\Delta F = \frac{(R_{CS}\ value-our\ value)}{(R_{CS}\ value)} \times 100$. Observe that they are in the range 4.7–10.5%.

4 Conclusion

In this paper, we formulated an optimization model to manage patients admissions, hospital rooms, and surgeries. Schedules with planned patient admissions, patient-to-room assignments and planned surgeries are developed using the matheuristic FiNeMath, which is based on solving the optimization formulation. Preliminary results achieved on a set of benchmark instances, point out that our approach is promising. Currently, we are working on improving the FiNeMath efficiency by developing suitable solution destroying phases to reduce computational times.

Acknowledgements We would like to thank Eugenio Rende who created a tool for computing input cost matrices.

References

1. Hulshof, P., Kortbeek, N., Boucherie, R., Hans, E., Bakker, P.: Taxonomic classification of planning decisions in health care: a structured review of the state of the art in OR/MS. Health Syst. **1**, 129–175 (2012)
2. Demeester, P., Souffriau, W., De Causmaecker, P., Vanden Berghe, G.: A hybrid tabu search algorithm for automatically assigning patients to beds. Artif. Intell. Med. **48**(1), 61–70 (2010)
3. Ceschia, S., Schaerf, A.: Local search and lower bounds for the patient admission scheduling problem. Comput. Op. Res. **10**(38), 1452–1463 (2011)
4. Ceschia, S., Schaerf, A.: Modeling and solving the patient admission scheduling problem under uncertainty. Artifi. Intell. Med. **56**(3), 199–205 (2012)
5. Ceschia, S., Schaerf, A.: Dynamic patient admission scheduling with operating room constraints, flexible horizons, and patient delays. J. Sched. **19**(4), 377–389 (2016)
6. Vancroonenburg, W., Della Croce, F., Goossens, D., Spieksma, F.C.R.: The red blue transportation problem. Eur. J. Op. Res. **237**(3), 814–823 (2014)
7. Guido, R., Solina, V., Conforti, D.: Offline Patient Admission Scheduling Problems. In: Sforza A., Sterle C. (eds) Optimization and Decision Science: Methodologies and Applications. ODS 2017. Springer Proceedings in Mathematics & Statistics, vol 217. Springer, Cham (2017)
8. Lusby, R.M., Schwierz, M., Range, T.M., Larsen, J.: An adaptive large neighborhood search procedure applied to the dynamic patient admission scheduling problem. Artif. Intell. Med. **74**, 21–31 (2016)
9. Guido, R., Groccia, M.C., Conforti, D.: An efficient matheuristic for offline patient-to-bed assignment problems. Eur. J. Op. Res. **268**(2), 486–503 (2018)

Equilibria on Networks with Uncertain Data—A Comparison of Different Solution Approaches

Joachim Gwinner and Friedemann Sebastian Winkler

Abstract This contribution is concerned with Wardrop traffic equilibria. As is well known these equilibria can be formulated as variational inequalities over a convex constraint set. Here we consider uncertain data that can be modeled as probabilistic. We survey different solution approaches to this class of problems, namely the expected value formulation, the expected residual minimization formulation, and the approach via random variational inequalities. To compare these solution approaches we provide and discuss numerical results for a 12 node network as a test example.

Keywords Wardrop traffic equilibrium · Uncertain data · Probabilistic approaches · Unfairness measure

1 Introduction

This contribution is concerned with Wardrop traffic equilibria. As is well known from the classic papers of Dafermos [3] and Smith [15] these equilibria can be formulated as finite-dimensional variational inequalities over a convex constraint set. For more recent research on the interplay between traffic equilibria and variational inequality theory we refer to [4, 7, 13].

Here we consider uncertain data that can be modeled as probabilistic. We survey different solution approaches to this class of problems, namely the expected value (EV) formulation introduced in [6], the expected residual minimization formulation

Dedicated to Professor H. Walk

J. Gwinner (✉)
Department of Aerospace Engineering, Institute of Mathematics,
Universität der Bundeswehr München, Neubiberg, Germany
e-mail: joachim.gwinner@unibw.de

F. S. Winkler
Universität der Bundeswehr München, Neubiberg, Germany
e-mail: winkler@me-lrt.de

© Springer Nature Switzerland AG 2018

P. Daniele and L. Scrimali (eds.), *New Trends in Emerging Complex Real Life Problems*, AIRO Springer Series 1,
https://doi.org/10.1007/978-3-030-00473-6_31

(ERM) introduced in [2], and the approach via random variational inequalities (RVI) introduced in [9, 10]. To compare these solution approaches we provide and discuss numerical results from the bachelor thesis [16] for a 12 node network as a test example. While for small networks the non-iterative algorithm given in [13], which has recently improved by branch and bound in [14], is applicable, here we use standard iterative solvers for variational inequalities and nonlinear complementarity problems taken from [5]. As a measure to evaluate these different approaches we consider an index of unfairness.

We do not consider sample path solution [8]; for a comparison between this method and the RVI approach we can refer to [11].

2 The Stochastic Variational Inequality Problem

Let us begin with the deterministic traffic equilibrium problem formulated as the variational inequality: Find a vector $x \in S$ (a closed convex set of finite dimension) such that

$$F(x)^T(y - x) \geq 0 \quad \forall y \in S\,.$$

Traffic equilibrium problems that arise in real-world applications often involve uncertainty. Influences like weather changes may have a significant effect on the network's congestion. Here we model such uncertainties as probabilistic. They may occur both in the defining function F and in the set S of the problem. We assume that there are l random variables that have to be considered and group those in a random vector ω. This leads to the stochastic variational inequality (SVI): Find $x \in S(\omega)$ such that

$$F(x,\omega)^T(y - x) \geq 0 \quad \forall y \in S(\omega)\,,$$

where $S(\omega) \subset \mathbb{R}^n$ is closed convex for any random parameter $\omega \in \Omega$ and where $(\Omega, \mathscr{F}, \mathscr{P})$ is a probability space with $\Omega \subset \mathbb{R}^l$ and a given probability distribution $\mathscr{P}$.

In the special case $S(\omega) = \mathbb{R}^n_+$ the problem reduces to the stochastic nonlinear complementarity problem (SNCP): Find a vector x such that

$$0 \leq F(x,\omega) \perp x \geq 0\,.$$

When F is an affine function of x for any $\omega \in \Omega$, i.e.

$$F(x,\omega) = M(\omega)x + q(\omega)\,,$$

where $M(\omega) \in \mathbb{R}^{n\times n}$, $q(\omega) \in \mathbb{R}^n$, the SNCP reduces to a stochastic linear complementarity problem SLCP.

It is obvious that—apart from the special deterministic case—there will be no single vector x that satisfies the SVI or the SNCP or the SLCP for all possible

$\omega \in \Omega$. Instead some deterministic problem or a sequence of those has to be constructed to satisfy the conditions approximately. Here the question arises as to which deterministic model should be used to achieve the best results. Over the last years various approaches have been studied, some of which we will survey and compare then.

3 Various Solution Approaches

In this section we first describe shortly the EV and ERM approaches for SNCP, then the ERM approach for SVI and finally sketch the main idea of the RVI approach.

The expected value (EV) formulation introduced in [6] and the expected residual minimization formulation (ERM) introduced in [2] are two deterministic formulations for the SNCP. The EV formulation is to solve a single nonlinear complementarity problem with the expectation $E[F(x, \omega)]$ as deterministic substitute. The ERM formulation is to minimize a residual function. A version of the ERM formulation using NCP functions (ERM-CP approach) is to find an optimal solution of

$$\text{minimize} f(x) := E\big[\|\Phi(x, \omega)\|^2\big], \quad x \in \mathbb{R}^n_+ ,$$

where

$$\Phi(x, \omega) = \begin{pmatrix} \varphi(x_1, F_1(x, \omega)) \\ \vdots \\ \varphi(x_n, F_n(x, \omega)) \end{pmatrix}$$

is the residual function and $\varphi : \mathbb{R}^2 \to \mathbb{R}$ is an NCP function that satisfies

$$\varphi(a, b) = 0 \Rightarrow a \geq 0, b \geq 0, ab = 0$$

In [17] the min function $\varphi_1(a, b) = \min(a, b)$ and the FB function $\varphi_2(a, b) = \sqrt{a^2 + b^2} - (a + b)$ have been used.

One of the problems of the NCP-ERM approach is that the min-function or FB-function are coercive but not convex in general. Hence a solution calculated with this model might not be globally optimal. There is an alternative due to [1] (ERM-VI approach) that uses the regularized gap function f_α and the D-gap function g_α given by

$$f_\alpha(x, \omega) := \max_{y \in S(\omega)} \big[F(x, \omega)^T (x - y) - \frac{1}{2\alpha} \|y - x\|^2\big],$$
$$g_\alpha(x, \omega) := f_\alpha(x, \omega) - f_{1/\alpha}(x, \omega) .$$

Then in the affine case with

$$S(\omega) = \{y \in \mathbb{R}^n \;:\; A(\omega)y = b(\omega),\, y \geq 0\}$$

the two proposed residual functions to be minimized are ($\tau > 0$ fixed)

$$\begin{aligned}\theta_\alpha^R(x) &:= E\big[f_\alpha(x,\omega) + \tau \|A(\omega)x - b(\omega)\|\big],\\ \theta_\alpha^D(x) &:= E\big[g_\alpha(x,\omega)\big]\end{aligned}$$

which both are shown in [1] to be convex for large enough α.

In the previous approaches (see also [12] for a more detailed survey) we have always found a single solution vector that was in some way optimal for all possible values of the random variables. The RVI approach (see [9, 10] for details) goes a different way. It is a numerical procedure that provides a solution function that depends on the random variables. A single solution vector can then still be obtained by using the expected value of this function. This approach has some advantages, for example we can calculate the variance and compare it to the interval of the random variables.

4 A Test Traffic Network

We consider a network $(\mathscr{N}, \mathscr{A}, \mathscr{W})$ consisting of the set $\mathscr{N}$ of 12 nodes, the set $\mathscr{A}$ of 26 (directed) arcs and the set $\mathscr{W}$ of 3 origin destination (OD) pairs: $(1, 8)$, $(6, 4)$, $(12, 2)$. The topology of the network is described by the node-arc incidence matrix E with the entries

$$E_{n,a} = \begin{cases} 1 \text{ if } a \text{ starts in } n\\ -1 \text{ if } a \text{ ends in } n\\ 0 \text{ else}\end{cases}$$

for $n \in \mathscr{N}$, $a \in \mathscr{A}$. Further for every OD pair $w \in \mathscr{W}$ there is a demand d_w at its destination that equals the supply at its origin. This gives rise to demand vectors D^w with the components

$$D_n^w = \begin{cases} d_w & \text{if } n \text{ is origin of } w\\ -d_w & \text{if } n \text{ is destination of } w\\ 0 & \text{else.}\end{cases}$$

The total flow x_a in an arc a composes of partial flows x_a^w resulting from the different OD pairs $w \in \mathscr{W}$,

$$x_a = \sum_{w \in \mathscr{W}} x_a^w.$$

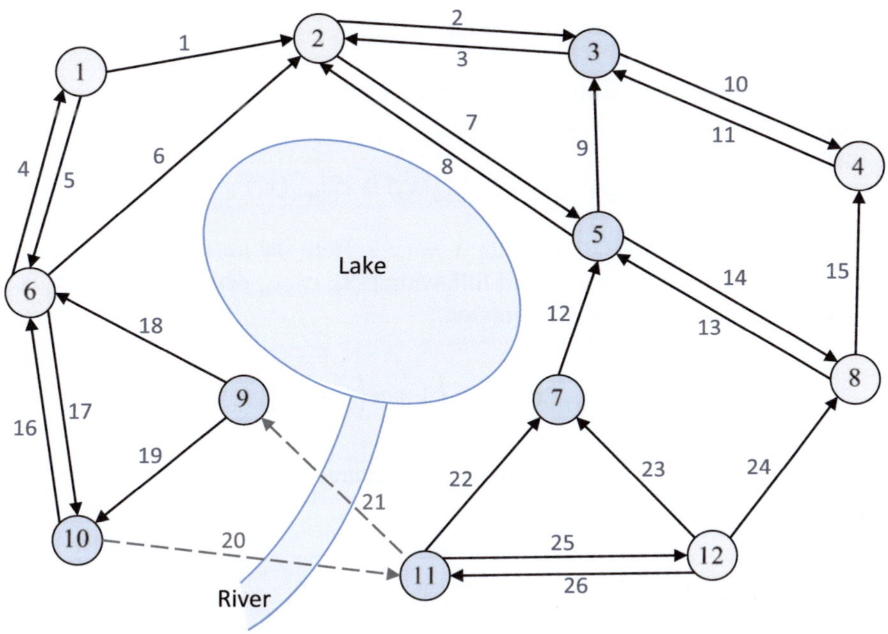

Fig. 1 12 node network with 3 OD pairs: (1, 8), (6, 4), (12, 2)

These part flows are grouped in vectors $x^w = (x_a^w)_{a\in\mathscr{A}}$ for each $w \in \mathscr{W}$. Then Wardrop's principle with mass conservation leads to the variational inequality: Find $x = (x^w)_{w\in\mathscr{W}} \in S$ such that

$$F(x)^T(\sum_{w\in\mathscr{W}} z^w - \sum_{w\in\mathscr{W}} x^w) \geq 0, \forall z \in S,$$

where the feasible set S and the vector cost function F are defined respectively by

$$S = \{x = (x^w)_{w\in\mathscr{W}} : Ex^w = D^w, x^w \geq 0\ (\forall w \in \mathscr{W})\},$$
$$F(x) = (c_a(\sum_{w\in\mathscr{W}} x^w))_{a\in\mathscr{A}}$$

with given cost function c_a on each arc $a \in \mathscr{A}$.

There is a stochastic influence on the arcs 20 and 21 (Fig. 1) via the random arc cost,

$$c_R^a = a\omega_c + b$$

with a, b constant and ω_c exponentially distributed. There is also a stochastic influence on the travel demands for each OD pair,

$$d_w = a_w + \omega_d b_w$$

with a_w, b_w constant and ω_d log-normal distributed.

The total arc travel cost is then given by

$$c_a(x) = c_I^a(x) + c_R^a(x_a) + c_{\text{GBPR}}^a(x_a)$$

depending on the total arc flow vector $x = (x_a)$. Here the traffic flow on an arc a can have a negative influence c_I^a and following [18], c_{GBPR}^a denotes the generalized Bureau of Public Roads (GBPR) function,

$$c_{\text{GBPR}}^a(x^a) = c_0^a\left(1 + \alpha\left(\frac{x^a}{c_1^a}\right)^\beta\right)$$

with $\alpha = 0.15$ and $\beta = 2.0$, c_0^a, c_1^a fixed for each arc a.

5 Numerical Results

The traffic network was simulated with different values and weights of the random parameters in a series of numerical experiments. The following results are the path flows for one realization. It is representative in a way that the other results produced showed very similar outcomes.

Table 1 Path flows calculated with the different approaches

Path	ERM-CP	ERM-VI	EV	RVI
1	164.9879	166.4100	162.0012	163.4000
2	47.4648	46.2170	49.4313	47.8210
3	58.3731	60.3030	59.0241	58.6970
4	85.1028	79.2660	81.3520	82.7460
5	23.2451	34.0920	24.7344	23.1180
6	45.6531	39.2730	46.3221	46.6580
7	130.2841	109.1500	132.9165	131.2300
8	49.7389	76.1050	48.6997	50.1100
9	0	0.5109	0	0
10	0	1.0676	0	0
11	126.2697	123.1200	122.6562	125.3900
12	73.7208	65.0660	71.5271	72.7900
13	11.8699	11.5180	13.4434	11.2530
14	26.9040	32.8970	27.9059	26.0160

Table 2 Nodes and arcs of all used paths

Path	Nodes	Arcs
1	1-2-5-8	1-7-14
2	1-6-10-11-12-8	5-17-20-25-24
3	6-1-2-3-4	4-1-2-10
4	6-2-3-4	6-2-10
5	6-10-11-12-8-4	17-20-25-24-15
6	6-10-11-7-5-3-4	17-20-22-12-9-10
7	12-11-9-10-6-2	26-21-19-16-6
8	12-11-9-6-2	26-21-18-6
9	12-11-7-5-2	26-22-12-8
10	12-11-7-5-3-2	26-22-12-9-3
11	12-7-5-2	23-12-8
12	12-8-4-3-2	24-15-11-3
13	12-8-5-2	24-13-8
14	12-7-5-3-2	23-12-9-3

Table 1 states the calculated path flows for all paths that were used in at least one of the solutions of the different approaches. Table 2 specifies the used paths by stating the used nodes and arcs.

To evaluate the numerical results obtained by the different approaches ERM-CP, ERM-VI, EV, and RVI we need a performance measure. Here we focus to an index of unfairness.

Note that for each fixed $\omega \in \Omega$ the Wardrop equilibrium reflects the fairness to all users with the same OD pair, since the travel cost for each used route connecting the same OD pair is equal or less than any unused route. However, in the uncertain case, the travel cost for any flow pattern connecting the same OD pair is not necessarily the same. For a fixed $\omega \in \Omega$ the unfairness of a feasible flow pattern for an OD pair $w \in \mathscr{W}$ is measured by

$$C_w^{\text{unfair}}(y, \omega) = \frac{C_w^{\max}(y, \omega)}{C_w^{\min}(y, \omega)},$$

where $C_w^{\max}$, $C_w^{\min}$ depending on the path flow vector y are the largest, respectively the smallest travel cost of routes being used, which connect the OD pair w. Thus, the expected unfairness of the outcome for the whole network under uncertainty is given by

$$\text{unf} := E[\frac{1}{|\mathscr{W}|} \sum_{w \in \mathscr{W}} C_w^{\text{unfair}}].$$

This index of unfairness together with the computed total cost are listed for the different approaches below (Table 3).

Table 3 Comparison for the different approaches

	ERM-CP	ERM-VI	EV	RVI
Unfairness	1.0093	1.0170	1.0240	1.0003
Total cost e+005	2.6756	2.6814	2.6464	2.6440

It can be seen that the RVI formulation has the lowest unfairness and the EV formulation the highest. As the total traffic flow is higher in the ERM solutions it is unavoidable that the total cost is also higher than in the case of EV and RVI.

References

1. Agdeppa, R.P., Yamashita, N., Fukushima, M.: Convex expected residual models for stochastic affine variational inequality problems and its application to the traffic equilibrium problem. Pac. J. Optim. **6**, 3–19 (2010)
2. Chen, X., Fukushima, M.: Expected residual minimization method for stochastic linear complementarity problems. Math. Oper. Res. **30**, 1022–1038 (2005)
3. Dafermos, S.: Traffic equilibrium and variational inequalities. Transp. Sci. **14**, 42–54 (1980)
4. Daniele, P., Giuffrè, S.: Random variational inequalities and the random traffic equilibrium problem. J. Optim. Theory Appl. **167**, 363–381 (2015)
5. Facchinei, F., Pang, J.-S.: Finite-dimensional variational inequalities and complementarity problems. Springer Series in Operations Research, vol. II. Springer, New York (2003)
6. Fang, H., Chen, X., Fukushima, M.: Stochastic R0 matrix linear complementarity problems. SIAM J. Optim. **18**, 482–506 (2007)
7. Giannessi, F., Maugeri, A. (eds.): Variational Inequalities and Network Equilibrium Problems. Plenum Press, New York (1995)
8. Gürkan, G., Özge, A.Y., Robinson, S.M.: Sample-path solution of stochastic variational inequalities. Math. Program. **84**, 313–333 (1999)
9. Gwinner, J., Raciti, F.: On a class of random variational inequalities on random sets. Numer. Funct. Anal. Optim. **27**, 619–636 (2006)
10. Gwinner, J., Raciti, F.: Random equilibrium problems on networks. Math. Comput. Model. **43**, 880–891 (2006)
11. Jadamba, B., Khan, A.A., Raciti, F.: Regularization of stochastic variational inequalities and a comparison of an Lp and a sample-path approach. Nonlinear Anal. **94**, 65–83 (2014)
12. Lin, G.-H., Fukushima, M.: Stochastic equilibrium problems and stochastic mathematical programs with equilibrium constraints: a survey. Pac. J. Optim. **6**, 455–482 (2010)
13. Maugeri, A.: Convex programming, variational inequalities, and applications to the traffic equilibrium problem. Appl. Math. Optim. **16**, 169–185 (1987)
14. Raciti, F., Falsaperla, P.: Improved noniterative algorithm for solving the traffic equilibrium problem. J. Optim. Theory Appl. **133**, 401–411 (2007)
15. Smith, M.J.: The existence, uniqueness and stability of traffic equilibrium. Transp. Res. **138**, 295–304 (1979)
16. Winkler, F.S.: Variational inequalities and complementarity problems for stochastic traffic user equilibria, Bachelor Thesis, Universität der Bundeswehr München and Kyoto University (2011)
17. Zhang, C., Chen, X.: Stochastic nonlinear complementarity problem and applications to traffic equilibrium under uncertainty. J. Optim. Theory Appl. **137**, 277–295 (2008)
18. Zhang, C., Chen, X., Sumalee, A.: Robust Wardrops user equilibrium assignment under stochastic demand and supply: expected residual minimization approach. Transp. Res. Part B **45**, 534–552 (2011)

Construction of Discrete Time Graphs from Real Valued Railway Line Data

Steven Harrod

Abstract Railway timetables are frequently modeled as discrete time expanded graphs. The selection of the magnitude of the discrete time unit can significantly alter the structure of the graph and change the solutions generated. This paper presents a method for generating improved mappings of real railway track segments to discrete arc graphs given a chosen discrete time unit. The results show that the dimensions of the generated graph are not monotonic and a range of values should be evaluated.

Keywords Railway timetable · Discrete optimization · Railway operations

1 Introduction

Frequently, railway timetabling problems are formulated as discrete time expanded graphs. The movements of the trains however, are measured in real valued time. In most formulations, feasible solutions require that train run times be lengthened or rounded up to the nearest discrete time unit, resulting in some increase in travel time and reduction in railway line capacity. It should be explained to the reader new to railways, that nearly all railways divide their rail networks into sections called "blocks". Train movement authorization is given according to these blocks, and a true microscopic model of a railway would represent each of these blocks as an arc. These blocks can be as small as 100 m.

The number of alternative train paths, the density of the graph, and the complexity of the problem all increase as the size of the discrete time unit shrinks. Frequently, these timetabling problems are very large, consisting of tens of thousands of discrete arcs, each of which is represented in a mathematical program by a binary decision variable. One way to reduce the complexity of these problems is to select a larger discrete time unit, with a subsequent increase in the model approximation error. This

S. Harrod (✉)
Technical University of Denmark, 2800 Kongens Lyngby, Denmark
e-mail: stehar@dtu.dk

© Springer Nature Switzerland AG 2018
P. Daniele and L. Scrimali (eds.), *New Trends in Emerging Complex Real Life Problems*, AIRO Springer Series 1,
https://doi.org/10.1007/978-3-030-00473-6_32

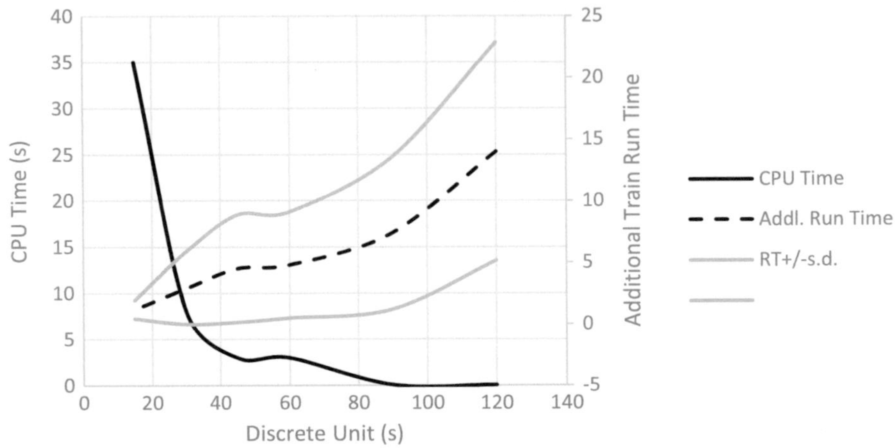

Fig. 1 Performance of range of discrete time units in Caimi et al. [2]

paper demonstrates a method of optimizing the discrete arc graph for a given time unit magnitude.

The design of these discrete arc graphs is one of many tasks in a class of problems variously referred to as "train routing" problems or "timetabling problems" (TTP). Harrod [7] provides a detailed survey on mathematical models of this class. Various prior papers in the literature have selected a discrete time unit according to environmental conditions, business rules, or their judgment, but, with one exception, the choice is not discussed at length and usually is limited to a single sentence. Examples of studies that apply a discrete time unit are Mills et al. [9], Brännlund et al. [1], Caprara et al. [4], Şahin [11], Harrod [6], and Lusby et al. [8].

Caimi et al. [2] presents a problem of the Berne, Switzerland station area. The paper describes assigning run times to trains as the ceiling function of the run time divided by the discrete time unit. A range of discrete time values between 15 and 120 s are tested on one problem scenario, and the results are shown in Fig. 1. The tradeoff between computation time and accuracy can be clearly seen. "Addl. Run Time" is the average additional movement time for each train path through the station due to the rounding up of the real valued run time to discrete time. The nominal run time through the station is 250 s. The bounding lines"RT $\pm$ s .d." are drawn one half of the standard deviation from the "Addl. Run Time" value. In this case, the selected discrete time unit of 90 s is approximately the headway between trains, minimizes the computation time, and is approximately in the midrange of the induced error in run times and capacity.

2 A Method for Constructing a Discrete Time Network

This section proposes a method for distilling a complex real-valued railway network structure into a smaller, discrete valued graph. Given a discrete time unit magnitude as a starting point, the method generates a discrete graph by merging adjacent track segments into longer segments to be represented by the graph arcs. The objective of the process is to minimize the deviation between the real valued travel time on the arcs and the assigned discrete travel time on the arcs.

The error created by the difference between the assigned integer time value and the original real valued movement time is called here "induced" model error. The chosen time unit either needs to minimize induced error over the average of all trains, or over a favored group of trains based on some objective criteria. Not all induced error is bad. Federal Railroad Administration [5] recommends that all simulated train times be increased by 7% to compensate for operating delays not accounted for in train simulations, and some additional induced error may also serve as schedule slack to protect against stochastic delays.

There are two, sometimes opposing, objectives in this method. First, to minimize the number of blocks $|B|$ so that the number of arcs or decision variables in the model is minimized. Second, to dimension the blocks so that the resulting discrete train travel time is a close match to the real valued train time and minimizes induced error. A number of parameters support this process, such as s, the minimum number of track segments to combine, which can define a default train separation or headway. If, for example, the track segments represent signalled track segments, and the rules dictate a two block separation (which is very common), then $s = 3$ will ensure a set of blocks B which will maintain this headway. Another factor to consider in merging track segments is the rolling minimum operating headway. Track segments should not be combined in such a way that they create large variation in the arc travel times, and thus create a bottleneck and reduce the overall flow rate of the line.

Figure 2 displays an example set of track blocks to be combined. The top picture depicts a length of track with signals and two occupying trains, separated by red and yellow signals. The source data is represented below by arcs (a) with travel times labeled. The bottleneck on this route is the segment with travel time $t = 5$, so the maximum flow on this route is one train each 5 time units (because only one train may occupy each segment at a time). The middle set of arcs (b) shows the effect of combining the first three track segments. The flow is not affected. The bottom set of arcs (c) makes a poor choice in combining the last two track segments and reduces the maximum flow to one train each 6 time units.

The combination of track segments into model track blocks is determined by problem (P), which is a simple set partitioning exercise. Refer to Table 1 for explanation of the set notation. The first component of the objective is a tie-breaker. In the event that more than one combination of track blocks offers the same objective value, the one with the least number of members is preferred. The selection of a coefficient of 0.001 is arbitrary, within a range. It should be small to insure that the second component of the objective is the dominant decision maker, but it should not

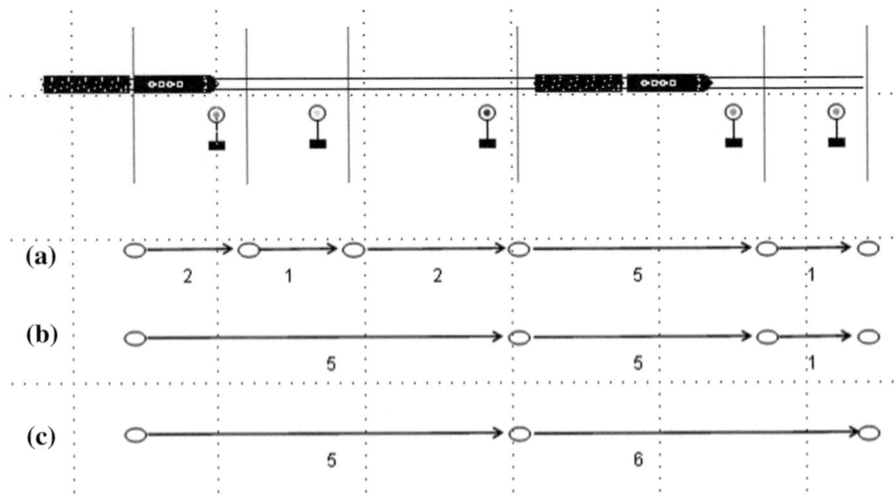

Fig. 2 Example track segment combination process

be too small, as extreme magnitude differences in coefficients can lead to numerical difficulties for integer program solvers [3].

The second component of the objective is the sum of the induced error for the source train data considered. All real valued train times are rounded up to the next integer discrete time value, thus assuring operational feasibility of the resulting model solution timings. The induced error α is the difference of the real value equivalent of the discrete time value and the source real valued train time. The error is then approximately the discrete time unit (μ) minus the modulo of the real time (t) divided by the time unit (Eq. 1).

$$\mu - (t \mod \mu) \tag{1}$$

All combinations of track segments are enumerated in set Ω, whose members are a couple (i, k) where i is the number of track segments combined and k is the first track segment index. For example, set member $(3, 5)$ defines the combination of track segments $\{5, 6, 7\}$, set member $(4, 3)$ defines the combination of track segments $\{3, 4, 5, 6\}$, set member $(5, 8)$ defines the combination of track segments $\{8, 9, 10, 11, 12\}$, etc. The size of the combination of track segments is limited by s, which may determine a minimum physical headway, a user selected maximum combination size l, and the maximum of the rolling minimum operating headway or bottleneck time described earlier. The single constraint requires that any solution cover all source track segments. This problem should be solved for a wide range of values of u (the discrete time unit value, see Table 1), and then the total induced error and number of model track blocks $|B|$ calculated and compared for each value prior to making a final selection.

Table 1 Components of Problem 2

Component	Type	Description
$x_{i,k}$	Binary variable	Represents the selection of segment set (i, k) as a model network track block
u	Parameter	Real value of discrete time unit
$t_{r,i,k}$	Parameter	Total real travel time for train r on segment set (i, k)
$\alpha_{r,i,k}$	Parameter	Induced error for train r on segment set (i, k) for given discrete time unit, $\alpha_{r,i,k} = u - mod(t_{r,i,k}, u)$
h	Parameter	Ruling minimum headway or maximum of rolling value of minimum operating headway
s	Parameter	Minimum number of track segments to combine into one model track block
l	Parameter	Maximum number of track segments to combine into one model track block
Γ	Set	The set of trains considered
Θ	Set	The *ordered* set of source track segments or blocks, ordered by network sequence (original data)
Ω	Set	The set of possible derived blocks, each member, (i, k), maps to a contiguous subset of Θ, starting at position k in set Θ and including i consecutive members (track segments) $\Omega = \left\{ i \in \{s \dots l\}, k \in \{1 \dots \lvert\Theta\rvert\} \middle\vert k \leq (\lvert\Theta\rvert - i + 1), \max_r t_{r,i,k} \leq h \right\}$
Δ_θ	Set	The set of derived or merged blocks from Ω that contain the indicated original source track segment θ $\Delta_\theta = \{(i, k) \in \Omega \vert k \leq \theta < k + i\}$

(P)

$$\min \sum_{(i,k)\in\Omega} 0.001 x_{i,k} + \sum_{r\in\Gamma,(i,k)\in\Omega} \alpha_{r,i,k} x_{i,k} \tag{2}$$

s.t.

$$\sum_{(i,k)\in\Delta_\theta} x_{i,k} = 1 \qquad \forall\theta \in \Theta \tag{3}$$

$$x \in \{0, 1\}$$

To date, the author has solved these directly using commercial solvers such as Cplex without difficulty. What is more difficult is managing the input data for these problems. A formal database structure is valuable for managing this data and the logical relationships between various data entries. The initial objective is to format the track network into a series of minimum dimension units for pre-processing. Each unit has dimensions of start location, end location, capacity (number of tracks), connected track segments at each end, and is tabulated as a record in the database. A suitable database for these records is displayed in Fig. 3. An initial impulse might be to record the track in one mile or kilometer units, but most signal systems do not allow the control of trains in these increments, so other dimensions or division points should be used. Most North American railways operate a fixed installation signalling system, with train control points fixed at the location of color light or position signals, so the locations of these signals are the best starting points for track segment records.

3 Application to a Real Test Case: The BNSF Transcontinental Railway, Winslow to Flagstaff, Arizona

This double track railway is at the midpoint of the journey between Chicago and Los Angeles, and is the dominant traffic lane for BNSF. In addition to the heaviest freight traffic on the BNSF network, the line hosts one daily passenger train in each direction, Amtrak's Southwest Chief between Chicago and Los Angeles. Winslow is a crew change point, and both Winslow and Flagstaff are station stops for the Southwest Chief. Between them lies 54 miles of double track through remote lands and a Navajo Indian Reservation.

Track network data and train timing data are supplied by BNSF. Signals are installed on this line approximately every 2–3 miles, so with source track segments of the same length, there are approximately 21 segments between Winslow and Flagstaff (approximate depending on one's interpretation of the signaling system). Train timing data is provided as computer simulations of a variety of trains. For each train timing, data is supplied westbound (WB: Winslow to Flagstaff), eastbound (EB: Flagstaff to Winslow), and for wet and dry rail in each direction, for a total of 4 timings for each train. Wet rail timings are longer than dry rail timings, because the wet rail limits acceleration and braking, so these timings are used as the more robust of the two choices. A two block separation is presumed. That is, each train is presumed to be trailed by a red signal, a yellow signal, and finally a green signal, and so $s = 3$. An arbitrarily large value of $l = 7$ is applied.

The train types considered are limited to the G and X class freight trains and the Amtrak Southwest Chief. Calculated over a 3 track segment rolling horizon, the bottleneck time for a freight train is approximately 25 min and occurs westbound around Darling, Arizona (milepost 326). Eastbound the bottleneck time is only 14 min, and at the same location. The bottleneck time for Amtrak is only 12 min, at the same location, and in both directions. This mix of trains was arbitrarily chosen to demonstrate a variety of train types. The typical train mix on this line is actually much more diverse, and will vary according to season, day of week, and time of day. The methods presented here may be applied to any specific scenario.

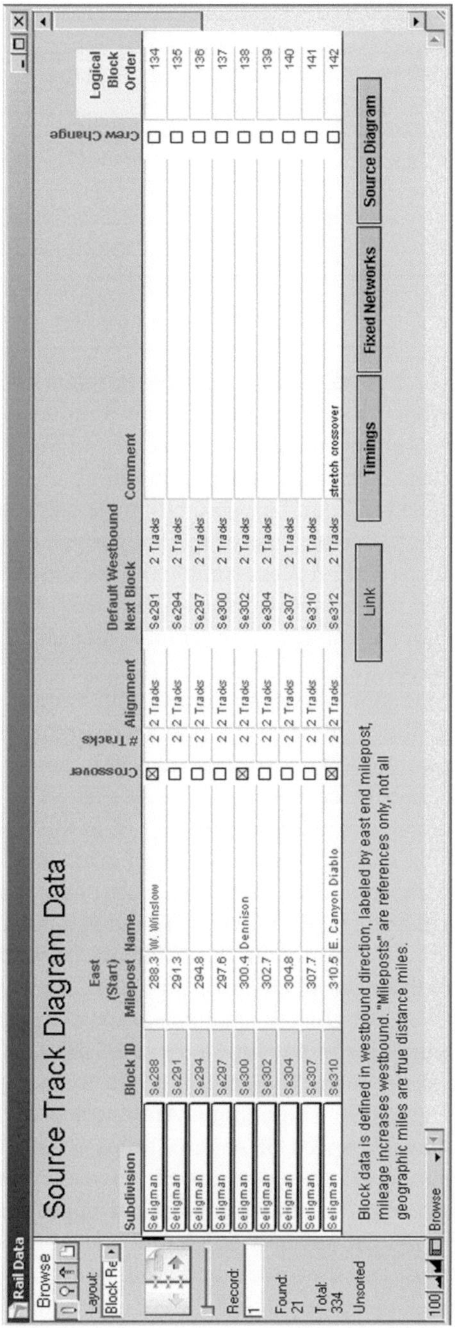

Fig. 3 Track network database

Table 2 Discrete timings for $u = 4.5$

Milepost	Block no.	True miles	Freight		Amtrak	
			WB	EB	WB	EB
288 (Winslow)	1	16.6	6	4	3	3
304	2	10.0	4	3	2	2
314	3	8.8	4	2	2	2
323	4	9.5	5	3	3	3
333 (Flagstaff)	5	11.3	6	4	4	4

Model network track blocks are determined using Problem 2 for discrete time units in 0.5 min increments from 1 to 20, based upon the train timings of the dominant freight trains only. The solution statistics are presented in Fig. 4. The induced error is displayed as a percentage of the real valued train timing. Again, all integer valued train timings are determined by rounding up (ceiling function) the real valued train timings, and the induced error is the sum of the integer valued train timings minus the real valued train timings. The problem complexity is estimated to be proportional to the arc count, presented as a complexity factor, $(1/u)|B|$. That is, the formulation complexity is a function of the number of geographic arcs and the granularity of the time horizon to be analyzed.

Candidate discrete time unit values are tagged with vertical lines in Fig. 4. The first candidate, a discrete time unit of $u = 3$, offers a desirable induced error of 6%, but a complexity factor of 1.67. At a discrete time unit value of 3.5 min, not only does the error increase, but the complexity increases as well. This is because the optimal number of model blocks increases from 5 to 6 at this discrete unit size. The next candidate unit size is $u = 4.5$, which offers an error of 9% and a complexity factor of 1.11, or approximately a 33% reduction in arc count for an admittedly 50% increase in error. This net error is still below 10% and is a practical level providing some schedule slack and compensating for the difference between simulated timings and expected timings. In this data set the trade-off between problem complexity and induced error becomes increasingly less favorable as the discrete unit size increases.

The resulting train timings in discrete time units of 4.5 min are presented in Table 2. The Amtrak trains are faster than the general freight trains by a factor of nearly 2 (compare column Freight/WB with column Amtrak/WB in Table 2), in spite of the fact that the top allowable speed of Amtrak trains is not twice that of freight trains as a rule. The authorized passenger speed between Winslow and Flagstaff is approximately 79 mph, and for general freight it is approximately 45 mph, or a ratio of approximately 1.75. The superior acceleration and braking properties of the Amtrak train allow it to navigate the route much faster than the general freight trains. Also note how the method has homogenized the arc dimensions. Block 1 is twice as many miles as block 3, but in 3 of 4 columns it is only 50% greater in travel time.

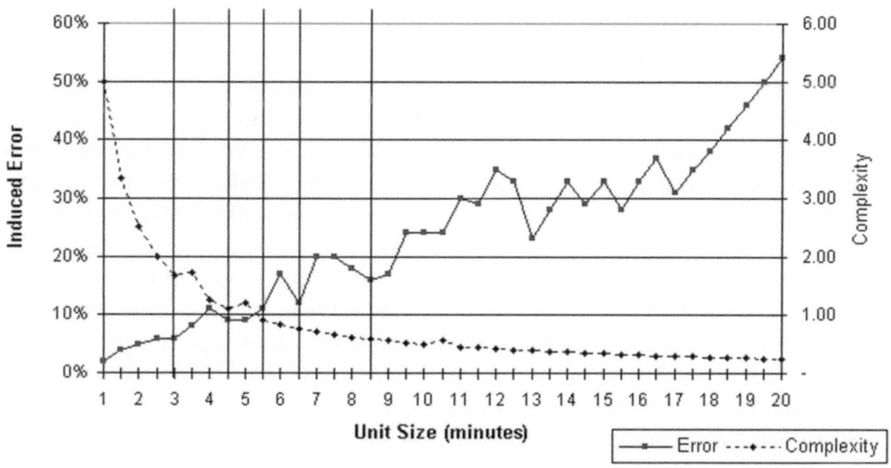

Fig. 4 Induced error and complexity as a function of discrete time unit

Reflecting back to the discussed [2] in the introduction, the results of this specific example demonstrate some similarities. First of all, Figs. 1 and 4 display the same trends, but recall also that Caimi et al. finds that the headway time between trains is a good heuristic for the preferred discrete time unit. In this example the limiting train headway on the BNSF line is not known, but it may be estimated by established methods described in [10]. In this case, the dominant freight traffic runs at 45 mph, a two block signal separation is expected, and the blocks are 2–3 miles long. The trains should thus typically operate with a five mile separation or about 6.66 min, which is a little higher than the result suggested by Fig. 4, but a good initial value. The advantage of this method is that clear guidance can be obtained quickly and with direct evidence, without the necessity of actually constructing and testing alternative models.

4 Conclusion

This paper introduced the application of discrete time units in scheduling problems for railways, and cited [2] as an example of the range of values that could be selected and their impact on the problem complexity and accuracy. Caimi et al. obtains its results by trial and error on a complex railway scheduling problem. This paper offers a mathematical program for generating prospective time unit values specific to a given railway line and train performance. The results and application are comparable. This research could be further extended with more detailed, large problem examples.

The choice of discrete time unit must not be arbitrary, because small changes in unit size can have large effects on the model's representation of and authenticity to actual operations. The problem demonstrated here is not of a large enough size for practical application, but the limitation at the moment is not the capability of solvers such as Cplex, but the time necessary to collect and structure the data. There are approximately 119 signaled track segments between Winslow and Needles, and the westbound freight journey time is 432 min. Using the principles described here, this network could be abstracted to a graph of 24 track blocks and a 4 min discrete time unit, offering approximately an 8% induced error and a complexity factor of 6.0.

The method described here provides a fast process for approximating a real valued set of sequential railway track segments as a discrete arc graph. Multiple discrete time magnitudes may be evaluated and compared on their induced error and resulting graph complexity. The progression of the graph development as the discrete time unit increases is not linear. In some cases a larger time unit offers reduced complexity without incurring larger induced error. Further investigation of this method could evaluate the robustness of the actual train timetabling solutions produced under different discrete time unit values.

References

1. Brännlund, U., Lindberg, P., Nôu, A., Nilsson, J.: Railway timetabling using lagrangian relaxation. Transp. Sci. **32**(4), 358–369 (1998)
2. Caimi, G., Burkolter, D., Herrmann, T., Chudak, F., Laumanns, M.: Design of a railway scheduling model for dense services. Netw. Spat. Econ. **9**, 25–46 (2009)
3. Camm, J.D., Raturi, A.S., Tsubakitani, S.: Cutting big m down to size. Interfaces **20**(5), 61–66 (1990)
4. Caprara, A., Fischetti, M., Toth, P.: Modeling and solving the train timetabling problem. Oper. Res. **50**(5), 851–861 (2002)
5. Federal Railroad Administration: Railroad corridor transportation plans. Technical report, Office of Railroad Development, RDV-10, Federal Railroad Administration, Washington, DC (2005)
6. Harrod, S.: Modeling network transition constraints with hypergraphs. Transp. Sci. **45**(1), 81–97 (2011)
7. Harrod, S.: A tutorial on fundamental model structures for railway timetable optimization. Surv. Oper. Res. Manag. Sci. **17**, 85–96 (2012)
8. Lusby, R., Larsen, J., Ryan, D., Ehrgott, M.: Routing trains through railway stations: a new set-packing approach. Transp. Sci. **45**(2), 228–245 (2011)
9. Mills, R., Perkins, S., Pudney, P.: Dynamic rescheduling of long-haul trains for improved timekeeping and energy conservation. Asia-Pac. J. Oper. Res. **8**, 146–165 (1991)
10. Parkinson, T., Fisher, I.: TCRP report 13, rail transit capacity. Technical report, Transportation Research Board, Washington D.C (1996)
11. Şahin, G.: New combinatorial approaches for solving railroad planning and scheduling problems. Ph.D. Thesis, University of Florida, Gainesville, FL (2006)

An Integer Linear Programming Formulation for Routing Problem of University Bus Service

Selin Hulagu and Hilmi Berk Celikoglu

Abstract The initial phase of our work, concentrating on the formulation of a staff service bus routing problem (SSBRP), is motivated by a real life problem of a university at a multi-centric metropolitan city. In order to improve the overall cost efficiency of the existing staff service bus operation system of the Technical University of Istanbul (ITU) we ultimately aim to find a set of staff service bus routes that provides transportation to and from four campuses for its eligible academics and administrative staff currently using service buses. An integer linear programming formulation for the SSBRP for the single campus case is presented.

Keywords Optimization · School bus routing problem · Traffic and transportation

1 Introduction

The summarized study concentrates on the formulation of a staff service bus routing problem for which an environmentally friend approach is adopted. Our work is motivated by a real life problem of around 2500 users composed of academics and administrative staff, around 1000 bus stations for a heterogeneous bus fleet and four campuses of the Technical University of Istanbul (ITU), where the highway network of operation spreads to two sides of Istanbul, the Asian and the European sides divided by the Bosphorus, and is mostly congested during morning and evening commutes. The increases on vehicle operating costs, together with the increases on urban traffic congestion and the consequent increased traveling times for services bus transport necessitate the cost and time effective planning of the scheduling and the routing of our staff service bus fleet. Therefore, for the cost-effective operation

S. Hulagu · H. B. Celikoglu (✉)
Department of Civil Engineering, Technical University of Istanbul, Ayazaga Campus, 34469 Maslak, Sariyer, Istanbul, Turkey
e-mail: celikoglu@itu.edu.tr

S. Hulagu
e-mail: hulaguselin@itu.edu.tr

© Springer Nature Switzerland AG 2018

P. Daniele and L. Scrimali (eds.), *New Trends in Emerging Complex Real Life Problems*, AIRO Springer Series 1,
https://doi.org/10.1007/978-3-030-00473-6_33

and management of ITU staff service bus fleet, we ultimately seek the optimization of the lines and the routes of service buses considering a number of measures on the vehicle operating costs, traveling times, and the environmental impacts. In the present paper, we summarize the initial phase of our work that is on the formulation of our objective as an Integer Linear Program (ILP) of the vehicle routing problem considering a single campus. We present the preliminary findings as we initially seek the solution with some simplifications within a number of scenarios and assumptions to minimize the overall total cost. While the optimized routes as the after plan for the entire operation have been being evaluated in comparison to the existing routing and scheduling plan currently in operation considering the measures sought to be minimized, i.e., fuel consumption, total cost, traveling time, and exhaust emissions, content of the present paper covers the initiation of the evaluation considering solely the routing for a single campus aiming to minimize solely the total operating cost.

2 Relevant Literature

A great number of studies on both the formulation and the solution of several variants of the Vehicle Routing Problem (VRP), that are clearly classified and summarized in the work by Toth and Vigo [21], do exist. In the following, a review on the VRP with specific reference to the School Bus Routing Problem (SBRP), analogous to the Staff Service Bus Routing Problem (SSBRP) we handle, is summarized.

The bus routing problem can simply be formulated to optimize the scheduling for a fleet of busses to pick up users from various stops and to deliver them to their destinations subject to a number of constraints, i.e., capacities of busses at the fleet, route length of busses, traveling time of users in busses, allocation of each bus stop to single a route, and etc. Park and Kim [17] present a review on SBRP decomposing the content of the works published up to 2010 into sub-problems of bus stop selection (e.g. [18]), bus route generation (e.g. [2]), school bell time adjustment (e.g. [12]), and bus route scheduling (e.g. [19]) following a data preparation step.

Approaches to solving the SBRP that has been being studied since the work by Newton and Thomas [16] are problem specific and have been generally adopted to solve formulations for combined problem of scheduling and routing of crew and vehicle (e.g. [5]), the rural postman problem (e.g. [11]), and the multi-objective vehicle routing problem (e.g. [13]).

Considering the decomposition proposed by Desrosiers et al. [9], the SBRP can be solved throughout four steps following data preparation: bus stop selection where users are assigned to stops; bus route generation; adjustment of school bell time; and route scheduling. Most approaches consider separately and sequentially these steps as sub-problems due to the complexity and the size of the overall problem though sub-problems are not independent. However, most of the relevant works published consider solely some components of the five-step SBRP, frequently the routing and route scheduling of vehicles, since bus stop locations and the opening and closing hours of schools are subject to the policies of local and/or nation-wide authorities such as the board of education, the municipality of the city, and etc. In this aspect a single sub-problem or a combination of its sub-problems has been extensively studied.

Newton and Thomas [16] specified a network to hold all the bus stops and then partitioned it into smaller feasible routes where each route could be covered by a single bus. Employing a clustering algorithm to group bus stops serving 5 schools in Indiana, Angel et al. [1] aimed to generate routes by finding minimum length routes among clusters so that the constraints of the problem are satisfied. Bennett and Gazis [3] studied the problem of bus route generation with 256 bus stops and around 30 routes in Toms River, NJ alternating the objective functions adopted, i.e., minimizing the total user-miles and employing a modified version of the Savings Algorithm of Clarke and Wright [7]. Bodin and Berman [4] followed a 3-opt procedure, including a look-ahead feature and a bus stop splitter, to generate an initial traveling salesman tour that is further partitioned into feasible routes considering the BSRP of a densely populated suburban area with around 13000 students and 25 schools. Given a set of routes of around 100 that deliver all students from their assigned bus stops of around 35 to their schools Swersey and Ballard [20] aimed to find the minimum number of buses to cover these routes by formulating an integer program and solving heuristically the problem of route scheduling using simple cutting planes. Employing a number of methods in land-use based generation of a set of routes; Desrosiers et al. [8, 10] studied the SBRP of 20,000 students and 60 schools in Montreal, Canada within an integer program formulation and a column generation solution approach. Bowerman et al. [6] proposed an approach composed of a multi-objective clustering algorithm to location based grouping of students at Wellington County, Ontario and a combination of a set covering algorithm and a traveling salesman problem algorithm to generate school bus routes and bus stops for each cluster.

3 Methodology: Problem Characteristics and Problem Formulation

The staff service bus transportation system of ITU operates as follows: In the morning, users are picked up at a bus stop that is within a considerable walking distance of their residence. The service bus covers the rest of the bus stops remaining on its route and terminates its trip at a campus where each route serves solely a single campus. In the afternoon, the process is reversed and users are dropped off at the bus stops where they were picked up in the morning.

While we define a number of optimization criteria to evaluate within scenarios the desirability of a particular set of staff service bus routes considering around 110 lines to serve four campuses in conjunction with the policies of the university board that are ‘Number of routes’, ‘Total bus route length’, ‘Load balancing’, ‘Length balancing’, and ‘User walking distance’, we consider the first two for the sample single campus case as summarized in the present paper. In the following we present an ILP formulation of the SSBRP for ITU service bus operation system, considering the routing from and to a single campus.

The SSBRP involves the routing of a number of buses that are operated to transport the staff of ITU between the campus and their houses where each bus is assigned to a single path for the pick-up and the drop-off purposes with a limited seat capacity and a restriction on the maximal traveling distance. Pick-ups and drop-offs are realized at bus stops, each of which is visited only once by a single bus. The objective is therefore to find the minimum number of service buses to serve all the staff and their corresponding routes, so as to minimize the total cost of the overall operation.

Our problem formulation follows a complete graph $G = (V, A)$ with V nodes and A arcs. Assuming that $M = \{1, 2, \ldots, m\}$ is the set of service busses, $V : \{2, 3, \ldots, n\}$ is the set of bus stops, and $V' : \{1, 2, 3, \ldots, n, n + 1\}$ is the set of all routing points that includes the campus node, $\{1\}$, and the depot node, $\{n + 1\}$, the distances between each node pair ij is characterized by a symmetric distance matrix, where d_{ij} represents the distance required to be traversed from node i to node j, and vice versa as well, for $i \neq j$ and $\forall i,\ j \in V'$. Each node $i \in V$ is assigned a number of staff to be picked-up or dropped-off, q_i. The SSBRP therefore turns out to determine m node paths connected to the single origin point as the campus, node 1, so that the total capacity on each path does not exceed the capacity constrain, Q, and the total length of each path does not exceed the bus route length constrain, D. The important decision variable for the problem we concentrate on is binary, x_{ij}, and it determines whether or not an arc sources from node i and sinks node j in the solution as given by (1):

$$x_{ij} = \begin{cases} 1, \text{ if arc } (i, j) \text{ is traversed} \\ 0, \qquad\qquad\qquad \text{otherwise} \end{cases} \quad \text{for } i \neq j \text{ and } \forall i,\ j \in V' \tag{1}$$

In order to obtain the optimal solution to the SSBRP we have adopted a node based approach for formulating the model with polynomial sized binary variable and constraints, in which model variables are selected considering explicitly the nodes of the graph, $G = \left(V', A\right)$. Assuming that: c_{ij} is the cost of traversing from node i to node j and is a function of the distance with the unit distance cost, α, to change the distance matrix to the cost matrix: $c_{ij} = \alpha \cdot d_{ij}$; and f is the fixed unit cost of operation per bus, the general form of the objective to our integer program formulation is given by (2):

$$\min\left\{\sum_{i=1}^{n}\sum_{j=1}^{n}\left(\left(c_{ij} \cdot x_{ij}\right) + (f \cdot m)\right)\right\} \tag{2}$$

subject to the assignment constraints given through (3.1) to (5):

$$\sum_{i=2}^{n} x_{n+1,i} \leq m \tag{3.1}$$

$$\sum_{i=2}^{n} x_{i,1} \le m \tag{3.2}$$

$$\sum_{j=1}^{n} x_{ij} = 1, \quad \forall i \in V \tag{4}$$

$$\sum_{i=2}^{n+1} x_{ij} = 1, \quad \forall j \in V \tag{5}$$

in addition to constraints on capacity, i.e., given through (6) to (8), and distance, i.e., given through (9) to (13) to eliminate sub tours:

$$u_i - u_j + Q \cdot x_{ij} + \left(Q - q_i - q_j\right) \cdot x_{ji} \le Q - q_j, \quad i \ne j, \quad i, j = 2, 3, \ldots, n \tag{6}$$

$$u_i - \sum_{j=2}^{n} \left(q_j \cdot x_{ji}\right) \ge q_i, \quad j \ne i, \quad i = 2, 3, \ldots, n \tag{7}$$

$$u_i + (Q - q_i) \cdot x_{n+1,i} \le Q, \quad i = 2, 3, \ldots, n \tag{8}$$

$$v_j = v_i + d_{ij} \tag{9}$$

$$v_i - v_j + \left(D - d_{i1} + d_{ij}\right) \cdot x_{ij} + \left(D - d_{i1} - d_{ji}\right) \cdot x_{ji} \le D - d_{i1}, \quad i \ne j, \quad i, j = 2, 3, \ldots, n \tag{10}$$

$$v_i \ge \sum_{j=2}^{n} \left(d_{ji} \cdot x_{ji}\right), \quad j \ne i, \quad i = 2, 3, \ldots, n \tag{11}$$

$$u_j = u_i + q_j \tag{12}$$

$$v_i \le D - \left(D \cdot x_{n+1,i}\right), \quad i = 2, 3, \ldots, n \tag{13}$$

$$x_{ij} \in \{0, 1\} \quad \forall i, j \in V' \tag{14}$$

$$u_i \ge 0 \quad \wedge \quad v_i \ge 0, \quad i \in (2, 3, \ldots, n) \tag{15}$$

where: u_i is the total amount of staff picked up (or dropped off) by the service bus after visiting node i for $i \in V'$; v_i is the total length traveled from depot to node i (or from campus to node i) for $i \in V'$.

As the SSBRP is asymmetrical, and hence the distances should be considered accordingly, the ILP given above is formulated specific to depots to campus morning routings as represented by (3.1) and (3.2). The formulation for the campus to depots afternoon routings necessitates the substitution of the two constraints respectively by (16.1) and (16.2).

$$\sum_{i=2}^{n} x_{i,1} \le m \tag{16.1}$$

$$\sum_{i=2}^{n} x_{i,n+1} \leq m \tag{16.2}$$

Note that constraints (4) and (5) are the degree constraints that secure each of the nodes to be visited only once, where constraints (3.1) and (3.2) restrict the maximum number of service buses that depart from node 1 to arrive node $n + 1$ to m.

In order to impose the restrictions given from (6) to (8) that each service bus has to carry at most its capacity, Q, the Miller et al. based constraints [15] proposed by Kara et al. [14] are utilized, specifically: constraint (6) to eliminate the sub-tours, as in a capacitated VRP, by ensuring that u_i follows a step-wise function; and constraints (7) and (8) restrict the lower and the upper bounds for the total number of staff picked-up by the service bus after leaving node i, q_i. In a similar manner, while the constraints (9) and (10) eliminate the sub-tours by ensuring the function of the total length traveled to be stepwise, constraints (11) and (12) restrict the lower and the upper bounds for the total length covered by the service bus until node i, v_i.

Constraints (14) and (15) represent respectively that: x_{ij} is an integer decision variable that is binary; and u_i and v_i are decision variables that can take continuous values.

4 Preliminary Results and Discussion

In the present study we have considered the intercontinental routings, where all the routing points other than the campus take place on the Asian side while the campus is located on the European side, from and to a single campus, ITU Taskisla Campus, as a sample case. The optimization of the service bus operation system of ITU is, therefore partially sought over the network piece with arcs of around 365 km out of 6718 (see road network colored red in Fig. 1) and traversing throughout of around 119 planning zones out of 451 (see borders colored black in Fig. 1) within an ILP formulation of the SSBRP considering the features: the number of stops in addition to the campus and the depot; the cost of traveled distance within an affordable range suggested by the university board; the number of seats available at buses; the maximum distance to be covered by all routes; the fixed cost per bus; and a-priori info on staff assignment to stops.

As the current routing plan (see in Fig. 2) for each line is observed to visit the bus stops that are turned out to be specified upon each user's preference we have considered in each zone a centroid with inter-modal transfer alternatives as the pick-up or drop-off of all the staff accommodating in the zone of interest. Though such an aggregation would violate a reasonable walking distance from and to homes of staff, the approach fits well the ultimate aim of decreasing the number of lines currently being operated, of around 110 to serve four campuses, to its half due to the cuts-off in the budget and/or relevant funds.

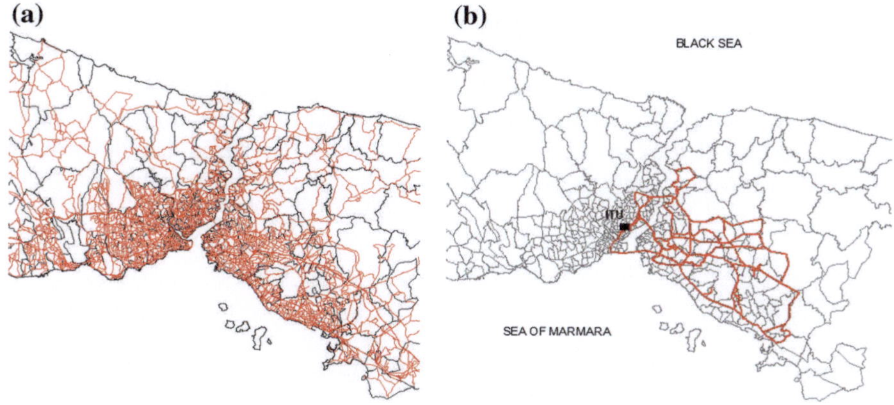

Fig. 1 **a** Istanbul road network with planning zones. **b** Network piece considered

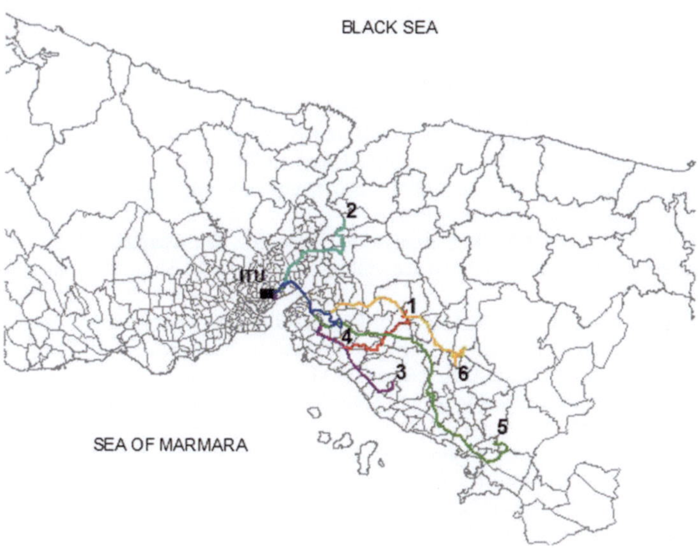

Fig. 2 Current routing plan for sample single campus

A GIS software is utilized to obtain the distances between each node pairs over the road network of Istanbul considering the urban freeways and urban arterials that serve to all zones with above mentioned centroids. The unit distance cost for each service bus is assumed to be 0.232 USD per kilometer with a fixed cost of 90288 USD per bus.

Model based exact solutions for 12 asymmetrical problems for 6 lines are sought with an open source code solver using Intel(R) Xeon(R) CPU E5-26500 processors each at 2.00 Ghz, and 32 GB of RAM. The optimal solution resulted in a total cost that corresponds to around 35% of distance cost reduction compared to the current routing plan. While we present in present paper the preliminary analyses aiming to minimize the overall operating cost to enhance the constraints service bus operating budget, our further research concentrates on extending the problem formulation to account for: (i) the pollution based routing incorporating either a number of constraints on air pollutants emitted by the buses that form the fleet or emission damage costs in the objective function; and (ii) the mixed loads from all the four campuses.

References

1. Angel, R.D., Caudle, W.L., Noonan, R., Whinston, A.: Computer-assisted school bus scheduling. Manage. Sci. **18**(6), 279–288 (1972)
2. Bektas, T., Elmastas, S.: Solving school bus routing problems through integer programming. J. Oper. Res. Soc. **58**(12), 1599–1604 (2007)
3. Bennett, B.T., Gazis, D.C.: School bus routing by computer. Transp. Res. **6**, 317–325 (1972)
4. Bodin, L.D., Berman, L.: Routing and scheduling of school buses by computer. Transportation Science **13**(2), 113–129 (1979)
5. Bodin, L.D., Golden, B., Assad, A., Ball, M.O.: Routing and scheduling of vehicles an crews: the state of the art. Comput. Oper. Res. **10**, 63–211 (1983)
6. Bowerman, R., Hall, B., Calamai, P.: A multi-objective optimization approach to urban school bus routing: formulation and solution method. Transp. Res. Part A **29**(2), 107–123 (1995)
7. Clarke, G., Wright, J.: Scheduling of vehicles from a central depot to a number of delivery points. Oper. Res. **12**(4), 568–581 (1964)
8. Desrosiers, J., Ferland, J.A., Rousseau, J.-M., Lapalme, G., Chapleau, L.: TRANSCOL: a multi-period school bus routing and scheduling system. TIMS Stud. Manage. Sci. **22**, 47–71 (1986)
9. Desrosiers, J., Ferland, J.A., Rousseau, J.-M., Lapalme, G., Chapleau, L.: An overview of a school busing system. In: Jaiswal, N.K. (ed.) Scientific Management of Transport Systems, pp. 235–243. North-Holland, Amsterdam (1981)
10. Desrosiers, J., Soumis, F., Desrochers, M., Sauvé, M.: Methods for routing with time windows. Eur. J. Oper. Res. **23**(2), 236–245 (1986)
11. Eiselt, H.A., Gendreau, M., Laporte, G.: Arc routing problems. Part II: The rural postman problem. Oper. Res. **43**, 399–414 (1995)
12. Fügenschuh, A.: Solving a school bus scheduling problem with integer programming. Eur. J. Oper. Res. **193**(3), 867–884 (2009)
13. Jozefowiez, N., Semet, F., Talbi, E.G.: Multi-objective vehicle routing problems. Eur. J. Oper. Res. **189**(2), 293–309 (2008)
14. Kara, I., Laporte, G., Bektas, T.: A note on the lifted Miller–Tucker–Zemlin subtour elimination constraints for the capacitated vehicle routing problem. Eur. J. Oper. Res. **158**(3), 793–795 (2004)
15. Miller, C.E., Tucker, A.W., Zemlin, R.A.: Integer programming formulations and traveling salesman problems. J. Assoc. Comput. Mach. **7**, 326–329 (1960)
16. Newton, R.M., Thomas, W.H.: Design of school bus routes by computer. Socio-Econom. Plann. Sci. **3**(1), 75–85 (1969)
17. Park, J., Kim, B.-I.: The school bus routing problem: a review. Eur. J. Oper. Res. **202**, 311–319 (2010)

18. Schittekat, P., Sevaux, M., Sörensen, K.: A mathematical formulation for a school bus routing problem. In: Proceedings of the IEEE 2006 International Conference on Service Systems and Service Management, Troyes, France (2006)
19. Spada, M., Bierlaire, M., Liebling, Th.M.: Decision-aiding methodology for the school bus routing and scheduling problem. Trans. Sci. **39**, 477–490 (2005)
20. Swersey, A.J., Ballard, W.: Scheduling school buses. Manage. Sci. **30**(7), 844–853 (1984)
21. Toth, P., Vigo, D.: Vehicle routing: problems, methods, and applications. MOS-SIAM series on optimization, vol. 18, 2nd ed. Philadelphia, SIAM (2014)

Testing Demand Responsive Shared Transport Services via Agent-Based Simulations

Giuseppe Inturri, Nadia Giuffrida, Matteo Ignaccolo, Michela Le Pira, Alessandro Pluchino and Andrea Rapisarda

Abstract In this paper, an agent-based model is presented to test the feasibility of different configurations of Demand Responsive Shared Transport (DRST) services in a real context. DRST services provide "just-in-time" mobility solutions by dynamically assigning a fleet of vehicles to passenger booking requests taking advantages of Information and Communication Technologies. First results show the impact of different route choice strategies on the system performance and can be useful to help the planning and designing of such services.

Keywords Shared mobility · Flexible transit · Dynamic ride sharing Mobility on demand · Agent-based model

---

G. Inturri (✉)
Department of Electric, Electronic and Computer Engineering, University of Catania, Via Santa Sofia 64, Catania, Italy
e-mail: ginturri@dica.unict.it

N. Giuffrida · M. Ignaccolo · M. Le Pira
Department of Civil Engineering and Architecture, University of Catania, Via Santa Sofia 64, Catania, Italy
e-mail: nadia.giuffrida@dica.unict.it

M. Ignaccolo
e-mail: matteo.ignaccolo@unict.it

M. Le Pira
e-mail: mlepira@dica.unict.it

A. Pluchino · A. Rapisarda
Department of Physics and Astronomy, University of Catania, Via Santa Sofia 64, Catania, Italy
e-mail: alessandro.pluchino@ct.infn.it

A. Rapisarda
e-mail: andrea.rapisarda@ct.infn.it

A. Pluchino · A. Rapisarda
Infn sezione di Catania, Via Santa Sofia 64, Catania, Italy

A. Rapisarda
Complexity Science Hub Vienna, Josefstädter Str. 39, 1080 Wien, Austria

© Springer Nature Switzerland AG 2018

P. Daniele and L. Scrimali (eds.), *New Trends in Emerging Complex Real Life Problems*, AIRO Springer Series 1,
https://doi.org/10.1007/978-3-030-00473-6_34

1 Introduction

This paper focuses on the potential contribution of innovative Demand Responsive Shared Transport (DRST) services provided by a fleet of vehicles, booked by users via mobile device applications, and scheduled in real-time to pick up and drop off passengers in accordance with their needs [1]. This flexible system stands between an expensive individual door-to-door ride service (like a taxi) and a typical transit service system.

From the operator's point of view, it is important to select the optimal strategy to assign vehicles to passengers' requests, so to perform high load factor and low driven distance (to reduce operation costs). From the users' point of view, it is important to minimize the additional time and distances they have to experience due to the shared ride and, possibly, to pay reduced fees with respect to an individual ride service.

In the last years, increasing attention has been paid on shared transport services. Optimization models have been proposed to solve dial-a-ride [2] or multiple depot vehicle scheduling problems [3]. More recently, simulation models have been used, e.g. to study the performance of dial-a-ride systems with fixed-route problems [4], the efficient scheduling of dynamic demand responsive transport (DRT) services [5], the dynamics of taxi-sharing systems [6], the effects of using a zoning vs. a no-zoning strategy and time-window settings [7].

In this respect, agent-based models (ABM) proved effective to reproduce complex social systems, and to overcome some limitations of "top-down" approaches, by reproducing the microinteraction among single autonomous agents in different context, e.g. among stakeholders involved in transport decision-making processes [8, 9]. In the field of ride sharing, they have been proposed to study taxi fleet operations [10], car sharing [11] and to investigate DRT services by developing an open-source simulation testbed [12]. Main benefits of ABM are the possibility to provide a realistic description of a system, capture emergent phenomena from the microinteraction among agents, and being flexible, in terms of number and type of modelled agents, agent behavior, degree of rationality, ability to learn and evolve, and rules of interactions [13].

In this paper, a new ABM is presented as a realistic environment where to simulate different DRST scenarios, with simple rules assigned to agents, in order to explore the transport demand and supply variables that make this service feasible and convenient. The model simulates the interaction between vehicles traveling along a semi-flexible route and users walking from their origins to stop-nodes to get a transport service to their destinations. The main novelty relies on the integration of a GIS-based demand model into the ABM that can be easily transferred to other contexts. The aim is to understand, starting from the micro-interaction between demand and supply agents (i.e. passengers and vehicles), the macroscopic behavior of the system so to monitor, via appropriate indicators, its performance and give suggestions on its planning, management and optimization. The city of Ragusa (Italy) is chosen as case study, where an innovative DRST has already been tested.

The remainder of the paper is organized as follows. Next section provides a detailed description of the ABM and its main components. Then, the case study with the main input variables is introduced. In the last section, preliminary results of the simulations are presented, and future research steps as well.

2 Model Description

The ABM has been implemented within NetLogo programming environment [14] to test the impact of different vehicle route choice strategies on the service efficiency and effectiveness.

The model can be described according to its main features, i.e. (i) transport network, (ii) demand model, (iii) agent (user and vehicle) dynamics, (iv) route choice strategies, (v) performance indicators.

Transport network. The network consists of a fixed route and three optional routes, composed of network nodes and links, stop-nodes and diversion-nodes. It reproduces the actual road network. Besides, a GIS map provides georeferenced socio-economic data about population and economic activities.

Demand model. A user group's[1] trip request is randomly generated with a negative exponential distribution according to a gravity model [15], with an average trip rate from an origin i to a destination j proportional to population density.

Agent (user and vehicle) dynamics. Agents' dynamics works as follows. If the distance between a trip origin (or destination) and the nearer service stop overcomes a prefixed threshold, the request is rejected. Otherwise, the user group moves to the nearest stop and waits until a vehicle (with available empty seats) reaches the stop. In this case, each user goes on board, and alights at the nearest stop to its required destination and it is recorded as a "satisfied" user. Conversely, if a prefixed maximum waiting time is overcome before a vehicle reaches the origin stop, the user group gives up and assumes the status of "unsatisfied".

As far as vehicle dynamic is concerned, a given number of vehicles with prefixed seat capacity is randomly generated at stops, and they start traveling along the fixed route at constant speed. At each stop, waiting users are loaded following the First-Come-First-Served queue rule, if the group size is not greater than the available empty seats. At each diversion node, a vehicle can shift to an optional route if there are waiting users or if on-board users have to alight along the route. The available vehicle seats are updated at each event of passenger loading/unloading.

[1]Each group has a maximum prefixed size according to vehicle's capacity.

Route choice strategies. All vehicles drive on the fixed route. At diversion nodes, a vehicle may drive on a flexible route in accordance with the assigned Route Choice Strategy (RCS). In this first version of the model, there are three RCSs:

- "Fully Random" (FR), vehicles can randomly decide whether to drive on optional routes;
- "All Vehicles drive on All flexible Routes" (AVAR), except for a random percentage of vehicles choosing, as for the FR strategy;
- "Each Vehicle is Assigned to a flexible Route" (EVAR), except for a random percentage of vehicles choosing, as for the FR strategy.

The randomness component has been introduced to add some "noise" to the system, since it has been demonstrated that this can have a beneficial role in increasing the efficiency of social and economic complex systems [16].

Performance indicators. The local interaction between passengers and vehicles determines system patterns that can be monitored via appropriate performance indicators. In this first version of the model, we decided to monitor 12 indicators, i.e.: (1) total number of served passengers (NP), (2) total driven distance [TDD (km)], (3) average passenger travel distance [APTD (km)], (4) average vehicle load factor (ALF), (5) average waiting time [AWT (min)], (6) average on-board time [AOBT (min)], (7) average total travel time [APTT (min)], (8) average vehicle speed [AVS (km/h)], (9) transport intensity [CI (km/pax)]—as ratio of total driven distance and number of passengers (TDD/NP), (10) total user travel time ([TPTT (h)]—including a penalty of 60 min for each unsatisfied user), (11) vehicle operation cost [OC (€)], and (12) total unit cost [TUC (€/pax)], evaluated according to the following equation, where VOT is the Value of Time:

$$TUC\left(\frac{€}{pax}\right) = \frac{TPTT(h) \cdot VOT\left(\frac{€}{h}\right) + OC(€)}{NP(pax)} \tag{1}$$

3 Case Study

The case study is Ragusa, a small-medium city (70,000 inhabitants) in the southeastern part of Sicily (Italy), where an innovative DRST service, called MVMANT,[2] has already been tested in 2016. Two distinct areas, i.e. the upper town, and the lower historical center of Ragusa Ibla, with a high touristic vocation, characterize the city. MVMANT has connected several park-and-ride facilities with the main destinations in Ragusa Ibla, which is scarcely connected to the center, offering a continuous service with midsize passenger vans.

Figure 1 shows the fixed (blue) and flexible (orange) routes, and census zones on the left colored according to population density (from light to dark green).

[2]https://www.mvmant.com/pilot-in-ragusa/.

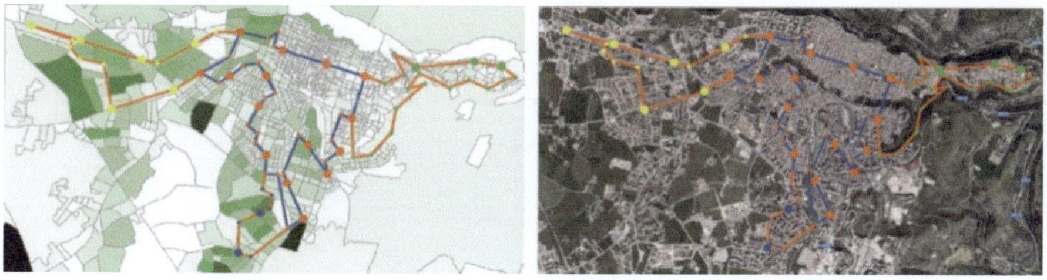

Fig. 1 Virtual map (left) and satellite map (right)

The main input variables of the system are:

- service variables, i.e. total simulation time (h), number of vehicles, vehicle maximum capacity (seats), vehicle average speed (km/h);
- demand variables, i.e. demand rate (request/hour), maximum number of passengers per group request, maximum waiting time (min);
- route choice strategy, i.e. FR, AVAR, EVAR, with a variable percentage of randomness for the last two strategies.

4 Preliminary Results and Conclusions

In order to test the model and the overall system performance, we performed several simulations by varying the number of vehicles and their capacity (with the same total capacity), and by considering different route choice strategies (i.e. FR, AVAR, and EVAR) with increasing levels of randomness.

Figure 2 (on the left) shows the total hours spent by all passengers (in yellow), the hours spent while waiting (in light blue), while on board (orange), the hours considering a penalty of 60 min for each "unsatisfied" user (grey), and the total number of transported passengers (in dark blue). It is worthy of notice that the total number of transported passengers decreases with 15 and 30 vehicles, since group requests with three passengers cannot be satisfied by vehicles with low capacity. On the right, Fig. 2 shows the total unit cost TUC (€/pax) by the number of vehicles. It is calculated by assigning a monetary value to each hour spent by the passengers in the system (10 €/h), adding the operation cost of vehicles (in the range of 0.5–1.0 €/km based on the vehicle size) and drivers' cost (20 €/h), and dividing by the number of passengers. It can be considered as a measure of the total system cost (demand and supply) for the mobility of one person. In this case, an optimal range can be identified in the range between 5 and 10 vehicles.

Comparing TUC for EVAR and AVAR strategies with variable randomness (Fig. 3), best results are found with EVAR and no randomness. This is because vehicle assignment to specific routes (EVAR) reduces the travelled distances by empty vehicles. By increasing randomness, the two strategies get closer in terms

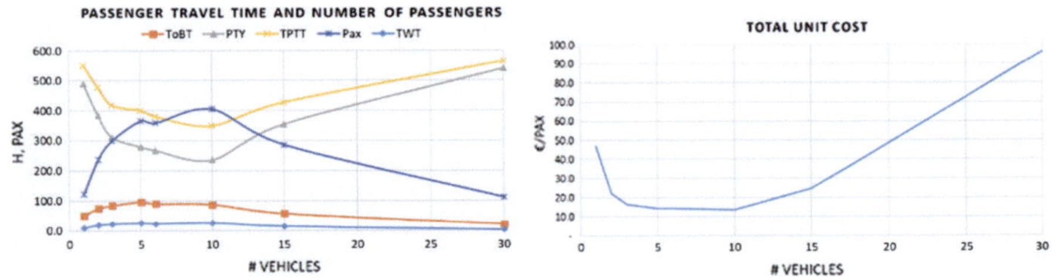

Fig. 2 Passenger travel time and number of passengers (on the left) and total unit cost TUC (on the right) (route choice strategy EVAR with 30% randomness; maximum group size = 3; total seat capacity = 30)

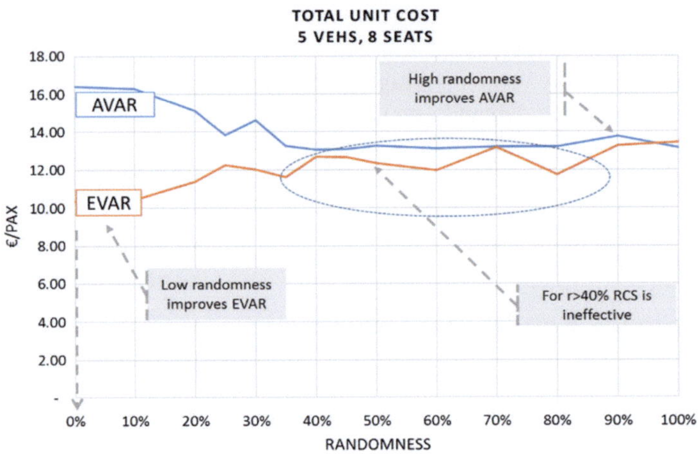

Fig. 3 Total unit cost by randomness (number of vehicles = 5; seat capacity = 8; maximum group size = 1)

of TUC. With randomness over 40%, the two strategies give the same results with intermediate and almost constant system performance. It can be concluded that a certain rate of randomness is beneficial for the AVAR strategy, since it implies that not all vehicles will simultaneously explore all the routes if demand is present (thus reducing the empty driven distance). Vice versa, EVAR strategy works better without randomness.

In conclusion, simulation results show that the service quality and performance considerably vary with the number and capacity of vehicles. In particular, given a fixed supply (in terms of total seat capacity), many vehicles with low capacity decrease passenger travel time (and cost) and increase the operator costs, while few high-capacity vehicles perform better from an operator's point of view. An optimal range can be found by considering a total unit cost accounting both for passenger travel time and vehicle operation cost.

Besides, simulation results with different vehicle dispatching strategies (from flexible to more fixed route strategies) show that assigning vehicles to specific routes reduces travelled distances by empty vehicles. If all vehicles are allowed to drive on all the routes, then a certain level of randomness in agent's choice is found to be beneficial for the system performance.

First results show that the model is able to reproduce different system configurations and to monitor, via appropriate indicators, its performance.

Further research will focus on: (a) comparing DRST with pure taxi and pure bus services; (b) testing other strategies to optimize the service (i.e. increase load factor, reduce vehicle-km), e.g. re-balancing/idle strategies; (c) testing reactive/adaptive agent behaviors for route choice strategies based on system states; (d) testing pricing strategies and public subsidies to increase the service effectiveness (in terms of satisfied demand); (e) testing the system performance with autonomous vehicles; (f) including elasticity of demand to price; (g) improving the demand model (e.g. including socio-demographics characteristics, data from surveys).

References

1. Ambrosino, G., Nelson, J.D., Romanazzo, M.: Demand responsive transport services: towards the flexible mobility agency. ENEA. Italian National Agency for New Technologies, Energy and the Environment. ISBN 88-8286-043-4 (2003)
2. Stein, D.M.: Scheduling dial-a-ride transportation systems. Transp. Sci. **12**(3), 232–249 (1978)
3. Bodin, L., Golden, B.: Classication in vehicle routing and scheduling. Networks **11**(2), 97–108 (1981)
4. Shinoda, K., Noda, I., Ohta, M., Kumada, Y., Nakashima, H.: Is dial-a-ride bus reasonable in large scale towns? Evaluation of usability of dial-a-ride systems by simulation. In: Multiagent for Mass User Support, pp. 105–119. Springer (2004)
5. Diana, M.: The importance of information flows temporal attributes for the efficient scheduling of dynamic demand responsive transport services. J. Adv. Transp. **40**(1), 23–46 (2006)
6. D'Orey, P.M., Fernandes, R., Ferreira, M.: Empirical evaluation of a dynamic and distributed taxi-sharing system. In: Proceedings of ITSC 2012, pp. 140–146. IEEE (2012)
7. Quadrifoglio, L., Dessouky, M., Ordonez, F.: A simulation study of demand responsive transit system design. Transp. Res. Part A **42**, 718–737 (2008)
8. Le Pira, M., Inturri, G., Ignaccolo, M., Pluchino, A.: Dealing with the complexity of stakeholder interaction in participatory transport planning. In: Zak, J., Hadas, Y., Rossi, R. (eds.) Advanced Concepts, Methodologies and Technologies for Transportation and Logistics; Advances in Intelligent Systems and Computing, vol. 572. Springer International Publishing. https://doi.org/10.1007/978-3-319-57105-8_3 (2018)
9. Marcucci, E., Le Pira, M., Gatta, V., Ignaccolo, M., Inturri, G., Pluchino, A.: Simulating participatory urban freight transport policy-making: accounting for heterogeneous stakeholders' preferences and interaction effects. Transp. Res. Part E **103**, 69–86 (2017)
10. Cheng, S.F., Nguyen, T.D.: Taxisim: a multiagent simulation platform for evaluating taxi fleet operations. In: Proceedings of the 2011 IEEE/WIC/ACM International Conferences on Web Intelligence and Intelligent Agent Technology, vol. 02, pp. 14–21 (2011)
11. Lopes, M.M., Martinez, L.M., de Almeida Correia, G.H.: Simulating carsharing operations through agent-based modelling: an application to the city of Lisbon, Portugal. Transp. Res. Procedia **3**, (2014)
12. Čertický, M., Jakob, M., Píbil, R.: Simulation testbed for autonomic demand-responsive mobility systems. In: Autonomic Road Transport Support Systems, pp. 147–164 (2016)
13. Bonabeau, E.: Agent-based modeling: methods and techniques for simulating human systems. PNAS **99**(Suppl 3), 7280–7287 (2002)

14. Wilensky, U.: NetLogo. Center for Connected Learning and Computer Based Modeling. Northwestern University, Evanston, IL (1999). In: http://ccl.northwestern.edu/netlogo/
15. de Dios Ortuzar, J., Willumsen, L.G.: Modelling Transport. Wiley (2011)
16. Pluchino, A., Rapisarda, A., Garofalo, C.: The Peter principle revisited: a computational study. Phys. A **389**(3), 467–472 (2010)

Production Control in a Competitive Environment with Incomplete Information

Konstantin Kogan and Fouad El Ouardighi

Abstract We consider an industry consisting of a large number of firms producing substitutable products and engaged in a dynamic Cournot-type competition. The firms are able to reduce their marginal production costs by accumulating their own experience as well as the experience spillovers from other firms. In particular, firms accumulate production experience through proprietary learning, which, however, depreciates over time. We determine steady-state Nash equilibrium policies that are based on the assumption that the firms do not have precise information about each competitor and therefore are unable to respond to a specific firm's dynamics. The firms, however, do react to overall industry trends. We show that in such a case, though the information used for production control is incomplete, in the long run the firms tend to the output they would converge to under complete information. We also find that industry growth due to more firms entering the market results in decreasing long-run equilibrium output of each firm when the depreciation of experience is higher than the rate of spillovers. Otherwise, the opposite result can emerge, i.e., the steady-state output will grow.

Keywords Production · Control · Differential games · Quantity competition

1 Introduction

We consider an industry consisting of firms competing by producing substitutable goods. Similar to a Cournot game [4], the firms choose product quantities (outputs) that determine the market price and thereby the firms' prinfofits. The firms are able

K. Kogan (✉)
Management Department, Bar-Ilan University, 52900 Ramat-Gan, Israel
e-mail: kogank@biu.ac.il

F. El Ouardighi
Operations Management Department, ESSEC Business School, Avenue Bernard Hirsch, 95021 Cergy Pontoise, France
e-mail: elouardighi@essec.fr

© Springer Nature Switzerland AG 2018
P. Daniele and L. Scrimali (eds.), *New Trends in Emerging Complex Real Life Problems*, AIRO Springer Series 1,
https://doi.org/10.1007/978-3-030-00473-6_35

to reduce their production marginal costs by accumulating their experience as well as through experience spillovers from their competitors (see, for example, the reviews on the effect of learning by doing in [1, 2, 5, 11, 14]). Due to its cumulative nature, production experience is interpreted herein as a stock variable that requires a dynamic approach. A similar interpretation is suggested by Jarmin [7] who develops an empirical multi-period model to study the intertemporal nature of learning by doing and spillovers.

Early analytical models that account for the effect of experience typically consider quantity-based competition over two periods, with proprietary learning-by-doing and/or spillovers (see, for example, [6, 8, 12]). Stokey [13] develops a more general model—a differential game with production dynamics and complete spillovers, i.e., the unit cost for any firm depends only upon cumulative industry production. Stokey characterizes equilibria and finds persistent free riding due to complete spillovers. To overcome tractability issues and determine explicit steady-state equilibrium in a similar problem, Miravete [10] assumes no spillovers and only fixed cost reduction due to accumulated output. Kogan et al. [9] provide an analytical solution for two competing firms with proprietary learning by doing and partial spillovers when the firms are able to account for the competitor's experience dynamics.

Unlike the cases that appear in the literature above and similar to what occurs in real life markets, we consider an industry that consists of a large number of firms which do not have full information about their competitors. In particular, not only the spillovers are partial, but also the firms are not able to account for the individual dynamics of each specific competitor. Instead the firms account for the overall industry dynamics when choosing the equilibrium time path of their output. We find the overall industry growth results in decreasing output and lower experience of individual firms (reduced competiveness) when the experience depreciation is higher than the rate of spillovers. Otherwise, the opposite result can emerge and the firms will become more competitive.

We proceed in the next section with the formulation of the problem. Section 3 is devoted to the Nash equilibria conditions while Sect. 4 focuses on steady-state (long-run) equilibrium policies. Section 5 summarizes our results.

2 The Model

Consider a mature duopoly with N manufacturers competing by producing substitutable goods. Specifically, the firms simultaneously choose a quantity to be produced per time unit (output) over an infinite planning horizon, $q_i(t) \geq 0$, $i = 1, 2, \ldots, N$. Similar to the Cournot competition approach, the market price P of the product at time t depends negatively on the total output at time t, $Q(t) = \sum_{i=1}^{N} q_i(t)$. That is, $P(Q(t)) = \alpha - Q(t)$, where α is the maximal price, $P(0) = \alpha$. The cumulative experience, $X_i(t)$, of a firm is due to its cumulative production, $\int_0^t q_i(s)ds$, and depreciation at the rate of ρ:

$$\dot{X}_i(t) = q_i(t) - \rho X_i(t),\ X_i(0) \geq 0,\ q_i(t) \geq 0,\ i = 1, 2, \ldots, N. \tag{1}$$

We assume each firm is unable to account for the experience dynamics of a specific competitor, but it is able to account for the entire industry experience, $X(t) = X(t) = \sum_{i=1}^{N} X_i(t)$, dynamics which with respect to (1) is determined by the overall output,

$$\dot{X}(t) = Q(t) - \rho X(t), \tag{2}$$

Equations (1) and (2) imply there is saturation in accumulating the production experience and this affects the firm's ability to reduce its unit production cost [3].

Similar to Jørgensen and Zaccour [8] and Fudenberg and Tirole [6], we assume a linear effect of proprietary learning-by-doing and spillovers on the firm's marginal production cost:

$$C(X_i(t), X(t)) = c - \gamma X_i(t) - \varepsilon X(t),\ i = 1, 2, \ldots, N, \tag{3}$$

where c is the initial unit production cost, γ is the rate of proprietary learning, and ε is the rate of learning from the industry experience spillover. Note, that calculation of the correct γ assumes reducing it by ε to eliminate the need to account for a spillover from firm i to itself, as $X(t)$ includes also X_i.

Given the discount rate δ of future profits, each firm maximizes its cumulative profit:

$$\max_{q_i(\cdot)} \int_0^{\infty} \left[\alpha - Q(t) - c + \gamma X_i(t) + \varepsilon X(t)\right] q_i e^{-\delta t} dt \tag{4}$$

subject to the constraint (3), $i = 1, 2, \ldots, N$.

Recalling our assumption that each firm is only able to respond to the overall industry behavior, formulation (1)–(4) presents a differential game where individual firms are playing against the entire industry when making a decision. In what follows we study time-dependent $\{q_i(t), t \geq 0\}$ Nash equilibrium output policies. Since such policies are weakly time-consistent, the equilibrium actions can be continued. This ensures that the players' commitments at any time along the equilibrium path are credible.

3 Equilibrium Policies

Using the maximum principle and omitting index t where it is obvious, the current value Hamiltonians for the differential game (1)–(4) are:

$$H_i = \left[\alpha - \sum_{j=1}^{N} q_j - c + \gamma X_i + \varepsilon X \right] q_i + \psi_i (q_i - \rho X_i) + \psi_i^X \left(\sum_{j=1}^{N} q_j - \rho X \right) \quad (5)$$

where the costate variables satisfy the following conditions:

$$\dot{\psi}_i = (\delta + \rho)\psi_i - \gamma q_i, \quad \lim_{t\to\infty} e^{-\delta t} \psi_i(t) = 0, \quad (6)$$

$$\dot{\psi}_i^X = (\delta + \rho)\psi_i^X - \varepsilon q_i, \quad \lim_{t\to\infty} e^{-\delta t} \psi_i^X(t) = 0. \quad (7)$$

Since the Hamiltonian H_i is concave in q_i, the equilibrium outputs are then found from

$$\frac{\partial H_i}{\partial q_i} = \alpha - 2q_i - \sum_{j \neq i} q_j - c + \gamma X_i + \varepsilon X + \psi_i + \psi_i^X = 0, \quad (8)$$

that is:

$$q_i = \frac{\alpha - \sum_{j \neq i}^{N} q_j - c + \gamma X_i + \varepsilon X + \psi_i + \psi_i^X}{2}. \quad (9)$$

If the initial experience of the firms is identical, the symmetric equilibrium output is:

$$q_i = \begin{cases} \frac{\alpha - c + \gamma X_i + \varepsilon X + \psi_i + \psi_i^X}{N+1}, & if\, \alpha - c + \gamma X_i + \varepsilon X + \psi_i + \psi_i^X > 0 \\ 0, & if\, \alpha - c + \gamma X_i + \varepsilon X + \psi_i + \psi_i^X \le 0. \end{cases} \quad (10)$$

From (10) we observe that the firm's equilibrium output policy is to produce proportionally to the firm's own current experience, X_i, to the entire industry current experience, X_i, as well as to the marginal profit ψ_i and ψ_i^X from the opportunity of increasing the firm's experience and the industry experience respectively by one more unit.

4 Steady-State Nash Equilibrium

To find an interior symmetric steady-state, we next set $\dot{X}_i = 0$ and $\dot{X} = 0$. Consequently, $q_i = \rho X_i$ for $i = 1, 2, \ldots, N$ and from (10) we have:

$$\frac{\alpha - c + X_i \gamma + \varepsilon X + \psi_i + \psi_i^X}{N+1} = \rho X_i \quad (11)$$

Differentiating (11) we find $\frac{\dot{\psi}_i + \dot{\psi}_i^X}{N+1} = 0$. To meet this equation, consider next

$$\dot{\psi}_i = 0 \text{ and } \dot{\psi}_i^X = 0,\ i = 1, 2, \ldots, N. \tag{12}$$

By accounting for (7) and (8), Eq. (12) lead to

$$\psi_i = \frac{\gamma q_i}{\delta + \rho} \text{ and } \psi_i^X = \frac{\varepsilon q_i}{\delta + \rho}. \tag{13}$$

Using (13), Eq. (11) transforms into

$$\frac{\alpha - c + X_i \gamma + \varepsilon X + \frac{\gamma q_i}{\delta+\rho} + \frac{\varepsilon q_i}{\delta+\rho}}{N+1} = \rho X_i. \tag{14}$$

Recalling $q_i = \rho X_i$, we have $\frac{\alpha - c + X_i\gamma + \varepsilon X + \left(\frac{\gamma}{\delta+\rho} + \frac{\varepsilon}{\delta+\rho}\right)\rho X_i}{N+1} = \rho X_i$, which, with respect to the symmetricity, $NX_i = X$, defines steady-state experience $X_i^{SS} = \frac{\alpha - c}{\rho(N+1) - \left(\gamma + \left(\frac{\gamma}{\delta+\rho} + \frac{\varepsilon}{\delta+\rho}\right)\rho\right) - N\varepsilon}$. The result is feasible, i.e., $X_i^{SS} \geq 0$, if $\rho(N+1) > \gamma + \left(\frac{\gamma}{\delta+\rho} + \frac{\varepsilon}{\delta+\rho}\right)\rho + N\varepsilon$. Defining next the minimal marginal cost c^m that the firm may achieve in the long run, $c^{SS} = c - \gamma X_i^{SS} - \varepsilon X^{SS} \geq c^m$, we conclude as follows.

Proposition 1 *If $\rho(N+1) > \gamma + \frac{\gamma+\varepsilon}{\delta+\rho}\rho + N\varepsilon$ and $c^{SS} \geq c^m$, the differential game (1)–(4) has a unique, symmetric steady-state Nash equilibrium characterized by positive experience, output and costates:*

$$X_i^{SS} = \frac{\alpha - c}{\rho(N+1) - \left(\gamma + \frac{\gamma+\varepsilon}{\delta+\rho}\rho\right) - N\varepsilon} \tag{15}$$

$$q_i^{SS} = \frac{\rho(\alpha - c)}{\rho(N+1) - \left(\gamma + \frac{\gamma+\varepsilon}{\delta+\rho}\rho\right) - N\varepsilon} \tag{16}$$

$$\psi_i^{SS} = \frac{\rho(\alpha - c)}{\rho(N+1) - \left(\gamma + \frac{\gamma+\varepsilon}{\delta+\rho}\rho\right) - N\varepsilon}\frac{\gamma}{\delta+\rho} \tag{17}$$

$$\psi_i^{X,SS} = \frac{\rho(\alpha - c)}{\rho(N+1) - \left(\gamma + \frac{\gamma+\varepsilon}{\delta+\rho}\rho\right) - N\varepsilon}\frac{\varepsilon}{\delta+\rho} \tag{18}$$

$i = 1, 2, \ldots, N.$ ■

From Eqs. (15)–(18) we readily conclude:

Corollary 1 *The higher the proprietary learning ability γ and the experience spillover ε, the greater the firms' steady-state output and experience.* ■

To find transient output dynamics, we differentiate q_i from (10) and substitute (1), (2), (6) and (7):

$$\dot{q}_i = \frac{\gamma(q_i - \rho X_i) + \varepsilon(Q - \rho X) + (\delta+\rho)\psi_i - \gamma q_i + (\delta+\rho)\psi_i^X - \varepsilon q_i}{N+1}.$$

Next using $\psi_i + \psi_i^X = (N+1)q_i - \alpha + c - \gamma X_i - \varepsilon X$ with respect to (10) we have

$$\dot{q}_i = \frac{-(\alpha - c)(\delta + \rho) + [(\delta + \rho)(N+1) + \varepsilon(N-1)]q_i - X_i((\gamma + \varepsilon N)(\delta + \rho) + \gamma\rho + \varepsilon\rho N)}{N+1}$$

and therefore

$$-(\alpha - c)(\delta + \rho) + [(\delta + \rho)(N+1) + \varepsilon(N-1)]q_i - \dot{q}_i(N+1) = X_i((\gamma + \varepsilon N)(\delta + \rho) + \gamma\rho + \varepsilon\rho N)$$
$$= X_i(\gamma + \varepsilon N)(\delta + 2\rho).$$

Consequently, after one more differentiation of $\dot{q}_i$ we obtain

$$\ddot{q}_i = \frac{[\delta(N+1) + \varepsilon(N-1)]\dot{q}_i - q_i((\gamma + \varepsilon N)(\delta + 2\rho) - \rho[(\delta + \rho)(N+1) + \varepsilon(N-1)]) - \rho((\alpha - c)(\delta + \rho)}{N+1}.$$

The solution to this equation with respect to the boundary conditions is

$$q_i = K_1 e^{\frac{1}{2}\left(A - \sqrt{A^2 - 4B}\right)t} + K_2,$$

where $A = \delta(N+1) + \varepsilon(N-1)$, $B = (\gamma + \varepsilon N)(\delta + 2\rho) - \rho[(\delta + \rho)(\mathrm{N}+1) + \varepsilon(N-1)]$, and K_1 and K_2 are integration constants.

We next show that $K_2 = q_i^{SS}$. Assume $B<0$, then K_2 is determined by $\dot{q}_i(\infty) = 0$. Differentiating (9) and requiring $\dot{q}_i = 0$, we have,

$$\gamma\dot{X}_i + \varepsilon\dot{X} + \dot{\psi}_i + \dot{\psi}_i^X = 0,$$

which with respect to (6) and (7) results in $\gamma(q_i - \rho X_i) + \varepsilon(Q - \rho X) + (\delta + \rho)\psi_i - \gamma q_i + (\delta + \rho)\psi_i^X - \varepsilon q_i = 0$. Differentiating the last expression we have, $-\gamma\rho\dot{X}_i - \varepsilon\rho\dot{X} + (\delta + \rho)\dot{\psi}_i + (\delta + \rho)\dot{\psi}_i^X = 0$ and therefore

$$-\gamma\dot{X}_i - \varepsilon\dot{X} + \frac{\delta + \rho}{\rho}\left(\dot{\psi}_i + \dot{\psi}_i^X\right) = 0.$$

Thus, we find that $\dot{\psi}_i + \dot{\psi}_i^X = 0$ and $\dot{X}_i + \dot{X} = 0$, which with respect to the symmetric conditions implies, $\dot{X}_i = 0$ and $\dot{X} = 0$. Next, according to (6), $(\delta + \rho)\psi_i - \dot{\psi}_i = \gamma q_i$, which after differentiating and recalling $\dot{q}_i = 0$ leads to $(\delta + \rho)\dot{\psi}_i - \ddot{\psi}_i = 0$. The only solution for ψ_i to this differential equation for t tending to infinity is a constant, i.e., $\dot{\psi}_i = \dot{\psi}_i^X = 0$. Accordingly, we have found that in the long run $\dot{q}_i = 0$ implies $\dot{X}_i = \dot{X} = 0$ and $\dot{\psi}_i = \dot{\psi}_i^X = 0$, which are the conditions employed to derive a unique steady state $\{q_i^{SS}, X_i^{SS}\}$ in Proposition 1. Consequently, $q_i(\infty) = K_2 = q_i^{SS}$ and the solution $q_i = K_1 e^{\frac{1}{2}\left(A - \sqrt{A^2 - 4B}\right)t} + q_i^{SS}$ converges asymptotically to the unique steady state q_i^{SS} as t tends to infinity if $B = (\gamma + \varepsilon N)(\delta + 2\rho) - \rho[(\delta + \rho)(\mathrm{N}+1) + \varepsilon(N-1)] <0$, which straightforwardly leads to the condition $\gamma + \frac{\gamma + \varepsilon}{\delta + \rho}\rho + N\varepsilon - \rho(N+1) <0$. Since this condition is identical to the condition stated in Proposition 1, we conclude as follows.

Proposition 2 *If an equilibrium steady state defined by Proposition 1 exists, i.e.,*

$$\gamma + \frac{\gamma + \varepsilon}{\delta + \rho}\rho + N\varepsilon < \rho(N+1),$$

then the firms converge to it and this state is assymptotically stable. ■

Note that from (10) and Proposition 1 we also observe the effect of myopic behavior of the firms if they ignore own experience dynamics, i.e., assume that $\psi_i = 0$ or the industry dynamics, i.e., $\psi_i^X = 0$ as stated below.

Proposition 3 *Myopic firms that ignore either their own dynamics or the industry dynamics or both are characterized by lower steady-state equilibrium experience and output. Furthermore, if the rate of spillover is greater than the rate of proprietary learning, $\varepsilon > \gamma$, a firm's long-run output more strongly reduced if the firm ignores industry dynamics rather than its own dynamics. Otherwise if $\varepsilon < \gamma$, then the effect on the firm's own experience dynamics is stronger.* ■

The effect of the industry size N on individual outputs can be found by differentiating X_i^{SS} with respect to N:

$$\frac{\partial X_i^{SS}}{\partial N} = -\frac{\alpha - c}{\left[\rho(N+1) - \left(\gamma + \frac{\gamma+\varepsilon}{\delta+\rho}\rho\right) - N\varepsilon\right]^2}(\rho - \varepsilon). \tag{19}$$

Consequently, if $\rho < \varepsilon$, then the derivative (19) will remain positive. This implies the following result.

Proposition 4 *If there exists a positive steady state and the spillover rate does not exceed the experience depreciation rate, $\rho > \varepsilon$, then the more firms comprising the industry, the lower the equilibrium steady-state output and the experience of each firm. Otherwise, if $\rho < \varepsilon$, the larger the industry, the greater the steady-state equilibrium experience and the output of each firm. In addition, the higher the proprietary learning ability of the firms, the stronger the respective change in the steady-state when N grows.* ■

Finally, we compare our results with the standard approach, assuming that each firm is able to account for the dynamics of every individual competitor rather than for the entire industry only. Then the Hamiltonian and the costates take the following form:

$$H_i = \left[\alpha - \sum_{j=1}^{N} q_j - c + \gamma X_i + \varepsilon \sum_{j=1}^{N} X_j\right] q_i + \psi_i(q_i - \rho X_i) + \sum_{j \neq i} \psi_i^j (q_j - \rho X_j) \tag{20}$$

$$\dot{\psi}_i = (\delta + \rho)\psi_i - (\gamma + \varepsilon)q_i, \quad \lim_{t\to\infty} e^{-\delta t}\psi_i(t) = 0, \tag{21}$$

$$\dot{\psi}_i^j = (\delta + \rho)\psi_i^j - \varepsilon q_i, \ \lim_{t\to\infty} e^{-\delta t}\psi_i^X(t) = 0. \tag{22}$$

The equilibrium outputs are then found from:

$$\frac{\partial H_i}{\partial q_i} = \alpha - 2q_i - \sum_{j\neq i} q_j - c + \gamma X_i + \varepsilon \sum_{j=1}^{N} X_j + \psi_i = 0 \text{ and}$$

$$q_i = \frac{\alpha - \sum_{j\neq i} q_j - c + \gamma X_i + \varepsilon \sum_{j=1}^{N} X_j + \psi_i}{2}.$$

Consequently, $q_i = \frac{\alpha - c + \gamma X_i + \varepsilon \sum_{j=1}^{N} X_j + \psi_i}{N+1}$ and similar to the derivation of Proposition 1, we find a steady state:

$$X_i^{SS} = \frac{\alpha - c}{\rho(N+1) - \left(\gamma + \frac{\gamma+\varepsilon}{\delta+\rho}\rho\right) - N\varepsilon} \text{ and } q_i^{SS} = \frac{\rho(\alpha - c)}{\rho(N+1) - \left(\gamma + \frac{\gamma+\varepsilon}{\delta+\rho}\rho\right) - N\varepsilon}. \tag{23}$$

We next readily conclude, when comparing our results in (15) and (23) with the next proposition.

Proposition 5 *Even if the firms are unable to account for the individual dynamics of their competitors and, instead, account for the overall industry dynamics, in the long run they accumulate the same experience and tend to the same output they would converge to under full information about their competitors.* ■

5 Conclusions

We address output competition between firms producing substitutable products for a single market. Assuming that each firm is unable to account for the dynamics of every single competitor but does respond to the entire industry trends, we find a symmetric, long-run Nash equilibrium along with its properties and conditions of existence. In particular, we find that the higher the proprietary learning ability of the firms and the industry experience spillover, the greater the firms' steady-state output and experience. Myopic firms that ignore either their own experience dynamics or the industry experience dynamics or both are characterized by lower steady-state equilibrium experience and output. On the other hand, the firms accumulate the same experience and tend to the same output in the long run if they account for the industry dynamics rather than individual competitors dynamics. That is, though the industry information used for production control is incomplete in terms of competitors dynamics, the firms tend to the long-run output they would attain under complete information.

We show that industry growth does not necessarily imply the firms' individual output growth. Specifically, if the spillover rate does not exceed the experience depreciation rate, then the more firms comprising the industry, the lower the equilibrium steady-state output and experience of each firm. Otherwise, the individual equilibrium outputs and experience can grow as more firms enter the market.

References

1. Arrow, K.J.: The economic implications of learning by doing. Rev. Econ. Stud. **29**(3), 155–173 (1962)
2. Cellini, R., Lambertini, L.: Dynamic R&D with spillovers: competition vs cooperation. J. Econ. Dyn. Control **33**(3), 568–582 (2009)
3. Cohn, D., Tesauro, G.: How tight are the Vapnik-Chervonenkis bounds? Neural Comput. **4**(2), 249–269 (1992)
4. Cournot, A.A.: Researches into the Mathematical Principles of the Theory of Wealth. Augustus M. Kelley Publishers, New York, 1971 (1838)
5. Dutton, J.M., Thomas, A.: Treating progress functions as a managerial opportunity. Acad. Manag. Rev. **9**(2), 235–247 (1984)
6. Fudenberg, D., Tirole, J.: Learning by doing and market performance. Bell J. Econ. **14**, 522–530 (1983)
7. Jarmin R.S.: Learning by doing and competition in the early rayon industry. CES, 93–4 (1993)
8. Jørgensen, S., Zaccour, G.: Optimal output strategies in a two-stage game with entry, learning-by-doing and spillovers. In: Petrosjan, A., Mazalov, V.V. (eds.) Game Theory and Applications. Nova Science Publishers, New York (2000)
9. Kogan, K., El Ouardighi, F., Herbon, A.: Production with learning and forgetting in a competitive environment. Int. J. Prod. Econ. (IJPE). **189**, 52–62 (2017)
10. Miravete, E.J.: Time-consistent protection with learning by doing. Eur. Econ. Rev. **47**(5), 761–790 (2003)
11. Schroeder, M.: Fractals, Chaos, Power Laws: minutes From An Infinite Paradise. Freeman, New York (1991)
12. Spence, A.M.: The learning curve and competition. Bell J. Econ. **12**(Spring), 49–70 (1981)
13. Stokey, N.L.: The dynamics of industry-wide learning. In: Heller, W.P., Starr, R.M., Starrett, D.A. (eds.). Essays in Honor of Kenneth J. Arrow, vol. 2. Cambridge, Cambridge University Press
14. Yelle, L.E.: The learning curve: historical review and comprehensive survey. Decis. Sci. **10**(2), 302–328 (1979)

Comparative Statics via Stochastic Orderings in a Two-Echelon Market with Upstream Demand Uncertainty

Constandina Koki, Stefanos Leonardos and Costis Melolidakis

Abstract We revisit the classic Cournot model and extend it to a two-echelon supply chain with an upstream supplier who operates under demand uncertainty and multiple downstream retailers who compete over quantity. The supplier's belief about retail demand is modeled via a continuous probability distribution function F. If F has the *decreasing generalized mean residual life (DGMRL)* property, then the supplier's optimal pricing policy exists and is the unique fixed point of the *mean residual life (MRL)* function. This closed form representation of the supplier's equilibrium strategy facilitates a transparent comparative statics and sensitivity analysis. We utilize the theory of stochastic orderings to study the response of the equilibrium fundamentals—wholesale price, retail price and quantity—to varying demand distribution parameters. We examine supply chain performance, in terms of the distribution of profits, supply chain efficiency, in terms of the *Price of Anarchy*, and complement our findings with numerical results.

Keywords Continuous distributions · Demand uncertainty
Generalized mean residual life · Comparative statics · Sensitivity analysis
Stochastic orders

C. Koki
Department of Statistics, Athens University of Economics and Business,
Patision 76, 104 34 Athens, Greece
e-mail: kokiconst@aueb.gr

S. Leonardos (✉) · C. Melolidakis
Department of Mathematics, National & Kapodistrian University of Athens,
Panepistimioupolis, 157 72 Athens, Greece
e-mail: sleonardos@math.uoa.gr

C. Melolidakis
e-mail: cmelol@math.uoa.gr

© Springer Nature Switzerland AG 2018
P. Daniele and L. Scrimali (eds.), *New Trends in Emerging Complex Real Life Problems*, AIRO Springer Series 1,
https://doi.org/10.1007/978-3-030-00473-6_36

1 Introduction

The global character of modern markets necessitates the study of competition models that capture two features: first, that retailers' cost is not constant but rather formed as the decision variable of a strategic, profit-maximizing supplier, and second that uncertainty about retail demand affects not only the retailers but also the supplier. Motivated by these considerations, in [10], we use a game-theoretic approach to extend the classic Cournot market in the following two-stage game: in the first-stage (acting as a Stackelberg leader), a revenue-maximizing supplier sets the wholesale price of a product under incomplete information about market demand. Demand or equivalently, the supplier's belief about it, is modeled via a continuous probability distribution. In the second-stage, the competing retailers observe wholesale price and realized market demand and engage in a classic Cournot competition. Retail price is determined by an affine inverse demand function.

Lariviere and Porteus [8] studied a similar model in which demand uncertainty affected a single retailer. They identified the property of *increasing generalized failure rate (IGFR)* as a mild unimodality condition for the *deterministic* supplier's revenue-function and then performed an extensive comparative statics and performance (efficiency) analysis of the supply chain at equilibrium. The properties of IGFR random variables were studied in a series of subsequent notes, [3, 9, 12].

In [10], we extended the work of [8] by moving uncertainty to the supplier and by implementing an arbitrary number of second-stage retailers. We introduced the *generalized mean residual life (GMRL)* function of the supplier's belief distribution F and showed that his *stochastic* revenue function is unimodal, if the GMRL function is decreasing—*(DGMRL)* property—and F has finite second moment. In this case, we characterized the supplier's optimal price as a fixed point of his *mean residual life (MRL)* function, see Theorem 1 below. Subsequently, we turned our attention to DGMRL random variables, examined their moments, limiting behavior, closure properties and established their relation to IGFR random variables, as in [3, 9, 12]. This study was done in expense of a comparative statistics and performance analysis, as the one in Sects. 3–5 of [8]. The importance of such an analysis is underlined among others in [1, 2, 6] and references therein.

1.1 Contributions—Outline

The present paper aims to fill this gap. Following the methodology of [8], we study the response of market fundamentals by utilizing the closed form characterization of the equilibrium obtained in [10]. Specifically, under the conditions of Theorem 1, the optimal wholesale price is the unique fixed point of the MRL function of the demand distribution F. This motivates the study of conditions under which two different markets, denoted by F_1 and F_2, can be ordered in the mrl-stochastic order, see [14].

The paper is organized as follows. In Sect. 2, we provide the model description and in Sect. 3, the existing results from [10] on which the current analysis is based. Our findings, both analytical and numerical are presented in Sect. 4. Section 5 concludes our analysis and discusses directions for future work.

1.1.1 Comparison to Related Works

Two-echelon markets have been extensively studied in the literature under different perspectives and various levels of demand uncertainty, see e.g. [5, 11, 15, 16]. In the present study, we depart from previous works by introducing the toolbox of stochastic orderings in the comparative statics analysis. The advantage of this approach is that we quantify economic notions, such as market size and demand variability, in various ways. Accordingly, we are able to challenge established economic intuitions by showing, for instance, that repsonses of wholesale prices to increasing market size or demand variability are not easy to predict, since they largely depend on the notion of variability that is employed.

2 The Model: Game-Theoretic Formulation

An upstream supplier produces a single homogeneous good at constant marginal cost, normalized to 0, and sells it to a set of $N = \{1, 2, \dots, n\}$ downstream retailers. The supplier has ample quantity to cover any possible demand and his only decision variable is the wholesale price r which he determines prior to and independently of the retailers' order-decisions. The retailers observe r—a price-only contract (there is no return option and the salvage value of the product is zero)—as well as the market demand parameter α and choose simultaneously and independently their order-quantities $q_i\,(r \mid \alpha)\,, i \in N$. They face no uncertainty about the demand and the quantity that they order from the supplier is equal to the quantity that they sell to the market (at equilibrium). The retail price is determined by an affine inverse demand function $p = (\alpha - q\,(r))^+$, where α is the *demand parameter* and $q\,(r) := \sum_{i=1}^{n} q_i\,(r)$ is the total quantity that the retailers release to the market.[1] Contrary to the retailers, we assume that at the point of his decision, the supplier has incomplete information about the actual market demand.

This supply chain can be represented as a two-stage game, in which the supplier acts in the first stage and the retailers in the second. A strategy for the supplier is a price $r \geq 0$ and a strategy for retailer i is a function $q_i : \mathbb{R}_+ \to \mathbb{R}_+$, which specifies the quantity that retailer i will order for any possible cost r. Payoffs are determined via the strategy profile $(r, \mathbf{q}\,(r))$, where $\mathbf{q}\,(r) = (q_i\,(r))_{i=1}^{n}$. Given cost r, the profit function $\pi_i\,(\mathbf{q}\,(r) \mid r)$ or simply $\pi_i\,(\mathbf{q} \mid r)$, of retailer $i \in N$, is $\pi_i\,(\mathbf{q} \mid r) = q_i\,(\alpha - q)^+ -$

[1]To simplify notation, we write q or $q\,(r)$ and q_i or $q_i\,(r)$ instead of the proper $q\,(r \mid \alpha)$ and $q_i\,(r \mid \alpha)$.

rq_i. For a given value of α, the supplier's profit function, π_s is $\pi_s(r \mid \alpha) = rq(r)$ for $0 \le r < \alpha$, where $q(r)$ depends on α via $\pi_i(\mathbf{q} \mid r)$.

To model the supplier's uncertainty about retail demand, we assume that after the pricing decision of the supplier, but prior to the order-decisions of the retailers, a value for α is realized from a continuous distribution F, with finite mean $\mathbb{E}\alpha < +\infty$ and nonnegative values, i.e. $F(0) = 0$. Equivalently, F can be thought of as the supplier's belief about the demand parameter and, hence, about the retailers' willingness-to-pay his price. We will use the notation $\bar{F} := 1 - F$ for the survival function and $\alpha_L := \sup\{r \ge 0 : F(r) = 0\} \ge 0$, $\alpha_H := \inf\{r \ge 0 : F(r) = 1\} \le +\infty$ for the support of F respectively. Under these assumptions, the supplier's payoff function π_s becomes stochastic: $\pi_s(r) = \mathbb{E}\pi_s(r \mid \alpha)$. All the above are assumed to be common knowledge among the participants in the market (the supplier and the retailers).

3 Existing Results

We consider only subgame perfect equilibria, i.e. strategy profiles $(r, \mathbf{q}(r))$ such that $\mathbf{q}(r)$ is an equilibrium in the second stage and $q_i(r)$ is a best response against any r for all $i = 1, 2, \ldots, n$. The equilibrium behavior of this market has been analyzed in [10]. To proceed with the equilibrium representation, we first introduce some notation.

3.1 *Generalized Mean Residual Life*

Let $\alpha \sim F$ be a nonnegative random variable with finite expectation $\mathbb{E}\alpha < +\infty$. The *mean residual life (MRL)* function $\mathrm{m}(r)$ of α is defined as

$$\mathrm{m}(r) := \mathbb{E}(\alpha - r \mid \alpha > r) = \frac{1}{\bar{F}(r)} \int_r^{\infty} \bar{F}(u)\,\mathrm{d}u, \quad \text{for } r < \alpha_H$$

and $\mathrm{m}(r) := 0$, otherwise, see, e.g., [4, 7] or [14]. In analogy to the *generalized failure rate (GFR)* function $\mathrm{g}(r) := r\mathrm{h}(r)$, where $\mathrm{h}(r) := f(r)/\bar{F}(r)$ denotes the hazard rate of F and the *increasing generalized failure rate (IGFR)* unimodality condition, defined in [8] and studied in [3, 9], we introduce, see [10], the *generalized mean residual life (GMRL)* function $\ell(r)$, defined as $\ell(r) := \frac{\mathrm{m}(r)}{r}$, for $0 < r < \alpha_H$. If $\ell(r)$ is *decreasing*, then F has the *(DGMRL) property*. The relationship between the (IGFR) and (DGMRL) classes of random variables is studied in [10].

We will use the notation DMRL for a random variable X with a decreasing mean residual life function $\mathrm{m}(r)$ and IFR for a random variable X with increasing failure rate $\mathrm{h}(r)$. We say that X_1 is smaller than X_2 in the *mean residual life order*, denoted as $X_1 \le_{\mathrm{mrl}} X_2$, if $\mathrm{m}_1(r) \le \mathrm{m}_2(r)$ for all r, see [14]. Of course $\mathrm{m}_1(r) \le \mathrm{m}_2(r)$ if and only if $\ell_1(r) = \ell_2(r)$ for all $r > 0$. Similarly, X_1 is smaller than X_2 in the usual stochastic (hazard rate) order, denoted as $X_1 \leq_{\text{st}} X_2$ ($X_1 \leq_{\text{hr}} X_2$), if $\bar{F}_1(r) \leq \bar{F}_2(r)$ ($\mathrm{h}_1(r) \leq \mathrm{h}_2(r)$) for all r. The $\leq_{\text{hr}}$-order implies the $\leq_{\text{mrl}}$-order. However, neither of the orders $\leq_{\text{st}}$ and $\leq_{\text{mrl}}$ imply the other.

3.1.1 Market Equilibrium

Using this terminology, we can express the supplier's optimal pricing strategy in terms of the MRL function and formulate sufficient conditions on the demand distribution, under which a subgame perfect equilibrium exists and is unique.

Theorem 1 ([10]) *Assume that the supplier's belief about the unknown, nonnegative demand parameter, α, is represented by a continuous distribution F, with support inbetween α_L and α_H with $0 \leq \alpha_L < \alpha_H \leq \infty$.*

(a) If an optimal price r^* for the supplier exists, then r^* satisfies the fixed point equation

$$r^* = \mathrm{m}\left(r^*\right) \tag{1}$$

(b) If F is strictly DGMRL and $\mathbb{E}\alpha^2$ is finite, then in equilibrium, the optimal price r^* of the supplier exists and is the unique solution of (1).

Expressing (1) in terms of the GMRL function $\ell(r)$, the supplier's optimal price r^* can be equivalently written as the solution of equation $\ell(r^*) = 1$.

4 Comparative Statics

The closed form expression of (1) provides the basis for an extensive comparative statics and sensitivity analysis on the distribution parameters of market demand. To understand the market-equilibrium behavior under different demand (distribution) characteristics, we employ (1) and the rich theory of *stochastic orders*, [4, 7, 14]. Based on π_i, π_s and Theorem 1, the market fundamentals at equilibrium are given in Table 1.

Here, Π_s^* refers to the supplier's *realized*—not expected—profit, i.e. $\Pi_s^* := \pi_S(r^* \mid \alpha)$. From Table 1, it is immediate that the total quantity q^* that is sold to the market and the retail price p^* are monotone in r^*. Accordingly, we restrict attention on the behavior of r^* as the distribution parameters vary.

To obtain a meaningful comparison between different markets, we assume throughout equilibrium uniqueness. Hence, unless stated otherwise, we consider only strictly DGRML distributions with finite second moment. Since the DGMRL is particularly inclusive, see [3, 10] and finiteness of the second moment of the demand is naturally to assume, we do not consider them as restrictive. Still, since these conditions are only sufficient and not necessary, the analysis applies to any other setting that guarantees equilibrium existence and uniqueness.

Table 1 Market fundamentals in equilibrium

Notation and expression	Definition
$r^* = \mathrm{m}(r^*)$	Wholesale price
$q^* = \frac{n}{n+1}(\alpha - r^*)^+$	Total quantity sold to the market
$p^* = \alpha - q^*$	Retail price
$\Pi_s^* = \frac{n}{n+1}(\alpha - r^*)^+ r^*$	Realized supplier's profit
$\Pi_i^* = \left(\frac{1}{n+1}(\alpha - r^*)^+\right)^2$	i-th retailer's profit, $i = 1, \ldots, n$

4.1 *Wholesale Price Determinants*

Although immediate from Theorem 1, the next Lemma showcases the importance of the characterization in (1). Let X_1 and X_2 denote two markets (or two instances of the same market) with demand distributions F_1, F_2. As stated above, X_1, X_2 are assumed to be nonnegative, strictly DGMRL random variables with finite second moment. We then have

Lemma 1 *Let X_1, X_2 denote the demand in two different market instances. If $X_1 \leq_{mrl} X_2$, then $r_1^* \leq r_2^*$.*

Proof Since $X_1 \leq_{\mathrm{mrl}} X_2$, we have that $\mathrm{m}_1(r) \leq \mathrm{m}_2(r)$ for all $r > 0$, by definition. Hence, by (1), $r_1^* = \mathrm{m}\left(r_1^*\right) \leq \mathrm{m}_2\left(r_1^*\right)$, which implies that $\ell_2\left(r_1^*\right) \geq 1$. Since $\ell_2(r)$ is strictly decreasing by assumption, this implies that $\ell_2(r) > 1$ for all $r < r_1^*$. Since r_2^* is the unique solution of $\ell\left(r_2^*\right) = 1$, this in turn implies that $r_2^* \geq r_1^*$.

Hence, the supplier charges a larger wholesale price in a market that is larger in the $\leq_{\mathrm{mrl}}$-order. Based on Lemma 1, the task of studying the behavior of the wholesale price r^* largely reduces to finding sufficient conditions that imply—or that are equivalent to—the $\leq_{\mathrm{mrl}}$-order. Such conditions can be found in [14], and are studied below.

4.1.1 Re-estimating Demand

We start with the response of the equilibrium wholesale price r^* to transformations that intuitively correspond to a larger market. Let X denote the random demand in an instance of the market under consideration. Let $c \geq 1$ be a positive constant. Moreover, let Z denote an additional source of demand that is independent of X. Let r_X^* denote the equilibrium wholesale price in the initial market and r_{X+Z}^* the equilibrium wholesale price in the market with random demand $X + Z$. How does r_X^* compare to r_{cX}^* and to r_{X+Z}^*?

While the intuition that the larger markets cX and $X + Z$ will give rise to higher wholesale prices is largely confirmed, see Theorem 2, the results do not hold in full generality and one needs to pay attention to some technical details. For instance, since

DGMRL random variables are not closed under convolution, see [10], the random variable $X + Z$ may not be DGMRL. This may lead to a multiplicity of equilibrium prices in the $X + Z$ instance, irrespectively of whether X is DGMRL or not. To focus on the economic interpretation of the comparative static analysis and to avoid an extensive discussion on the technical conditions, we assume that Z is a random variable such that the market $X + Z$ has again a unique wholesale equilibrium price. However, we consider this assumption as a restriction to the applicability of statement (ii) of Theorem 2.

Theorem 2 *Let $X \sim F$ be a nonnegative DGMRL random variable with finite second moment which describes the demand distribution in a market instance.*

(i) If $c \geq 1$ is a positive constant, then $r_X^* \leq r_{cX}^*$.
(ii) If Z is a nonnegative random variable with finite second moment, independent of X such that $X + Z$ remains strictly DGMRL, then $r_X^* \leq r_{X+Z}^*$.

Proof The proof of (i) follows directly from the preservation property of the $\leq_{\text{mrl}}$-order that is stated Theorem 2.A.11 of [14]. Specifically, since $c\text{m}(r/c)$ is the mrl function of cX, we have that for all > 0

$$c\text{m}(r/c) = r \cdot \frac{\text{m}(r/c)}{r/c} = r \cdot \ell(r/c) \geq r \cdot \ell(r) = \text{m}(r)$$

where the inequality follows from the assumption that X is DGMRL. Hence, $X \leq_{\text{mrl}} cX$ which by Lemma 1 implies that $r_X^* \leq r_{cX}^*$.

Statement (ii) is more involved since r_{X+Z}^* may not be unique in general. However, under the assumption that $X + Z$ remains strictly DGMRL, we may adapt Theorem 2.A.11 of [14] and obtain the claim in a similar fashion to part (i). Although Theorem 2.A.11 is stated for DMRL random variables, the proof extends in a straightforward way to DGMRL random variables.

Another way to treat the possible multiplicity of equilibrium wholesale prices in the $X + Z$ market and the fact that $X + Z$ may not be DGMRL is the following. Since, X is strictly DGMRL, we know that $r < \text{m}_X(r)$ for all $r < r_{X_1}^* = \text{m}_{X_1}\left(r_{X_1}^*\right)$. Together with $X_1 \leq_{\text{mrl}} X_1 + Z$, this implies that for all $r < r_{X_1}^*$, the following holds: $r < \text{m}_{X_1}(r) \leq \text{m}_{X_1+Z}(r)$, and hence that $r_{X_1+Z}^* \geq r_{X_1}^*$ for any r_{X+Z}^* such that (1) holds. Hence, in this case, we can compare the r_X^* with every r_{X+Z}^* separately and obtain that r_X^* is less than any possible equilibrium wholesale price in the market r_{X+Z}^*. However, as mentioned above, we prefer to restrict attention to markets that preserve equilibrium uniqueness.

4.1.2 Closure Properties

Next, we turn our attention to operations that preserve the $\leq_{\text{mrl}}$-order. Let X_1, X_2 denote two different instances of the market, i.e., two different demand distributions or beliefs about it, such that $X_1 \leq_{\text{mrl}} X_2$. In this case, we know that $r_1^* \leq r_2^*$. We are interested in determining transformations of X_1, X_2 that preserve the $\leq_{\text{mrl}}$-order and hence, by Lemma 1, the ordering $r_1^* \leq r_2^*$. Again, to avoid technicalities, we assume that X_1, X_2 are such that Theorem 1 applies, i.e., that they are nonnegative, strictly DGMRL and have finite second moment.

Theorem 3 *Let $X_1 \sim F_1, X_2 \sim F_2$ denote the demand in two different market instances, such that $X_1 \leq_{mrl} X_2$. Then,*

(i) If ϕ is an increasing convex function, then $r^*_{\phi(X_1)} \leq r^*_{\phi(X_2)}$.
(ii) If Z is a nonnegative, IFR random variable with finite second moment, independent of X_1, X_2 such that $X_1 + Z$ and $X_2 + Z$ remain strictly DGMRL, then $r^*_{X_1+Z} \leq r^*_{X_2+Z}$.
(iii) If $X_p \sim F_1 + (1 - p) F_2$ is strictly DGMRL for some $p \in (0, 1)$, then $r^*_{X_1} \leq r^*_{X_p} \leq r^*_{X_2}$.

Proof Statements (i) through (iii) follow directly from Theorems 2.A.19, Lemma 2.A.8 and Theorem 2.A.18 respectively. The assumption that the transformed random variables remain strictly DGMRL ensures equilibrium uniqueness.

If instead of $X_1 \leq_{\text{mrl}} X_2$, X_1 and X_2 are ordered in the weaker $\leq_{\text{hr}}$-order, i.e., if $X_1 \leq_{\text{hr}} X_2$ and Z is DMRL (instead of merely IFR), then Lemma 2.A.10 of [14] implies that statement (ii) of Theorem 3 remains true. Formally,

Corollary 1 *Let $X_1 \sim F_1, X_2 \sim F_2$ denote the demand in two different market instances, such that $X_1 \leq_{hr} X_2$. If Z is a nonnegative, IFR random variable with finite second moment, independent of X_1, X_2 such that $X_1 + Z$ and $X_2 + Z$ remain strictly DGMRL, then $r^*_{X_1+Z} \leq r^*_{X_2+Z}$.*

Following the exposition of [14], the above collection of statements can be extended to incorporate more case-specific results.

Although Theorems 2 and 3 are immediate *once* Theorem 1 and Lemma 1 have been established, their implications in terms of the economic intuitions are non-trivial. In particular, both Theorems imply that if the supplier reestimates upwards her expectations about the demand then she will charge a higher price. However, this intuitive conclusion depends on the conditions that imply the $\leq_{\text{mrl}}$-order and does not hold in general, as discussed in Sect. 4.1.4 below.

4.1.3 Market Demand Variability

The response of the equilibrium wholesale price to increasing (decreasing) demand variability is less straightforward. There exist several notions of stochastic orders that compare random variables in terms of their variability and depending on which we employ, we may derive different results. First, we introduce some notation.

Variability or Dispersive Orders: Let $X_1 \sim F_1$ and $X_2 \sim F_2$ be two nonnegative random variables with equal means, $\mathbb{E}X_1 = \mathbb{E}X_2$, and finite second moments. If

$\int_r^{+\infty} \bar{F}_1(u)\,du \le \int_r^{+\infty} \bar{F}_2(u)\,du$ for all $r \ge 0$, then X_1 is said to be smaller than X_2 in the *convex order*, denoted by $X_1 \le_{cx} X_2$. If F_1^{-1} and F_2^{-1} denote the right continuous inverses of F_1, F_2 and $F_1^{-1}(r) - F_1^{-1}(s) \le F_2^{-1}(r) - F_2^{-1}(s)$ for all $0 < r \le s < 1$, then X_1 is said to be smaller than X_2 in the *dispersive order*, denoted by $X_1 \le_{disp} X_2$. Finally, if $\int_{F_1^{-1}(p)}^{\infty} \bar{F}_1(u)\,du \le \int_{F_2^{-1}(p)}^{\infty} \bar{F}_2(u)\,du$ for all $p \in (0, 1)$, then X_1 is said to be smaller than X_2 in the *excess wealth order*, denoted by $X_1 \le_{ew} Y$. Shaked and Shanthikumar [14] show that $X \le_{disp} Y \implies X \le_{ew} Y \implies X \le_{cx} Y$ which in turn implies that $\text{Var}(X) \le \text{Var}(Y)$. Further insights and motivation about these orders are provided in Chap. 3 of [14].

Less Variability implies Lower Wholesale Price: Under our assumptions the $\le_{mrl}$-order is not implied by the $\le_{cx}$-order. Hence, the $\le_{cx}$-order is not enough to conclude that wholesale prices are ordered according to the respective market variability, i.e., that less (more) variability gives rise to lower (higher) wholesale prices. However, if we restrict attention to the $\le_{ew}$ and $\le_{disp}$ orders, then more can be said. Recall that α_L denotes the left end of the support of a variable X. Accordingly, we will write α_{Li} to denote the left end of the support of variable X_i for $i = 1, 2$.

Theorem 4 *Let $X_1 \sim F_1, X_2 \sim F_2$ be two nonnegative, DGMRL random variables with $\alpha_{L1} \le \alpha_{L2}$ which denote the demand in two different market instances. If either X_1, X_2 or both are DMRL and $X_1 \le_{ew} X_2$, then $r_1^* \le r_2^*$.*

Theorem 4 follows directly from Theorem 3.C.5 of [14]. Based on its proof, the assumption that at least one of the two random variables is DMRL (and not merely DGMRL) cannot be relaxed. Belzunce et al. [4] argue about the restricted applicability of the $\le_{ew}$-order due to the difficulty in the evaluation of incomplete integrals of quantile functions and provide useful characterizations of the $\le_{ew}$-order to remedy this problem.

A result of similar flavor can be obtained if we use the $\le_{disp}$ order instead. Again, the condition that both X_1 and X_2 are DGMRL does not suffice and we need to assume that at least one is IFR.

Theorem 5 *Let $X_1 \sim F_1, X_2 \sim F_2$ be two nonnegative, DGMRL random variables which denote the demand in two different market instances. If either X_1, X_2 or both are IFR and $X_1 \le_{disp} X_2$, then $r_1^* \le r_2^*$.*

Theorem 5 follows directly from Theorem 3.B.20 (b) of [14] and the fact that the $\le_{hr}$-order implies the $\le_{mrl}$-order. Again, more case specific results can be drawn from the analysis of [14].

The main insight that we get from Theorems 4 and 5 is that less (more) variability implies lower (higher) wholesale prices. This is in sharp contrast with the results of [8] and sheds light on the effects of demand uncertainty. If uncertainty affects the retailer, then the supplier charges a higher price and captures all supply chain profits as variability reduces. Contrarily, if uncertainty falls to the supplier, then the supplier charges a lower price as variability increases. In this case, the supplier captures a lower share of system profit, see also (2) below.

The above cases correspond to two extremes: in [8] uncertainty falls solely to the retailer, whereas in the present analysis uncertainty falls solely to the supplier.

Based on the aggregate findings for the two cases, the following question naturally arises: is there a way to distribute demand uncertainty among supplier and retailers to mitigate its adverse effects and to evenly distribute supply chain profits among market participants? Answering this question exceeds the scope of the comparative statics analysis. However, it highlights a interesting direction for future work.

Parametric families of distributions: To further elaborate on the effects of relative variability on the wholesale price, we may compare our analysis with the approach of [8]. Given a random variable X with distribution F, [8] consider the random variables $X_i := \delta_i + \lambda_i X$ with $\delta_i \geq 0$ and $\lambda_i > 0$ for $i = 1, 2$. They conclude that in this case, the wholesale price is dictated by the coefficient of variation, $CV_i = \frac{\sqrt{\text{Var}(X_i)}}{\mathbb{E}X_i}$. Specifically, if $CV_2 < CV_1$, then $r_1^* < r_2^*$, i.e., in their model, a lower CV, or equivalently a lower relative variability, implies a higher price.

To establish a similar comparison, we utilize the comprehensive comparison of parametric distributions in terms of stochastic orders that is provided in Sect. 2.9 of [4]. For instance, consider two normal random variables $X_1 \sim N\left(\mu_1, \sigma_1^2\right)$ and $X_2 \sim N\left(\mu_2, \sigma_2^2\right)$. By Table 2.2 of [4], if $\sigma_1 < \sigma_2$ and $\mu_1 \leq \mu_2$, then $X_1 \leq_{\text{mrl}} X_2$ and hence, by Lemma 1, $r_1^* < r_2^*$. However, by choosing σ_i and μ_i appropriately, we can achieve an arbitrary ordering of relative variability, i.e. of CV_1 and CV_2. The reason is that the conclusions from this approach are obscured by the fact that changing μ_i for $i = 1, 2$, does not only affect the CV_i's but also the central location of the demand distributions. In this sense, the approach using dispersive orders seems more appropriate because, under the assumption that $\mathbb{E}X_1 = \mathbb{E}X_2$, it isolates the effect of the variability of the distribution on wholesale prices via stochastic orderings.

4.1.4 Stochastically Larger Market

It is well known that the usual $\leq_{\text{st}}$-order does not imply nor is implied by the $\leq_{\text{mrl}}$-order, see [14]. This implies that in a stochastically larger market, the supplier may still charge a lower price, which is in line with the intuition of [8] that "size is not everything" and that prices are driven by different forces. Such an example is provided below. Let

$$f\left(r; \omega, \kappa, \phi\right) := \frac{\kappa\left(\kappa^2 + \omega^2\right)}{\kappa^2 \cos\left(\omega\phi\right) + \kappa^2 + \kappa\omega \sin\left(\omega\phi\right) + \omega^2} \cdot e^{-\kappa r}\left(\cos\left(\omega\left(r - \phi\right)\right) + 1\right)$$

for $r \geq 0$, denote the densities of a parametric family of exponentially decaying sinusoids. For $(\omega, \kappa, \phi) = (0, \kappa, 0)$, f corresponds to the exponential distribution with parameter κ. Figure 1 depicts the survival functions $\bar{F}, \bar{G}$, the log-survival ratio $\log\left(\bar{F}/\bar{G}\right)$ and the optimal wholesale prices r_F^* and r_G^* for F corresponding to $(\omega, \kappa, \phi) = (\pi, 0.8, 1.2)$ and G to $(\omega, \kappa, \phi) = (0, 0.9, 0)$. Since the log-survival ratio remains throughout positive, we infer that $G \leq_{\text{st}} F$. However, as shown in the graph below $r_F^* = 1.0299 < r_G^* = 1.1114$. Although, both functions have a unique fixed point, F is not DMRL (nor DGMRL). Several simulations have not provided a conclusive answer to whether stochastic dominance implies also a larger price if we restrict to the DMRL (or DGMRL) class of random variables.

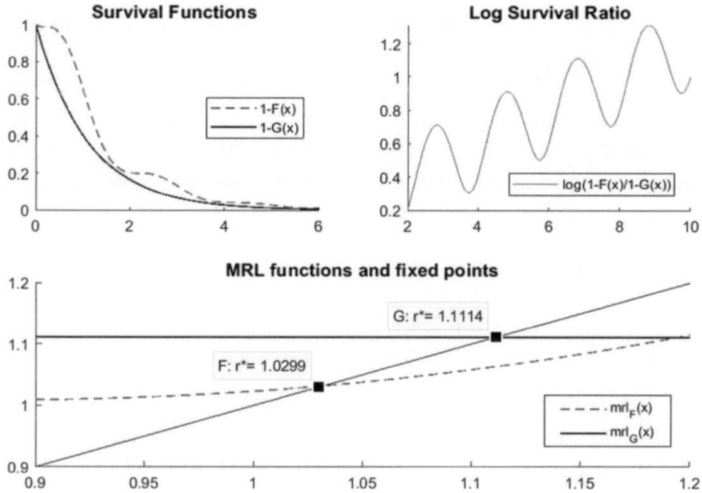

Fig. 1 F stochastically dominates G, however $r_G^* > r_F^*$

4.2 *Supply Chain Performance*

We measure the supply chain performance in terms of the ratio $\sum_{i=1}^{n} \Pi_i^* / \Pi_s^*$ which describes the division of the *realized* system profit between retailers and suppliers. If $\alpha \leq r^*$, then there is no transaction and the profits of all participants are equal to zero. For $\alpha > r^*$, we have that

$$\frac{\sum_{i=1}^{n} \Pi_i^*}{\Pi_s^*} = \frac{n\left(\frac{1}{n+1}\left(\alpha - r^*\right)\right)^2}{\frac{n}{n+1} r^*\left(\alpha - r^*\right)} = \frac{1}{n+1}\left(\frac{\alpha}{r^*} - 1\right) \tag{2}$$

Hence, the division of realized profit between supplier and retailers depends on the number n of retailers and the wholesale price r^*. Specifically, for a given realized demand α, as n or r^* increase, the supplier captures a larger share of the system profits.

4.2.1 Supply Chain Efficiency

As a benchmark, we will first determine the equilibrium behavior and performance of an integrated supply chain. Let π_I denote the profit of an integrated firm. The integrated firms' decision variable is now the retail price r, and hence its expected profit is given by $\pi_I(r) = r\mathbb{E}(\alpha - r)^+ = r\mathrm{m}(r)\bar{F}(r)$. By the same argument as in the proof of Theorem 1, π_I is maximized at $r^* = \mathrm{m}(r^*)$. In particular, the equilibrium price of both the integrated and non-integrated supplier is the same. Hence, the integrated firm's realized profit in equilibrium is equal to $\Pi_I^*(r^* \mid \alpha) = r^*(\alpha - r^*)^+$.

In a similar fashion to [13], we define the realized *Price of Anarchy (PoA)* of the system as the worst-case ratio of the realized profit of the centralized supply chain, Π_I^*, to the realized aggregate profit of the decentralized supply chain, $\Pi_D^* := \Pi_s^* + \sum_{i=1}^n \Pi_i^*$. To retain equilibrium uniqueness, we restrict attention to the class $\mathcal{G}$ of nonnegative DGMRL random variables. If the realized demand α is less than r^*, then both the centralized and decentralized chain make 0 profits. Hence, we define the PoA as: PoA $:= \sup_{F\in\mathcal{G}} \sup_{\alpha>r^*} \frac{\Pi_I^*}{\Pi_D^*}$. We then have

Theorem 6 *The PoA of the system is given by*

$$PoA = 1 + \mathcal{O}(1/n) \tag{3}$$

Proof A direct substitution in the definition of PoA yields:

$$\text{PoA} := \sup_{F\in\mathcal{G}} \sup_{\alpha>r^*} \left\{ \frac{(n+1)^2}{n} \cdot \left(n + \frac{\alpha}{r^*}\right)^{-1} \right\} \tag{4}$$

Since $\left(n + \frac{\alpha}{r^*}\right)^{-1}$ decreases in the ratio α/r^*, the inner sup is attained asymptotically for $\alpha \to r^*$. Hence, PoA $= \sup_{F\in\mathcal{G}} \left\{ \frac{(n+1)^2}{n} \cdot (n+1)^{-1} \right\} = 1 + \frac{1}{n}$.

Theorem 6 implies that the supply chain becomes less efficient as the number of downstream retailers increases. Although the PoA provides a useful worst-case scenario, for a fixed F and a realized demand α, it is also of interest to study the response of the ratio Π_I^*/Π_D^* to different wholesale prices. Specifically, for any given value of α, and fixed F, Π_I^*/Π_D^* increases as the wholesale price increases. Hence, a higher wholesale price corresponds to worst efficiency for the decentralized chain.

Together with the observation that with a higher wholesale price, the supplier captures a larger share of the system profits, this motivates—from a social perspective—the study of mechanisms that will lead to reduced wholesale prices for fixed demand levels and fixed market characteristics (number of retailers and demand distribution). Such a study falls not within the topic of the present analysis but constitutes a promising direction for future research.

5 Conclusions

Along with [10], the present study provides a probabilistic and economic analysis that aims to extend the work of [3, 8, 9, 12]. The characterization, under mild conditions, of the supplier's optimal pricing policy as the unique fixed point of the MRL function of the demand distribution, provides a powerful tool for a multifaceted comparative statics analysis. Theorems 2 and 3 demonstrate how stochastic orderings, coupled with this characterization, provide predictions of the response of the wholesale price in a versatile environment of various demand transformations. Based on a numerical example, Sect. 4.1.4 confirms [8]'s intuition that prices are driven by different forces than market size. In Sects. 4.2 and 4.2.1, we show that number of second stage retailers and wholesale prices have a direct impact on supply chain performance and efficiency. A more extended version of the present study is subject of ongoing work.

References

1. Acemoglu, D., Jensen, M.K.: Aggregate comparative statics. Games Econ. Behav. **81**, 27–49 (2013). https://doi.org/10.1016/j.geb.2013.03.009
2. Athey, S.: Monotone comparative statics under uncertainty. Q. J. Econ. **117**(1), 187–223 (2002). http://www.jstor.org/stable/2696486
3. Banciu, M., Mirchandani, P.: Technical note—new results concerning probability distributions with increasing generalized failure rates. Oper. Res. **61**(4), 925–931 (2013). https://doi.org/10.1287/opre.2013.1198
4. Belzunce, F., Martinez-Riquelme, C., Mulero J.: An Introduction to Stochastic Orders (Chapter 2—Univariate Stochastic Orders). Academic Press (2016). https://doi.org/10.1016/B978-0-12-803768-3.00002-8
5. Bernstein, F., Federgruen, A.: Decentralized supply chains with competing retailers under demand uncertainty. Manag. Sci. **51**(1), 18–29 (2005). https://doi.org/10.1287/mnsc.1040.0218
6. Jensen, M.K.: Distributional comparative statics. Rev. Econ. Stud. **85**(1), 581–610 (2018). https://doi.org/10.1093/restud/rdx021
7. Lai, C.-D., Xie, M.: Stochastic Ageing and Dependence for Reliability. Springer Science+Business Media, Inc. (2006). ISBN-10: 0-387-29742-1
8. Lariviere, M., Porteus, E.: Selling to the newsvendor: an analysis of price-only contracts. Manuf. Serv. Oper. Manag. **3**(4), 293–305 (2001). https://doi.org/10.1287/msom.3.4.293.9971
9. Lariviere, M.: A note on probability distributions with increasing generalized failure rates. Oper. Res. **54**(3), 602–604 (2006). https://doi.org/10.1007/978-1-4615-4949-9_8
10. Leonardos, S., Melolidakis, C.: Selling to Cournot Oligopolists: Pricing under Uncertainty & Generalized Mean Residual Life (2017). arXiv:1709.09618
11. Pan, K., Lai, K.K., Liang, L., Leung, S.: Two-period pricing and ordering policy for the dominant retailer in a two-echelon supply chain with demand uncertainty. Omega **36**(4), 919–929 (2009). https://doi.org/10.1016/j.omega.2008.08.002
12. Paul, A.: A note on closure properties of failure rate distributions. Oper. Res. **53**(4), 733–734 (2005). https://doi.org/10.1287/opre.1040.0206
13. Perakis, G., Roels, G.: The price of anarchy in supply chains: quantifying the efficiency of price-only contracts. Manag. Sci. **53**(8), 1249–1268 (2007). https://doi.org/10.1287/mnsc.1060.0656
14. Shaked, M., Shanthikumar, G.: Stochastic Orders. Springer Series in Statistics, New York (2007). ISBN 0-387-32915-3
15. Wu, C.-H., Chen, C.-W., Hsieh, C.-C.: Competitive pricing decisions in a two-echelon supply chain with horizontal and vertical competition. Int. J. Prod. Econ. **135**(1), 265–274 (2012). https://doi.org/10.1016/j.ijpe.2011.07.020
16. Yang, S.-L., Zhou, Y.-W.: Two-echelon supply chain models: considering duopolistic retailers different competitive behaviors. Int. J. Prod. Econ. **103**(1), 104–116 (2006). https://doi.org/10.1016/j.ijpe.2005.06.001

Speeding-up the Exploration of the 3-OPT Neighborhood for the TSP

Giuseppe Lancia and Marcello Dalpasso

Abstract A move of the 3-OPT neighborhood for the Traveling Salesman Problem consists in removing any three edges of the tour and replacing them with three new ones. The standard algorithm to find the best possibble move is cubic, both in its worst and average time complexity. Since TSP instances of interest can have thousands of nodes, up to now it has been impossible to use the 3-OPT local search on anything other than quite small instances. In this paper we describe an alternative algorithm whose average complexity appears to be quadratic and which allowed us to use 3-OPT on instances with several thousand nodes. The algorithm is based on a rule for quickly choosing two out of three edges in a good way, and then completing the choice in linear time. To this end, the algorithm uses max-heaps as a suitable data structure.

Keywords Traveling Salesman Problem · 3-OPT · K-OPT · Local search
Average running time

1 TSP and the K-OPT Neighborhood

The Traveling Salesman Problem (TSP) calls for finding the shortest Hamiltonian cycle (tour) in a complete graph $G = (V, E)$ of n nodes, weighted on the arcs. We consider the symmetric TSP, i.e., the graph is undirected and the distance $c(i, j)$ between two nodes i and j is the same irrespective of the direction in which we

G. Lancia (✉)
DMIF, University of Udine, Udine, Italy
e-mail: giuseppe.lancia@uniud.it

M. Dalpasso
DEI, University of Padova, Padova, Italy
e-mail: marcello.dalpasso@unipd.it

© Springer Nature Switzerland AG 2018

P. Daniele and L. Scrimali (eds.), *New Trends in Emerging Complex Real Life Problems*, AIRO Springer Series 1,
https://doi.org/10.1007/978-3-030-00473-6_37

traverse an edge. A tour is identified by a permutation of vertices $(v_1, \ldots, v_n)$. The length of a tour T, denoted by $c(T)$, is the sum of the lengths of the edges of the tour. More generally, for any set F of edges, we denote by $c(F)$ the value $\sum_{e \in F} c(e)$.

Local search (LS) [1, 6] is often a very effective way to tackle hard combinatorial optimization problems, including the TSP. Assume the problem is $\min_{x \in X} f(x)$. Given a map which associates to every solution x a set $N(x)$ called its *neighborhood*, in LS we start at any solution x^0, set $s := x^0$, and look for a solution $x^1 \in N(s)$ better than s. If found, we replace s with x^1 and iterate the same search. We continue this way until we get to a solution $s = x^i$ such that $f(s) = \min\{f(x)|x \in N(s)\}$. We say that s is a *local optimum*. Replacing x^i with x^{i+1} is called performing a *move* of the search. The total number of moves performed from x^0 to the local optimum is called the *length of the convergence*. If $x \in N(s)$ and $f(x) < f(s)$ we say that the move from s to x is an *improving move*. There are two main strategies of LS, namely *first-improvement* and *best-improvement*. In the first-improvement strategy, x^{i+1} is the first solution that we find in $N(x^i)$ such that $f(x^{i+1}) < f(x^i)$. In best-improvement, x^{i+1} is such that $f(x^{i+1}) = \min\{f(x)|x \in N(x^i) \wedge f(x) < f(x^i)\}$. Neither one of these strategies is better than the other on all instances.

A popular LS neighborhood for the TSP is the so-called K-OPT. Let $K \in \mathbb{N}$ be a constant. A K-OPT move on a tour T consists in first removing a set R of K edges and then inserting a set I of K edges so as $(T \setminus R) \cup I$ is still a tour. A K-OPT move is improving if $c(I) < c(R)$, while it is *best improving* if $c(R) - c(I)$ is the maximum over all possible choices of R, I.

The first use of K-OPT dates back to 1958 with the introduction of 2-OPT in [3]. In 1965 Lin [5] described the 3-OPT neighborhood, and experimented with the $\Theta(n^3)$ algorithm. The instances which could be tackled at the time were fairly small ($n \leq 150$). In 1968, Steiglitz and Weiner [8] described an improvement over Lin's method which made it 2 or 3 times faster, but still cubic in nature.

The exploration of the K-OPT neighborhood, for a fixed K, might be considered "fast" from a theoretical point of view, since there is an obvious polynomial algorithm (complete enumeration, of time $\Theta(n^K)$). However, in practice, complete enumeration makes the use of K-OPT impossible already for $K = 3$, if n is large enough. For a given tour of, say, $n = 5000$ nodes, the time required to try all 3-OPT moves, on a reasonably fast computer, is more than an hour, let alone converging to a local optimum. For this reason, 3-OPT has never been really adopted for the heuristic solution of TSP instances of interest.

An important recent result in [4] proves that, under a widely believed hypothesis similar to the $P \neq NP$ conjecture, it is impossible to find the best 3-OPT move with a worst-case algorithm of time $O(n^{3-\epsilon})$ for any $\epsilon > 0$ so that complete enumeration is, in a sense, optimal. However, this gives us little consolation when we are faced with the problem of applying 3-OPT to a large TSP instance. In fact, for complete enumeration the average case and the worst case coincide, and we might wonder if there exists a better practical algorithm, much faster than complete enumeration on the majority of instances but still $O(n^3)$ in the worst case. The algorithm described in this paper is such an example.

1.1 Our Contribution

The TSP is today very effectively solved, even to optimality, by using sophisticated mathematical programming based approaches, such as Concorde [2]. No matter how ingenious, heuristics can hardly be competitive with these approaches when the latter are given enough running time.

However, heuristics such as local search have a great quality: they are simple (to understand, to implement and to maintain) and, in general, very fast, so that they can overcome their simplemindedness by being able to sample a huge amount of good solutions in a relatively small time. Of course if too slow, we lose all the interest in using a LS approach despite its simplicity.

This is exactly the situation for the 3-OPT. The goal of our work has been to show that, with a clever implementation of the search for improving moves, it can be actually used since it can become orders of magnitude faster than its standard implementation.

Let us give a flavour of the type of results that we can achieve. Assume we have an average PC and 5 hours of time which we want to devote to local search starting from as many random tours as possible. Assume $n = 1000$ and each convergence goes through 500 intermediate solutions. With the enumerative approach we could sample only one local optimum and would stop while halfway through the second convergence. With our method we would sample more than 300 local optima.

As we will see, another advantage of our method is that while we are approaching the local optimum, our method becomes faster in finding (if it exists) an improving move or the best improving move. The brute force approach, on the other hand, takes constant time for finding the best improving move, or, if it is looking for the first improvement, it does in fact become slower, since the number of candidates to try before finding an improvement becomes larger and larger near the local optimum.

2 Notation

Let $G = (V, E)$ be a complete graph on n nodes, and $c : E \mapsto \mathbb{R}^+$ be a cost function for the edges. Without loss of generality, $V = \{0, 1, \ldots, \bar{n}\}$, where $\bar{n} = n - 1$. We will describe an effective strategy for finding either the best improving or any improving move for a given current tour $(v_1, \ldots, v_n)$ which, wlog, we assume to be $T = (0, 1, \ldots, \bar{n})$.

We will be using modular arithmetic frequently. For convenience, for each $x \in V$ and $t \in \mathbb{N}$ we define

$$x \oplus t := (x + t) \mod n, \qquad x \ominus t := (x - t) \mod n.$$

We define the *forward distance* $d^+(x, y)$ from node x to node y as the unique $t \in \{0, \ldots, n - 1\}$ such that $x \oplus t = y$. Similarly, we define the *backward distance* $d^-(x, y)$ from x to y as the $t \in \{0, \ldots, n - 1\}$ such that $x \ominus t = y$.

Finally, the *distance* between any two nodes x and y is defined by

$$d(x, y) := \min\{d^+(x, y), d^-(x, y)\}$$

A 3-OPT move is fully specified by two sets, i.e., the set of removed and of inserted edges. We call a *removal set* any set of three tour edges, i.e., three edges of type $\{i, i \oplus 1\}$. A removal set is identified by a triple $S = (i_1, i_2, i_3)$ with $0 \le i_1 < i_2 < i_3 \le \bar{n}$, where the edges removed are $R(S) := \{\{i_j, i_j \oplus 1\} : j = 1, 2, 3\}$. We call any such triple a *selection*. A selection S is *complete* if $d(i_j, i_h) \ge 2$ for each $j \neq h$, otherwise we say that S is a *partial* selection. Complete selections are more important than partial selections, since there is only a quadratic number of partial selections but a cubic number of complete ones.

Let S be a selection and $I \subset E$ with $|I| = 3$. If $(T \setminus R(S)) \cup I$ is still a tour then I is called a *reinsertion set*. Given a selection S, a reinsertion set I is *pure* if $I \cap R(S) = \emptyset$, and *degenerate* otherwise. Finding the best 3-OPT move when the reinsertions are constrained to be degenerate is $O(n^2)$ (in fact, 3-OPT degenerates to 2-OPT in this case). Therefore, the most computationally expensive task is to determine the best move when *the selection is complete and the reinsertion is pure*. We refer to this kind of moves as *true* 3-OPT. Thus, in the remainder of the paper we will focus on true 3-OPT moves.

2.1 Reinsertion Schemes

Let $S = (i_1, i_2, i_3)$ be a complete selection. When the edges $R(S)$ are removed from a tour, the tour gets broken into 3 consecutive segments which we can label by $\{1, 2, 3\}$ (segment j ends at node i_j). Since the selection is pure, each segment is indeed a path of at least one edge. A reinsertion set patches back the segments into a new tour. If we adopt the convention to start always a tour with segment 1 traversed clockwise, the reinsertion set: (i) determines a new ordering in which the segments are visited along the tour and (ii) may cause some segments to be traversed counterclockwise. In order to represent this fact we use a notation called a *reinsertion scheme*. A reinsertion scheme is a signed permutation of $\{2, 3\}$. The permutation specifies the order in which the segments 2, 3 are visited after the move. The signing $-s$ tells that segment s is traversed counterclockwise, while $+s$ tells that it is traversed clockwise. For example, the third reinsertion set depicted in Fig. 1 is represented by the reinsertion scheme $< +3, -2 >$ since from the end of segment 1 we jump to the beginning of segment 3 and traverse it forward. We then move to the last element of segment 2 and proceed backward to its first element. Finally, we close the cycle by going back to the first element of segment 1.

There are potentially $2^2 \times 2! = 8$ reinsertion schemes, but for some of these the corresponding reinsertion sets are degenerate. A scheme for a pure reinsertion must not start with $+2$, nor end with "$+3$", nor be $< -3, -2 >$. This leaves only 4 possible schemes, let them be $r_1, \ldots, r_4$.

Clearly, there is a bijection between reinsertion schemes and reinsertion sets. If r is a reinsertion scheme, we denote by $I(r)$ the corresponding reinsertion set. The enumeration of all true 3-OPT moves can be done as follows: (i) We consider, in turn, each reinsertion scheme $r_1, \ldots, r_4$; (ii) Given r_j, we consider all complete selections $S = (i_1, i_2, i_3)$, obtaining the moves defined by $(R(S), I(r_j))$. The cost of step (ii) by complete enumeration is $\Theta(n^3)$. In the remainder of the paper we will focus on a method for lowering significantly, in practice, the complexity of this step.

2.2 The Number of Complete Selections

For space reasons we state the following theorem without proof. For generality, we consider the complete selections of K-OPT, i.e., k-tuples $(i_1, \ldots, i_n)$ of increasing indices with $d(i_j, i_{j\oplus 1}) > 1$ for two consecutive indices.

Theorem 1 *For each $K = 2, \ldots, \lfloor n/2 \rfloor$ the number of complete K-OPT selections is*

$$\binom{n-K+1}{K} - \binom{n-K-1}{K-2}$$

Corollary 1 *The number of complete 3-OPT selections is*

$$\binom{n-2}{3} - (n-4) = \frac{n^3 - 9n^2 + 20n}{6}$$

From the corollary, and knowing that there are 4 pure reinsertion schemes for each 3-OPT complete selection, we can compute the number $T_3(n)$ of true 3-OPT moves. For example, it is $T_3(1{,}000) = 660{,}680{,}000$, while $T_3(5{,}000) = 83{,}183{,}400{,}000$ and $T_3(10{,}000) = 666{,}066{,}800{,}000$ giving a striking example of why the exploration of the 3-OPT neighborhood would be totally impractical unless some effective strategies were adopted.

3 Speeding-up the Search: The Basic Idea

Our method can be used to find either the best improving selection or any improving selection. In the rest of the paper we will focus on the Best-Improvement case, since it is the harder of the two. The changes needed in order to adopt the method for a First-Improvement search are trivial.

According to the enumerative strategy outlined in the previous section, suppose we have fixed a reinsertion scheme r, and want to find the best selection for it. Our goal is to provide an alternative, much faster, way to do it than the following, classical, "nested-for" approach over all indices:

```
for ( i1 = 0; i1 ≤ n̄ − 4; i1++ )
    for ( i2 = i1 + 2; i2 ≤ n̄ − 2 − P(i1 = 0); i2++ )
        for ( i3 = i2 + 2; i3 ≤ n̄ − P(i1 = 0); i3++ )
            evaluateMove(i1, i2, i3, r); [* check if move is improving. possibly update best *]
```

(The expression $\mathcal{P}(A)$, given a predicate A returns 1 if A is true and 0 otherwise).

Our idea for speeding-up the search is based on this consideration. Suppose there is a magic box which knows all the pairs of indices that belong to some best improving selection, and that we can inquire the box by specifying two labels (e.g., "i_2" and "i_3", etc.). The box, in time $O(1)$ would return us a pair of values (e.g., v_2 and v_3) such that there exists at least one best improving 3-OPT move in which the two specified indices have those particular values. At this point we could enumerate the values for the missing index and determine the best completion possible. This way, finding a best improving 3-OPT move would take $\Theta(n)$ time rather than $\Theta(n^3)$.

The bulk of our work has then been to simulate, heuristically, a similar magic box, i.e., a data structure that can be queried and should return two out of the three indices of a best improving selection much in a similar way as described above. In our heuristic version, the box, rather than returning a pair of indices that are certainly in a best improving solution, returns a pair of indices that *are likely to be in a best improving solution*. As we will see, this can already greatly reduce the number of possible selections candidate to be best improving. In order to assess the likelihood of two specific indices to be in a best solution, we will use suitable two-arguments functions described in the next sections.

3.1 The Fundamental Quantities τ^+ and τ^-

We define two functions of $V \times V$ into $\mathbb{R}$ which, loosely speaking, will be used to determine, for each pair of indices of a selection, the contribution of that pair to the value of a move. The rationale is that, the higher the contribution, the higher the probability that a particular pair is in a best selection.

The two functions are called $\tau^+()$ and $\tau^-()$. For each $a, b \in \{0, \ldots, \bar{n}\}$, we define: (1) $\tau^+(a, b)$ to be the difference between the cost from a to its successor and to the successor of b, and (2) $\tau^-(a, b)$ to be the difference between the cost from a to its predecessor and to the predecessor of b:

$$\tau^+(a, b) = c(a, a \oplus 1) - c(a, b \oplus 1), \qquad \tau^-(a, b) = c(a, a \ominus 1) - c(a, b \ominus 1)$$

Clearly, each of these quantities can be computed in time $O(1)$, and computing their values for a subset of possible pairs can never exceed time $O(n^2)$.

4 A 3-Phase Procedure for Searching the 3-OPT Neighborhood

The pure 3-OPT reinsertion schemes are four (see Fig. 1), namely: $r_1 =< +3, +2 >$; $r_2 =< -2, -3 >$; $r_3 =< +3, -2 >$; and $r_4 =< -3, +2 >$.

Notice that r_3 and r_4 are symmetric to r_2. Therefore, we can just consider r_1 and r_2 since all we say about r_2 can be applied, *mutatis mutandis*, to r_3 and r_4 as well.

Given a reinsertion scheme r, the cost $\Delta(i_1, i_2, i_3)$ of a move with selection $S = (i_1, i_2, i_3)$ is the difference between the cost of the removed edges $\{\{i_1, i_1 \oplus 1\}, \{i_2, i_2 \oplus 1\}, \{i_3, i_3 \oplus 1\}\}$ and the cost of the reinsertion set $I(r)$. A key observation is that we can break-up the function $\Delta()$, that has $\Theta(n^3)$ possible arguments, into a sum of functions of *two parameters each* (therefore, $\Theta(n^2)$ possible arguments). That is, we'll have

$$\Delta(i_1, i_2, i_3) = f^1(i_1, i_2) + f^2(i_2, i_3) + f^3(i_1, i_3) \quad (1)$$

for suitable functions $f^1(), f^2(), f^3()$, each representing the contribution of a particular pair of indices to the value of the move. The domains of these functions are subsets of $\{0, \ldots, \bar{n}\} \times \{0, \ldots, \bar{n}\}$ which limit the valid input pairs to values obtained from two specific elements of a selection. Let $\mathcal{S}$ be the set of all complete selections. For $a, b \in \{1, 2, 3\}$, let us define

$$\mathcal{S}_{ab} := \{(x, y) : \exists(v_1, v_2, v_3) \in \mathcal{S} \text{ with } v_a = x \text{ and } v_b = y\} \quad (2)$$

Then the domain of f^1 is $\mathcal{S}_{12}$, the domain of f^2 is $\mathcal{S}_{23}$ and the domain of f^3 is $\mathcal{S}_{13}$.

Below, we describe these functions for r_1 and r_2 (remember that r_3 and r_4 are symmetric to r_2):

[r_1 :] We have $I(r) = \{\{i_1, i_2 \oplus 1\}, \{i_2, i_3 \oplus 1\}, \{i_1 \oplus 1, i_3\}\}$ (see Fig. 1) and

$$\Delta(i_1, i_2, i_3) = \tau^+(i_1, i_2) + \tau^+(i_2, i_3) + \tau^+(i_3, i_1)$$

$$f^1 : (x, y) \in \mathcal{S}_{12} \mapsto \tau^+(x, y); \quad f^2 : (x, y) \in \mathcal{S}_{23} \mapsto \tau^+(x, y);$$
$$f^3 : (x, y) \in \mathcal{S}_{31} \mapsto \tau^+(x, y).$$

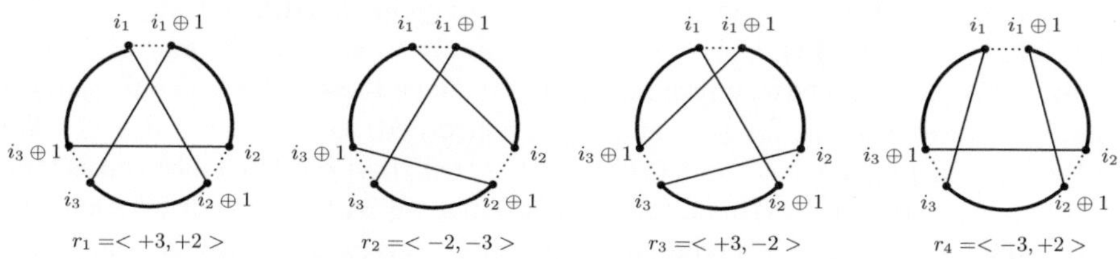

Fig. 1 The pure reinsertion schemes of 3-OPT

[r_2 :] We have $I(r) = \{\{i_1, i_2\}, \{i_2 \oplus 1, i_3 \oplus 1\}, \{i_1 \oplus 1, i_3\}\}$ (see Fig. 1) and

$$\Delta(i_1, i_2, i_3) = \tau^+(i_1, i_2 \ominus 1) + \tau^-(i_2 \oplus 1, i_3 \oplus 2) + \tau^+(i_3, i_1)$$

$$f^1 : (x, y) \in \mathcal{S}_{12} \mapsto \tau^+(x, y \ominus 1); \;\; f^2 : (x, y) \in \mathcal{S}_{23} \mapsto \tau^+(x \oplus 1, y \oplus 2);$$
$$f^3 : (x, y) \in \mathcal{S}_{31} \mapsto \tau^+(x, y).$$

The two-parameter functions f^1, f^2, f^3 can be used to discard very quickly from consideration all triples such that no two of the indices give a sufficiently large contribution to the total. Better said, we keep in consideration only candidate triples for which at least one contribution of two indices is large enough. Assume we want to find the best selection and we currently have a selection $S^* = (\bar{v}_1, \bar{v}_2, \bar{v}_3)$ of value $V := \Delta(S^*)$ (the current "champion"). We make the trivial observation that for a selection (i_1, i_2, i_3) to beat S^* it must be

$$\left(f^1(i_1, i_2) > \frac{V}{3}\right) \vee \left(f^2(i_2, i_3) > \frac{V}{3}\right) \vee \left(f^3(i_1, i_3) > \frac{V}{3}\right)$$

These are not exclusive, but possibly overlapping conditions, which we will consider in turn with a three-phase algorithm. For $j = 1, 2, 3$, in the j-th phase, we will restrict our search to the selections (i_1, i_2, i_3) which satisfy the j-th condition. Furthermore, we will not enumerate those selections with a complete enumeration, but rather *from the most promising to the least promising*, stopping as soon as we realize that no selection of the phase has still the possibility of being the best selection overall.

The best data structure for performing this kind of search (which hence be used for our heuristic implementation of the magic box described in Sect. 3) is the *Max-Heap*. A heap is perfect for taking the highest-valued elements from a set, in decreasing order. It can be built in linear time with respect to the number of its elements and has the property that the largest element can be extracted in logarithmic time, while still leaving a heap.

We then build a a max-heap H^1 in which, for each $(x, y) \in \mathcal{S}_{12}$ such that $f^1(x, y) > V/3$, we put the triple $(x, y, f^1(x, y))$. The heap is organized according to the value of f^1. Assume H^1 has L elements. Building the heap has cost $O(L)$, and then extracting the element of maximum $f^1()$ value, while keeping the heap structure, has cost $O(\log L)$. We will now start extracting the elements from the heap. Let us denote by $(x[j], y[j], f^1(x[j], y[j]))$ the j-th element extracted. Heuristically, we might say that $x[1]$ and $y[1]$ are the most likely values that the pair of indices i_1 and i_2 can take in a best improving selection, since these values give the largest possible contribution (as far as i_1 and i_2 are concerned) to the move value (1). We will keep extracting the maximum $(x[j], y[j], f^1(x[j], y[j]))$ from the heap as long as $f^1(x[j], y[j]) > V/3$. This does not mean that we will extract all the elements of H^1, since the value of V could change (namely, increase) during the search and hence the extractions might terminate before the heap is empty.

Each time we extract the heap maximum, we have that $x[j]$ and $y[j]$ are two possible indices (i.e., i_1 and i_2) out of three for a candidate selection to beat S^*. With a linear-time scan, we can search the third missing index (i.e., i_3) and see if we get indeed a better selection than S^*. To find i_3 we run a for-cycle with $y[j]+2 \leq i_3 \leq \bar{n}$, checking each time if $(x[j], y[j], i_3)$ is a better selection than S^*. Whenever this is the case, we update S^*. Notice that we also update V so that the number of elements still in the heap for which $f^1(x, y) > V/3$ may decrease considerably. Say that, overall, M elements are extracted from the heap. Then the cost of the phase is $O(n^2 + L + M(n + \log L))$ which, on our tests, was is in practice growing as a quadratic function of n. Worst-case, this is $O(n^3)$ like complete enumeration but, as we will show in our computational experiments, it is *much* smaller in practice. This is because the Mn triples which are indeed evaluated for possibly becoming the best selection have a much bigger probability of being good than a generic triple, since two of the three indices are guaranteed to help the value of the move considerably.

After this phase of probing the heap H^1, we run an analogous phase in which we probe a heap H^2 containing all the triples $(x, y, f^2(x, y))$ for which $(x, y) \in \mathcal{S}_{23}$ and $f^2(x, y) > V/3$. For each element $(x[j], y[j])$ extracted from the heap, we look for the missing index i_1, with $0 \leq i_1 \leq x[j]-2$. Notice that the value V determining which elements belong to H^1 is the value of the current best solution, i.e., it is not the value that V had at the start of the previous phase, but the value it had at its end.

A third and final phase probes a heap H^3 containing all the triples $(x, y, f^3(x, y))$ for which $(x, y) \in \mathcal{S}_{13}$ and $f^3(x, y) > V/3$, each time looking for the missing index i_2 in the range $x[j]+2 \leq i_2 \leq y[j]-2$.

5 Computational Results

Here we describe some preliminary computational results. More experiments and a full detailed discussion of the computational experience are delayed to a journal version of this paper.

In a first set of experiments we have tested the idea on random instances in which the edge costs are uniformly distributed over the interval $[1, \ldots, n^2]$. The following table refers to 10 random graphs for each choice of n from 1,000 to 3,000 with increments of 500. For each instance, we generated a random tour and looked for the best 3-OPT move both with the complete enumeration (C) and our heap-based method (H):

Times are in seconds on a Intel Pentium G645 @2.9 GHz. We put the rows of avg next to each other and it is easy to see that the speedup of our method goes from about 100× to more than 300× for increasing n. This behavior was true for all instances tried. By fitting the growth of the time required to perform a move from a random permutation over a larger set of graph sizes than what shown here, we determined the quadratic polynomial function $t(n) = An^2 + Bn + C$ microseconds, with $A = 0.097$, $B = 9.18$ and $C = 0$.

	$n = 1000$	$n = 1500$	$n = 2000$	$n = 2500$	$n = 3000$
Min C	39.416	146.936	396.748	805.064	1452.312
Max C	39.936	147.624	385.872	810.576	1463.624
Avg C	39.712	147.272	392.300	807.712	1457.236
Avg H	0.312	0.688	1.672	2.448	3.336
Min H	0.184	0.560	1.560	2.060	2.500
Max H	0.372	0.748	1.684	2.936	4.624

When n gets large, comparing the two approaches becomes impractical since the time for complete enumeration gets too long. We can however abort complete enumeration after, say 10,000,000 triples and estimate the final time quite accurately as $\bar{t} \times T_3(n)/10{,}000{,}000$, where $\bar{t}$ is the time at abortion and $T_3(n)$ was given after Corollary 1. This way, for example, we can show that the time for finding the best 3-OPT move from a random solution when $n = 5{,}000$ is more than 10 h, while with our method (which we carry out without aborting) it is about 30 s.

The same type of behavior is true for instances from the TSPLIB [7], although the speedup percentage seems lower than for uniform-cost random graphs. We sampled 94 instances of TSPLIB and performed a best move from a random starting solution (10 times). The instance sizes went from $n = 21$ to $n = 2392$. For small n it is practically impossible to distinguish the running times and our method starts to emerge as a clear winner for $n \geq 100$. The larger instances are reported below:

Instance	u1817	d2103	u2152	u2319	pr2392
n	1817	2103	2152	2319	2392
Avg C	261.560	425.936	460.812	600.008	654.436
Avg H	3.500	2.624	6.124	3.312	6.040

6 Conclusions and Future Directions

We have described a practical method for finding the best 3-OPT move which exhibits, empirically, a quadratic running time. A (probably quite hard) question to study would be to mathematically prove that the expected running time of this algorithm on random instances is lower than cubic. A direction of future research would also be to assess the effectiveness of 3-OPT, now that we can use it, on the TSP, i.e., to study the quality of 3-OPT local optima (for both best- and first- improvement LS).

Finally, we are developing similar ideas to speed-up the search of the 4-OPT neighborhood, which, from preliminary results, appear very promising.

References

1. Aarts, E., Lenstra, J.K. (eds.): Local Search in Combinatorial Optimization, 1st edn. Wiley, Inc., New York, NY, USA (1997)
2. Applegate, D.L., Bixby, R.E., Chvatl, V., Cook, W.J.: The Traveling Salesman Problem: A Computational Study. Princeton University Press (2006)
3. Croes, G.A.: A method for solving traveling-salesman problems. Oper. Res. **6**(6), 791–812 (1958)
4. de Berg, M., Buchin, K., Jansen, B., Woeginger, G.J.: Fine-grained complexity analysis of two classic TSP variants. In: 43rd International Colloquium on Automata, Languages, and Programming, ICALP 2016, 11–15 July 2016, Rome, Italy, pp. 5:1–5:14 (2016)
5. Lin, S.: Computer solutions of the traveling salesman problem. Bell Syst. Tech. J. **44**(10), 2245–2269 (1965)
6. Papadimitriou, C.H., Steiglitz, K.: Combinatorial Optimization: Algorithms and Complexity, vol. 01. Prentice Hall (1982)
7. Reinelt, G.: TSPLIB—a traveling salesman problem library. ORSA J. Comput. **3**, 376–384 (1991)
8. Steiglitz, K., Weiner, P.: Some improved algorithms for computer solution of the traveling salesman problem. In: Proceedings of the 6th annual Allerton Conference on System and System Theory, pp. 814–821. University of Illinois, Urbana (1968)

The Green Vehicle Routing Problem with Occasional Drivers

Giusy Macrina and Francesca Guerriero

Abstract This paper introduces a new variant of the green vehicle routing problem with crowd-shipping. The company has an own mixed fleet composed of conventional combustion engine and electric vehicles. In addition, ordinary people named "occasional drivers" are available to deliver items to some customers on their route. The objective is to minimize the sum of routing costs of conventional and electric vehicles, by including fuel consumption cost and energy consumption cost, and occasional drivers' compensation. We describe an integer linear programming formulation for the problem and we also provide a comprehensive analysis on several indicators, such as routing costs and polluting emissions. The results show how the use of occasional drivers may lead not only to more convenient solutions, but also to highly interesting scenarios in a green perspective.

Keywords Green vehicle routing problem · CO_2 emissions · Electric vehicles Crowd-shipping · Occasional drivers

1 Introduction

Global environmental pollution is a critical issue facing the whole population. Among the other logistic sectors, transportation is one of the major producer of polluting emissions. Introducing environmental aspects in classic logistics is necessary to protect the health of people, by reducing negative externalities. Several researchers have started to develop "green" solutions for the classic vehicle routing problem (VRP). In the last years, this problem, named green vehicle routing problem (GVRP), has been widely studied. The goals of the GVRP are twofold: on one hand some authors

---

G. Macrina (✉) · F. Guerriero
Department of Mechanical, Energy and Management Engineering, University of Calabria, 87036 Rende CS, Italy
e-mail: giusy.macrina@unical.it

F. Guerriero
e-mail: francesca.guerriero@unical.it

© Springer Nature Switzerland AG 2018
P. Daniele and L. Scrimali (eds.), *New Trends in Emerging Complex Real Life Problems*, AIRO Springer Series 1,
https://doi.org/10.1007/978-3-030-00473-6_38

focused on pollution reduction in VRP with conventional vehicles, by introducing the pollution routing problem (PRP) (e.g. see: [3, 5, 8, 9]), on the other hand a large part of works concerns the use of alternative fuel vehicles, in particular electric vehicles (EVs), in the VRP (E-VRP) (e.g. see: [6, 7, 11, 12]). The main goal of the E-VRP is to introduce the EVs in transportation planning and reduce their energy consumption. Even if the reduction of polluting emissions and the use of alternative fuel vehicles are important challenges for transportation companies, however, a large number of cars continues to travel on the road, producing polluting emissions and generating congestion. Ordinarily, these cars are underused and could be better exploited. This is the main goal of "Crowd-shipping": to allow deliveries, that usually are performed by companies, to ordinary people, named occasional drivers (ODs). Crowd-shipping becomes a new opportunity and a creative solution to exploit underused assets, i.e. ordinary cars, by optimizing the costs of same-day and last-mile deliveries and reducing the environmental impacts. Several big on-line retailers have started to use Crowd-shipping, such as Walmart, DHL and Amazon. Crowd-shipping requires a platform to connect ODs to customers, thus, when a customer sends a request of fast delivery on computer/phone application, if an OD accepts to make the delivery, the platform sends him all the information necessary to perform the service. Arslan et al. [2] presented an overwiew on benefits of Crowd-shipping. They considered a peer-to-peer platform, taking into account the possibility of using traditional vehicles and ad hoc vehicles. Archetti et al. [1] introduced the vehicle routing problem with occasional drivers (VRPOD). The authors supposed that the company can make deliveries not only by using its own fleet, but also some ODs. Some works extended the problem presented in Archetti et al. [1], in particular Macrina et al. [10] considered multiple deliveries for ODs and time windows for both ODs and customers, whereas Dahle et al. [4] assumed that the availability of ODs is uncertain, thus they introduced stochasticity on ODs and supposed to have some stochastic information about their appearance.

The main goal of this work is to combine the features of GVRP with Crowd-shipping, and to analyse new innovative green solutions for vehicle routing, thus we present the green vehicle routing problem with ODs (GVRPOD). We consider the availability of a mixed fleet of internal combustion commercial vehicles (ICCVs) and EVs, named "regular vehicles", and a certain number of ODs. We formulate a mathematical model to represent this new problem as a variant of the GVRP, then we perform analysis in order to evaluate how using ODs may lead not only to more convenient solutions, but also more eco-friendly transportation plans. The remainder of this paper is organized as follows: in Sect. 2 we present the mathematical model for the GVRPOD, in Sect. 3 we describe our preliminary numerical study to highlight the benefit of using ODs in GVRP, finally, Sect. 4 summarizes the conclusions.

2 Mathematical Model for the GVRPOD

We present an integer linear programming formulation for the GVRPOD. Let C be the set of customers, and R the set of recharging stations. Let R' be the set of recharging stations and their copies to allow multiple visits. Let $s \in R$ be the depot node in which each route of regular vehicles starts and ends. We duplicate the depot, thus we denote with t the destination node. Let $W = C \cup R \cup \{s, t\}$ and $W' = C \cup R' \cup \{s, t\}$. Let K be the set of available ODs and V the set of their v_k destinations. We define the node set as $N' = C \cup R' \cup \{s, t\} \cup V$. The $GVRPOD$ is formulated on a graph $G = (N', A)$, where A is the set of arcs. In what follows, we indicate with the superscripts C and E the conventional and electric vehicles, respectively; as long as with the superscript OD the ODs. All the parameters and decision variables used to formulate mathematically the problem under study are summarized in Table 1.

Every customer $i \in C$ must be served only once, either by a regular vehicle (EV or ICCV) or by an OD. For EVs we consider a constant energy consumption proportional to the distance travelled. Every recharging station is characterized by a recharging speed, the EV can be charged at recharge stations on the roads or at depot during the night, at a lower price. Partial battery recharging is allowed and we also take into account that recharging the last 10% of battery may require long times, whereas recharging up to 90% may damage battery. In order to define the energy consumption and CO_2 emissions for the ICCVs, we consider the estimation made by Ubeda et al. [13]

We model the GVRPOD as follows:

$$\text{Minimize} \quad w^r \sum_{i \in R'} \sum_{j \in W'} g_{ij} + w^e \Big(\sum_{(i,j) \in A} \pi d_{ij} x_{ij}^E - \sum_{i \in R'} \sum_{j \in W'} g_{ij} \Big) \tag{1}$$

$$+ w^f \sum_{(i,j) \in A} f(u_i^C) d_{ij} x_{ij}^C + \sum_{(i,j) \in A} c_{ij}^E d_{ij} x_{ij}^E + \sum_{(i,j) \in A} c_{ij}^C d_{ij} x_{ij}^C$$

$$+ \sum_{k \in K} \sum_{i \in C \cup \{s\}} \sum_{j \in C} \rho c_{ij}^{OD} r_{ij}^k - \sum_{k \in K} \sum_{j \in C} c_{s v_k}^{OD} r_{sj}^k$$

$$\text{subject to} \quad \sum_{j \in W'} x_{ij}^E \leq 1 \qquad i \in R' \tag{2}$$

$$\sum_{j \in W' \setminus \{s\}} x_{ij}^E - \sum_{j \in W' \setminus \{t\}} x_{ji}^E = 0 \qquad i \in W' \tag{3}$$

$$\sum_{j \in W \setminus \{s\}} x_{ij}^C - \sum_{j \in W \setminus \{t\}} x_{ji}^C = 0 \qquad i \in W \tag{4}$$

$$\sum_{j \in W'} x_{sj}^E \leq n^E \tag{5}$$

$$\sum_{j \in W} x_{sj}^C \leq n^C \tag{6}$$

$$u_j^E \geq u_i^E + q_j x_{ij}^E - Q^E (1 - x_{ij}^E) \qquad i \in W' \setminus \{s, t\}, j \in W' \setminus \{s\} \tag{7}$$

Table 1 Parameters and variables of GVRPOD

Parameters		
q_i	Demand customer $i \in C$	kg
s_i	Service time customer $i \in C$	h
$[e_i, l_i]$	Time windows node $i \in W'$	h
d_{ij}	Distance to travel on the arc $(i, j) \in A$	km
t_{ij}	Time to travel on the arc $(i, j) \in A$	h
c_{ij}^E	EVs travel cost to transit on the arc $(i, j) \in A$	€/km
c_{ij}^C	ICCVs travel cost to transit on the arc $(i, j) \in A$	€/km
$c_{ij}^O D$	ODs travel cost to transit on the arc $(i, j) \in A$	€/km
T	Maximum duration of a regular route/end of depot time window	h
Q^E	Maximum capacity of EVs	kg
Q^C	Maximum capacity of ICCVs	kg
Q_k^{OD}	Maximum capacity of OD $k \in K$	kg
B^E	Maximum battery capacity of EVs	KWh
ρ_i	Recharging speed of recharging station $i \in R'$	KWh/h
π	Energy consumption rate	KWh/km
w^e	Recharging cost at depot	€/KWh
w^r	Recharging cost at recharging station	€/KWh
w^f	Diesel cost	€/L
Variables		
x_{ij}^E	Binary variable equal to one only if the EV traverses the arc $(i, j) \in A$	
x_{ij}^C	Binary variable equal to one only if the ICCV traverses the arc $(i, j) \in A$	
g_{ij}	Energy recharged by the EV at the recharging station $i \in R'$ for travelling to $j \in W'$	KWh
u_i^C	Amount of load left in the ICCV after visiting node $i \in W'$	kg
u_i^E	Amount of load left in the EV after visiting node $i \in W'$	kg
τ_j	Arrival time of a regular vehicle to the node $j \in W'$	h
r_{ij}^k	Binary variable equal to one only if OD $k \in K$ traverses arc $(i, j) \in A$	
f_i^k	Arrival time of OD $k \in K$ at customer $i \in C$	h
w_i^k	Available capacity of the OD $k \in K$ after visiting customer $i \in C$	kg

$$u_j^C \geq u_i^C + q_j x_{ij}^C - Q^C(1 - x_{ij}^C) \qquad i \in W \setminus \{s,t\}, j \in W \setminus \{s\} \quad (8)$$

$$u_j^E \leq Q^E \qquad j \in W' \quad (9)$$

$$u_j^C \leq Q^C \qquad j \in W \quad (10)$$

$$u_s^E = 0 \quad (11)$$

$$u_s^C = 0 \quad (12)$$

$$\tau_j \geq \tau_i + (t_{ij} + s_i)x_{ij}^E - M(1 - x_{ij}^E) \qquad i \in C, j \in W' \quad (13)$$

$$\tau_j \geq \tau_i + (t_{ij} + s_i)x_{ij}^C - M(1 - x_{ij}^C) \qquad i \in W, j \in W \quad (14)$$

$$\tau_j \geq \tau_i + t_{ij}x_{ij}^E + \frac{1}{\rho_i} g_{ij} - M(1 - x_{ij}^E) \qquad i \in R', j \in W' \quad (15)$$

$$e_j \leq \tau_j \leq l_j \qquad j \in W' \quad (16)$$

$$z_{ij} \leq (z_{hi} + g_{ij}) - \pi d_{ij} x_{ij}^E + M(1 - x_{ij}^E) + M(1 - x_{hi}^E) \qquad h \in W' i, \in W' \setminus \{s\}, \; j \in W', i \neq j, i \neq h, j \neq h \quad (17)$$

$$z_{sj} \leq B^E - \pi d_{sj} x_{sj}^E + M(1 - x_{sj}^E) \qquad j \in W' \quad (18)$$

$$g_{ij} \leq B^E - z_{hi} + M(1 - x_{ij}^E) + M(1 - x_{hi}^E) \qquad i \in R' \setminus \{s\}, h \in W', j \in W' \quad (19)$$

$$z_{ij} \geq 0.1B^E \qquad i \in R', j \in W' \quad (20)$$

$$g_{ij} \leq 0.9B^E \qquad i \in R', j \in W' \quad (21)$$

$$\sum_{j \in C \cup \{v_k\}} r_{ij}^k - \sum_{h \in C \cup \{s\}} r_{hi}^k = 0 \qquad i \in C, k \in K \quad (22)$$

$$\sum_{j \in C \cup \{v_k\}} r_{sj}^k - \sum_{j \in C \cup \{s\}} r_{jv_k}^k = 0 \qquad k \in K \quad (23)$$

$$\sum_{k \in K} \sum_{j \in C \cup \{v_k\}} r_{sj}^k \leq |K| \quad (24)$$

$$\sum_{j \in C} r_{sj}^k \leq 1 \qquad k \in K \quad (25)$$

$$w_j^k \geq w_i^k + d_i r_{ij}^k - Q_k(1 - r_{ij}^k) \qquad j \in C \cup \{v_k\}, i \in C \cup \{s\}, k \in K \quad (26)$$

$$w_s^k \leq Q_k \qquad k \in K \quad (27)$$

$$f_i^k + t_{ij} r_{ij}^k - \alpha(1 - r_{ij}^k) \leq f_j^k \qquad i \in C, j \in C, k \in K \quad (28)$$

$$f_i^k \geq e_{v_k} + t_{si} \qquad i \in C, k \in K \quad (29)$$

$$f_{v_k}^k \leq l_{v_k} \qquad k \in K \quad (30)$$

$$f_i^k + t_{iv_k} r_{iv_k}^k - \alpha(1 - r_{iv_k}^k) \leq f_{v_k}^k \qquad i \in C, k \in K \quad (31)$$

$$e_i \leq f_i^k \leq l_i \qquad i \in C \quad (32)$$

$$\sum_{j \in C \cup \{t\}} x_{ij}^E + \sum_{j \in C \cup \{t\}} x_{ij}^C + \sum_{h \in C \cup \{v_k\}} \sum_{k \in K} r_{ih}^k = 1 \qquad i \in C \quad (33)$$

$$x^E_{ij}, x^C_{ij} \in \{0, 1\} \qquad i \in W', j \in W' \qquad (34)$$

$$u^E_i, u^C_i, \tau_i \geq 0 \qquad i \in W' \qquad (35)$$

$$g_{ij} \geq 0 \qquad i \in R', j \in W'. \qquad (36)$$

$$r^k_{ij} \in \{0, 1\} \qquad (i, j) \in A, k \in K \qquad (37)$$

$$0 \leq w^k_i \leq Q_k \qquad i \in C \cup \{s, v_k\}, k \in K \qquad (38)$$

$$f^k_i \geq 0 \qquad i \in C \cup \{s, v_k\}, k \in K. \qquad (39)$$

The objective function is composed of 7 terms. The first one represents the recharging cost when an EV is recharged to a recharging station. The second one is the energy consumption cost, we suppose that the EVs were recharged to the depot during the night and we do not consider the energy recharged during the route. The third term is the total diesel cost, whereas the fourth and fifth ones are the routing cost of EVs and ICCVs, respectively. The sixth term is the cost of compensation of the OD k for the delivery service with $\rho \geq 0$, the seventh one is the cost of the OD k when it does not perform the delivery service. Constraints (2)–(21) are linked to regular vehicles. Constraints (2) ensure that a recharging station can be visited at most once, whereas constraints (3) and (4) are the flow conservations constraints for regular vehicles. Constraints (5)–(6) impose a maximum number of available EVs and ICCVs, respectively. Capacity constrains are defined by (7)–(12), constraints (13)–(15) determine the arrival time at each node, whereas (16) are the time windows constraints. Constraints (17)–(18) are associated with the EVs' energy consumption and ensure that the maximum battery capacity is not exceeded, and (19) represent the partial battery charging. Constraints (20)–(21) define the state of charge of the battery. Constraints (22)–(32) are linked to ODs. In particular, constraints (22)–(23) are the flow conservation constraints. Constraints (24) ensure that at most $|K|$ ODs are used to serve the customers, whereas conditions (25) impose that an OD may perform only one route. Constraints (26)–(27) are the capacity constraints, whereas constraints (28)–(29) are the time windows constraints and also define the time at which the ODs are available to make deliveries, constraints (31) define the arrival time at the destination node v_k. Customers' time windows are defined by constraints (32). Constraints (33) ensure that each customer is visited at most once, either by a regular vehicle (EVs or ICCVs) or by an OD. Finally, conditions (34)–(39) define the domains of variables.

3 Numerical Study

In this section we conduct a preliminary computational study to assess the validity of the proposed GVRPOD model. The goal of this study is to show how the use of ODs may be not only profitable in terms of cost reduction, but also it may reduce polluting emissions. With this purpose, we modelled and then solved a classical GVRP and then we compared the results obtained with those achieved with GVRPOD. All the test are conducted using an Intel 2.60 GHz processor and 16 GB of RAM. The model was implemented in Java and solved with CPLEX. We conducted the preliminary computational study on different small-size instances. We started from a set of 30 instances based on Solomon VRPTW instances, proposed by Schneider et al. [12] for their E-VRP with time windows. The authors modified these instances by randomly introducing the recharging stations. Given a E-VRPTW instance, with the customers locations identified by the coordinates (x_i, y_i), we randomly generated the destinations for the ODs, in the square with lower left hand corner $(\min_i\{x_i\}, \min_i\{y_i\})$ and upper righthand corner $(\max_i\{x_i\}, \max_i\{y_i\})$, (see: [1]). We then set the number of ICCVs and EVs and, if necessary, recalculated a reasonably time window to ensure feasibility. Our comparative analysis is divided into two phases, in the first one we focus on profitability of using ODs, thus we compare the GVRP and the GVRPOD in terms of costs, in the second one we focus on green aspects and we make a comparison in terms of polluting emissions. For all the experiments, we use the parameters setting reported in Table 2. We fixed the battery capacity for EVs to 10 KWh and $\pi = 0.125$. Considering the technologies introduced by Felipe et al. [7] (slow, medium and fast), we suppose that all the recharging stations support the medium technology, hence the recharging speed is fixed at 20,000 [KWh/h] and $w^r = 0.4$ [€/KWh], whereas the EVs can be recharged with the slow technology during the night at depot and $w^e = 0.17$ [€/KWh]. Fuel cost $w^f = 0.8$ [€/L], and to calculate ODs compensation ρ is set to 1.10.

Table 3 presents the comparison results for each GVRPOD instance against GVRP. Each table has 9 columns, the first one reports instances name, then we have four columns referred to the GVRPOD and three to the GVRP, which show the cost of the solution and the number of employed vehicles, ICCVs, EVs and ODs (only for GVRPOD), respectively. The last column is the GAP on cost, calculated as $(\text{Objective}_{\text{GVRPOD}} - \text{Objective}_{\text{GVRP}})/\text{Objective}_{\text{GVRP}}$. The results summarized in Table 3a clearly show that for instances with five customers the use of ODs is highly competitive. The GAP is on average 14% and there is a reduction in terms of cost for the majority of the considered instances. We have the same trend for instances with 10 and 15 customers, indeed, Table 3b, c show that the GAP is on average 24% and 23%, respectively; in addition, overall we have a reduction in terms of cost and number of employed ICCVs. However, the use of ODs becomes more interesting if we analyse the reduction in terms of polluting emissions, due to the decrease of the number of ICCVs used in transportation plans. Table 3 shows that the GVRPOD

Table 2 Parameters setting

$\|C\|$	$\|R\|$	n^E	n^C	Q^E	Q^C	$\|K\|$	Q_k^{OD}
5	[1–3]	1	1	80.00	80.00	2	[20–30]
10	[2–4]	2	2	80.00	80.00	3	[20–40]
15	[2–5]	3	3	80.00	80.00	3	[20–40]

Table 3 Results for the GVRPOD and GVRP

GVRPOD					GVRP			
Test	Cost	#ICCVs	#EVs	#ODs	Cost	#ICCVs	#EVs	GAP (%)
(a) Results for instances with $\|C\| = 5$								
C101C5	210.63	1	1	1	252.32	1	1	17
C103C5	148.15	0	1	2	165.72	1	1	11
C206C5	161.62	1	1	2	226.93	1	1	29
C208C5	169.50	0	1	1	193.36	1	1	12
R104C5	132.37	0	1	1	138.30	1	1	4
R105C5	123.62	1	1	1	151.34	1	1	18
R202C5	105.29	0	1	2	128.84	0	1	18
R203C5	169.38	0	1	1	196.40	1	1	14
RC105C5	236.41	1	1	0	236.41	1	1	0
RC108C5	237.54	1	1	2	257.89	1	1	8
RC204C5	181.96	1	1	0	181.96	1	1	0
RC208C5	106.54	0	1	2	186.12	1	1	43
AVG	165.25	0.50	1.00	1.25	192.96	0.92	1.00	14
(b) Results for instances with $\|C\| = 10$								
C101C10	337.56	1	2	2	378.51	1	2	11
C104C10	210.18	0	2	3	310.03	1	2	32
C202C10	245.32	0	2	3	294.36	1	2	17
C205C10	233.28	0	2	3	261.83	1	2	11
R102C10	210.74	1	1	2	248.68	1	2	15
R103C10	125.49	0	2	2	190.88	0	2	34
R201C10	166.74	0	2	3	218.28	1	2	24
R203C10	141.16	0	1	3	235.17	0	2	40
RC102C10	254.82	1	1	3	400.43	2	2	36
RC108C10	305.86	0	2	2	346.44	2	0	12
RC201C10	212.90	0	2	3	313.55	1	2	32
RC205C10	259.66	0	2	3	346.29	1	2	25
AVG	225.31	0.25	1.75	2.67	295.37	1.00	1.83	24
(c) Results for instances with $\|C\| = 15$								
C106C15	180.835	0	2	3	274.7025	0	3	34.2
C208C15	284.8813	0	3	3	358.1721	1	3	20.5
R209C15	246.9075	0	3	3	325.9467	1	3	24.2
R102C15	287.69	0	2	3	381.4325	2	3	24.6
RC103C15	331.9534	1	2	3	395.3261	2	3	16.0
RC202C15	325.2825	0	2	3	413.1033	1	2	21.3
AVG	276.26	0.17	2.33	3.00	358.11	1.17	2.83	23

Table 4 GVRP versus GVRPOD: polluting emissions

Test	ε_{GVRP}	ε_{GVRPOD}	$\Delta(\varepsilon)(\%)$
$\|C\| = 5$	92.17	19.01	−79
$\|C\| = 10$	73.15	35.20	−52
$\|C\| = 15$	47.08	11.22	−76

uses less ICCVs than GVRP, this has a great impact on environment. The GVRPOD uses a number of ICCVs that is the 45% and the 75% less than the ICCVs used by the GVRP for instances with five and 10 customers, respectively; up to the 85% less for instances with 15 customers. Since the amount of polluting emissions are strongly related to the number of fuel vehicles on route, and we know that the ODs will travel whether performing the delivery or not, the use of ODs become a strategic choice. To calculate the amount of polluting emissions, we use the function $\varepsilon(u)$ defined in Sect. 2. In particular, the total amount of emissions is calculated as $\varepsilon = \sum_{(i,j)\in A} \varepsilon(u_i^C) d_{ij} x_{ij}^C$. Table 4 summarizes the results obtained, focusing on polluting emissions. The first column gives the class of instances, the second and the third ones show the average of polluting emissions obtained solving the GVRP and GVRPOD, respectively. The last column is the polluting emissions variation $\Delta(\varepsilon)$ calculated as $100 \times (\varepsilon_{GVRP} - \varepsilon_{GVRPOD})/\varepsilon_{GVRP}$. Looking at results in Table 4, it is clear that the use of ODs highly impacts on polluting emissions reduction. We have a reduction of about the 79% of emissions for instances with five customers, 52% for instances with 10 customers and of about the 73% for instances with 15 customers. Our preliminary computational study clearly shows the advantages obtained by using ODs in routing plan, not only in terms of total cost reduction, but also in terms of environmental impact, that is, polluting emissions and traffic congestion reduction.

4 Conclusion

In this work we have proposed a new variant of the green vehicle routing problem, introducing the crowd-shipping. In this problem, named green vehicle routing problem with occasional drivers (GVRPOD), the company has to serve some customers in an urban area with a mixed fleet of vehicles, composed of conventional vehicles and electric vehicles, and may also use some occasional drivers (ODs). The ODs are ordinary people who decide to deliver something to other people on their route, for a small compensation. We have modelled the GVRPOD, then we have shown the benefits to use the ODs in transportation planning, not only in terms of costs reduction but also in terms of environmental impacts.

Acknowledgements This work was supported by MIUR "PRIN 2015" funds, project: Transportation and Logistics in the Era of Big Open Data - 2015JJLC3E_003 - CUP H52F15000190001.

References

1. Archetti, C., Savelsbergh, M., Speranza, M.G.: The vehicle routing problem with occasional drivers. Eur. J. Oper. Res. **254**, 472–480 (2016)
2. Arslan, A.M., Agatz, N., Kroon, L., Zuidwijk, R.: Crowdsourced delivery: a dynamic pickup and delivery problem with ad-hoc drivers. In: Erasmus Research Institute of Management ERIM Paper, Social Science Research Network (2016). http://papers.ssrn.com/sol3/papers.cfm?abstracs_id=2726731
3. Bektaş, T., Laporte, G.: The pollution-routing problem. Transp. Res. Part B **45**, 1232–1250 (2011)
4. Dahle, L., Andersson, H., Christiansen, M.: The vehicle routing problem with dynamic occasional drivers. In: Lecture Notes in Computer Science. 10572 LNCS, pp. 49-63 (2017)
5. Demir, E., Bektaş, T., Laporte, G.: An adaptive large neighborhood search heuristic for the pollution-routing problem. Eur. J. Oper. Res. **223**, 346–359 (2012)
6. Erdoğan, S., Miller-Hooks, E.: A green vehicle routing problem. Transp. Res. Part E **48**(1), 100–114 (2012)
7. Felipe, A., Ortuño, M.T., Righini, G., Tirado, G.: A heuristic approach for the green vehicle routing problem with multiple technologies and partial recharges. Transp. Res. Part E. **71**, 111–128 (2014)
8. Franceschetti, A., Honhon, D., Van Woensel, T., Bektaş, T., Laporte, G.: The time-dependent pollution-routing problem. Transp. Res. Part B. **56**, 265–293 (2013)
9. Koç, Ç., Bektaş, T., Jabali, O., Laporte, G.: The fleet size and mix pollution-routing problem. Transp. Res. Part B **70**, 239–254 (2014)
10. Macrina, G., Di Puglia Pugliese, L., Guerriero, F., Laganà, D.: The vehicle routing problem with occasional drivers and time windows. In: Sforza, A., Sterle, C. (eds.) Optimization and Decision Science: Methodologies and Applications. Springer Proceedings in Mathematics & Statistics, vol. 217, pp. 577–587. ODS, Sorrento, Italy, Springer (2017)
11. Montoya, A., Guéret, C., Mendoza, J.E., Villegas, J.G.: The electric vehicle routing problem with nonlinear charging function. Transp. Res. Part B. **103**, 87–110 (2017)
12. Schneider, M., Stenger, A., Goeke, D.: The electric vehicle routing problem with time windows and recharging stations. Transp. Sci. **48**(4), 500–520 (2012)
13. Ubeda, S., Faulin, J., Serrano, A., Arcelus, F.J.: Solving the green capacitated vehicle routing problem using a tabu search algorithm. Lect. Notes Manag. Sci., J. Manuf. Syst. **6**, 141–149 (2014)

Using Cryptography Techniques as a Safety Mechanism Applied to Components in Autonomous Driving

Antonino Mondello and Alberto Troia

Abstract Many applications are being developed that adopt a new emerging technology inspired by biological structures in nature to solve real-life problems; this approach involves implementations based on artificial neural networks (ANNs), deep learning, and other forms of artificial intelligence (AI). Autonomous driving is one area where these AI implementations can be applied; however, with it brings several uncertainties, including the safety and security of the implementation. The intent of this paper is to provide a new perspective in using cryptography as a methodology to implement safety in the hardware that incorporates AI technology in automotive while addressing at the same time classical problems due to physical and software failures.

Keywords Artificial intelligence · Machine learning · Deep learning Neural network · Genetic algorithm · Neuron · Gene · HASH · HMAC · SHA256 Digest · Weight matrix · Secure storage · Memory · Automotive Autonomous driving

1 Introduction

Modern electronic systems may be affected by systematic or random errors. While a systematic error is a kind of marginality and/or bug that can be easily reproduced and solved during the system validation phase, random errors are more difficult to detect and solve. Several standards exist, such as the ISO26262, that are trying to formalize processes and procedures to reduce at very low level the risk of failures; however,

A. Mondello (✉)
Micron Technology Inc., Catania, Italy
e-mail: amondell@micron.com

A. Troia (✉)
Micron Technology Inc., Munich, Germany
e-mail: atroia@micron.com

© Springer Nature Switzerland AG 2018

P. Daniele and L. Scrimali (eds.), *New Trends in Emerging Complex Real Life Problems*, AIRO Springer Series 1,
https://doi.org/10.1007/978-3-030-00473-6_39

unfortunately, a failure event can always occur. If a critical event happens, a robust system must be able to detect it and place the entire system in a safe condition while also executing rescue procedures.

Current literature covers safety and security as different topics, this paper has the purpose to link both the concepts, providing a unique option to address the issues—based on modern cryptography—to detect and correct certain errors that might affect data inside a storage device.

2 Safety and Security Issues of a Memory System

A storage device cannot be considered simply a container of data; it must have the capability to ensure data integrity against a wide range of physical or human sources of errors. Data inside a storage device can be threatened by:

Bit flips: One or more bits randomly change their values due to a physical defect or internal circuit malfunction. The bit flips failure rate is typically mitigated by using proper error correction techniques internal to the storage device. If the number of bits flipped exceed the capability of the adopted error correction scheme, the errors cannot be detected, or worse, new errors can be injected when attempting to correct the errors. This may cause unexpected system behavior that could affect the safety of the system;

Bus communication errors: System/component noise affects data transmitted through the internal/external buses. In this case, error detection and correction techniques can be implemented, but with the same limitation discussed above;

Memory content changes: Intentional, un-authorized changes of data by the direct action of hackers or by malicious software. This type of issue emerged in recent years with the internet of things revolution. Data inside the memory can also be modified by a system software bug; such kind of changes are not intentional, but the effect might be unpredictable and, in some cases, if the data changed are the vital for the systems, it could be similar to a hacker attack;

Memory replacement: An extreme technique to gain control of the system by replacing original storage components with non-genuine components. This kind of attack is fed by a parallel marketing of non-original components.

A storage device cannot be considered simply a container of data; it must have the capability to ensure data integrity against a wide range of physical or human threats.

3 Secure Memories

The data threats introduced in the previous section can be fixed by using new techniques; the usage of such techniques requires a reinterpretation of the problem posed under a different perspective: the cryptographic one.

Memory usage and data protection: Only authorized users can modify and/or read stored data. This topic implies that the storage device command set should be authenticated; in other words, the device must be able to detect commands coming from unauthorized hosts and avoid an attempt to reuse commands already used by the authorized entities;

Memory data attestation: The storage device should be able to attest the data stored. Data attestation check should be performed at each power cycle or reset of the system, and on each host request.

Memory identification: The memory should be able to prove its identity to a requester to avoid un-authorized component replacement;

To guarantee these features, the component must implement at least two kinds of cryptographic primitives: the HASH function and the MAC function, described in the following paragraphs.

3.1 The Authentication Problem

Consider the simple system structure depicted in Fig. 1, where a host, i.e. a processor, is asking for a command execution—such as read, program or erase data—to a component like a storage device. The system design wants only the authorized host to execute such commands.

A possible solution for this problem is to implement a command authentication mechanism based on certain cryptographic functions called Message Authentication Code (MAC) [1, 2]; such functions are used to calculate what is called signature of the command by using the formula:

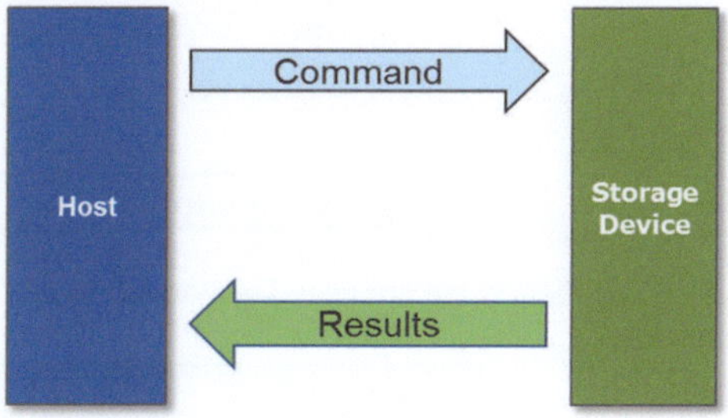

Fig. 1 Storage device and host

$$Signature = MAC(key, Command) \tag{1}$$

The key is secret information shared between the host and storage device in a secure environment like the system factory. The security of the system is based on the impossibility to calculate the message signature without knowing the secret key. The MAC function is well known to community, since it is not representing the secret; a function, can be promoted to MAC function, when the properties described in the next paragraph are matched.

Such a signature is used in conjunction with the following communication protocol to ensure system authentication:

Host: Sends a packet of information consisting of commands to be executed and the related signature (Fig. 2) calculated with the formula (1) and using the secret key.

Storage device: From the received message, takes the command field and locally calculates the signature by using the same formula (1) and its copy of the secret key, then compares the signature calculated with the one received; if they don't match, the message is discarded; otherwise, it is considered authentic and executed, and the storage device sends the command result and the related signature to the host (Fig. 3).

The host: verifies the authenticity of the received message by locally calculating the signature—with its secret key copy—and comparing it with the one received.

This approach is very robust in guaranteeing authentication, but it has a problem—it is prone to a kind of channel side attack called *replay attack*. In fact, if someone intercepts the message, by using a bus sniffer, they can reuse it later because it is correctly signed by the authorized sender. A solution to this problem can be found by introducing in the communication packet a field called freshness (see Fig. 4).

The freshness is a piece of information that, by definition, changes at each command/command transaction.

There are many ways to define the freshness: it can be the transmission timestamp, a random number, or a simple number that is incremented at each transition and with the property not decremented in any way. For that reason, it is called a monotonic counter (MC) approach. By using the MC as a measure of freshness, the transmission protocol is modified in the following way:

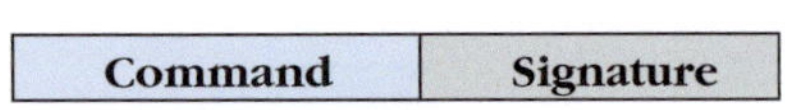

Fig. 2 Authenticated message packed

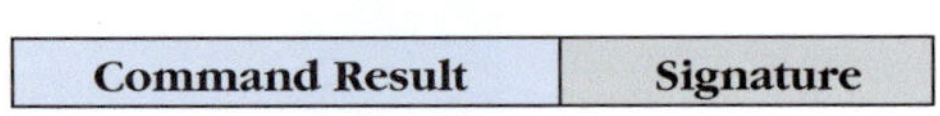

Fig. 3 Authenticated command response packed

Fig. 4 Message packed structure with freshness

Fig. 5 Response packed with freshness

Host: sends the message as in Fig. 4, where signature is calculated with the formula:

$$Signature = MAC\left[key, (Command|MC)\right] \tag{2}$$

Storage device: Takes from the received message the monotonic counter C value and verifies that it has a value greater than the one used in the last transmission. If it is, it calculates and verifies the signature received by comparing it with the local one calculated by using its secret key. If the check is successful, the message is considered authentic and the storage device executes the command. It then increments the internal MC value and sends the command result—the new MC value and the related signature—to the host (Fig. 5).

The host: Verifies the authenticity of the command result by checking that the MC value is greater than the last value and locally recalculating and verifying the signature received by using its own local secret key copy.

This protocol can ensure the authenticity in commands and results in exchanges between the host and storage device.

Two components that use this kind of transmission protocol implement what is called a secure command set. If the key is maintained secret, there is no way to use the system components without knowing the key.

A secure command set that includes program, erase, read, or configuration changes commands ensure the system is used only by someone who knows the secret key.

3.2 MAC Function Requirements

There are several functions that can be used as MAC functions to generate the signature of a message; these functions must satisfy some key requirements:

- "Easy" to be calculated: Given a message X, the calculus of MAC(key, X) is not computationally difficult;
- "Hard" to be inverted: Given a MAC(key, X), it must be "impossible" to determine the key value by knowing X and the MAC value; in other words, the calculation $\left[\mathrm{MAC(key,\ X)}\right]^{-1}$ must be computationally infusible;
- "Negligible" collision probability: Given two messages $\mathrm{X} \neq \mathrm{Y} \rightarrow \mathrm{MAC(key,\ X)} \neq \mathrm{MAC(key,\ Y)}$

The cryptographic strength of the MAC together with the secrecy of the key guarantees the authenticity and the integrity of the messages exchanged.

A powerful MAC function used in many cryptographic systems is the HMAC-SHA256. Such a function is described in [3, 4] and is very robust in the fact that it satisfies all the requirements defined for MAC function. HMAC-SHA256 can process a message up to 2^{64} bits in length and produce a 256-bit signature. As of today, there is no known possibility to invert this MAC and no collision conditions have been identified.

3.3 The Data Attestation Problem

A storage device such as a flash memory or RAM must guarantee the genuineness of stored data. In some cases, it is also requested of nonvolatile memories the capability to recover corrupted data with a genuine copy to guarantee system functionality in case of a hacker attack or severe system malfunction.

The changing of stored data can be detected using some cryptography tools including hash functions. A hash function is a function that is able to map data of arbitrary size to a "short" fixed size called HASH or digest of the data. Once a digest is calculated on a genuine data pattern, the result is called a golden digest. This data is stored in an area that is not user-accessible and is compared on demand with the current digest calculated at the moment of the request. The result of this comparison enables the system to understand if the array content is genuine or was accidentally or intentionally modified. If requested, by using an authenticated command, the memory can provide the hash result as a command result by using the authenticated protocol defined in the previous paragraph. A minimum set of authenticated commands for attestation purposes can include the ones that define the secure memory area to be hashed, generate the golden digest, and verify the current digest value of the secure area versus the golden one.

3.4 HASH Function Requirements

A function is eligible as a hash function if it is:

- "Easy" to be calculated: Given a message X, the calculus of HASH(X) has low computational complexity
- "Hard" to be inverted: $[\text{HASH}(X)]^{-1}$ must be impossible to be calculated
- "Negligible" collision probability: Given two messages $X \neq Y \rightarrow \text{HASH}(X) \neq \text{HASH}(Y)$.

Cryptographic literature proposes many hash functions [1, 2]. A common one is the SHA256 described in [3].

The implementation of an authenticated command able to calculate a hash of a selected memory area and compare the result to a golden value, enables a powerful mechanism for data attestation.

In the next paragraphs we will show how to use such kind of methodology to implement an Artificial Neural Network with the capability to attest its integrity.

4 Secure ADAS System Based on Neural Networks

In recent years, vehicle makers introduced autonomous driving systems. Due to the complexity of such systems, they were based on artificial intelligence (AI) implemented by using Artificial Neural Networks (ANN) [5].

In brief, an ANN is a set of cells called neurons that are interconnected to each other using a set of connections, and organized in layers, as depicted in Fig. 6.

The neural signals processed by the generic neuron m is sent to the neuron k after a multiplication by certain numerical constants called synaptic weights (w_{mn}). The constant b_m, not always present ($b_m = 0$) is called the *bias* of the neuron (Fig. 7).

The relationship between neuron inputs and the output is given by the formula:

$$a_m = f_m\left[b_m + \sum_{k=1}^{R} w_{mk} \cdot p_k\right] \quad (3)$$

where: $f_m(n)$ is a real function, called *activation function*, it can be, in principle, different for each neuron; in practice, they use only a few different kinds of functions. The most common activation functions are described in Table 1.

The process that permits us to define the bias and synaptic weight is called the *learning process*. As showed by [5] page 14-3, a quite general model of an ANN

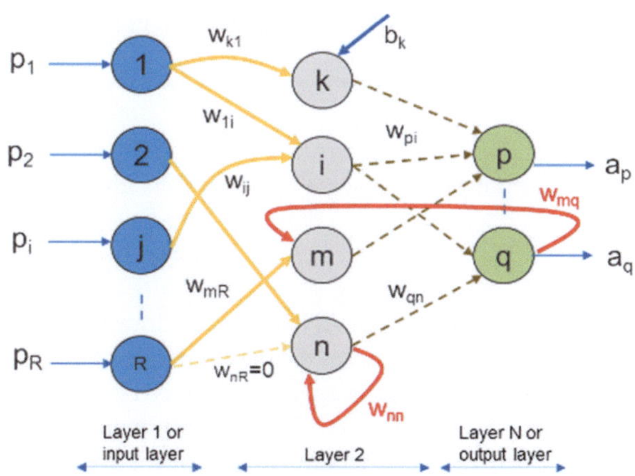

Fig. 6 Generic artificial neural network structure

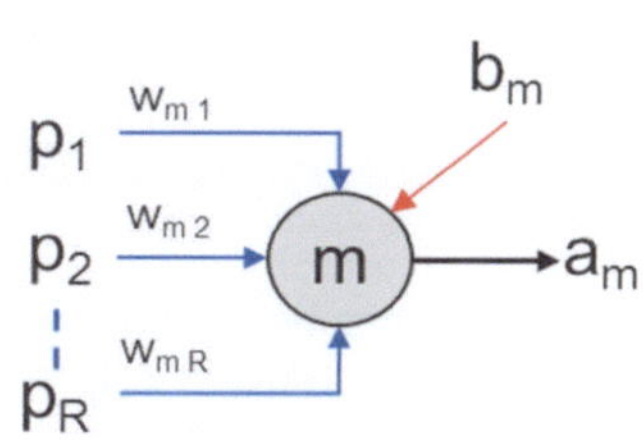

Fig. 7 Neuron structure

Table 1 Different kinds of activation functions

Hard limiter	Linear	Log-sigmoid
$f(t) = \begin{cases} b \text{ if } t \geq 0 \\ a \text{ if } t < 0 \end{cases}$	$f(t) = a \cdot t$	$f(t) = \frac{1}{1+e^{-t}}$

Table 2 Matrices needed to describe RANNs

$P = \{p_i^m\}$	Input vector of network; the inputs can be connected to all layers
$N^m = \{n_i^m\}$	Input vector of neurons of layer m
$A^m = \{a_i^m\}$	Vector is the output vector of layer m
$B^m = \{b_i^m\}$	Vector of neurons bias of layer m
$LW^{m,l} = \{lw_{i,j}^{m,l}\}$	Matrix with the synaptic weight from the layer l to the layer m. The element $lw_{i,j}^{m,l}$ is the weight between neuron i of layer m and the neuron j of layer l
$IW^{m,l} = \{iw_k^{m,l}\}$	Matrix of synaptic weight associated to the network input l to the layer m. The element $iw_k^{m,l}$ is the weight between input l to the neuron k of layer m

can be described by using matrix formalism, by defining the sets of matrices listed in table Table 2.

The input vector $N^m(t)$ of each neuron present in the *m-th* layer can be written as:

$$N^m(t) = \sum_l \sum_d LW^{m,l}(d) \cdot a^l(t-d) + \sum_l \sum_d LW^{m,l}(d) \cdot p^l(t-d) + b^m \quad (4)$$

The output of the layer m of network is given by:

$$A^m(t) = f^m(n^m(t)) \quad (5)$$

From this overview of ANNs, we can draw the conclusion that the implementation of a neural network requires the manipulation of large amounts of data stored in matrix form (see Table 2). Such data must be stored in a nonvolatile memory device and must be written, read, and updated in a secure manner. This challenge will be addressed in the next paragraphs where we describe a possible hardware implementation.

5 Cryptography as Usage to Improve Safety

The ANNs can be implemented by using any methodology known in the field. Whatever hardware is used—GPU, CPU, FPGA etc.—it must implement and calculate the formulas introduced in the previous paragraph. Such formulas are based on matrices that contain a large amount of data stored in a storage device. What we would like to discuss here is the usage of cryptography as a methodology to understand if the storing and reading of the data associated with the ANN is correct or was modified by random errors or hacker attacks.

Suppose to store the matrices: $IW^{m,l}$, $LW^{m,l}$, b^m in a specific area of a nonvolatile storage device, in fixed or floating-point notation based on the kind of implementation; before using such matrices, the calculus unit can request the storage device to attest to the values stored; the storage device activates the internal hash engine and calculates:

$$Digest = \text{HASH}\left(IW^{m,l} \middle| LW^{m,l} \middle| b^m\right) \quad \forall m, l \tag{6}$$

To practically perform such calculus, the elements of each matrix involved is read row by row from the first column to the last one. The read data forms a long sequence of numbers that are processed by the HASH engine.

To better clarify this, consider the simple ANN depicted in Fig. 8 with just four neurons, two inputs and outputs, the non-zero elements of the above matrices are represented directly on the figure.

By reading the matrices, we have the sequence of numbers:

$$\begin{aligned} S =&iw_1^{1,1},\ iw_2^{1,1},\ iw_1^{2,1},\ iw_1^{2,1},\ lw_{1,1}^{2,1},\ lw_{1,2}^{2,1},\ lw_{2,1}^{2,1},\ lw_{2,2}^{2,1},\ lw_{1,1}^{1,2},\ lw_{1,2}^{1,2},\\ &lw_{2,1}^{1,2},\ lw_{2,2}^{1,2},\ b_1^1,\ b_2^1,\ b_1^2,\ b_2^2 \end{aligned} \tag{7}$$

Which are the ones involved in the HASH calculus:

$$Digest = \text{HASH(S)} \tag{8}$$

The result of the comparison of the above calculated digest and the golden one previously calculated enables the system to check the integrity of the ANN, this can indicate that there are no accidental or malicious changes. The presence of an authenticated commands set repeats the integrity check every time and defines some policy to update the ANN in a secure way.

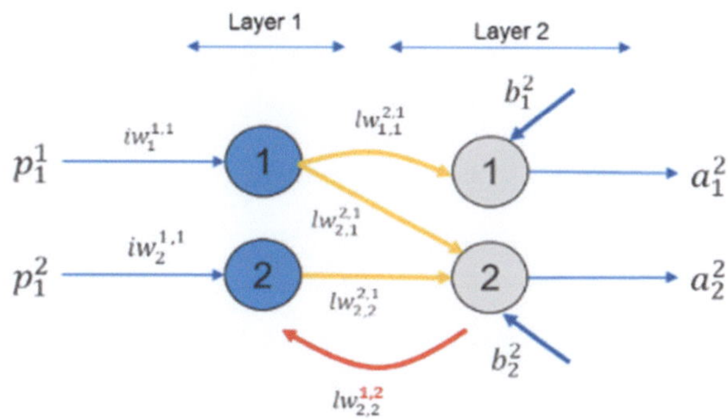

Fig. 8 Simple ANN

6 Conclusions

This paper presented a new interpretation of the use of cryptography as a security and safety mechanism. The security and safety of the system, implementing this methodology, is guaranteed by two cryptographic primitives: MAC and HASH; the described method enables a robust system against a wide range threats.

References

1. Yaschenko, V.V.: Cryptography: An Introduction. AMS (2002)
2. Ferguson, N. et al.: Cryptography Engineering. Wiley (2010)
3. Fips-180-2. SHA-256: Secure Hash Algorithm [PDF]. http://csrc.nist.gov/publications/fips/fips180-2/fips180-2withchangenotice.pdf
4. Fips-198-1. HMAC-SHA-256: Hash Based Message Authentication Code [PDF]. http://csrc.nist.gov/publications/fips/fips198-1/FIPS-198-1_final.pdf
5. Hagan, M.T. et al.: Neural Network Design, 2nd ed. ISBN 978-0971732117

Simplifying the Minimax Disparity Model for Determining OWA Weights in Large-Scale Problems

Thuy Hong Nguyen

Abstract In the context of multicriteria decision making, the ordered weighted averaging (OWA) functions play a crucial role in aggregating multiple criteria evaluations into an overall assessment supporting the decision makers' choice. Determining OWA weights, therefore, is an essential part of this process. Available methods for determining OWA weights, however, often require heavy computational loads in real-life large-scale optimization problems. In this paper, we propose a new approach to simplify the well-known minimax disparity model for determining OWA weights. We use the binomial decomposition framework in which natural constraints can be imposed on the level of complexity of the weight distribution. The original problem of determining OWA weights is thereby transformed into a smaller scale optimization problem, formulated in terms of the coefficients in the binomial decomposition. Our preliminary results show that the minimax disparity model encoded with a small set of these coefficients can be solved in less computation time than the original model including the full-dimensional set of OWA weights.

Keywords Ordered weighted averaging · OWA weights determination · Binomial decomposition framework · k-additive level · Large-scale optimization problems

1 Introduction

In many disciplines, decision makers have to deal with problems involving the aggregation of multicriteria evaluations and producing overall assessments within the application context. The ordered weighted averaging function (OWA) introduced by Yager [23] is a fundamental aggregation function in decision making theory. One of the main motivations behind selecting the OWA functions for multicriteria aggrega-

T. H. Nguyen (✉)
Department of Industrial Engineering, University of Trento,
Via Sommarive 9, 38123 Povo, TN, Italy
e-mail: hongthuy.nguyen@unitn.it

© Springer Nature Switzerland AG 2018

P. Daniele and L. Scrimali (eds.), *New Trends in Emerging Complex Real Life Problems*, AIRO Springer Series 1,
https://doi.org/10.1007/978-3-030-00473-6_40

tion is their flexibility in providing a general class of weighted aggregation functions bounded by the *min* and the *max* functions.

The determination of appropriate OWA weights is a very important object of study when applying OWA functions in the context of decision making. Among the various methods that can be found in the literature, namely [21, 22, 24], and more recently [2, 7, 15], we can distinguish the following methodological categories: (a) methods based on learning OWA weights from data, (b) methods based on weight-generating functions, and (c) methods based on the characteristic measures of *orness* and *disparity*.

In this paper we focus on the problem of determining a special class of OWA functions based on the disparity measure under a given degree of orness since these two measures can lead to OWA weights associated with different attitudinal characters of decision makers, see [16, 23]. Depending on their weighting structure, OWA functions reflect different preferences of decision makers, from the optimistic to the pessimistic attitude. The attitudinal character of decision makers is measured by their orness, which takes values in the unit interval. The maximum (minimum) orness value is attained when decision makers are purely optimistic (purely pessimistic). On the other hand, the disparity evaluates the non-uniformity of the OWA weights. In the case of the arithmetic mean the disparity takes its minimal value.

In the literature several methods have been introduced to obtain the optimal weights by using the disparity measure. After the pioneering work of O'Hagan [17] on the maximal entropy method and the variance-based methods of Yager [25] and Fullér and Majlender [9], Wang and Parkan [20] proposed the minimax disparity method in which the objective is to minimize the maximum absolute difference between two adjacent weights. Liu [14] proved the equivalence of the solutions of the minimum variance approach suggested by Fullér and Majlender [9] and the minimax disparity model proposed by Wang and Parkan [20] under a given degree of orness. Extensions of disparity-based models for determining OWA weights are presented in [1, 8, 10, 18, 19]. In this paper, we focus on the minimax disparity model since it has recently received a great deal of interest in the literature and is easy to solve due to its simple linear programming formulations.

The usual academic instances of the minimax disparity model focus on solving problems with small dimensions ($n = 3, 4, 5, 6$). However, in applied operational research, optimization problems are often much more complex and require a heavy computational demand when there are hundreds or thousands of variables. In order to overcome the complexity of high-dimensional problems, we consider the binomial decomposition framework, proposed by Calvo and De Baets [6], see also Bortot and Marques Pereira [4] and Bortot et al. [3, 5], which refers to the k-additive framework introduced by Grabisch [11–13]. This framework allows us to transform the original problem, expressed directly in terms of the OWA weights, into a problem in which the weights are substituted by a new set of coefficients. In this transformed representation, we can consider only a reduced number of these coefficients, associated with the first k-additive levels of the OWA function, and we can set the remaining coefficients to zero. Preliminary experiments show that the solution found for the transformed model can still be a good approximated solution for the original model, while the computational demand in high-dimensional problems can be significantly reduced.

The remainder of this paper is organized as follows. In Sect. 2 we briefly review the OWA functions and their representation in the binomial decomposition framework. Section 3 reviews the recent development of the minimax disparity model for determining OWA weights. In Sect. 4 we recall the minimax disparity model and reformulate it in terms of the coefficients in the binomial decomposition framework. We illustrate our approach for dimension $n = 10, 20, 30, 40$. Finally, Sect. 5 contains some conclusive remarks.

2 OWA Functions and the Binomial Decomposition Framework

In this section we consider a point $\boldsymbol{x} \in \mathbb{R}^n$, with $n \geq 2$. The increasing reordering of the coordinates of $\boldsymbol{x}$ is denoted as $x_{(1)} \leq \cdots \leq x_{(n)}$. We now introduce the definition of the OWA function and its characterizing measures.

Definition 1 An *Ordered Weighted Averaging (OWA) function* of dimension n is an averaging function $A : \mathbb{R}^n \longrightarrow \mathbb{R}$ with an associated weighting vector $\boldsymbol{w} = (w_1, \ldots, w_n) \in [0, 1]^n$, such that $\sum_{i=1}^n w_i = 1$ and

$$A(\boldsymbol{x}) = \sum_{i=1}^{n} w_i \, x_{(i)}. \tag{1}$$

Different OWA functions are classified by their weighting vectors. The OWA weights are characterized by two measures called orness and disparity. In the following part we review these two measures and their properties.

Definition 2 Consider an OWA function with an associated weighting vector $\boldsymbol{w} = (w_1, \ldots, w_n) \in [0, 1]^n$ such that $\sum_{i=1}^n w_i = 1$, two characterizing measures called *orness* and *disparity* are defined as

$$Orness(\boldsymbol{w}) = \frac{1}{n-1} \sum_{i=1}^{n} (i-1) w_i, \qquad Disparity(\boldsymbol{w}) = \max_{i \in \{1,\ldots,n-1\}} |w_i - w_{i+1}|. \tag{2}$$

Yager [23] introduced the orness measure to evaluate the level of similarity between the OWA function and the *or (maximum)* operator. On the other hand, the disparity measure, as proposed by Wang and Parkan [20], is defined as the maximum absolute difference between two adjacent weights. Its value shows how unequally multicriteria evaluations are taken into account in the aggregation process. Both OWA characterizing measures are bounded in the unit interval. Three special OWA weighting vectors are $\boldsymbol{w}_* = (1, 0, \ldots, 0)$, $\boldsymbol{w}_A = (\frac{1}{n}, \ldots, \frac{1}{n})$ and $\boldsymbol{w}^* = (0, \ldots, 0, 1)$. For these vectors we have the orness equal to 0, 0.5, 1 and disparity equal to 1, 0, 1, respectively.

In the following we recall the binomial decomposition of the OWA functions proposed by Calvo and De Baets [6], see also Bortot and Marques Pereira [4].

Definition 3 The *binomial OWA functions* $C_j : \mathbb{R}^n \longrightarrow \mathbb{R}$, with $j = 1, \ldots, n$, are defined as

$$C_j(\boldsymbol{x}) = \sum_{i=1}^{n} w_{ji}\, x_{(i)} = \sum_{i=1}^{n} \frac{\binom{n-i}{j-1}}{\binom{n}{j}}\, x_{(i)} \qquad j = 1, \ldots, n \tag{3}$$

where the binomial weights w_{ji}, $i, j = 1, \ldots, n$, are zero when $i + j > n + 1$, according to the usual convention that $\binom{p}{q} = 0$ when $p < q$, with $p, q = 0, 1, \ldots$.

Theorem 1 (Binomial decomposition) *Any OWA function* $A : \mathbb{R}^n \longrightarrow \mathbb{R}$ *can be written uniquely as*

$$A(\boldsymbol{x}) = \alpha_1 C_1(\boldsymbol{x}) + \alpha_2 C_2(\boldsymbol{x}) + \cdots + \alpha_n C_n(\boldsymbol{x}) \tag{4}$$

where the coefficients α_j, $j = 1, \ldots, n$, *are subject to the boundary and monotonicity conditions,*

$$\alpha_1 = 1 - \sum_{j=2}^{n} \alpha_j \geq 0 \quad \textit{and} \quad \sum_{j=2}^{n} \Big[1 - n \frac{\binom{i-1}{j-1}}{\binom{n}{j}} \Big] \alpha_j \leq 1 \qquad i = 2, \ldots, n \tag{5}$$

The detailed proof of Theorem 1 is given in Bortot and Marques Pereira [4].

The binomial decomposition (4) expresses the linear combination between the OWA weights and the coefficients α_j and can be written as the linear system

$$\begin{cases} w_1 &= w_{11}\alpha_1 + w_{21}\alpha_2 + \cdots + w_{n-1,1}\alpha_{n-1} + w_{n1}\alpha_n \\ w_2 &= w_{12}\alpha_1 + w_{22}\alpha_2 + \cdots + w_{n-1,2}\alpha_{n-1} \\ \ldots. & \\ w_n &= w_{1n}\alpha_1 \end{cases} \tag{6}$$

where the binomial weights $w_{ji} = \frac{\binom{n-i}{j-1}}{\binom{n}{j}}$, $i, j = 1, \ldots, n$, and the coefficients α_j, $j = 1, \ldots, n$, are subject to the conditions (5).

We notice that the coefficient matrix of the linear system is composed of the binomial weights $w_{ji} = \frac{\binom{n-i}{j-1}}{\binom{n}{j}}$, $i, j = 1, \ldots, n$, with the first $n - j + 1$ weights being positive and non-linear decreasing, and the last $j - 1$ weights equal to zero. Hence there always exists a unique vector of coefficients alpha satisfying the linear system which is triangular and whose coefficient matrix is full rank and invertible.

3 The Minimax Disparity Model for Determining OWA Weights

In this section, we briefly review the recent development of the specific class of OWA functions whose weights are determined by the minimax disparity methods. In 2005 Wang and Parkan [20] revisited the maximum entropy method introduced by O'Hagan [17] and proposed the minimax disparity procedure to determine the OWA weights in the convex optimization problem

$$\begin{aligned} &\text{Min.} \quad \left\{ \max_{i\in\{1,\dots,n-1\}} |w_i - w_{i+1}| \right\} \\ &\text{s.t.} \quad \sum_{i=1}^{n} w_i = 1, \quad \text{Orness}(\boldsymbol{w}) = \eta = \frac{1}{n-1}\sum_{i=1}^{n}(i-1)w_i, \quad 0 \le w_i \le 1 \end{aligned} \tag{7}$$

where $0 \le \eta \le 1$ stands for the orness of the weighting vector.

The objective function is non-linear due to the absolute difference between two adjacent weights. In order to overcome this non-linearity, the authors introduced a new variable called $\delta = \max_{i\in\{1,\dots,n-1\}} |w_i - w_{i+1}|$ and described the original problem equivalently as

$$\begin{aligned} &\text{Min.} \quad \delta \\ &\text{s.t.} \quad \sum_{i=1}^{n} w_i = 1, \qquad \text{Orness}(\boldsymbol{w}) = \eta = \frac{1}{n-1}\sum_{i=1}^{n}(i-1)w_i, \\ &\qquad w_i - w_{i+1} - \delta \le 0, \quad w_i - w_{i+1} + \delta \ge 0, \quad 0 \le w_i \le 1. \end{aligned} \tag{8}$$

4 The Minimax Disparity Model and the Binomial Decomposition

In Sect. 3, we have reviewed minimax disparity methods for determining OWA weights. The empirical results of those methods are obtained for small dimensions ($n = 3, 4, 5, 6$). In real-life scenarios, we usually encounter large-scale optimization problems. In this context, the optimization problems formulated directly in terms of the OWA weights require high computational resources. Our objective is to make the minimax disparity methods for determining OWA weights more tractable in high-dimensional problems.

In the binomial decomposition framework, any OWA function can be expressed uniquely into a linear combination of the coefficients α_j, $j = 1, \dots, n$, and the binomial OWA functions as described in Sect. 2. The positive aspect of the binomial decomposition framework is the possibility of using k-additive levels in which a certain number of coefficients $\alpha_1, \ldots, \alpha_k$, where $k \leq n$, are used in the process of OWA weights determination while the remaining coefficients are set to zero. Using less coefficients can markedly reduce the computation time.

We now transform the minimax disparity model (8) into a problem in which the weights are substituted by a set of coefficients α_j, $j = 1, \ldots, n$,

$$
\begin{aligned}
&\text{Min. } \delta \\
&\text{s.t. } \sum_{i=1}^{n} \alpha_j = 1, \qquad \text{Orness}(\boldsymbol{\alpha}) = \eta = \sum_{j=1}^{n} \frac{n-j}{(n-1)(j+1)} \cdot \alpha_j, \\
&\sum_{j=1}^{n-i+1} w_{ji}\alpha_j - \sum_{j=1}^{n-i} w_{j,i+1}\alpha_j - \delta \leq 0 \qquad i = 1, \ldots, n-1, \\
&\sum_{j=1}^{n-i+1} w_{ji}\alpha_j - \sum_{j=1}^{n-i} w_{j,i+1}\alpha_j + \delta \geq 0 \qquad i = 1, \ldots, n-1, \\
&\sum_{j=2}^{n} \left[1 - n\frac{\binom{i-1}{j-1}}{\binom{n}{j}}\right]\alpha_j \leq 1 \qquad i = 1, \ldots, n
\end{aligned} \tag{9}
$$

where $0 \leq \eta \leq 1$ stands for the orness of the weighting vector. Notice that the first and the last constraints correspond to the boundary and the monotonicity conditions of the OWA weighting vector $\boldsymbol{w} = (w_1, ..., w_n) \in [0, 1]^n$ with $\sum_{i=1}^{n} w_i = 1$.

In Table 1 we report the empirical results of our proposed model with full-dimension coefficients alpha ($k = n$) for the case $n = 10$. The coefficients alpha for the central orness value 0.3, 0.4, 0.5, 0.6 and 0.7 have high sparsity. In particular, the sparsity of coefficients alpha is 90% for orness $\eta = 0.5$ and 80% for orness $\eta = 0.3, 0.4, 0.6, 0.7$. This suggests that by using a smaller k-additive level ($k < n$) one can exploit the sparsity of the model and speed up the performance of

Table 1 The coefficients alpha of our proposed method for $n = 10$

Orness η	1	0.9	0.8	0.7	0.6	0.5	0.4	0.3	0.2	0.1	0
α_1	10	4.3	2.71	1.98	1.49	1	0.51	0.02	0	0	0
α_2	−45	−5.4	−1.93	−0.98	−0.49	0	0.49	0.98	0	0	0
α_3	120	0	0	0	0	0	0	0	0	0	0
α_4	−210	0	0	0	0	0	0	0	3	0	0
α_5	252	12.6	0	0	0	0	0	0	0	0	0
α_6	−210	−16.8	0	0	0	0	0	0	−9	0	0
α_7	120	2.4	0	0	0	0	0	0	13.71	8.4	0
α_8	−45	9	1.29	0	0	0	0	0	−9.64	−13.5	0
α_9	10	−6.5	−1.57	0	0	0	0	0	3.43	7.5	0
α_{10}	−1	1.4	0.5	0	0	0	0	0	−0.5	−1.4	1

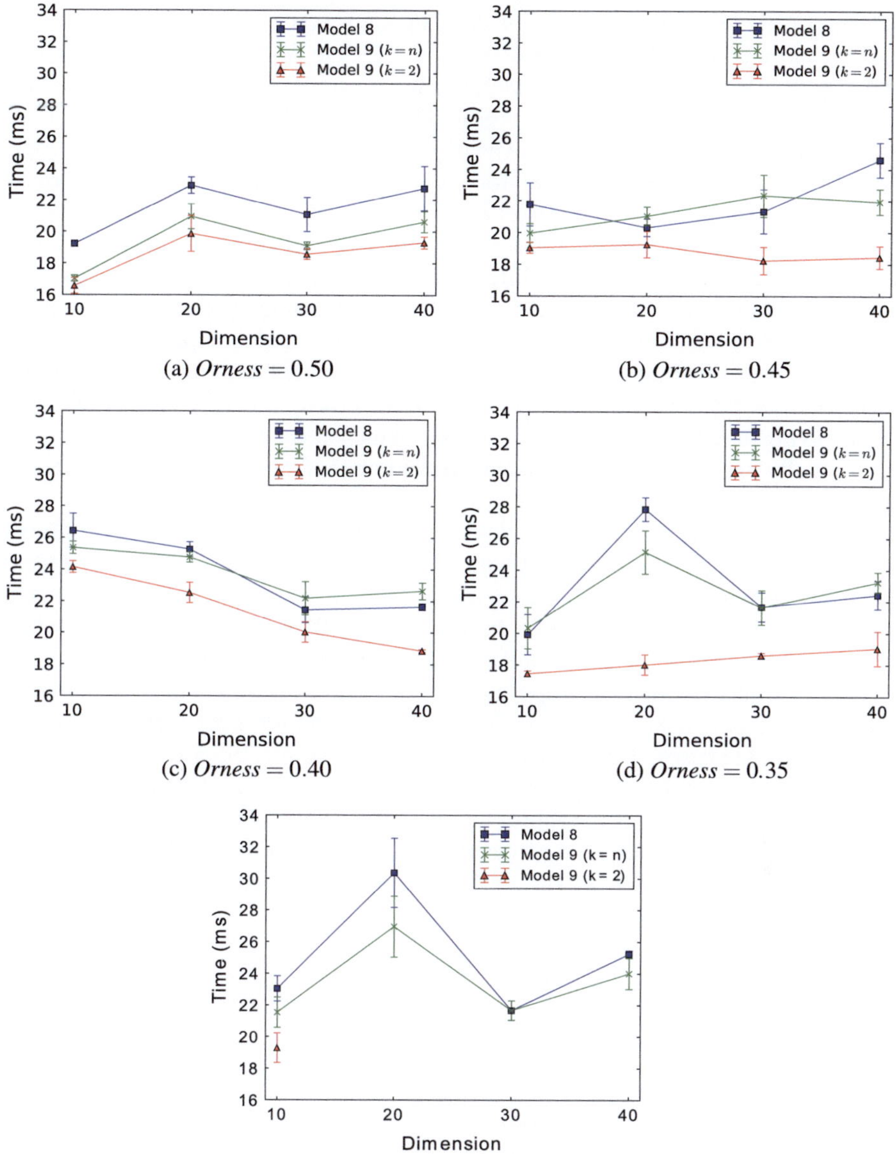

(a) *Orness* = 0.50
(b) *Orness* = 0.45
(c) *Orness* = 0.40
(d) *Orness* = 0.35
(e) *Orness* = 0.30

Fig. 1 Computation time of the original model and our proposed model with various degrees of orness for $n = 10, 20, 30, 40$

solvers. In Fig. 1 the performance of the CPLEX solver is shown, with respect to the original method, and our proposed method for the two k-additive levels $k = n$ and $k = 2$. The graph shows the average running time (out of 300 runs), including the standard error, for various orness degrees $\eta = 0.50, 0.45, 0.40, 0.35, 0.30$ and dimensions $n = 10, 20, 30, 40$. The proposed method with full-dimension coefficients alpha facilitates the solver for some orness degrees. Notably, for orness $\eta = 0.5$, this method is always faster the conventional one due to its high sparsity (90%). For the other degree of orness, the proposed model with full-dimension alpha and the conventional model do not differ significantly. Even though both models have the same number of constraints and variables, the constraints of coefficients alpha related to the monotonicity condition in model (9) are more complex than those in model (8). As a result, the full-dimension proposed model requires more computation time than the conventional method for some orness values. However, we note that applying the proposed method with lower-dimensional coeffcients alpha ($k < n$) substantially improves the performance. The number of variables used in our model reduces by $(1 - \frac{k}{n})\%$.

As shown in Fig. 1 the solver spends less time to obtain precisely the optimal weights with our proposed method with 2-additive level ($k = 2$). When the orness value differs significantly from the central orness, for instance $\eta = 0.30$, the model with the 2-additive level is adequate for identifying OWA weights for $n = 10$ while a larger number of coefficients alpha is needed for $n = 20, 30, 40$.

Experiments show that the proposed approach with a reduced number of variables can assist decision makers in finding OWA weights faster for some degrees of orness. In the remaining cases, decision makers can exploit the flexibility of the model and choose the k-additive level that provides the best trade-off between the computational demand and the accuracy of the approximated and possibly suboptimal OWA weights. As an example, we consider our proposed model with the orness value equal to 0.2. If the k-additive level increases from 3 to 10, we obtain better objective values as

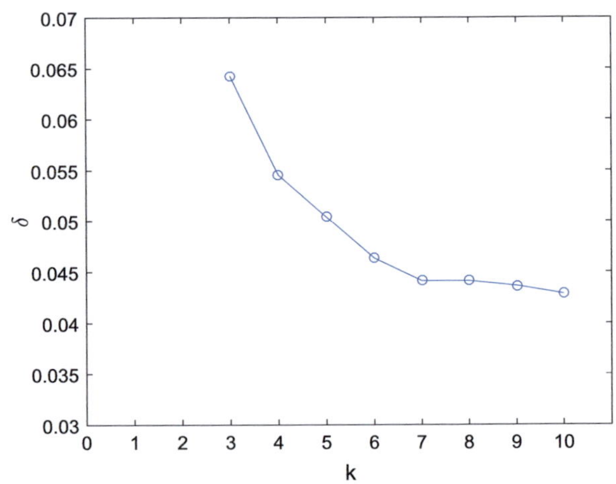

Fig. 2 The objective value δ corresponding to the k-additive levels (for the cases $k = 1, 2$ there is no solution of coefficients alpha). Reproduced with permission from arXiv:1804.06331

expected (see Fig. 2). However it is evident that $k = 7$ leads to the best trade-off between the accuracy of the optimal value and the dimensionality reduction of the optimization problem.

5 Conclusions

This paper proposes a new methodology for determining OWA weights in large-scale optimization problems where the cost of computation of optimal weights is very high. Our model allows the optimization of the OWA weights to be transformed into the optimization of the coefficients in the binomial decomposition framework, considering the k-additive levels in order to reduce the complexity of the proposed model. Empirical results show that a small set of the coefficients in the binomial decomposition can efficiently model the full-dimensional set of the OWA weights.

However, there are still some issues that need to be addressed in future research. For instance, applying the proposed method to non-linear OWA weight determination models may significantly reduce the computation time. It is necessary to develop an algorithm to identify which k-additive level in the set $\{1, \ldots, n\}$ gives the best trade-off between accuracy and computational complexity according to the specific applications.

Acknowledgements The author would like to thank Ricardo Alberto Marques Pereira and Silvia Bortot for their helpful remarks on the manuscript. The author would like to thank to Andrea Mariello for his practical advice on the experiments. The author is grateful to the anonymous referees for their valuable suggestions.

References

1. Amin, G.R., Emrouznejad, A.: An extended minimax disparity to determine the OWA operator weights. Comput. Ind. Eng. **50**(3), 312–316 (2006)
2. Beliakov, G., Bustince Sola, H., Calvo, T.: A Practical Guide to Averaging Functions, Studies in Fuzziness and Soft Computing, vol. 329. Springer, Heidelberg (2016)
3. Bortot, S., Fedrizzi, M., Marques Pereira, R.A., Nguyen, T.H.: The binomial decomposition of generalized Gini welfare functions, the S-Gini and Lorenzen cases. Inf. Sci. (to appear)
4. Bortot, S., Marques Pereira, R.A.: The binomial Gini inequality indices and the binomial decomposition of welfare functions. Fuzzy Sets Syst. **255**, 92–114 (2014)
5. Bortot, S., Marques Pereira, R.A., Nguyen, T.H.: The binomial decomposition of OWA functions, the 2-additive and 3-additive cases in n dimensions. Int. J. Intell. Syst. 187– 212 (2018)
6. Calvo, T., De Baets, B.: Aggregation operators defined by k-order additive/maxitive fuzzy measures. Int. J. Uncertain. Fuzz. **6**(6), 533–550 (1998)
7. Carlsson, C., Fullér, R.: Maximal entropy and minimal variability OWA operator weights: a short survey of recent developments. In: Collan, M., Kacprzyk, J. (eds.) Soft Computing Applications for Group Decision-Making and Consensus Modeling, pp. 187–199. Springer, Heidelberg (2018)
8. Emrouznejad, A., Amin, G.R.: Improving minimax disparity model to determine the OWA operator weights. Inf. Sci. **180**(8), 1477–1485 (2010)

9. Fullér, R., Majlender, P.: On obtaining minimal variability OWA operator weights. Fuzzy Sets Syst. **136**(2), 203–215 (2003)
10. Gong, Y., Dai, L., Hu, N.: An extended minimax absolute and relative disparity approach to obtain the OWA operator weights. J. Intell. Fuzzy Syst. **31**(3), 1921–1927 (2016)
11. Grabisch, M.: Alternative representations of discrete fuzzy measures for decision making. Int. J. Uncertain. Fuzz. **5**(5), 587–607 (1997)
12. Grabisch, M.: Alternative representations of OWA operators. In: Yager, R.R., Kacprzyk, J. (eds.) Recent Developments in the Ordered Weighted Averaging Operators: Theory and Practice, Studies in Fuzziness and Soft Computing, vol. 265, pp. 73–85. Springer, Heidelberg (1997)
13. Grabisch, M.: k-order additive discrete fuzzy measures and their representation. Fuzzy Sets Syst. **92**(2), 167–189 (1997)
14. Liu, X.: The solution equivalence of minimax disparity and minimum variance problems for OWA operators. Int. J. Approx. Reason. **45**(1), 68–81 (2007)
15. Liu, X.: A review of the OWA determination methods: classification and some extensions. In: Yager, R.R., Kacprzyk, J., Beliakov, G. (eds.) Recent Developments in the Ordered Weighted Averaging Operators: Theory and Practice, pp. 49–90. Springer, Heidelberg (2011)
16. Mezei, J., Brunelli, M.: A closer look at the relation between orness and entropy of OWA function. In: Collan, M., Kacprzyk, J. (eds.) Soft Computing Applications for Group Decision-making and Consensus Modeling, Studies in Fuzziness and Soft Computing, pp. 201–211. Springer, Heidelberg (2018)
17. O'Hagan, M.: Aggregating template or rule antecedents in real-time expert systems with fuzzy set logic. In: Twenty-Second Asilomar Conference on Signals, Systems and Computers, vol. 2, pp. 681–689. IEEE (1988)
18. Sang, X., Liu, X.: An analytic approach to obtain the least square deviation OWA operator weights. Fuzzy Sets Syst. **240**, 103–116 (2014)
19. Wang, Y.M., Luo, Y., Hua, Z.: Aggregating preference rankings using OWA operator weights. Inf. Sci. **177**(16), 3356–3363 (2007)
20. Wang, Y.M., Parkan, C.: A minimax disparity approach for obtaining OWA operator weights. Inf. Sci. **175**(1–2), 20–29 (2005)
21. Xu, Z.: An overview of methods for determining OWA weights: research articles. Int. J. Intell. Syst. **20**(8), 843–865 (2005)
22. Xu, Z.S., Da, Q.L.: An overview of operators for aggregating information. Int. J. Intell. Syst. **18**(9), 953–969 (2003)
23. Yager, R.R.: On ordered weighted averaging aggregation operators in multi-criteria decision making. IEEE Trans. Syst. Man Cybrn. **18**(1), 183–190 (1988)
24. Yager, R.R.: A general approach to criteria aggregation using fuzzy measures. Int. J. Man Mach. Stud. **39**(2), 187–213 (1993)
25. Yager, R.R.: On the inclusion of variance in decision making under uncertainty. Int. J. Uncertain. Fuzz. **04**(05), 401–419 (1996)

Fleet Size and Mix Pickup and Delivery Problem with Time Windows: A Novel Approach by Column Generation Algorithm

M. N. Tchoupo, A. Yalaoui, L. Amodeo, F. Yalaoui and P. Flori

Abstract In this paper, a new efficient column generation algorithm is proposed to tackle the Fleet Size and Mix Pick-up and Delivery Problem with Time Windows (FSMPDPTW). This work is motivated by fleet sizing for a daily route planning arising at a Hospital centre. Indeed, a fleet of heterogeneous rented vehicles is used every day to pick up goods to the locations and to deliver it to other locations. The heterogeneous aspect of the fleet is in term of fixed cost, capacity, and fuel consumption. The objective function is the minimization of the total fixed cost of vehicles used and the total routing cost. The problem is modelled as a set partitioning problem, and an efficient column-generation algorithm is used to solve it. In the resolution, the pricing problem is decomposed in sub-problems, such that each vehicle type has its own sub-pricing problem. A mixed integer linear program, a powerful labelling algorithm and the regret heuristic is provided to solve the pricing sub-problems. Based on Li and Lim's benchmark (altered Solomon's benchmark) for demands and from Lui and Shen's benchmark for vehicles types; a new set of benchmarks is proposed to test the efficiency of the propound algorithm.

Keywords Transportation logistics · Pickup and delivery with time windows
Heterogeneous fleet · Column-generation algorithm

M. N. Tchoupo (✉) · A. Yalaoui · L. Amodeo · F. Yalaoui
LOSI, ICD, UMR, CNRS, University of Technology of Troyes, Troyes 6281, France
e-mail: moise.noumbissitchoupo@utt.fr

A. Yalaoui
e-mail: alice.yalaoui@utt.fr

L. Amodeo
e-mail: lionel.amodeo@utt.fr

F. Yalaoui
e-mail: farouk.yalaoui@utt.fr

P. Flori
Hospital of Troyes, 101 Av. Anatole France, Troyes, France
e-mail: pauline.flori@ch-troyes.fr

© Springer Nature Switzerland AG 2018

P. Daniele and L. Scrimali (eds.), *New Trends in Emerging Complex Real Life Problems*, AIRO Springer Series 1,
https://doi.org/10.1007/978-3-030-00473-6_41

1 Introduction

In Pickup and Delivery Problem with Time Windows (PDPTW), the demands are paired and every pair is a request which must be satisfied in the same route and by the same vehicle. Each request consists of delivering goods from a predefined location (pickup customer) to other one (delivery customer). In the literature, a lot of works in PDPTW didn't consider the fleet dimensioning in heterogeneous fleet case.This problem can be described as the problem of designing least cost routing plan to satisfy a set of transportation requests by a given heterogeneous fleet of vehicles. Each route starts and ends at the depot. Each customer must be served within a given time window. The amount of goods must not exceed the capacity of the vehicle. The service time indicates how long it will take for the pickup or delivery to be performed. A vehicle is allowed to arrive at a location before the beginning of its time windows, and in this case must wait until the start of the time window. For each request, the pickup customer must be visited before the corresponding delivery customer with the same vehicle and the same route, but not necessarily immediately after. In this problem, an unlimited fleet of vehicles with a fixed capacity is considered. The problem is NP-hard because it contains the PDPTW which is NP-Hard [2]. In this work, it is assumed that the matrix of distance and the matrix of time satisfies the triangular inequality.

The Fleet Size and Mix Vehicle Routing Problem (FSM) was introduced by [4] in which the vehicles fleet is considered as heterogeneous and unlimited. The FSM with fixed and variable costs was introduced by [3]. A classification and a review of the literature on the Fleet Size and Mix Vehicle Routing Problem with time Windows (FSMTW) is provided by [5]. Li and Lim [6] proposed a tabu-embedded simulated annealing algorithm to solve the Pickup and Delivery Problem with Time windows (PDPTW). They proposed six newly-generated different data sets with various distribution properties. An adaptive large neighbourhood search heuristic was proposed by [11]. For a survey on pickup and delivery problems until 2008, see [9].

Recently, many extensions of PDPTW problems motivated by real life problem were discussed in literature. Xu et al. [12] proposed a set partitioning formulation to solve a practical heterogeneous fleet pickup and delivery problem with constraints on unloading sequence. They also take into account practical aspects such as: multiple time windows, multiple depots, compatibility constraints between carrier and vehicles, orders and vehicles, and between orders. Qu and Bard [10] introduces a variant of the heterogeneous pickup and delivery problem in which the capacity of each vehicle can be modified by reconfiguring its interior to satisfy different types of customer demands. They developed a two-phase heuristic that makes use of ideas from greedy randomized adaptation search procedures with multiple starts. Noumbissi Tchoupo et al. [8] developed a Bender's decomposition algorithm for PDPTW with heterogeneous fleet (HVRPPDTW) to minimize the sum of vehicle fixed costs and the total routing cost.

In this paper a new variant of the PDPTW problem is considered. It is an extension of PDPTW by considering re-sizing of the heterogeneous vehicle fleet. Because

PDPTW is an NP hard problem, FSMPDPTW is also an NP hard problem. In Sect. 2, a set partition formulation is proposed to model it. Solution methodology to solve this NP hard problem, is presented in Sect. 3. A mixed integer linear program is proposed in Sect. 3.1; in Sect. 3.2 is presented different ways to reduce the search space, and a labelling algorithm is proposed in Sect. 3.3. In Sect. 3.4, a fast heuristic algorithm is proposed to quickly generate columns with negative reduced cost. Section 4 is dedicated to numerical experiment.

2 Problem Formulation

This section presents a set partitioning formulation for FSMPDPTW to model the problem, based on the set partitioning formulation of [1]. The formulation uses a directed graph $G = (V', E)$, where $V' = \{0, \ldots, 2n\}$ is a set of $2n + 1$ nodes and E is a set of edges. Node 0 represents the depot and the set $V = V' \setminus \{0\}$ corresponds to the $2n$ customers. Let $P = \{p_1, \ldots, p_n\}$ the set of n pickup demands and $D = \{d_1, \ldots, d_n\}$ the set of n delivery demands. K is the set of vehicles types. Q^k is the capacity, f^k the fixed cost and v^k the cost per unit of distance of a vehicle type k $\forall k \in K$. The travel time (respectively distance) to get to the location of demands j from the location of demand i is noted t_{ij} (respectively d_{ij}), s_i is the time service required at the node i, q_i the amount of goods to pickup or deliver for node i, e_i the earlier time at which the service may begin at node i and l_i the latest time at which the service may begin at node i. For $i \in \{1, \ldots, n\}$ pickup demand p_i and its corresponding d_i delivery demand comply the equation :

$$q_{p_i} = -q_{d_i} \tag{1}$$

The set of edges E is formally defined by Eq. (2).

$$(i, j) \in E \iff \begin{cases} e_i + s_i + t_{ij} \leq l_j \quad \text{and} \\ \exists k \in K /\ q_i + q_j \leq Q^k \quad \text{and} \quad |q_i| \leq Q^k \quad \text{and} \quad |q_j| \leq Q^k \end{cases} \tag{2}$$

A route $r = (0, i_1, \ldots, i_l, 0)$ performed by a vehicle of type k, is a cycle starting and ending at the depot 0, passing by the customers $\{i_1, \ldots, i_l\} \subset V$, and complying the constraints of time windows, capacity of a vehicle of type k and the paring constraints. Pairing constraints means that if a pickup is served in a route, then the corresponding delivery is served by the same route and after the pickup demand.

Let $\mathscr{R}^k$ be the set of index of all feasible routes of vehicle type $k \in K$ and let $\mathscr{R} = \bigcup_{k \in K} \mathscr{R}^k$. For each route $r \in \mathscr{R}^k$, a routing cost c_r^k is associated. $\mathscr{R}_r^k$ is the subset of customers visited by the route $r \in \mathscr{R}^k$. Furthermore, let λ_ρ be a binary variable equal to 1 if and only if $\rho \in \mathscr{R}$ is chosen in the solution. Note that for all $\rho \in \mathscr{R}$, there is $vh_\rho \in K$ and $r_\rho \in \mathscr{R}^{k_\rho}$ such that: $\rho = r_\rho$. The set partitioning formulation of the problem is the following:

$$(MP)\ \ Minimize\quad \sum_{\rho\in\mathscr{R}}(f^{vh_\rho}+c_{r_\rho}^{vh_\rho})\lambda_\rho \tag{3}$$

$$\sum_{\rho\in\mathscr{R}/i\in\rho}\lambda_\rho=1,\qquad (\pi_i)\quad \forall i\in P; \tag{4}$$

$$\lambda_\rho\in\{0,1\},\qquad \forall\rho\in\mathscr{R}. \tag{5}$$

In this formulation, the objective function (3) minimizes the total cost. Constraints (4) specifies that each pickup demand $p\in P$ must be covered once by one route, and involves that the corresponding delivery is served in the same route and by the same vehicle. $\forall i\in P,\ \ \pi_i$ is the dual variable associate to constraint (4).

3 Solution Methodology

The master problem (*MP*) has an exponential number of variables ($|\mathscr{R}|=|K|2^{2n+1}$ variables) and $n=|P|$ constraints. The linear relaxation of (*MP*) is easier to solve than (*MP*) and if the optimal solution of linear relaxation is integer, then it is the optimal solution of (*MP*). For large networks, the number of variables in the master problem becomes prohibitive, and it is not possible to generate all the variables of the master problem. The idea of the *column generation* is to work only with a sufficiently meaningful subset of variables, considered in the *restricted master problem* (*RMP*). In the rest of the paper, the linear relaxation of (*RMP*) is noted by (*LRMP*). In column generation, an iteration consists firstly of optimizing the restricted master problem in order to determine the current optimal objective function value $\bar{z}$ and the dual multipliers π. Secondly, it consists of finding, if it exists, a variable λ_ρ with negative reduced cost. If the cost of an optimal solution of the pricing problem is positive, there is no improving variable and the solution found by the linear relaxation of the master problem is optimal. Otherwise, the founded variable is added to the (*LRMP*) and the process is repeated. When the optimal solution of (*LRMP*) is not integer, in order to obtain integer solutions to the original problem, (*RMP*) is solved. In this case, it is not possible to know if the solution obtained by the (*RMP*) is the optimal solution of the (*MP*).

3.1 Pricing Problem Formulation

The problem studied allows to subdivide the pricing problem in $|K|$ sub-pricing problems, with a problem for each vehicle type. Given the dual multipliers π of the (*LRMP*) and a type $k\in K$ of vehicle, the sub-pricing problem is modelled by a linear integer program. Given the decisional variables:

- x_{ij} is a binary variable equal to 1 if the vehicle k travels from the location of demand i to the location of demand j;
- T_i is the time at which vehicle k begins the service at node i;
- $Q_{i,j}$ is the number of item in the vehicle k on the arc (i, j).

The mixed integer linear program is defined by:

$$(PR_k) \qquad \text{Minimize:} \quad f^k + \sum_{(i,j)\in A} (v^k d_{ij} - \pi_i^*) x_{ij} \tag{6}$$

$$\sum_{j:(i,j)\in A} x_{ij} \leq 1, \qquad \forall i \in P; \tag{7}$$

$$\sum_{j:(i,j)\in A} x_{ij} = \sum_{j:(n+i,j)\in A} x_{n+i,j}, \qquad \forall i \in P, \tag{8}$$

$$\sum_{i\in P} x_{0i} \leq 1; \tag{9}$$

$$\sum_{i\in P} x_{0i} = \sum_{i\in D} x_{i0}, \tag{10}$$

$$\sum_{j:(i,j)\in A} x_{ij} = \sum_{j:(j,i)\in A} x_{ji}, \qquad \forall i \in P \cup D; \tag{11}$$

$$\sum_{j:(i,j)\in A} x_{ij} = \sum_{j\in P} x_{0j}, \qquad \forall i \in P \cup D, \tag{12}$$

$$Q_{ij} \leq Q^k x_{ij}, \qquad \forall (i, j) \in A; \tag{13}$$

$$\sum_{i\in P} Q_{0i} = \sum_{i\in D} Q_{i0}; \tag{14}$$

$$\sum_{j:(i,j)\in A} Q_{ij} - \sum_{j:(i,j)\in A} Q_{ji} = q_i \sum_{j:(i,j)\in A} x_{ij}, \qquad \forall i \in P \cup D; \tag{15}$$

$$T_i + (l_i + s_i + t_{ij}) x_{i,j} \leq T_j + l_i, \qquad \forall (i, j) \in A; \tag{16}$$

$$T_i + (s_i + T_{i,n+i}) \sum_{j:(i,j)\in A} x_{ij} \leq T_{n+i}, \qquad \forall i \in P; \tag{17}$$

$$x_{ij} \in \{0, 1\}, T_i \in [e_i, l_i], Q_{ij} \in \left[0, Q^k\right] \ \forall i, j \in N. \tag{18}$$

The objective function (6) minimizes the reduced cost of a path. Constraints (7)–(9) assure that each pickup demand is served at most once and the corresponding delivery demand is served in the same route. Constraints (10)–(12) guarantee that the route starts and ends at the depot. Constraints (13)–(16) assure the respect of time windows and the capacity constraints. Constraint (17) assures that every delivery demand is satisfied after the corresponding pickup demand but not necessary immediately after the pickup demand. For a given vehicle type $k \in K$, the solution of the sub-problem of pricing associated is a path starting at the depot and ending at the depot, under pairing, precedence, time window and capacity constraints. The proposed model is based on a graph. So in the next section, the acceleration techniques are provided to reduce the search space and speed up the resolution of problem.

3.2 Reduction of the Search Space

In this subsection, is presented different ways to reduce the search space by suppressing infeasible arcs in A. Considering the graph $G = (V', A)$, Dumas et al. [2] suggested a set of infeasible arcs for PDPTW. It is proposed to add:

- $(p_i, p_j) \in A$ and $(p_j, d_i) \in A$ are infeasible if $e_{p_i} + t_{p_i p_j} + t_{p_j d_i} > l_{d_i}$,
- $(p_i, d_j) \in A$ and $(d_j, d_i) \in A$ are infeasible if $e_{p_i} + t_{p_i d_j} + t_{d_j d_i} > l_{d_i}$,
- $(p_i, p_j) \in A$ is infeasible if $q_{p_i} + q_{p_j} > Q^k$ for all $k \in K$.

In the following, all infeasible arcs above are removed from $G = (V', A)$. Also in the aim to speed up the resolution of the pricing problem, a criteria is proposed to identify and remove the requests whose associate dual value is not interesting compare to cost required to complete it.

Proposition 1 *Let $i_0 \in P$ and $k \in K$; if:*

$$m_{i_0} v^k > \pi_{i_0}, \ \textit{then}, \sum_{j:(i_0,j)\in A} x_{i_0 j} = 0;$$

with $m_{i_0} = min_{i,j\in N} z_{i,j,i_0} + min_{i,j\in N} z_{i,j,n+i_0}$, *and* $\forall i, j, b \in N, \ z_{i,j,b} = d_{i,b} + d_{b,j} - d_{i,j}$, *if* $(0, i, b, j)$ *is a feasible path, and equal to infinity otherwise.*

In the next subsection, a labelling algorithm is proposed to solve the sub-pricing problem to optimality.

3.3 Labelling Algorithm

In our study case, the sub-pricing problem is the search of a path with a negative reduce cost. The idea in labelling algorithm is to evaluate all possible paths in the graph. To speed up the generation and the evaluation of paths, dominance criteria are used. In this study, it is assumes that the matrix of distance and the matrix of time verify triangular inequality. In the proposed labelling algorithm, labels are extended from the source to the pickup demands. Each label, is constituted by:η the last node of the partial path, t the departure time from the last node, qt the total load after visiting the last node, c the accumulated cost, $\mathscr{D}$ the set of the delivery demands not yet completed such that the corresponding pickup demand has been completed, $\mathscr{P}$ the set of the pickup demands completed, such that the corresponding delivery demand has been completed. The reference of each parent label is also stored. The notations η_L, $t_L, qt_L, c_L, \mathscr{D}_L, \mathscr{P}_L$ are used to refer to the attributes of the label L. A label L can be extended along arc (η_L, p_j) (p_j is a pickup demand) if:

1. $(\eta_L, p_j) \in A$ and $p_j \notin \mathscr{P}$,
2. $qt_L + q_{p_j} \leq Q^k$ (pricing problem associated to the type of vehicle $k \in K$).

A label L can be extended along arc (η_L, d_j) (d_j is a delivery demand) if:

$$(\eta_L, d_j) \in A \text{ and } d_j \in \mathscr{D}.$$

Proposition 2 *A label L' is dominated by label L if:*

1. $\eta_L = \eta_{L'}$,
2. $t_L \leq t_{L'}$,
3. $c_L \leq c_{L'}$,
4. $\mathscr{P}_{L'} \subseteq \mathscr{P}_L$,
5. $\mathscr{D}_L \subseteq \mathscr{D}_{L'}$.

3.4 Heuristic Algorithm for a Sub-pricing Problem

In this section, fast heuristic is proposed to accelerate the resolution of the (*LRMP*). The heuristic proposed in this section are based on the well know *greedy heuristic*. At each iteration of the algorithm, one request is chosen to be insert. Let $diffCost_{\{i\},\rho}$ denote the change in objective value induced by inserting pickup demand i and its delivery demand $i + n$ into the route ρ and at the positions that decrease the most the objective value. If the demand i cannot be inserted in route ρ, then the cost is considered as infinity. Finally the pickup demand p (and its delivery demand) inserted is the one such that:

$$p = argmin_{\ i \in U}\ diffCost_{\{i\},\rho}$$

Where U is the set of unplanned requests. This process continues until no more request can be inserted. This heuristic were inspired by heuristics described in [11].

4 Computational Experiments

The algorithms have been implemented on eclipse, the programming language was Java and the experiments have been carried out on a 2.6 GHz and 8 Go of RAM.

Table 1 Computational results for instances with a clustered distribution of demands

Inst	Cost		LB	Time (min)		Gap	Inst	Cost		LB	Time (min)		Gap
	Lab	MILP		Lab	MILP			Lab	MILP		Lab	MILP	
lc101	*2057.26*	2064.95	2014	1	5	2	lc201	1950.49	1967.06	1762	63	120	10
lc102	*2241.81*	2296.28	2187	35	57	5	lc202	2076.02	2575.95	1908	96	120	8
lc103	*1800.61*	1813.87	1754	62	105	3	lc203	1972.22	2390.52	1639	133	120	17
lc104	*1642.48*	1642.48	1497	95	120	9	lc204	1800.37	2210.03	1449	60	120	20
lc105	*2108.24*	2108.24	2106	17	35	1	lc205	1451.86	1792.03	1390	61	120	4
lc106	*2100.34*	2106.23	2061	56	81	2	lc206	2083.75	2144.84	1746	92	120	16
lc107	*2247.07*	2247.07	2168	53	120	4	lc207	1531.84	1752.52	1415	60	120	8
lc108	*2078.23*	2087.66	2074	75	120	1	lc208	1509.06	1817.73	1417	90	120	6
lc109	*1702.41*	1702.41	1684	72	108	1							

Table 2 Computational results for instances with a uniform distribution of demands

Inst	Cost		LB	Time (min)		Gap	Inst	Cost		LB	Time (min)		Gap
	Lab	MILP		Lab	MILP			Lab	MILP		Lab	MILP	
lr101	*3050.76*	3050.76	3050	0.1	0.1	0	lr201	2340.53	2340.53	2139	61	120	9
lr102	*2713.17*	2713.17	2713	22	42	0	lr202	2823.71	2823.71	2053	64	120	27
lr103	*2343.15*	2223.84	2223	31	84	0	lr203	3181.84	3037.15	1895	80	120	40
lr104	*1733.56*	1733.56	1730	86	120	0.2	lr204	2848.01	3163.52	1704	60	120	40
lr105	*2649.53*	2651.06	2613	0.1	1	1.4	lr205	2635.37	3042.47	1999	60	120	24
lr106	*2343.66*	2343.66	2335	5	25	0.3	lr206	3339.36	3182.30	2100	77	120	37
lr107	*1837.26*	1837.26	1837	52	68	0	lr207	2692.2	3250.98	1601	106	120	41
lr108	*1775.42*	1775.42	1775	60	78	0	lr208	2600.08	3016.7	1601	60	120	38
lr109	*2227.15*	2256.99	2181	4	18	3.3	lr209	3105.56	3133	1921	82	120	38
lr110	*2217.79*	2217.79	1995	23	52	10	lr210	2692.60	3133.29	1915	84	120	29
lr111	*1936.11*	1936.11	1867	38	87	3.5	lr211	3098.91	3197	1624	60	120	48
lr112	*1881.31*	1881.31	–	120	120	–							

Table 3 Computational results for instances with a semi-clustered distribution of demands

Inst	Cost		LB	Time (min)		Gap	Inst	Cost		LB	Time (min)		Gap
	Lab	MILP		Lab	MILP			Lab	MILP		Lab	MILP	
lrc101	*2841.29*	2841.29	2841	1	0.1	0	lrc201	2850.87	2850.87	2614	61	120	9
lrc102	*2546.89*	2546.89	2537	6	19	0.4	lrc202	3173.58	3025.69	2505	60	120	21
lrc103	*2172.36*	2172.36	2114	31	99	2.6	lrc203	3511.22	3187.89	2134	73	120	39
lrc104	*1920.75*	1920.75	1890	91	109	1.6	lrc204	2954.55	2729.7	1719	60	120	42
lrc105	*2774.13*	2774.13	2748	1	2	0.9	lrc205	2903.55	3120.42	2426	63	120	16
lrc106	*2619.42*	2619.42	2522	1	3	3.7	lrc206	2578.4	2990.31	2261	62	120	12
lrc107	*2219.31*	2219.31	2183	6	19	1.6	lrc207	2874.87	2874.87	2265	74	120	26
lrc108	*2038.63*	2038.63	2014	31	100	1.2	lrc208	3581.08	3794.06	2121	71	120	41

To test the effectiveness of the proposed algorithm, the generation of the instances with reasonable size up to 100 demands is obtained by coupling Li and Lim's Benchmarks for the PDPTW and Lui and Shen's [7] benchmark for the set of types of vehicles. Computational results are summarized in the Tables 1, 2 and 3. And the solution gap is defined by (100 $\times$ (solution value-lower bound)/lower bound) (commonly use). **Inst** is the name of the instance, **Lab** design the labelling algorithm, **LB** is the lower bound obtained by Lab. And the instances in **bold** was solved to optimality.

In all configurations, the resolution of the pricing problem by labelling algorithm outperformed the resolution by the Mixed integer linear program (**MILP**). It is best in quality of solution and in computational time. The labelling algorithm allows to have a lower bound for 98% of the instances. And in average, the gap between the upper bound found by and the proposed column generation algorithm is 12%.

5 Conclusion

In this paper, an effective column generation algorithm is proposed to tackle a new combination of Fleet Size and Mix Vehicle Routing Problem and Pickup and Delivery Problem with Time Windows. A set partitioning formulation is provided to model it and a based-column generation algorithm was proposed to solve it. A labelling algorithm, the heuristics and the acceleration techniques are provided to speed up the resolution. A new set of benchmarks is proposed. The proposed algorithms are able to find an optimal solution for 21% of the instances, and for the instances with small time windows, it returns a solution close to optimal solution (with a gap of 2% on average). The experiments show that the designed labelling algorithm is better (quality and computational time) than the proposed mixed integer linear program. In the future, it would be also interesting to extend the problem by considering: multi-depot or multi-dimensional loading.

References

1. Baldacci, R., Toth, P., Vigo, D.: Exact algorithms for routing problems under vehicle capacity constraints. Ann. Oper. Res. **175**(1), 213–245 (2010)
2. Dumas, Y., Desrosiers, J., Soumis, F.: The pickup and delivery problem with time windows. Eur. J. Oper. Res. **54**(1), 7–22 (1991)
3. Ferland, J.A., Michelon, P.: The vehicle scheduling problem with multiple vehicle types. J. Oper. Res. Soc. **39**(6), 577–583 (1988)
4. Golden, B., Assad, A., Levy, L., Gheysens, F.: The fleet size and mix vehicle routing problem. Comput. Oper. Res. **11**(1), 49–66 (1984)
5. Koç, Ç., Bektaş, T., Jabali, O., Laporte, G.: Thirty years of heterogeneous vehicle routing. Eur. J. Oper. Res. **249**(1), 1–21 (2016)
6. Li, H., Lim, A.: A metaheuristic for the pickup and delivery problem with time windows. Int. J. Artif. Intell. Tools **12**(02), 173–186 (2003)

7. Liu, F.-H., Shen, S.-Y.: The fleet size and mix vehicle routing problem with time windows. J. Oper. Res. Soc. **50**(7), 721–732 (1999)
8. Noumbissi Tchoupo, M., Yalaoui, A., Amodeo, L., Yaloui, F., Lutz, F.: Problême de collectes et livraisons avec fenêtres de temps et flotte homogène (2016)
9. Parragh, S.N., Doerner, K.F., Hartl, R.F.: A survey on pickup and delivery problems. J. für Betr. Swirtschaft **58**(1), 21–51 (2008)
10. Qu, Y., Bard, J.F.: The heterogeneous pickup and delivery problem with configurable vehicle capacity. Transp. Res. Part C: Emerg. Technol. **32**, 1–20 (2013)
11. Ropke, S., Pisinger, D.: An adaptive large neighborhood search heuristic for the pickup and delivery problem with time windows. Transp. Sci. **40**(4), 455–472 (2006)
12. Xu, H., Chen, Z.-L., Rajagopal, S., Arunapuram, S.: Solving a practical pickup and delivery problem. Transp. Sci. **37**(3), 347–364 (2003)

Designing the Municipality Typology for Planning Purposes: The Use of Reverse Clustering and Evolutionary Algorithms

Jan W. Owsiński, Jarosław Stańczak and Sławomir Zadrożny

Abstract The paper presents the preliminary results of a study, meant to determine the typology of the Polish municipalities (roughly 2500 in number), oriented at planning and programming purposes. An initial typology of this kind was elaborated by the geographers, based on a number of individual features, as well as location-related characteristics. This typology is "re-established" or "approximated" via the "reverse clustering" approach, elaborated by the authors, consisting in finding the parameters of the clustering procedure that yield the results the closest to the initial typology, for the set of municipalities, described by the definite set of variables. Altogether, one obtains the clusters (types, classes) of municipalities that are possibly similar to the original ones, but conform to the general clustering paradigm. The search for the clustering that is the most similar to the initial typology is performed with an evolutionary algorithm. The paper describes the concrete problem, the approach applied, its interpretations and conclusions, related to the results obtained.

Keywords Clustering · Reverse clustering · Municipalities · Typology · Planning Spatial planning

1 Introduction: The Generic Problem

Several decades ago the issue of "regionalisation", meaning the "optimum division of space into coherent parts for planning, programming and policy purposes" was very much on the scientific and political agenda. This was linked with both the emergence

J. W. Owsiński (✉) · J. Stańczak · S. Zadrożny
Systems Research Institute, Polish Academy of Sciences, Newelska 6, 01-447 Warsaw, Poland
e-mail: owsinski@ibspan.waw.pl

J. Stańczak
e-mail: Jaroslaw.Stanczak@ibspan.waw.pl

S. Zadrożny
e-mail: zadrozny@ibspan.waw.pl

© Springer Nature Switzerland AG 2018

P. Daniele and L. Scrimali (eds.), *New Trends in Emerging Complex Real Life Problems*, AIRO Springer Series 1,
https://doi.org/10.1007/978-3-030-00473-6_42

of the territorial planning and management, and the political need to address the grassroots initiatives. Among the methodologies tried out, clustering was one of the most frequent and important. Since that time the focal points changed a lot and the issue apparently ceased to be of such importance.

Yet, the very problem remained: if a country, like Poland, consists of, say, 2500 municipalities, highly differentiated in terms of area, population, socio-economic function and economic power—often by several orders of magnitude (!)—and largely self-governed, then it becomes extremely important to design the policies and strategies not for all of them, and not for each of them, but for their definite ("optimally designed") classes.

It is assumed that very similar, if not the same, policies, rules and strategies would be designed by the central authority, and negotiated with the body of the municipalities, for the so designed classes. So, one should account in this design for the essential socio-economic, demographic and resource-environmental aspects of the units considered.

Thus, notwithstanding the shifts in fashions and scientific focuses, studies are performed of delimitation of geographical areas that should be subject to diversified treatment in policy terms. An instance of the aspect that ought to be accounted for is the economic viability of the municipalities. This is also the case of Poland—a high share of remote rural municipalities cannot cope with ageing, poverty, lack of jobs, outflow of the young etc. At the other extreme are the dense urban cores, featuring also population outflow, but, simultaneously, increasing density and intensity of business and intellectual activity, with sky-rocketing real estate prices.

The question arises: are we able to assess and to improve upon the expert-provided spatial divisions, produced for policy purposes? What are the respective methodological proposals? How do they work?

2 The Concrete Question—And the Way to Answer It

2.1 *The Typology of Municipalities*

In the light of the above, we consider the typology of municipalities, elaborated for the territory of Poland by the specialists from the Institute of Geography and Spatial Organization of the Polish Academy of Sciences in Warsaw. This typology was elaborated in two versions, the basic one consisting of ten municipality classes, as shown in Table 1. Municipalities were classified into these categories, with the idea of applying diversified policies and strategies to them. Now: given that the future of millions of people in such units depends on these policies, and hence the classification, how can we assert it is "correct"?

Table 1 Functional typology of Polish municipalities

Functional types	Number of units		Population		Area	
	Number	%	in ‘000	%	‘000 km^2	%
1. Urban functional cores of provincial capitals	33	1.3	9557	24.8	4.72	1.5
2. Outer zones of urban functional areas of provincial capitals	266	10.7	4625	12.0	27.87	8.9
3. Cores of the urban areas of subregional centres	55	2.2	4446	11.6	3.39	1.1
4. Outer zones of urban areas of subregional centres	201	8.1	2409	6.3	21.38	6.8
5. Multifunctional urban centres	147	5.9	3938	10.2	10.39	3.3
6. Communes having pronounced transport function	138	5.6	1448	3.8	20.06	6.4
7. Communes having pronounced non-agricultural functions	222	9.0	1840	4.8	33.75	10.8
8. Communes with intensive farming function	411	16.6	2665	6.9	55.59	17.8
9. Communes with moderate farming function	749	30.2	5688	14.8	93.83	30.0
10. Communes featuring extensive development	257	10.4	1878	4.9	41.59	13.3
Totals for Poland	2479	100.0	38,495	100.0	312.59	100.0

Source Śleszyński and Komornicki [7]

2.2 *The Reverse Clustering Approach*

The proposal that we forward here is as follows: let us estimate, how precisely this expert-provided classification can be re-created with the possibly broadly conceived clustering approach, representing a systematic and methodologically coherent perspective. This is exactly what the “reverse clustering” approach is about (see, e.g. [5]). Namely, given a set of objects X, indexed i, $i = 1, \ldots, n$, represented by vectors x_i, $x_i = (x_{i1}, \ldots, x_{ik}, \ldots, x_{im})$, and its partition P into subsets (clusters) A_q, $q = 1, \ldots, p$, we wish to determine the clustering procedure that would yield the partition most similar to P. The clustering procedure that we seek is represented by a vector Z, including: (1) the algorithm applied (progressive merger, k-means, density, …), (2) the parameters of the algorithm (e.g. the cluster number in k-means), (3) the weights of variables, and (4) the distance function (as represented, say, by the Minkowski exponent).

Thus, having the initial partition P, we look for some P^*, produced by a clustering procedure, that would be most similar to P. The criterion we use is the basic or derivative of the Rand index (counting the pairs in the same cluster in P and in P^* and in a different cluster in the two partitions).

The search in the space of feasible Z is performed with an algorithm, developed by one of the authors [6], a two-level evolutionary procedure, where the operators applied (basically, crossover and mutation, with definite excursions allowed) are also subject to selection and evolution, depending upon their previous performance for a given thread of individuals. The choice of the evolutionary algorithm was made in view of the cumbersome landscape of optimisation (shape of the objective function and the constraints). At this stage, the issue of computational burden was of secondary importance.

The "optimum" clustering procedure found provides a rationale for the partition, indicating the particular instances of difference that should be considered in a particular manner. At the same time, it might be a confirmation for the initial partition P, even if definite divergences occur.

In the study here illustrated we referred to three basic, and most popular clustering paradigms, namely: k-means (see [4, 8]), classical hierarchical aggregation (see [2, 3]), and DBSCAN density algorithm (see [1]).

3 The Reverse Clustering Exercise and Its Preliminary Results

3.1 The Preliminaries

The analysis here presented in its preliminary stage, addresses the classification, mentioned before, of the Polish municipalities. Since the particular classes address millions of inhabitants, we would very much like to know, to what extent they can be reconstructed with the possibly well fitted clustering procedures, meaning that they can be—hypothetically—more objectively justified.

The study consisted in (1) establishment of the list of variables, characterising the municipalities, (2) the processing of values of these variables so as to facilitate the running of the consequent procedure, (3) running of the "reverse clustering" procedure, so as to minimise the gap between the partition P and the one obtained from the clustering procedure, P^*.

Table 2 shows the variables, selected on the substantive basis to characterise the municipalities. Given the complexity of the setting—the variables, the spatial circumstances, the criteria—no wonder there appears a wide discrepancy between the categorisations obtained, as shown in an exemplary contingency table, Table 3.

Besides the sheer complexity, there is a difference as to how the two typologies were produced—for the original one, the procedure involved several steps, at which various criteria were applied, some hardly directly quantifiable (e.g. the role of

Table 2 Variables describing municipalities, accounted for in the study

No.	Variable	No.	Variable
1	Population density	10	Share of registered employed persons
2	Share of agricultural land	11	Number of businesses per 1000 inhabitants
3	Share of overbuilt areas	12	Average employment in a business indicator
4	Share of forests	13	Share of businesses from manufacturing and construction
5	Share of population in excess of 60 years	14	Number of pupils and students per 1000 inhabitants
6	Share of population below 20 years	15	Number of students in above primary schools per 1000 inhabitants
7	Birthrate for the last 3 years	16	Own revenues of the municipality per capita
8	Migration balance rate for the last 3 years	17	Share of revenue from Personal Income Tax in municipal revenues
9	Average farm acreage indicator	18	Share of expenditures into social care in total of expenditures from municipal budget

Source Own elaboration

transport), the criteria differing among the municipality types, distinguished on the preceding steps of the procedure, so that one could easily expect the kind of results, illustrated in Table 3.

Table 3 (and that despite low coincidence with the initial partition: only 46%) confirms the general outline of the initial partition. So, out of 10 initial classes, 5 are represented by the “reconstructed” ones, gathering the majority of items from the initial ones (classes 1, 3, 4, 8, and 9), and 3 classes, containing the biggest shares (even if at a par with some other ones) of the items in the class (nos. 2, 6 and 7). The remaining two classes, 5 and 10, must be subject to separate investigations. The last one, no. 10, is replaced in the results by the single-item class (an outlier), an extremely rich, apparently peripheral unit, where a large-scale lignite strip mine is located, with accompanying power plant and other enterprises.

The (partial) confirmation of the general pattern of the original typology is given by the dispersion of units in the solution among the original types. So, e.g., the original type 1 is spread in the solution among the new types 1, 3 and 5, corresponding to urban centres. Very similarly, the original type 3 is spread among the new types of alike character, 3 and 5. Some doubts may arise when we look in the other direction, namely the distribution of units in the new types among the original ones (e.g. rural farming units falling—even if occasionally—together with the urban centres). It should be emphasised that the diversity of the rural areas in Poland (even the new type 10 put apart) is indeed very high, partly justifying the assignment of some of them to other types.

Table 3 Illustration of contingency between initial classification and clustering result—the best partition, obtained for the k-means algorithm

Classes: Obtained: Given:	1	2	3	4	5	6	7	8	9	10	Sum off diagonal
1	**21**	0	11	0	1	0	0	0	0	0	12
2	25	**88**	8	88	5	16	15	7	13	0	177
3	3	0	**35**	0	17	0	0	0	0	0	20
4	5	8	2	**103**	1	13	31	8	28	0	96
5	10	0	46	4	**46**	37	2	0	2	0	101
6	0	0	2	22	2	**34**	16	29	33	0	104
7	5	0	9	8	3	18	**104**	40	33	1	117
8	0	0	0	22	0	38	0	**376**	56	0	116
9	0	1	1	75	6	67	39	137	**339**	0	326
10	1	0	3	7	3	23	97	40	88	**0**	261
Sums:	70	97	117	329	84	246	306	637	592	**1**	2479

Source Own elaboration

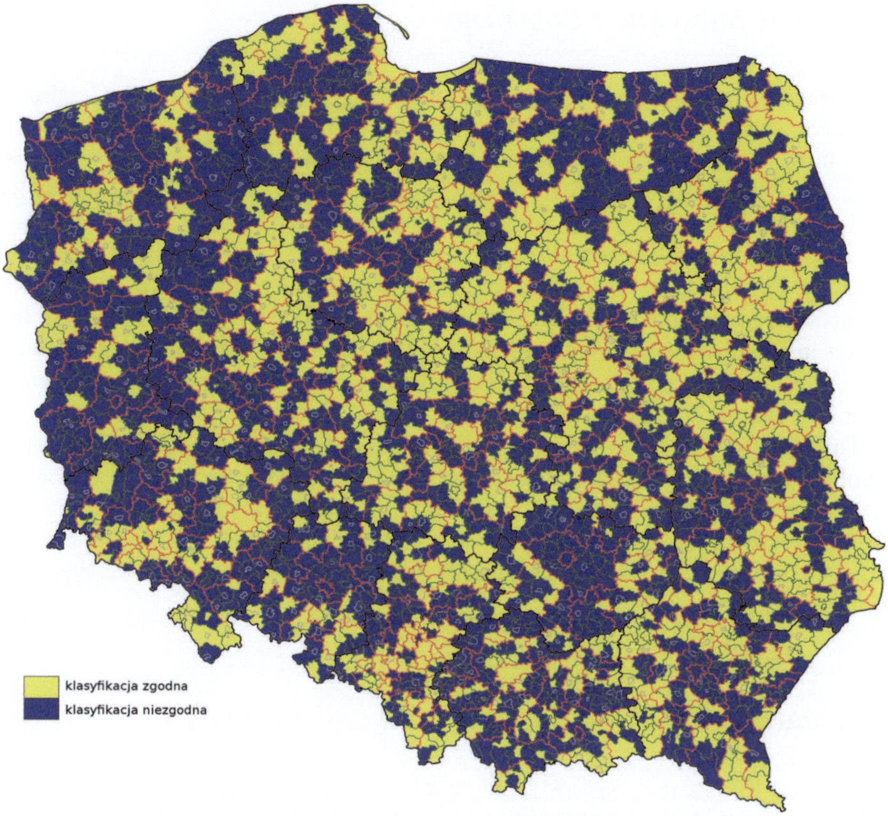

Fig. 1 Municipalities classified in the same types in original data and solution (yellow) and in different ones (blue) *Source* Own elaboration

Figure 1 shows the areas classified in the same types (yellow) and in different types (blue). The tendency towards formation of coherent areas is an indication of the consistent character of the results and the methodology applied (the data NOT containing any location aspect).

The results shown here come from the k-means-like algorithm, the general hierarchical aggregation algorithm (the Lance-Williams formula providing some of the parameters optimised) with DBSCAN also being tried out. The respective optimisation concerned altogether more than 20 variables, playing different roles in the optimisation exercise (e.g. variable weights vs. the cluster number or Lance-Williams parameters).

3.2 Some Substantive Conclusions

The results shown here provide a rich material for the substantive considerations. Take, e.g., the last class obtained in the solution—an outlier, as mentioned, constituted by a single municipality, with features quite exceptional on the national scale. Then, the initial class 5, definitely, highly "fuzzy", no wonder hardly unequivocally rendered by clustering.

In general, the clustering results confirm the overall character of the original typology, but also, very pungently indicate the potential doubts, or even need for corrections, in the initial classification. This is particularly well illustrated by the entries in Table 3, such as (row/column): 1/3, 2/4, or 5/3. If one compares these with the verbal definitions of the initial classes, the apparent confusion becomes quite clear.

This study shall continue efforts oriented at possibly faithful rendition of the initial typology through the use of additional variables, forming the vector Z, variations in the treatment of the objective function (Rand index as of now), and reasons for the divergences between the original and obtained classifications. It is hoped that this will also yield indications as to the direction of potential corrections to the typology of municipalities.

References

1. Ester, M., Kriegel, H.-P., Sander, J., Xu, X.-W.: A density-based algorithm for discovering clusters in large spatial databases with noise. In: Simondis, E., Han, J., Fayyad, U.M. (eds.) Proceedings of the Second International Conference on Knowledge Discovery and Data Mining (KDD-96), pp. 226–231. AAAI Press (1996)
2. Lance, G.N., Williams, W.T.: A generalized sorting strategy for computer classifications. Nature **212**, 218 (1966)
3. Lance, G.N., Williams, W.T.: A general theory of classification sorting strategies. 1. Hierarchical systems. Comput. J. **9**, 373–380 (1967)
4. Lloyd, S.P.: Least squares quantization in PCM. Bell Telephone Labs Memorandum, Murray Hill, NJ (1957); reprinted in IEEE Transactions Information Theory, IT-28 (1982), **2**, 129–137
5. Owsiński, J.W., Kacprzyk, J., Opara, K., Stańczak, J., Zadrożny, S.Ł.: Using a reverse engineering type paradigm in clustering: an evolutionary programming based approach. In: Torra, V., Dalbom, A., Narukawa, Y. (eds.) Fuzzy Sets, Rough Sets, Multisets and Clustering. Dedicated to Prof. Sadaaki Miyamoto. Studies in Computational Intelligence, vol. 671. Springer (2017)
6. Stańczak, J.: Biologically inspired methods for control of evolutionary algorithms. Control Cybern. **32**(2), 411–433 (2003)
7. Śleszyński, P., Komornicki, T.: Functional classification of Poland's communes (gminas for the needs of the monitoring of spatial planning (in Polish with English summary). Przegląd Geograficzny **88**, 469–488 (2016)
8. Steinhaus, H.: Sur la division des corps matériels en parties. Bulletin de l'Academie Polonaise des Sciences **IV**(C1.III), 801–804 (1956)

A Software for Production-Transportation Optimization Models Building

E. Parra

Abstract Corporations are using large mixed integer mathematical programming (MIP) models in strategic, medium term and short term planning. To build these models it is necessary a mathematical programming language and a set of optimizers. Some expert in the particular business must code the variables, constraints and objective function in the equations that reflect the actual problem. The person who do this must be both an expert in the field where company operates and a mathematical expert to write the mathematical model. A software to do this job easy for the planner (non-mathematical expert) is introduced. The software uses only some intuitive codes and data obtained from different sources. The purpose of this model builder software is to generate MIP supply chain optimization models. A previous version of the software has been used by large Spanish company for both medium term detailed planning and to analyze strategic investments.

Keywords MIP models · Mathematical programming software
Supply chain optimization

1 Introduction

Industrial size mathematical programming models for joint global optimization of supply, manufacturing and distribution operations is possible with today's personal computers and the availability of very good commercial optimizers. Large companies are reporting the use of MIP models. Only three examples as a sample: IBM [1], BMW [2] or the beverage industry [3]. The optimization software industry offers very good products to build and solve these models.

The main challenge is how to build such large models without an optimization expert. Above examples and many more needs a specialized expert or an expensive

E. Parra (✉)
Department of Economics, University of Alcala, Madrid, Spain
e-mail: enrique.parra@uah.es

© Springer Nature Switzerland AG 2018

P. Daniele and L. Scrimali (eds.), *New Trends in Emerging Complex Real Life Problems*, AIRO Springer Series 1,
https://doi.org/10.1007/978-3-030-00473-6_43

consultancy job to design the model, implemented with some language, i.e.: [4–9] for example, linked with an optimizer [10].

2 Supply Chain Optimisation

Large corporations optimize production and logistic decisions jointly. The size of the problem to be solved is huge: even a case with 10 factories, 5 products, 20 storage facilities, 50 clients, 12 months drives to a mode with thousand variables and constraints Also, it is common the existence of threshold quantities when we observe the actual cases: either an activity it's not been made or it has to be made beneath a minimum and finally in production its common the appearance of mixtures that must fulfil some properties. In that case semi-continuous variables appears in the mathematical formulation. To design this type of MIP models, each company must need expert help; the expert will use any of the commercial software available or will design an ad hoc software. Each change must be managed by the mathematical programming expert.

The software used can be categorized in two groups: optimizers and modelling languages. The first one, optimizers, solves the problem, typically in matrix format. The programmer is the responsible to fill the matrix accessing to databases and using the accepted formulation. Of course, the software could allow some flexibility in the problem formulation. The modelling languages help in model equations writing and, even, in the connections with databases that contain the necessary parameters for model building.

Table 1 shows a production and transportation optimization model written in an algebraic meta-language similar to commercial ones. Algebraic languages code the equations of a mathematical model … once written. The model is developed by an expert both in the "business" and mathematical modelling.

Then, the model is solved and the solution is translated to the practical planning decisions. It is necessary consultancy forever, because some tuning or model maintenance will be needed.

A different way to do this is introduced. The software described later builds the mathematical model using a set of intuitive codes instead of the explicit definitions of the elements of the model (variables, constraints …) to do the model building task easy for the planner. Follows the last version features of this software.

Table 1 Optimization formulation with an algebraic meta-language

```
TITLE OptPlan;
 INDEXES
     Product := (A1, A2, A3);  month:= (Jan, Feb, Mar, Apr);
     Fact := (p1, p2, p3, p4);  FactOrigin := Fact;  FactDest:= Fact;
 DATA
     SalePrice[product] := (120.00, 100.00, 115.00);
     Demand[Fact, product, mes] := DATAFILE("Demand.dat");
     CosProd[Fact, product]      := DATAFILE("CosProd.dat");
     RatProd[Fact, product]      := DATAFILE("RatProd.dat");
     DaysAvaProd [month] := (23, 20, 23, 22);
     CosInv[Fact, product] := DATAFILE("CosInv.dat");
     CapStock[Fact] := (800, 400, 500, 400);
     CostTra[FactOrigin, FactDest]:= DATAFILE ("CostTra.dat");
  VARIABLES
     Prod[Fact, product, month];  Stock[Fact, product, month];
     Sales[Fact, product, month];
     Shipping[product,month, FactOrigin, FactDest] WHERE (FactOrigin <> FactDest);
  COMPUTATIONS
        TotalRev  := SUM(Fact, product, month: SalePrice * Sales);
        TotalCosProd := SUM(Fact, product, month: CosProd* Prod);
        TotalCosInv := SUM(Fact, product, month: CosInv * Inventario);
        TotalCostShipping :=
         SUM(product,month, FactOrigin,FactDest: CostShipping * Shipping);
        TotalCost     := TotalCosProd + TotalCosInv + TotalCostShipping;
  MODEL
     MAX Profit= TotalRev - TotalCost;
  SUBJECT TO
    CapProd[Fact,Month]:SUM(product: Prod/RatProd) <=DaysAvaProd;
    Balance[Fact, product, month]:
      Prod + Stock[month-1]
     + SUM(FactOrigin:Shipping[FactOrigin, FactDest:=Fact])
     = Sales + Stock + SUM(FactDest:Shipping[FactOrigin:=Fact,FactDest]);
     MaxStock[Fact, month]: SUM(product: Stock)  <=  CapStock;
  LIMITS
    Sales < Demand;
```

3 MIP Models Building Software

3.1 Introduction

TPOS is the last version of a software that allows an easy way to represent, for a user who is not an expert in mathematical programming. The software can build a wide range of planning models for networks to maximize the production and distribution variable costs margin. The system generates and solves large industrial models (approx. 100,000 variables and 10,000 equations), and the model building process is based in attribute definition for nodes and arcs in a multiproduct and multi-period network.

A set of "codes" drive the model building phase, even the network structure. Depending on the content and number of keywords a different model is generated. User is autonomous: with the same software can afford different situations and purposes. TPOS allows building different models starting with some intuitive codes written by a person not necessary expert in mathematical programming but that it is an expert in the manufacturing and logistic system.

This software allows the generation of multi-product and multi-period models by the usage of networks where each node can be defined as a factory, a storage facility, a client, a transshipment facility…

3.2 Mathematical Model

The network is based in **N** nodes (external suppliers, factories, transshipment sites, destinations, customers) connected, by transport arcs (A), **P** products and **T** planning horizon periods (periods can be of variable length). They are the members, respectively, from NODES, PRODUCTS y PERIODS sets. Let i, i′∈ N (Nodes), j, j′ ∈ P (Products), k ∈ T (Periods).

The generated optimization model purpose is to find the optimal value of the following *variables*:

1. *Pr(i, j, k)*. Quantity produced of product j at node i during period k.
2. *TR(i, j, j′, k)*. Quantity transformed from product j into product j′ at node i during period k.
3. *X(i, l, j, k)*. Quantity (if it is generated as a continuous variable) transported from origin node i to destination node l of product i during period k, or (see below), if it is generated as binary variable, indicates than node l must be provided from node I for all products.
4. *CS(i, j, k)*. Quantity consumed (sales = demand satisfaction) of product j at node i during period k.
5. *OUT(i, j, k)*. Quantity of product j leaving node i during period k.
6. *EF(i, j, k)*. Quantity stocked at the end of period k, of product j at node i.
7. *MX(i, j, j′, k)*. Quantity of ingredient j to be used in the final product j′ by blending in period k at node i.

Objective function:

Maximize the variable margin: Revenues from product sales – Variable Costs (sum of supply, manufacturing, blending, stock and transportation costs)

TPOS generates mixed integer models for to two options: (1) X variables can be binary in order to force the model to select the same origin-destination arc for all the products required at a prefixed node type or (2) semi-continuous variables are used to force the outflows from a node to a threshold level or zero.

Follows the mathematical model built by TPOS and an example of how the user use codes to build it without an explicit equation formulation.

Constraints (keywords for model building are mentioned later, mathematical details are avoided):

1. Balance for Initial stock, productions, transformations, consumption, shipments and final stock for each node (i), product (j) and period (k).
2. Quantities Pr (i, j, k) are produced in node i, from a lower limit, for each product j and during each period k (keyword PROD) with a unitary cost. These quantities can be seen like supplies at the node. This is the only way to generate material in the model.
3. Lower limit and upper limit exists for joint production (sum for all products) for each period y node.
4. Lower and upper limit exists for a product production in the whole horizon and at each node.
5. Any product can be transformed into other with a yield, a unitary cost and lower and upper limits. Different limits can be included.
6. Any product can be stocked from a period to the next one; there exists stock limits for each pair product/node, and also a total for each node. At the beginning of the planning horizon, the nodes have an initial stock by product.
7. Sum of all product stocks at node i in each of the periods are limited.
8. Lower and upper stock limits for each node i, product j and period k.
9. Lower and upper demand limit, for each period, product and node; the revenues are the product of a price and the demand. This is the way to represent the sales. They are the only usual income from the model.
10. Lower and upper transportation limits for each arc (from node i to node i′), product j and period k. This is the way to model the transportation between nodes.
11. Lower and upper limits for the inflow to a node. To avoid very low shipping it is possible to impose a threshold level defining the outflow variable as a semi continuous variable.
12. If outflows are modelled as semi continuous variable, new variables must be generated, in addition to OUT(i, j, k) for the outflow from node, the model will include three auxiliary variables, SS1(i, j, k), SS2(i, j, k), BS(i, j, k); the last one it is binary (BS(i, j, k) $\in \{0, 1\}$) and some constraints are reformulated.
13. In case that X must be binary all the equations must be changed to substitute X by the term Demand (i, j, k). X(i, i′, j, k). In addition to this, new constraints are generated.

14. It is possible, as it was mentioned above, to force that all the products will be received in a node from the same origin. This is made by adding some constraints (remember X(i, i′, j, k) is binary).
15. Blending. It's possible to blend several products at a node to obtain other product with certain limits in properties. It's not allowed a successive blending (pooling problem) to avoid non-linear relations.

3.3 User Codes

For the end user, the system is a black box; the user feed it with data, very intuitive for him/her, and gets a solution in very easy format. This is the most important TPOS feature.

Table 2 shows the type of data and the keywords used to build a model. Table 3 shows an example of a model with the codes in Table 1.

The data for any type of model can be read from databases, spreadsheets, etc. Using this data, the model is generated and solved by the software [10]. Optimal solution can be analyzed using other software because solution is reported in user language or for be included in some database system (Multidimensional databases are recommended).

Table 2 Codes for defining models

1. General: "TITLE" (titles), "COSTALM" (Stock cost),
2. Network Structure: "NODES" (multiple node models), "PRODUCTS" (multiproduct), "PERIODS" (multiple period), "NET" (transportation arcs), “JOINTS” (all products together in the same arc)
3. Stocks: "STOC" (joint limits), "STOI" (initial stock and limits), "FS" (stock limits).
4. Production: "PROD" (limits and production cost), "PRODC", "PRODP" (limits)
5. Demand: "DEM" (limits and price for the sales)
6. Transport: "INF" (inflow limits), "OUT" (outflow limits, could be semi-continuous), "X" (arc/products/periods limits)
7. Conversions / Transformations: “TR” (feasibility, limits and costs for transformation from one product to another), “TR1”, “TR2”, “TR3”, “TR4” (other limits)
8. Blending: “MX” (possibilities, limits & costs of blend some products to get another), “MC”, “CA”, “CP”, (blending data)
9. Special constraints: "RE", "COE”. Any other constraint over the variables.

Table 3 A model coded in TPOS

Title y other data	TITLE", "Test" "tons" "euros"
	"CostAlm", 1
Products	"PRODUCTS",2
	"PrG","G"
	"PrE","E"
Periods	"PERIODS",2
	"Jan18","1",31
	"Feb18","2",28
Nodes	"NODES",19
. Factories	"Factory 1","FACT1", (Node parameters)
	"Factory 2","FACT2",(Node parameters)
. Warehouses	"Wareh 1"," Ware1 ",(Node parameters)
	"Wareh 2","Ware2",(Node parameters)
. Clients	"Client 01","CLI01",(Node parameters)
.	... definition of the 15 customers like nodes
	"Client 15","CLI15",(Node parameters)
Initial Stock	"STOI","******",(initial)
Stocks limits	"STOC","*****",(min , max)
	"EF" "*******" (min , max)
Production (total)	"PRODC","******",(min , max)
Production (joint)	"PRODP","******",(min , max)
Inflows Limits	"INF","*******",(min , max)
Outflow Limits	"OUT","*******",(min , max)
	"OUT","WARE2G1",(min , max)
Production limits	"PROD","*******",(min , max)
	"PROD","FACT1G*",(min , max)
	….rest of limits
Transformation PrG into PrE	
	"TR","FACT1GE*",(min , max, yield, cost)
	"TR","FACT2GE*",(min , max, yield, cost)
Customer demand	"DEM","*******",(min, max, price)
	"DEM","CLI01**",(min, max, price)
	.... rest of demands
	"DEM","CLI15**",(min, max, price)
Network and cost	"NET","FACT1","WARE1",(min, max, cost)
	"NET","FACT1","WARE2",(min, max, cost)
	... rest of transportation cost
	« NET","WARE2","CLI15",(min, max, cost)
Transportation contraint	"X" "*****CLI15**", (min, max)
	"X" "FACT1WARE2E1" , (min, max)

4 Conclusions

A software for MIP models generation is introduced. This software allows building (and solving linked with an optimizer) of mixed integer mathematical programming for optimal joint multilevel production and transport planning in large companies. It is possible to represent complex networks: several factories, storage facilities, clients, transportation arcs between them, several products on the network, blending and/or transformation between products in a node, several periods (days, weeks, months). The objective is to maximize the global variable margin of the firm.

This system has been used for factory supply planning, for factory to customer product distribution planning, for investment analysis, etc. always with the same tool. The models are built using easy codes instead of the mathematical formulation of equations.

References

1. Degbotse, A., Denton, B.T., Fordyce, K., Milne, R.J., Orzell, R., Wang, C.: IBM blends heuristics and optimization to plan its semiconductor supply chain. Interfaces **43**(2), 130–141 (2013)
2. Fleischmann, B., Ferber, S., Henrich, P.: Strategic planning of BMW's global production network. Interfaces **36**(3), 194–208 (2006)
3. Guimarães, L., Amorim, P., Sperandio, F., Moreira, F., Almada-Lobo, B.: Annual distribution budget in the beverage industry: a case study. Interfaces **44**(6), 605–626 (2014)
4. AIMMS (2018). https://aimms.com
5. AMPL (2018). www.ampl.com
6. CPLEX (2018). https://www.ibm.com/analytics/data-science/prescriptive-analytics/cplex-optimizer
7. GAMS (2018). https://www.gams.com
8. LINGO (2018). https://www.lindo.com
9. XPRESS (2018). http://www.fico.com/en/products/fico-xpress-optimization
10. Fourer, R.: Linear programming: software survey. OR MS Today Informs. **42**, 3 (2015)

Modelling Local Search in a Knowledge Base System

Tu-San Pham, Jo Devriendt and Patrick De Causmaecker

Abstract In this paper we present how the basic building blocks of local search approaches—problem constraints, neighbourhood moves, objective function, move evaluations—can be modelled declaratively using FO (·), an extension of first order logic. We extend the Knowledge Base System IDP with three built-in local search heuristics, namely first improvement, best improvement and tabu search, which take those building block specifications as input and execute local search accordingly. To demonstrate the framework, three neighbourhood moves for three different problems are modelled and tested.

Keywords Local search · Metaheuristics · Knowledge base system

1 Introduction

A Knowledge Base System (KBS) is a computer program that reasons and uses a general knowledge base to solve different complex problems. The IDP system [4] is an experiment on building such a KBS using the rich formal language FO (·) [7] for expression of the knowledge. FO (·) is an extension of first order logic with types, cardinality, arithmetics, and recursiveness to elegantly model of a wide range of concepts. The knowledge specified in FO (·) is called a *knowledge base* (KB), which IDP uses as input to a set of builtin *inference methods* to solve a large variety of tasks.

For example, given a KB on professors teaching courses, students studying, resources available and rules to be respected, the IDP system can generate schedules at the start of the year using the *model expansion* inference; verify correctness of hand-made and revised schedules using the *model checking* inference; generate a reason of why a certain schedule is not consistent using the *explain unsat* inference; proving invariants on the knowledge using the *theorem proving* inference—all using the same KB. The ability to solve many tasks distinguishes IDP from other program-

T.-S. Pham (✉) · J. Devriendt · P. De Causmaecker
KU Leuven, Leuven, Belgium
e-mail: san.pham@kuleuven.be

© Springer Nature Switzerland AG 2018

P. Daniele and L. Scrimali (eds.), *New Trends in Emerging Complex Real Life Problems*, AIRO Springer Series 1,
https://doi.org/10.1007/978-3-030-00473-6_44

ming languages or modelling systems, which are suited for solving only one task. IDP has been used in application domains such as interactive configuration [14, 15], machine learning and data mining [2] or access control systems [3].

Important classes of problems appearing in many applications are constraint satisfaction and combinatorial optimization, which also are the focus of this paper. IDP's inferences to solve such problems are *model expansion*, and its variant, *optimization*. The current back-end engine for these two inferences is MINISAT(ID) [5]. As a CP-SAT-based solver, it shows limited performance on typical operations research problems, such as the assignment problem [8], or on large real-world problems such as routing and scheduling. Local search on the other hand, is a well-studied combinatorial optimization technique, which has been succesfully applied to solve such problems.

In this paper, we specify local search neighbourhoods in FO $(\cdot)$ and extend the IDP system with a local search back-end. Doing so, we hope to arrive at a declarative, modular and reusable formalization of local search heuristics. Given a problem, as long as the problem's specifications and its necessary components (how to get valid moves, how to evaluate delta objective, how to create the neighbour solution given a current solution and a move) can be modelled in FO $(\cdot)$, the local search can be simulated in IDP.

The contribution of this work is two-fold. Firstly, formalizing local search neighbourhoods allows us to benefit from the existing functionality of the underlying system (in this case, IDP). E.g., we can automatically generate an initial feasible solution using the *model expansion* inference; we can debug neighbourhoods by using the *explainunsat* inference (which gives the reason why a solution is not feasible). Those benefits from the underlying system also distinguish this work from related work such as constraint-based local search systems (Comet [11], Oscar [12]) and Localsolver [1], which are particularly dedicated to solving optimization problems and are based on a set of built-in libraries of neighbourhood moves and combinators to construct local search heuristics for a set of problems. Secondly, the ultimate goal of a knowledge base system is to tackle all types of problems efficiently—including those for which the current state-of-the-art employs local search. Our work is a step towards introducing local search in a knowledge base environment.

The rest of the paper is organized as follows. Section 2 gives a brief introduction of IDP and FO $(\cdot)$. Section 3 shows how neighbourhood moves and move evaluations are formalized in FO $(\cdot)$, and how three local search heuristics are simulated using these formalizations. Experimental results are reported in Sect. 4. Conclusions and future work close the paper in Sect. 5.

2 IDP and FO $(\cdot)$

The IDP system [4] is a Knowledge Base System (KBS) equipped with (1) FO $(\cdot)$, a high level language which allows users to specify knowledge of a problem domain, and (2) a set of *inferences* to solve a wide range of problems around this knowledge.

A thorough introduction to IDP and FO $(\cdot)$, including examples and demos, can be found at https://dtai.cs.kuleuven.be/software/idp. Many of IDP's inferences are fundamentally different (e.g. theorem proving and model checking), therefore a specific solving approach is applied for each inference method. For *model expansion* and its variant *optimization*, IDP employs a two-phase *ground-and-solve* strategy where a FO $(\cdot)$ specification is grounded to a set of constraints in Extended Conjunctive Normal Form (ECNF) [6], which is then handled by MINISAT(ID) [5], a CP-SAT-based solver.

A specification (or modelling) in IDP consists of different components. The three most important components are *vocabularies* specifying the types and variables used, *theories* specifying problem constraints in FO $(\cdot)$, and *structures* representing both input data and solutions.

As a running example we employ the travelling salesman problem (TSP). Given a set of nodes and a distance function between them, the TSP consists of finding the shortest Hamiltonian cycle—the *tour*—visiting all cities. Here, we give a model for the TSP in IDP which we also made available online at http://goo.gl/TTv85c.

Example 2.1 (TSP)

The four components for the TSP modelling in IDP are as follows:

1. The vocabulary $Vproblem$ specifying the parameters ($Node$, $Distance$) and the variables ($Path$, $Reachable$) of the problem.
2. The theory $Tproblem$ over $Vproblem$, which is a conjunction of FO $(\cdot)$ formulas specifying the constraints of the problem:

$$\forall x: \exists! y: Path(x, y).$$
$$\forall x: \exists! y: Path(y, x).$$
$$\{\ Reachable(Depot).$$
$$Reachable(x) \leftarrow \exists y: Reachable(y) \wedge Path(y, x).\}$$
$$\forall x: Reachable(x).$$

   The first two lines of the theory in the example represent the flow constraints. Line three and four inductively define the $Reachable$ predicate, starting from the depot, and inductively adding neighbouring nodes according to the links present in $Path$. The last line then states that all nodes must belong to $Reachable$, forming a subtour elimination constraint—if a node is not reachable from the depot, it belongs to a subtour different from the one containing the depot.
3. The term Obj over $Vproblem$ representing the total travelling distance and serving as the objective function.
4. The (partial) structure S over $Vproblem$ describing type and parameter values. In S, the assignment to problem parameters (in this case, $Node$ and $Distance$) represents an instance of the problem, while assignments to variable symbols (in this case, $Path$ and $Reachable$) could form a solution for that instance. IDP distinguishes between parameters and variables simply by the fact that variables are unassigned in the initial structure.

Given these components, the inference *model expansion* can be applied to expand a (partial) structure S (an input instance) into a (complete) structure respecting the constraints in theory $Tproblem$ (a feasible solution); the inference *minimization* can be applied to obtain the minimal solution according to the objective function Obj. Besides, users can specify a *query* and apply *querying* inference to retrieve information from the model.

3 Modelling Local Search

Local search is a heuristic method to solve an optimization problem by moving from solution to solution in the search space using *neighbourhood moves*. A neighbourhood move is a function which maps a solution to another. Given a set of neighbourhood moves and some initial feasible solution, a local search algorithm explores the search space by enumerating the neighbours. When a neighbour satisfies an acceptance criterion, it becomes the starting point of a new iteration. This loop ends when a stop condition is met, and the solution with the best objective value is returned. Metaheuristics are local search-based heuristics, which guide a subordinate heuristic to escape from local optima.

In this section, the formalization of local search heuristics' building blocks in FO $(\cdot)$ is presented. We further describe how IDP is extended with three local search heuristics—first improvement, best improvement and tabu search.

3.1 Modelling Local Search Neighbourhood Moves

To model local search neighbourhood moves for a problem, several components are added to the problem specification. The essential part of the modelling is the function mapping a current solution to the corresponding neighbour solution given a neighbourhood move. This function is modelled in a theory *Tnext* which is built over two vocabularies *Vmove* representing a move and *Vnext* representing a neighbour solution. A query *queryGetDeltaObj* is added to allow a user-defined move evaluation. Valid moves are obtained by a query *queryGetValidMoves*. The above components are illustrated using the running example of the TSP as follows.

Example 3.1 In this example, we model the *2-opt* move for the TSP. In order to model the move, the problem specification is extended by introducing the predicate $Reach$ to represent the order of nodes in the solution. For example, $Reach(A, C)$ holds means that on the path starting from the depot, node A appears before C. To represent moves and neighbour solutions, two vocabularies $Vmove$ and $Vnext$ are added to the modelling. $Vmove$ consists of 4 constants A, B, C and D representing the four nodes involved in the 2-opt move—(A, B) and (C, D) are two edges in the solution in this specific order. $Vnext$ consists of the predicate $next\_dec\_Path(Node, Node)$, which represents a neighbour solution. The function mapping a current solution to its neighbour is defined in theory *Tnext* as below:

$$\{next\_dec\_Path(A, C).$$
$$next\_dec\_Path(B, D).$$
$$next\_dec\_Path(x, y) \leftarrow dec\_Path(x, y) \wedge Reach(y, A) \wedge y = minimum(Node).$$
$$next\_dec\_Path(x, y) \leftarrow dec\_Path(x, y) \wedge Reach(D, x) \wedge D = minimum(Node).$$
$$next\_dec\_Path(y, x) \leftarrow dec\_Path(x, y) \wedge Reach(B, x) \wedge Reach(y, C) \wedge y \neq minimum(Node).\}$$

Theory *Tnext* consists of a definition describing how to create a neighbour solution given a solution and a move. Roughly, it states that the new solution consists of 2 edges (A, C) and (B, D), the segment from the depot to A, the segment from D to the depot and the reversed segment from B to D. By applying *model expansion* on *Tnext* and a structure with assignments to dec_Path, $Reach$ and A, B, C, D, an interpretation of $next\_dec\_Path$, which represents a neighbour solution.

To complete the modelling, two queries are declared. Query $queryGetDelObj$ evaluates a move by calculating the delta between the total travelling time of the current solution and its neighbour: $\Delta = d_{AC} + d_{BC} - (d_{AB} + d_{CD})$. Query $queryGetValidMoves$ gets all valid moves from a given solution.

3.2 Metaheuristics Framework

In this section, we describe how IDP takes formal neighbourhood move specifications as input and combines them with the built-in heuristics to perform local search. The common mechanism is as follow. Firstly, an initial feasible solution is obtained using IDP's model expansion on *Tproblem*. Next, a set of valid moves is computed by solving the query *queryGetValidMoves* on the current solution. For each valid move, the query *queryGetDelObj* is applied to evaluate the move. If the move is accepted, the corresponding neighbour is computed by performing model expansion on $Tnext$. The obtained neighbour solution now functions as the current solution whose neighbourhood is further investigated in the next iterations.

As already mentioned, three local search heuristics are provided: first improvement, best improvement and tabu search. First improvement search and best improvement search are two simple heuristics which iterate through all valid moves and proceed with a move once the first improvement or the best improvement is found. Tabu search [9] is a more advanced metaheuristic which keeps a list of forbidden moves to avoid revisiting a recently visited solution. These heuristics take the neighbourhood move specifications described in Sect. 3.1 as input and run the corresponding local search heuristic accordingly. By way of illustration, we describe the first improvement local search procedure in Algorithm 1.

Given a problem, as long as the problem along with the necessary components can be modelled in FO $(\cdot)$, a local search can be simulated using the framework.

Algorithm 1: First improvement search

```
input  : instance S
output: solution bestSol
iniS ← model expansion on (S, Tproblem);
curSol ← iniS;
bestSol ← iniS;
repeat
    moves ← evaluate queryGetValidMoves on curSol;
    foreach move ∈ moves do
        delObj ← evaluate queryGetDelObj on move and curSol;
        if delObj < 0 then
            neighbour ← create new solution from move and curSol by applying model
            expansion on Tnext;
            bestSol ← neighbour;
            curSol ← neighbour;
            break;
        end
    end
until no improvement found or timeout;
```

4 Experiments

The main purpose of the experiment is to demonstrate how the framework is utilized to model and solve different problems. We model and test three different neighbourhood moves for three different problems as follows: (1) The TSP and the 2-opt move neighbourhood as presented in Sect. 3; (2) *The assignment problem* which consists of finding a bijection between a set of agents and a set of tasks that minimizes the assignment cost. A simple neighbourhood with moves that swap the assignment of two agents is modelled; (3) *The colouring violations problem* which consists of finding a graph colouring which minimizes the number of adjacent nodes sharing the same colour. The neighbourhood consists of moves that assign a different colour to a node. The modelling and executable code for all problems in this section can be found at http://github.com/tusanpham/LocalSearchInIDP.git.

We modelled each of these problems and neighbourhoods in FO $(\cdot)$. Those specifications are then taken as input by IDP to perform the three built-in local search heuristics (first improvement, best improvement and tabu search). Additionally, for each problem, we compare the results with IDP's builtin minimization inference.

Instances for the TSP were obtained from the TSPLIB [13], instances for the assignment problem were taken from [8], and instances for the colouring violations problem were taken from Michael Trick's operations research page [10]. A time limit of 300 seconds were imposed on all experiments. The experiment was conducted on an Intel Core i7-5600 cpu with 8GiB of RAM, running Ubuntu 14.04 64-bit. The IDP system we extended was IDP 3.6.

In Table 1, results of some representative instances from the three problems are reported. The first two columns show the best objective value and running time

Table 1 Results of the three test problems

	Instance	IDP minimization		Best improvement		First improvement		Tabu search	
		Best sol	Time	Best sol	Time	Best sol	Time	Best sol	Time
TSP	br17.atsp	39	301.88	39	3.83	39	4.37	39	3.66
	eil51.tsp	995	301.98	**441**	46.94	461	40.83	**441**	51.39
	ft70.atsp	64,305	302.12	47,078	300.11	61,173	352.98	**46,155**	300.77
	ftv33.atsp	2433	301.97	1658	300.02	2047	301.38	**1577**	300
	pr76.tsp	516,157	301.93	**113,187**	195.61	114,353	124.28	**113,187**	200.45
	rbg443.atsp	–	–	**8169**	399.09	8126	373.1	**8169**	403.96
	swiss42.tsp	2420	301.74	**1378**	35.48	1418	40.34	**1378**	35.56
Assignment	21	**52**	300	65	40.19	63	39.6	60	301.92
	25	67	300	76	83.96	**60**	57.3	**60**	304.29
	29	142	300	95	97.93	89	81.76	**74**	303.96
	33	127	300	**90**	176.47	95	148.89	**90**	301.02
	37	337	300	111	250.48	141	168.34	**103**	305.07
Colouring	anna.col	**2**	300	342	301.71	118	102.96	304	333.27
	homer.col	**641**	29.62	999	300	815	302.42	999	300
	le450_25a.col	**533**	300	1000	300	694	308.341	1000	300
	miles1500.col	8344	300	8967	300	**3032**	305.971	8868	300
	mulsol.i.4.col	**246**	300	917	308.01	384	303.383	917	300.06
	queen7_7.col	**0**	118.59	6	154.73	6	115.004	**0**	303.03
	zeroin.i.3.col	537	300	854	307.96	**286**	306.847	854	306.21

of IDP's minimization while the next columns contain results obtained by the best improvement, first improvement and tabu search respectively.

We firstly discuss the results on the TSP problem. It is apparent from the table that all three local search heuristics outperform IDP's minimization routine, especially in big instances. This suggests that at least for the TSP, a general solver can benefit from domain knowledge which can be modelled easily using a descriptive language, as we did with the 2-opt neighbourhood.

The results of the assignment problem show a clear difference between the three heuristics. Tabu search outperforms the remaining two heuristics on most instances, which is easily explained by the strong mechanism of tabu search to prevent repeating moves. This result highlights the benefit of modularity offered by our framework. Given a single neighbourhood formulation, it is easy to experiment with different heuristics before committing to the best one.

Finally, the results from the colouring violations problem, in contrast to the two above problems, show a win for the IDP's minimization routine over the three heuristics. This can be explained by the CP-SAT backend behind IDP's minimization being more suitable for this particular problem compared to local search with a simple swap move. This shows the benefit of having several back-ends in a single engine.

In the TSP, besides performing local search, we also experiment the *explain unsat* reference to reason why a solution is not valid. The inference works well on instances up to 78 nodes and gives a readable explanation on which variable assignments lead to constraints violation. The usage of *explain unsat* is available online at http://goo.gl/U4hvkf.

5 Conclusions and Future Work

In this paper, we describe our work on a local search framework within the Knowledge Base System IDP using the formal language FO $(\cdot)$. Three local search heuristics are provided, taking a formal description of problem's constraints, neighbourhood moves, objective function, and move evaluations as the input. The framework is demonstrated through three different problems in the experiment section. The experiment is a proof-of-concept on modelling and automatically exploiting neighbourhoods of different problems, and gauges the performance of our proof-of-concept. It is also interesting that we have improved the IDP's performance on two problems (TSP and assignment problems) using very simple local search modelling. It shows that we can improve a general solver by supplying it with neighbourhood specifications.

Regarding future work, we plan to propose a complete framework which allows specifying multiple neighbourhoods and combining them to create more sophisticated local search heuristics. In addition, the use of a formal language in our work enables the potential of applying formal methods to local search. For example, an automatic theorem prover can be applied to prove the correctness of neighbourhoods.

References

1. Benoist, T., Estellon, B., Gardi, F., Megel, R., Nouioua, K.: Localsolver 1. x: a black-box local-search solver for 0–1 programming. 4OR: Q. J. Op. Res. **9**(3), 299–316 (2011)
2. Bruynooghe, M., Blockeel, H., Bogaerts, B., De Cat, B., De Pooter, S., Jansen, J., Labarre, A., Ramon, J., Denecker, M., Verwer, S.: Predicate logic as a modeling language: modeling and solving some machine learning and data mining problems with idp3. Theory. Pract. L. Program. **15**(6), 783–817 (2015)
3. Cramer, M., Van Hertum, P., Ambrossio, D.A., Denecker, M.: Modelling delegation and revocation schemes in IDP. arXiv:1405.1584 (2014)
4. De Cat, B., Bogaerts, B., Bruynooghe, M., Janssens, G., Denecker, M.: Predicate logic as a modelling language: the IDP system (2016). http://arxiv.org/abs/1401.6312v2
5. De Cat, B., Bogaerts, B., Devriendt, J., Denecker, M.: Model expansion in the presence of function symbols using constraint programming. In: 2013 IEEE 25th International Conference on Tools with Artificial Intelligence, Herndon, VA, USA, 4–6 November 2013, pp. 1068–1075. IEEE Computer Society (2013). http://dx.doi.org/10.1109/ICTAI.2013.159
6. De Cat, B.: Separating knowledge from computation: An FO(·) knowledge base system and its model expansion inference. Ph.D. thesis. KU Leuven, Leuven, Belgium (2014)
7. Denecker, M., Vennekens, J.: Building a knowledge base system for an integration of logic programming and classical logic. In: M. García de la Banda, E. Pontelli (eds.) ICLP, LNCS, vol. 5366, pp. 71–76. Springer (2008). http://dx.doi.org/10.1007/978-3-540-89982-2_12
8. Devriendt, J.: Exploiting symmetry in model expansion for predicate and propositional logic. Ph.D. thesis, Informatics Section, Department of Computer Science, Faculty of Engineering Science (2017). https://lirias.kuleuven.be/handle/123456789/564687. Denecker, Marc (supervisor)
9. Glover, F., Laguna, M.: Tabu search. In: Handbook of Combinatorial Optimization, pp. 3261–3362. Springer (2013)
10. Michael trick's operations research page. https://mat.gsia.cmu.edu/COLOR/instances.html (2017)
11. Michel, L., Van Hentenryck, P.: The Comet programming language and system. In: P. van Beek (ed.) CP, LNCS, vol. 3709, pp. 881–881. Springer (2005)
12. OscaR Team: OscaR: Scala in OR (2012). https://bitbucket.org/oscarlib/oscar
13. TSP instance library. www.iwr.uni-heidelberg.de/groups/comopt/software/TSPLIB95/ (2017)
14. Van Hertum, P., Dasseville, I., Janssens, G., Denecker, M.: The KB paradigm and its application to interactive configuration. Theory. Pract. L. Program. **17**(1), 91–117 (2017)
15. Vlaeminck, H., Vennekens, J., Denecker, M.: A logical framework for configuration software. In: Proceedings of the 11th ACM SIGPLAN conference on Principles and practice of declarative programming, pp. 141–148. ACM (2009)

A Hybrid Method for Cloud Quality of Service Criteria Weighting

Constanta Zoie Rădulescu and Marius Rădulescu

Abstract The Multi-Criteria Decision Making (MCDM) methods can be used for selection of a Cloud Services Provider (CSP). The most critical input of these methods is the assignment of criteria weights which can be based on subjective, objective, or a combination of weighting methods. In this paper a new hybrid method is proposed for Quality of Service (QoS) criteria analysis and weighting. The approach is based on a subjective weighting method and an objective weighting method. The hybrid method is applied in a case study. An analysis of causal relations and the degree of influence between QoS criteria based on DEMATEL method is presented.

Keywords Subjective weighting · Objective weighting · DEMATEL method Quality of service · Cloud service provider

1 Introduction

The services are made available by the Cloud Services Providers (CSPs) with a variety of Quality of Service (QoS) attributes (criteria). For analysis of the benefits and risks in adoption a CSP the analysis of QoS metrics plays an important role [1–3].

A decision to select a CSP, taking into account the user requirements is a multi-criteria problem. The set of alternatives is the set of CSPs and the set of criteria is the set of QoS criteria. In MCDM approaches, weights of criteria reflect the relative importance of criteria in the decision making process. The most critical input to the most MCDM methods is the assignment of criteria weights which can be based on subjective, objective, or a combination of weighting methods.

C. Z. Rădulescu (✉)
National Institute for R&D in Informatics, Bucharest, Romania
e-mail: radulescucz@yahoo.com

M. Rădulescu
Institute of Mathematical Statistics and Applied Mathematics, Bucharest, Romania
e-mail: mradulescu.csmro@yahoo.com

© Springer Nature Switzerland AG 2018

P. Daniele and L. Scrimali (eds.), *New Trends in Emerging Complex Real Life Problems*, AIRO Springer Series 1,
https://doi.org/10.1007/978-3-030-00473-6_45

Subjective weighting methods are based on the experts (decision makers) opinion while the emphasis of the objective methods is on the statistical evaluation of data provided by the decision matrix. The approach based on the objective weighting is particularly applicable for situations where reliable subjective weights cannot be obtained [4, 5]. Since in the most real problems, the decision maker's expertise and judgment should be taken into account, subjective weighting may be preferable. In the situations when such reliable subjective weights are difficult to be obtained, the use of objective weights is useful [6]. Each of these methods has its own advantages and disadvantages. The uncertainty in the decision maker judgments is the main disadvantage of the subjective methods, while the objective methods do not benefit from the expertise and experience of the experts [7]. Subjective and objective weighting methods are compared in many MCDM problems. Examples of subjective weighting methods are [7]: AHP method [8], Weighted Least-Square Method [9], Delphi method [10], Simple Multi-attribute Rating Technique (SMART) [11], SMARTER [12] and Simos' procedure [13], [14]. Examples of objective weighting criteria methods are: the entropy method [15], Shannon's entropy concept [16], multiple objective programming [17], principal component analysis [18] and the mean square deviation method [19].

In this paper a hybrid method is proposed for Quality of Service (QoS) criteria analysis and weighting. The hybrid method is a combination of a subjective weighting method—Decision-Making Trial and Evaluation Laboratory (DEMATEL) method and an objective weighting method—Shannon method. Based on cloud expert' evaluation, the initial QoS criteria weights are calculated with DEMATEL method. DEMATEL method divides the set of QoS criteria in cause and effect groups and builds the Influential Network Relation Map (INRM). Based on INRM an analysis of the important QoS criteria influencing the adoption of a cloud service can be achieved. The second method that is the Shannon method takes into account the criteria weights computed with DEMATEL and the decision matrix of CSPs evaluated by QoS criteria. The Shannon method uses the concept of information entropy. The final weights of QoS criteria are obtained. These QoS criteria weights can be used in the selection problem of a CSP. Finally the hybrid method is applied in a case study. An analysis of causal relations and the degree of influence between QoS criteria is presented.

2 The Hybrid Method for QoS Criteria Weighting

The hybrid method is a combination of DEMATEL, a subjective weighting method and Shannon method an objective weighting method.

The DEMATEL method was developed by the Science and Human Affairs Program of the Battelle Memorial Institute of Geneva between 1972 and 1976. It takes into account the subjective evaluation of experts and solves complex and interrelated problems [20, 21]. The aim of DEMATEL is to reveal direct/indirect causal relations (dependencies) among system variables.

The DEMATEL method compares the interaction relationship between variables and uses a matrix to calculate the direct and indirect causal relationships and the degree of influence between variables, especially using the Influential Network Relation Map (INRM) to express the causal relationships and the degree of influence between variables [22]. The DEMATEL method can solve problems visually and can isolate the related variables into cause and effect groups in order to improve the understanding of the causal relationships among these variables [23].

One of the objective weighting methods which has been proposed by researchers is the Shannon entropy method [16]. The entropy concept is used in various fields of science. The concept of Shannon's entropy has an important role in information theory and is used as a general measure of uncertainty [6]. In MADM the entropy weight describes how much different alternatives approach one another in respect to a certain criterion. The greater is the value of the entropy, the smaller is the entropy weight, then the smaller are the differences between different alternatives in this specific criterion, and the less information the specific criterion provides, and the less important this criterion becomes in the decision making process [5].
The hybrid method steps are:

Step 1. Selection of a set of m QoS criteria and the set of n CSPs.
Step 2. The measure scale for the relationship between m QoS criteria is considered across five levels. The scores 0, 1, 2, 3, and 4 represent 'no influence,' 'low influence,' 'medium influence,' 'high influence,' and 'very high influence,' respectively. For the QoS criteria selected a questionnaire for evaluation is defined.
Step 3. The cloud expert is asked to make sets of comparisons between QoS criteria from the selected set. The direct matrix is build:

$$A = \left(a_{ij}\right), i = 1, 2, \ldots, n, \; j = 1, 2, \ldots, n.$$

Step 4. Build the $m \times m$ normalized direct relation matrix N. N is obtained by normalizing matrix A. In N all principal diagonal elements are equal to zero:

$$s = \max\Big(\max_{1\le i\le m} \sum_{j=1}^{m} a_{ij}, \max_{1\le j\le m} \sum_{i=1}^{m} a_{ij}\Big), \quad N = A/s$$

Build the $m \times m$ QoS criteria total influence matrix $T = (t_{ij})$ as: $T = N(I - N)^{-1}$ where I is an identity matrix. Note that: $\lim_{h\to\infty} N^h = (0_{m\times m})$.
Step 5. Construct the QoS criteria Influential Network Relation Map (INRM) with the help of the vectors $\mathbf{y} = (y_1, y_2, \ldots, y_n)$ and $\mathbf{z} = (z_1, z_2, \ldots, z_n)$:

$$z_i = \sum_{j=1}^{m} t_{ij}, \quad i = 1, 2, \ldots, m \; y_j = \sum_{i=1}^{m} t_{ij}, \quad j = 1, 2, \ldots, m$$

The vector z is the sum of all columns vectors of T. The i-th entry of $\mathbf{z}$ represents the degree the i-th QoS criterion influences all other QoS criteria. The vector y is the sum of all vector rows of T. The j-th entry of $\mathbf{y}$ represents the degree the j-th QoS criteria are influenced by all other QoS criteria. The sum $(z_i + y_i)$ represents the degree of influence between the i-th criterion and the other QoS criteria. The difference $(z_i - y_i)$ represents the degree of causality between the i-th criterion and the other QoS criteria. If $(z_i - y_i)$ is positive, then the i-th QoS criterion influences other QoS criteria and shall be attributed as a cause type. If $(z_i - y_i)$ is negative, then the i-th QoS criterion is influenced by other QoS criteria and shall be attributed as an effect type. If $(z_i - y_i)$ has a negative value and the value of $(z_i + y_i)$ is very small, it means that the i-th QoS criterion is more independent, and there are less factors which will impact the QoS criterion. When $(z_i - y_i)$ has a positive value and the value of $(z_i + y_i)$ is very small, it means that the i-th QoS criterion is also independent, and can impact a few other QoS criteria.

If $(z_i - y_i)$ has a negative value and the value of $(z_i + y_i)$ is very big, it means that QoS criterion i is the core problem required to be solved. When $(z_i - y_i)$ has a positive value and the value of $(z_i + y_i)$ is big, it means that QoS criterion i is the driving factor of solving the core problem, and shall be listed as the priority.

Step 6. Build the DEMATEL QoS criteria weights as:

$$w_i^D = (z_i + y_i) / \sum_{j=1}^{m} (z_j + y_j), \quad i = 1, 2, \ldots, m$$

Step 7. The cloud expertis asked to make CSPs assessments using QoS criteria and is build an $n \times m$ average decision matrix $E = \left(e_{ij}\right)$.

Step 8. The normalized decision matrix is calculated:

$$P = (p_{ij}^w), \quad p_{ij}^w = e_{ij} / \sum_{k=1}^{n} e_{kj}, \quad i = 1, 2, \ldots, n$$

Then e_j^w is calculated: $e_j^w = -\frac{\sum_{i=1}^{n} p_{ij}^w \ln(p_{ij}^w)}{\ln(n)}, \ 0 \le e_j^w \le 1$

Define the criteria weight based on entropy concept:

$$w_j^o = \frac{(1 - e_j^w)}{\sum_{k=1}^{n} (1 - e_k^w)}, \quad j = 1, 2, \ldots, m$$

Step 9. The final form of the entropy weights is calculated:

$$w_j = \frac{w_j^D * w_j^o}{\sum_{k=1}^{m} w_j^D * w_j^o}, \quad j = 1, 2, \ldots, m$$

3 Case Study

The cloud QoS criteria are selected by an expert in the field of Cloud Computing. The QoS criteria are: availability, security, VM Cost, Service Response Time and usability. This QoS criteria selected are quantitative and qualitative criteria. The quantitative criteria are: VM cost and Service Response Time and. The qualitative criteria are: availability, security and usability. The expert's evaluation is made based on a DEMATEL method measure scale. The QoS criteria considered for evaluation and analysis are presented in Table 1.

The initial direct relation/influence matrix A is obtained. Build the normalized direct relation matrix N. The initial direct relation/influence matrix A and normalized

Table 1 QoS criteria

Nr. crt.	QoS criteria	Symbol	Description	QoS criteria type	Measure unit
1	Availability	C1	The time that a service or system is available. It is time a system or component is functional to the total time it is required or expected to function	max	0–10
2	Security	C2	Security criteria indicate the effectiveness of a CSP in controlling access to services, service data and physical facilities from which services are provided	max	0–10
3	VM Cost	C3	The cost of virtual machines	min	\$/h
4	Service response time	C4	The service response time is defined as the time it takes for any workload to place a request for work on the virtual environment and for the virtual environment to complete the request	min	s
5	Usability	C5	The level of support in using the service provided by the CSP. Support level is the extent of technical assistance provided for CSP to its customers	max	0–10

Table 2 The elements of INRM

QoS criteria	C1	C2	C3	C4	C5
z	3.439	2.783	3.164	2.574	1.202
y	2.219	2.567	3.101	2.541	2.735
z − y	1.220	0.216	0.064	0.033	−1.533
Impact	Cause	Cause	Cause	Cause	Effect
z+y	5.659	5.350	6.265	5.116	3.938
DEMATEL QoS criteria weights	0.215	0.203	0.238	0.194	0.150

matrix N are calculated. Then the QoS criteria total influence matrix T is build and the QoS criteria INRM (Fig. 1, Table 2).

The arithmetic mean of the array of the vector $z - \boldsymbol{y}$ is 0 and the arithmetic mean of the array of the vector $z+\boldsymbol{y}$ is 5.265. Starting from the means 0 and 5.265, the horizontal and vertical lines, divide the INRM into four quadrants, shown in Fig. 1. The horizontal axis $z+\boldsymbol{y}$ represent the prominence and the vertical axis $z - \boldsymbol{y}$ represents the relation. According to the analysis of Fig. 1, the high prominence and relation in first quadrant are C1 (Availability), C2 (Security) and C3 (VM cost). These QoS criteria have the highest interaction influence degree with other QoS criteria, and thus, they are the driving factors. The fourth quadrant has high prominence but low relation.

C4 (Service Response Time) criterion is in the second quadrant (low prominence but high relation). However, C4 criterion is very close to the fourth quadrant.

Other QoS criterion, C5 (Usability) is in third quadrant, with low prominence and relation. In the fourth quadrant there is no criterion.

Fig. 1 Influential network relation map—INRM

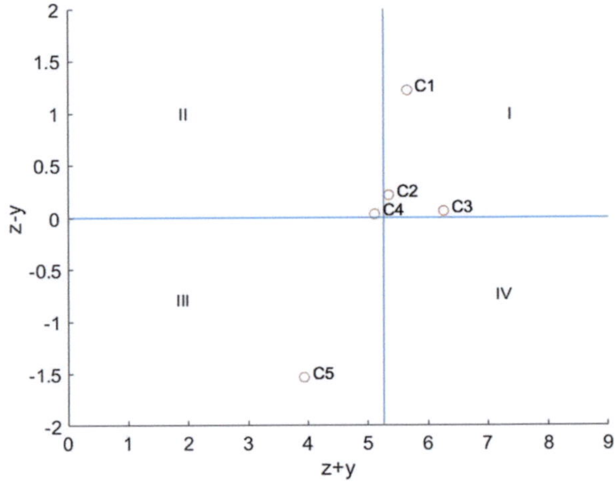

Table 3 Final QoS criteria weights

QoS criteria	DEMATEL QoS criteria weights	Shannon QoS criteria weights	Final QoS criteria weights
C1	0.215	0.072	0.072
C2	0.203	0.026	0.025
C3	0.238	0.456	0.506
C4	0.194	0.412	0.373
C5	0.150	0.034	0.024

The CSPs are selected for evaluation. The decision matrix E is build. The normalize decision matrix and Shannon QoS criteria weights are calculated.

Then the final form of the QoS criteria weights are calculated according to the Step 9 (Table 3).

The criterion with the highest weight in the DEMATEL subjective weighting method is C3 (VM cost). The second criterion is C1 (Availability). The criterion with the highest weight in the Shannon objective weighting method is C3—VM Cost and the second is C4—Service Response Time.

The method that combines the weights obtained by the DEMATEL subjective method with the weights obtained by the objective method Shannon considers that the highest weight has the criterion C3—VM Cost. Note that the final criteria weights are close to the Shannon criteria and the difference between the DEMATEL criteria weights and the Shannon criteria weights is lower than the difference between the final criteria weights and Shannon criteria weights.

4 Conclusions

The present paper calculates the cloud QoS criteria weights based on a combination of the subjective method DEMATEL and the objective method Shannon. With the help of DEMATEL method we investigate the causal relationships and degree of influence between a set of QoS criteria. The analysis provides important information to CSPs for promoting their cloud services and a better understanding of the most important QoS criteria that influences customers' decisions in the adoption of cloud services. Consequently CSPs can develop a more focused strategy for meeting customer demands.

Acknowledgements This research was supported by the project PN 18 19 01 01 and PN 18 19 05 01 from the Romanian Core Program of the Ministry of Research and Innovation.

References

1. Rădulescu, C.Z., Rădulescu, I.: An extended TOPSIS approach for ranking cloud service providers. Stud. Inf. Control **26**(2), 183–192 (2017)
2. Rădulescu, C.Z., Rădulescu, D.M., Harţescu, F.: A cloud service providers ranking approach, based on SAW and modified TOPSIS methods. In: Proceedings of the 16th International Conference on Informatics in Economy (IE 2017), Bucharest, Romania, pp. 7–12 (2017)
3. Rădulescu, C.Z., Balog, A., Rădulescu, D.M., Dumitrache, M.: A decision making framework for weighting and ranking criteria for cloud provider selection. In: Proceedings of the 20th International Conference on System Theory, Control and Computing (ICSTCC), October 13–15 (2016)
4. Deng, H., Yeh, C.H., Willis, R.J.: Inter-company comparison using modified TOPSIS with objective weights. Comput. Oper. Res. **27**, 963–973 (2000)
5. Wang, T.C., Da Lee, H.: Developing a fuzzy TOPSIS approach based on subjective weights and objective weights. Expert Syst. Appl. **36**, 8980–8985 (2009)
6. Lotfi, F.H., Fallahnejad, R.: Imprecise Shannon's entropy and multi attribute decision making. Entropy **12**, 53–62 (2010)
7. Ardakani, M.A., Milani, A., Yannacopoulos, S., Shokouhi, G.: On the effect of subjective, objective and combinative weighting in multiple criteria decision making: a case study on impact optimization of composites. Expert Syst. Appl. **46**, 426–438 (2016)
8. Saaty, T.L.: The Analytic Hierarchy Process. McGraw-Hill, New York, NY, USA (1980)
9. Chu, A.T.W., Kalaba, R.E., Spingarn, K.: A comparison of two methods for determining the weights of belonging to fuzzy sets. J. Optim. Theory Appl. **27**(4), 531–538 (1979)
10. Hwang, C.L., Lin, M.J.: Group Decision Making Under Multiple Criteria: Methods and Applications. Springer, Berlin (1987)
11. Edwards, W.: How to use multi attribute utility measurement for social decision-making. IEEE Trans. Syst. Man Cybern. **7**(5), 326–340 (1977)
12. Edwards, W., Barron, F.H.: SMARTS and SMARTER -improved simple methods for multi attribute utility measurement. Org. Behav. Human Decis. Process. **60**, 306–325 (1994)
13. Simos, J.: L'évaluation environnementale: Un processus cognitif négocié. Lausanne: DGF-EPFL, These de doctorat (1990)
14. Figueira, J., Roy, B.: Determining the weights of criteria in the ELECTRE type methods with a revised Simos' procedure. Eur. J. Oper. Res. **139**, 317–326 (2002)
15. Hwang, C.L., Yoon, K.: Multiple Attribute Decision Making: Methods and Applications. Springer, Berlin (1991)
16. Shannon, C.E., Weaver, W.: The Mathematical Theory of Communication. The University of Illinois Press, Urbana (1947)
17. Choo, E.U., Wedley, W.C.: Optimal criterion weights in repetitive multicriteria decision-making. J. Oper. Res. Soc. **36**, 983–992 (1985)
18. Croux, C., Filzmoser, P., Fritz, H.: Robust sparse principal component analysis. Technometrics **55**, 202–214 (2013)
19. Saaty, T.L.: The Analytic Hierarchy Process, p. 287. The Wharton School University of Pensilvania, United States of America (1991)
20. Gabus, A., Fontela, E.: Perceptions of the World Problematique: Communication Procedure, Communicating with Those Bearing Collective Responsibility. Battelle Geneva Research Center, Geneva, Switzerland (1973)
21. Wang, W.C., Lin, Y.H., Lin, C.L., Chung, C.H., Lee, M.T.: DEMATEL-based model to improve the performance in a matrix organization. Expert Syst. Appl. **39**, 4978–4986 (2012)
22. Gölcük, I., Baykasoglu, A.: An analysis of DEMATEL approaches for criteria interaction handling within ANP. Expert Syst. Appl. **46**, 346–366 (2016)
23. Lee, Y.C., Li, M.L., Yen, T.M., Huang, T.H.: Analysis of adopting an integrated decision making trial and evaluation laboratory on a technology acceptance model. Expert Syst. Appl. **37**, 1745–1754 (2010)

Cooperative Policies for Drug Replenishment at Intensive Care Units

Roberta Rossi, Paola Cappanera, Maddalena Nonato and Filippo Visintin

Abstract This paper addresses the effects of lateral transshipment within a drug inventory policy in a real case-study involving two Intensive Care Units. An extension of a previous developed integer linear programming model is proposed, which decides when, what and how much to order, ensuring orders regularity. Preliminary results on realistic instances suggest the potential advantage in terms of reduction of order occurrences while using excess stock efficiently and profitably. This analysis is a first step towards the introduction of the cooperative model within an optimization-simulation tool deployed in a rolling-horizon framework.

Keywords Cooperation · Lateral transshipment · Point-of-use drugs inventory Hospital logistics

1 Introduction and Problem Description

This study focuses on drugs replenishment problems in a real case-study involving two wards at a public hospital in Tuscany (Italy). Specifically, two Intensive Care Units (ICUs) are considered, although the approach proposed may be used whenever two wards use the same subset of drugs. In particular, we propose combined optimization-simulation policies to support nurses in drug inventory management.

R. Rossi (✉) · P. Cappanera
D.I.N.F.O., University of Florence, Florence, Italy
e-mail: roberta.rossi@unifi.it

P. Cappanera
e-mail: paola.cappanera@unifi.it

M. Nonato
D.E., University of Ferrara, Ferrara, Italy
e-mail: maddalena.nonato@unife.it

F. Visintin
D.I.E.F., University of Florence, Florence, Italy
e-mail: filippo.visintin@unifi.it

© Springer Nature Switzerland AG 2018

P. Daniele and L. Scrimali (eds.), *New Trends in Emerging Complex Real Life Problems*, AIRO Springer Series 1,
https://doi.org/10.1007/978-3-030-00473-6_46

Indeed, in the real case considered, nurses issue drug replenishment orders besides taking care of patients. This task is very time consuming due to (i) the lack of predictive/optimization tools and (ii) technological support, such as Automated Dispensing Machines (ADMs), RFID technology, etc [14], and (iii) patients conditions which can evolve dynamically.

Drugs replenishment management at hospital is challenging since it has different peculiarities with respect to industrial settings where optimization management modeling techniques are well studied and consolidated. While some aspects can be treated in a similar way, others require original solutions; for example, in this setting there are not neither cost-opportunity issues nor time-dependent costs of storage, priority is on patient care thus imposing drugs demand is always covered and backorders are not allowed. In such a context it is thus dramatically important to reduce working time nurses spend for tasks not directly related to patient care. We deal with a particular subset of drugs, namely antibiotics, which are crucial for at least two reasons: their availability is vital since strictly related to mortality and their improper use might give rise to drug-resistance. A reasoned, effective and careful drugs order management is indeed important. The criticality concerns above all demand irregularity. The process that regulates demand is as follows. According to guidelines [12] when clinicians suspect an infection is in progress, immediately a diagnostic test is sent to the microbiology laboratory to identify the microorganism that caused the possible infection and an empirical broad-spectrum therapy is started; test results arrive after several days and the therapy is corrected accordingly. As a consequence, the therapeutic plan for the same patient, within the same length of stay, may be modified several times: each therapy in the care plan has a well-defined pattern but is highly subject to updates required to follow changes in patient's clinical conditions. This results in intermittent drug consumption, which has to be added to beds availability which is limited in these wards compared to other hospital units.

In a previous work by the same authors [5], a hybrid push-first pull-second inventory strategy was proposed and compared with a classic periodic review policy (s, S), widely used in literature as a benchmark. Our hybrid policy is based upon a deterministic optimization model, preliminary presented in [3] and extended in [4], that assumes a forecasted demand is known. The model allows to schedule orders in a given planning period, incorporates stakeholders' perspectives, capacity constraints related to different storage units (shared among drugs or dedicated), controls budget while pursuing service regularity by imposing that the same quantity of drug is ordered each time an order of that drug is triggered. The policy has been tested through its integration with a discrete event simulation tool which generates actual daily demand and mimics real-time drug consumption at ward. Simulation implements the order schedule determined by optimization and triggers rush orders to face stock-outs as well as extra orders to restore safety stock levels. The combined optimization-simulation approach has been then deployed in a rolling horizon framework.

A wide computational testing done on a 365-day rolling horizon showed that the use of the hybrid policy is advantageous with respect to a (s, S) policy both in terms of stock-out events—that give rise to rush orders, and of regular orders. In addition, the hybrid policy seems to be quite robust with respect to the planning period length, with the number of rush orders that remains almost stable as the frequency with which the optimization model is run increases (every 7, 14 and 21 days).

However, using the hybrid policy has a possible drawback, i.e. the tendency to accumulate drugs in stock which may occur when forecasted demand differs from the realized one. In an attempt to preserve the advantages of the hybrid policy while limiting its possible drawbacks, in this work we investigate how the cooperation between two wards may be used to reduce stocks in excess and to increase efficiency. Indeed, a context where each ward adopts its own internal replenishment policy without coordinating with an entity of upper level in the logistic chain, seems to be a natural setting for cooperation between wards. Summarizing, cooperation may allow a more efficient management of drugs stocked at wards, both in terms of service quality and costs. When cooperation occurs, potential stock-outs may be covered with drugs borrowed by a neighbouring ward and thus promptly available for patient care which is the primary concern in this setting. Second, drug exchanges may also allow to reduce the stock level of those drugs that have been ordered as a consequence of a forecasted demand and not consumed. The motivation of this research is therefore to analyse methodologies and decision-making tools to support a collaboration between entities at the same level of the supply chain as an alternative to the classic order made to the entity of a higher level.

The paper is organized as follows. In Sect. 2 the related literature is briefly reviewed, in Sect. 3 the cooperative approach is introduced, while in Sect. 4 preliminary computational results are presented. Finally, some conclusions are drawn.

2 Literature Overview

Nowadays, efficient and optimized inventory and material logistics management in hospitals is an urgent problem to face with especially where there is scarcity of resources and huge economic cuts aimed at limiting wastes. Despite its importance, this problem has only been addressed in recent years as witnessed in [16]. An analysis of the literature was conducted in order to review how the problem of drug logistics is dealt with in a cooperative setting, namely where entities belonging to the same level of the supply chain—typically, two or more wards in a hospital, exchange products as an alternative to trigger an order to the entity of upper level—typically, the hospital's supplier. Even if the focus is on the health care setting, literature review includes also contributions from the industrial setting. We selected 170 papers relative to quantitative models and mainly published since 2009. It turned out that a still limited health care literature exists, especially concerning cooperative strategies. Relevant health care applications concern humanitarian emergencies studies, where a prompt delivery of drugs is essential [15].

In regards to the industrial setting, cooperation is usually referred with the term "lateral transshipment" and cooperative strategies are usually adopted to cope with unexpected demand peaks as an alternative to facilities enlargement or to add auxiliary warehouses to overcome limited capacity issues [7, 9]. Reasons to consider cooperation may also include delivery time and cost reduction. However, when transshipment costs are high or the interaction between entities of the same level is critical, adding stock capacity is the preferred solution [13].

Here we focus on quantitative optimization models. Cooperative strategies are incorporated into inventory/replenishment well consolidated mathematical methodologies: there, events triggering lateral transshipment are usually of two types according to [10]: (i) a product with not sufficient availability (stock-out) or for which the stock falls below a safety level is requested (the so called *reactive transshipment models*); (ii) a predefined time point or the occurrence of specific conditions trigger lateral transshipment (the so called *proactive models*). Usually these models are combined with periodic review reorder policies. The proposed approaches can be further classified according to whether they involve routing or not [6].

Different classes of constraints can be found in these cooperative models. Typical examples include: storage capacity constraints, quality of service issues, flow conservation, allocation and time windows constraints in which product deliveries can take place. We have paid particular attention to flow conservation constraints which play a central role in all the models with a lot-sizing structure where they are used to rule the consistency of product stock levels over the planning horizon. Flow variables in addition, allow to differentiate quantities transferred between entities at the same level according to lateral transshipment policies with respect to quantities transferred according to standard replenishment policies [1, 2, 6, 8, 11, 15].

3 Methodology

We present how the single ward model can be extended to include cooperation. Suppose we have two wards that, in case of need, can transship drugs to each other. When one of them incurs a stock-out for one or more drugs, it may ask for the drug to the other ward as an alternative to trigger a rush order. The idea is to use the formulation of the single ward as a building block to model the case of two cooperating wards. In addition to have a building block for each ward, the lateral transshipment events between them have to be managed by means of proper constraints. The optimization model used to define the building block relevant to a ward (described in [4]) is characterized by the following groups of constraints: (i) flow conservation constraints of each drug stock level; (ii) storage capacity constraints, both for dedicated and shared spaces; (iii) budget control; (iv) orders' regularity (v) principal stakeholder's perspective; (vi) variable domain.

Wards are indexed by r and both of them have their own variables and manage them by taking decisions as already described in the single ward model, regulating the order schedule ν^r_{fd} and the lot size of each drug Δ^r_f. The main parameters and variables relative to the cooperative model are summarized in Tables 1 and 2, while Fig. 1 provides a graphical representation of the structure of the complete model starting from the building blocks corresponding to each ward. The interaction

Table 1 Sets and parameters

R	Set of wards (indexed by r)
F	Set of drugs (indexed by f)
D	Ordered set of days (indexed by d)
q^r_{fd}	Demand of drug f on day d (number of doses) in ward r
U_f	Number of doses in each box of drug f

Table 2 Variables

s^r_{fd}	Stock level of drug f at the end of day d at ward r, expressed in number of doses
τ^{12}_d	Equal to 1 when a transshipment event from ward 1 to ward 2 occurs on day d
τ^{21}_d	Equal to 1 when a transshipment event from ward 2 to ward 1 occurs on day d
ζ^{12}_{fd}	Transshipped quantity of drug f from ward 1 to ward 2 on day d, expressed in number of doses
ζ^{21}_{fd}	Transshipped quantity of drug f from ward 2 to ward 1 on day d, expressed in number of doses
y^r_{fd}	Order quantity of drug f on day d at ward r, expressed in number of boxes
ν^r_{fd}	Equal to 1 when a standard order of drug f occurs on day d at ward r; 0 otherwise
υ^r_d	Equal to 1 when a standard order occurs on day d at ward r; 0 otherwise
Δ^r_f	Order quantity of drug f at ward r, expressed in number of boxes

between the two wards mainly impacts on the flow conservation constraints involving stocks levels s^r_{fd}. In particular, the flow conservation constraints exemplified for a typical working day in (1)–(3) from ward 1 point of view, impose that for each drug and on each day d of the planning period, the quantity in stock at the end of d ($s^1_{f,d}$) is equal to the residual of the previous day ($s^1_{f,d-1}$) plus the quantity potentially arrived on day d ($U^1_f y^1_{f,d-1}$) minus the current day demand ($q^1_{f,d}$) plus what received through lateral transshipment from the other ward ($\zeta^{21}_{f,d}$) minus what is sent by lateral transshipment to the other ward ($\zeta^{12}_{f,d}$)—see constraints (1). Clearly, for a same drug f, on a given d, lateral transshipment may occur in one direction at most and the quantity transshipped cannot exceed the demand of the receiving ward for drug f on the current day [see constraints (2)–(3)].

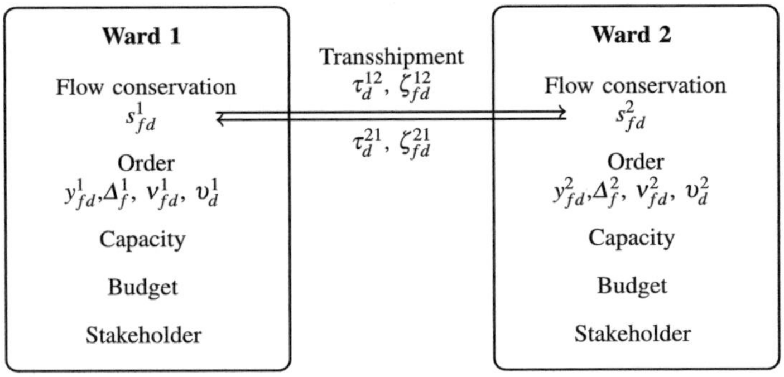

Fig. 1 Qualitative model constraints' block structure

$$s^1_{f,d} = s^1_{f,d-1} - q^1_{f,d} + U^1_f y^1_{f,d-1} + \zeta^{21}_{f,d} - \zeta^{12}_{f,d} \qquad \forall f \in F, \forall d \in D \tag{1}$$
$$\zeta^{21}_{f,d} \leq q^1_{f,d} \tau^{21}_d \qquad \forall f \in F, \forall d \in D \tag{2}$$
$$\zeta^{12}_{f,d} \leq q^2_{f,d} \tau^{12}_d \qquad \forall f \in F, \forall d \in D \tag{3}$$

Transshipment penalties contribute to the definition of the objective function and their role is experimentally investigated in Sect. 4 using 3 different values for them. The simulation model was coded integrating Rockwell Arena and Python. The main resources in the model are the wards beds, while the main entities flowing in the model are patient entities, nurse entities and orders entities. For the moment being, the model considers two wards each associated with a set of wards beds.

Sampling from suitable empirical distributions, each day, for each ward, and for each patient type, the model generates a number of arriving "patient entities". Each incoming patient entity triggers a script (coded in Python) that determines the patient therapy, i.e. the number of days s/he will spend in the ward and the number of doses of each drug s/he will be administered for each day. Once assigned with a therapy each patient seizes a ward bed. Such a bed will be released once the therapy is over. For each ward, every day, the model creates a "nurse entity". Such a nurse, checks the therapy associated with each patient in the ward and administers the drugs accordingly (one patient at time). The stock level associated with each administered drug automatically decreases upon administration. If the amount of drug in stock is smaller than needed, depending on the drug type, the nurse either: (1) issues a rush order to the pharmacy; or (2) issues a request to the other ward. If the request is accepted (2a) the nurse uses the drug from the other ward, and subsequently issues a regular order to restore an adequate stock level in both wards. If the request is rejected (2b) the nurse issues a rush order to the pharmacy as in (1).

In all cases after a time equal to the lead time (which is shorter for rush orders than for regular orders), the drug is restocked and the pending administration is performed. The decision of accepting or rejecting a request coming from another ward is taken based on several criteria (which vary according with the scenario) including: (i) the current stock of the requested drug; (ii) the pending orders for the requested drug;

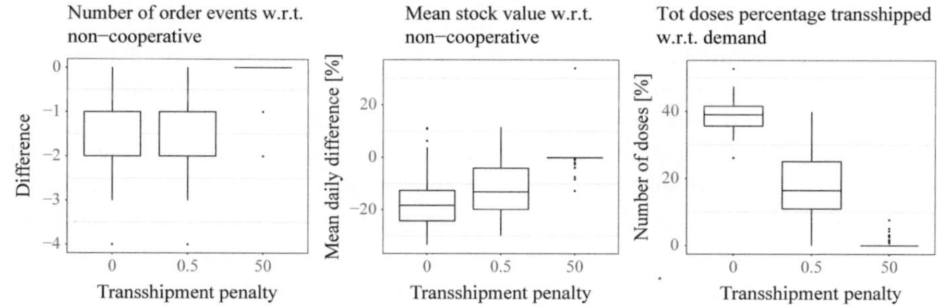

Fig. 2 Boxplots report the gap between *cooperative* and *standalone* policies over all instances and w.r.t. different penalties, concerning the number of order events (on the left), mean stock value percentage (in the middle), and percentage of drug doses covered by lateral transshipment w.r.t. demand (on the right)

(iii) the expected demand for the requested drug. For each ward, every N days (e.g. 7 days) the simulation model executes, in shell (i.e. without advancing the simulation clock), an algorithm (either an optimization model or a heuristics, depending on the scenario) that indicates for each drug, the number of boxes to order. These orders are issued immediately after the algorithm has produced the solution and fulfilled after the regular order lead time has elapsed.

In this study we have considered 18 drugs, 6 patient types, a ward capacity of 8 beds for both wards, a lead time for regular orders equal to 1 day and a lead time for rush order equal to 2 h.

4 Computational Results

The cooperative model has been compared to the non-cooperative version over a set of realistic demand instances generated through the empirical distributions retrieved from real data collected at the hospital. The MILP model was coded in Python and solved by the IBM ILOG CPLEX 12.7 solver on a MacBook Pro equipped with Intel Core i5 cpu. We considered a 14-days planning horizon. For each ward, one set of 50 demand instances has been generated and each instance has been solved as a stand alone. Then, the 100 instances have been coupled one by one to yield 50 Ward1-Ward2 pairs, and on each pair the cooperative model, allowing for transshipment, has been solved. Three variants corresponding to different transshipment penalties have been investigated, namely 0 (no penalty), 0.5, and 50 (high penalization variant), and the impact on (i) number of order events, (ii) demand coverage, (iii) mean stock value has been recorded. Figure 2 depicts the boxplots for each variant of the cooperative model penalty with respect to the stand alone version. As shown on the left, compared to the non-cooperative version (where there is an average of 4 order events) the number of order events decreases on average by 1 when the transshipment is granted without penalty as well as when the penalty is soft; when the penalty is high, the behaviour is similar to the stand alone policy. In the middle of Fig. 2 boxplots refer to stock value; the benefit of lateral transshipment in terms of daily reduction of stock value is remarkable for all penalty but the high variant. On the right of Fig. 2 the percentage of doses exchanged between the wards compared to the demand is shown, and it suggests that lateral transshipment allows to guarantee a good demand coverage (and service level), despite the decrease in the number of standard order events. As a whole, we can say that the number of order events benefits from the cooperative approach; carrying costs decrease as well, as much as the percentage of demand handled by transshipment increases.

5 Conclusions and Work in Progress

The aim of this research is to study the effects of cooperation for drug inventory management at point of use in a hospital setting. We seek for a strategy which allows a better use of the constrained resources, namely nurses time in terms of number of order events and money tight in stock. To meet this target we extend a previously proposed optimization model which schedules drug orders imposing regularity by allowing for cooperation between two wards. We investigate different transshipment cost variants; preliminary results comparing the cooperative policy to the stand alone case have been presented. It can be observed that 0.5 penalty is a good compromise between carrying costs and number of transshipment operations. In all cases, cooperation allows a consistent reduction of the number of order events and levels the tendency to accumulate stocks. Future research will assess the impact of lateral transshipment within the optimization-simulation tool deployed in a rolling-horizon framework already developed in [5]. Such an impact will be compared also with the maximum improvement that can be obtained when a unique shared stock for the two wards is considered. Indeed, lateral transshipment can be seen as a valid alternative to promptly react to shortages instead of triggering *rush orders*.

Acknowledgements This work has been partially supported by LINFA (Logistica INtelligente del FArmaco) project, funded by Regione Toscana under the call PAR FAS 2007–2013, Linea d'azione 1.1 Bando FAR FAS 2014.

References

1. Al-e-hashem, S.M.J.M., Rekik, Y.: Multi-product multi-period Inventory routing problem with a transshipment option: a green approach. Int. J. Prod. Econ. **157**, 80–88 (2014)
2. Azadeh, A., Elahi, S., Farahani, M.H., Nasirian, B.: A genetic algorithm-Taguchi based approach to inventory routing problem of a single perishable product with transshipment. Comput. Ind. Eng. **104**, 124–133 (2017)

3. Cappanera, P., Nonato, M., Rossi, R.: Empirical data driven intensive care unit drugs inventory policies. In: Proceedings of the Third International Conference on Health Care Systems Engineering, pp. 155–166. Springer Proceedings in Mathematics and Statistics (2017)
4. Cappanera, P., Nonato, M., Rossi, R.: Stakeholder involvement in drug inventory policies (2017) (submitted)
5. Cappanera, P., Nonato, M., Rossi, R., Visintin, F., Nanetti, L.: A hybrid push-first pull-second inventory policy for critical drugs inventory at ICUs (2018) (submitted)
6. Coelho, L.C., Cordeau, J.F., Laporte, G.: The inventory-routing problem with transshipment. Comput. Oper. Res. **39**(11), 2537–2548 (2012)
7. Feng, X., Moon, I., Ryu, K.: Warehouse capacity sharing via transshipment for an integrated two-echelon supply chain. Transp. Res. Part E Logist. Transp. Rev. **104**, 17–35 (2017)
8. Firouz, M., Keskin, B.B., Melouk, S.H.: An integrated supplier selection and inventory problem with multi-sourcing and lateral transshipments. Omega **70**, 77–93 (2017)
9. Lee, C., Park, K.S.: Inventory and transshipment decisions in the rationing game under capacity uncertainty. Omega **65**, 82–97 (2016)
10. Paterson, C., Kiesmller, G., Teunter, R., Glazebrook, K.: Inventory models with lateral transshipments: a review. Eur. J. Oper. Res. **210**(2), 125–136 (2011)
11. Peres, I.T., Repolho, H.M., Martinelli, R., Monteiro, N.J.: Optimization in inventory-routing problem with planned transshipment: a case study in the retail industry. Int. J. Prod. Econ. **193**, 748–756 (2017)
12. Rhodes, A., Evans, L.E., Alhazzani, W., Levy, M.M., Antonelli, M., Ferrer, R., Kumar, A.: Surviving sepsis campaign: international guidelines for management of sepsis and septic shock: 2016. Intensive Care Med. **43**, 304–377 (2017)
13. Rosales, C.R., Rao, U.S., Rogers, D.F.: Retailer transshipment versus central depot allocation for supply network design. Decis. Sci. **44**(2), 329–356 (2013)
14. Rosales, C.R., Magazine, M., Rao, U.: Point-of-use hybrid inventory policy for hospitals. Decis. Sci. **45**(5), 913–937 (2014)
15. Rottkemper, B., Fischer, K., Blecken, A.: A transshipment model for distribution and inventory relocation under uncertainty in humanitarian operations. Socio-econ. Plann. Sci. **46**(1), 98–109 (2012)
16. Volland, J., Fügener, A., Schoenfelder, J., Brunner, J.O.: Material logistics in hospitals: a literature review. Omega **69**, 82–101 (2017)

A Hybrid Metaheuristic for the Optimal Design of Photovoltaic Installations

Matteo Salani, Gianluca Corbellini and Giorgio Corani

Abstract We consider the Photovoltaic Installation Design Problem (PIDP) were photovoltaic modules must be organized in strings and wired to a set of electronic devices. The aim is to minimize installation costs and maximize power production considering "mismatch losses" caused by non-uniform irradiation (shading) and directly related to design decisions. We relate the problem to the known class of location routing problems and thanks to the existing knowledge on the problem, we design a route-first cluster-second heuristic. We propose an efficient machine learning approach to evaluate the installation performances accounting for mismatch losses. We prove that our approach is effective on real-world instances provided by our industrial partner.

Keywords Metaheuristic · Machine learning · Photovoltaic installation design

1 Introduction

Photovoltaic (PV) technology has progressed in the last decades, in particular for what concerns rising efficiencies and falling prices of components [8]. Anyway, the system design process still represents a bottleneck for the realization and capitalization of efficient PV installations [7].

Modern design support tools are still lacking of methods for a comprehensive evaluation of alternatives. One of the misregarded aspects concern the electrical performance (voltage and current) of PV installations where mismatch losses effects,

---

M. Salani (✉) · G. Corani
Dalle Molle Institute for Artificial Intelligence (IDSIA),
USI/DTI-SUPSI, Lugano, Switzerland
e-mail: matteo.salani@idsia.ch

G. Corbellini
Institute for Applied Sustainability to the Built Environment (ISAAC),
DACD-SUPSI, Porza, Switzerland

© Springer Nature Switzerland AG 2018
P. Daniele and L. Scrimali (eds.), *New Trends in Emerging Complex Real Life Problems*, AIRO Springer Series 1,
https://doi.org/10.1007/978-3-030-00473-6_47

due to different level of irradiation between cellsand modules, can have a signficant impact on energy yield up to 25% with respect to the optimal configuration [11].

In the context of an applied project, we collaborate with an industrial partner: InSun SA, an IT company based in Lugano, which provides a platform of services to the players of the photovoltaic value chain. The goal of the project is to enrich the inSun's platform with Machine Learning and Optimization components.

2 Photovoltaic Installation and Mismatch Losses

In a photovoltaic installation layout, modules are organized in strings (in series) and strings are assembled in parallel and connected to *Maximum Power Point Trackers* (MPPT) that are electronic components able to set the optimal operational point in current and voltage of PV strings. All strings connected to an MPPT must be composed of an equal number of modules. Strings are joined in junction boxes and then dedicated connection cables reach MPPTs which are physically located within *Inverters*, devices used to convert the energy produced from direct current to alternate current. Every inverter can host one or more MPPTs. Finally inverters are connected to the power grid.

When modules are facing the same environmental condition (irradiation and temperature) every module is working at its own best condition. When modules are in different conditions (mostly due to partial shading because of surrounding objects) some of them are not working at their global maximum power point. These partial shading phenomenon lead to additional non linear losses, called "mismatch losses" [3].

To estimate the mismatch losses of a given plant configuration, for every sun position when partial shading is occurring, a time consuming simulation must be carried out. In order to tackle the computational issue, we adopted a meta-model approach, approximating by a statistical model the mismatch losses.

Thus for each PV configuration, composed by a different number of PV strings s, we build a synthetic data set by simulating the physical model in a range of different conditions and we train a statistical predictor aimed at approximating the mismatch loss computed by the physical model. We use as inputs the ratio between diffuse irradiation and global irradiation and the proportion of shadow referring to each string, which thus yield $s + 1$ independent variables.

For each different number of string s, we train a random forest model [6, Chap. 15] as a statistical predictor. The random forest algorithm is widely recognized as one the most accurate machine learning approaches for function approximations [5], given its ability in capturing interaction between features and non-linearities.

Table 1 The random forest model reliably approximates the mismatch loss computed by the physical model on the different data sets for $s \in \{1, \ldots, 7\}$

	Number of strings						
	1	2	3	4	5	6	7
Number of data points	200	2300	17,700	20,000	20,000	20,000	20,000
Mean mismatch loss	0.11	0.13	0.14	0.15	0.16	0.16	0.16
Mean absolute error	1e−2	7e−3	4e−3	5e−3	5e−3	6e−3	7e−3
Correlation true/predicted	0.97	0.99	1.00	1.00	1.00	1.00	1.00

In order to validate the model we adopt 10-folds cross-validation for $s \in \{1, \ldots, 7\}$. As reported in Table 1, we obtained very good approximation in all data sets, with an absolute error smaller by at least one order of magnitude than the mismatch loss to be predicted and correlations larger than 0.95 between the values produced by the physical model and those estimated by the random forest. In most data sets, the error is *two* orders of magnitude smaller than the mismatch loss computed by the physical model and the correlation is 0.99. The data set that contains a single string is the most critical; the correlations between inputs and mismatch loss are so low in this setting that a linear regression yields a true/predicted correlation of only 0.53 while the random forest obtains a true/predicted correlation of about 0.95. This shows that the relation between input and output are strongly non-linear.

3 Photovoltaic Installation Design Problem

The photovoltaic installation design problem (PIDP) is described as follows: given a set M of homogeneous PV modules, a set T of MPPTs, a set I of inverters, a set S of irradiance samples, determine the inverter configuration, i.e. the MPPTs to be used among the available ones, for each MPPT, its configuration, i.e. the number and length of pv strings and the assignment of modules to MPPTs and their organization in strings. Finally, for a given MPPT and a given set of pv strings to be connected in parallel, the location of the junction where strings are assembled in parallel.

For each module $m \in M$, let p_m be its nominal power and c_m its position in a bi-dimentional space. For each tracker $t \in T$, let nm_t and ns_t be the maximum number of modules and the maximum number of strings that can be connected to it. Additionally, each string attached to the tracker $t \in T$ must be composed by no less than lm_t modules and no more than um_t modules. For each sample $s \in S$, let di_s and ii_s the direct and diffuse irradiances, respectively. We are also given a relative weight w_s of a given sample $s \in S$ as the fraction of time of an year that the corresponding irradiance sample is expected. For each module $m \in M$ and irradiation sample $s \in S$, we are given the status of the module $a_{n,s} \in (0, 1)$. It represents the power fraction of the module with respect to the nominal power we are expecting at irradiation sample s.

The PIDP has multiple objectives. The first objective, z_1, is to maximize the number of modules in the solution. As feasible MPPT configurations are limited there may be problem instances in which not all modules can be connected. The second objective, z_2, is the minimization of overall installation cables. The cable necessary to assemble the installation is composed of the cable necessary to form each PV string, plus the cable necessary to connect strings in parallel and the cable necessary to connect the strings to the MPPTs. The length of all cables is computed using the L1 norm as in general cables are secured to the hardware which is organized in a sort of a grid. The third objective, z_3, is the minimization of production losses due to mismatch effects. According to industrial requirements, we consider objecives lexicographically: given any two solutions r_1 and r_2, r_1 dominates r_2 if

$$z_1(r_1) > z_1(r_2) \vee (z_1(r_1) = z_1(r_2) \wedge z_2(r_1) < z_2(r_2) \wedge z_3(r_1) < z_3(r_2))$$

We observe that the PIDP can be conveniently cast to a multi-depot location routing problem (LRP) with some problem specific additional constraints. For brevity, we refer the reader to [9, 10, 13]. Variants and applications of the LRP are reviewed in [4], we mention [1] as authors addressed the LRP with profits, i.e. the problem in which some of the customers can be left unserved and [14] addressing the LRP with multiple objectives.

Briefly, we describe how an instance of the PIDP can be seen as an instance of the LRP and outline the few differences among the two problems. The set of customers is composed by the set of modules M. The set of potential depot nodes corresponds to the set of potential junction boxes.

The number of vehicles is unbounded and there are no fixed costs for opening a depot. The capacity of each vehicle is constrained from both above and below by um_t and lm_t. This is one of the differences between PIDP and LRP. In the latter, route capacity is constrained only from above. The cost matrix, can be conveniently described by the assembly distance between modules. Adjacent modules that can be connected directly by their own cables thus having distance equal to 0. We finally remark the second difference between PIDP: all PV strings connected to an MPPT must be composed of the same number of modules, in LRP terms, the number of customers associated with routes belonging to the same depot must be the same. This peculiarity is not addressed in any of the reviewed contributions.

Analysing the LRP literature, we focused on methods based on the route first cluster second principle introduced by Beasley [2] where the computational effort related to routing is concentrated on computing one or few "giant" TSP tours once and the clustering component is able to produce a population of non-dominated solutions.

4 Hybrid Heuristic

In order to solve the PIDP, we designed an hybrid heuristic. Each objective is treated separately by a different component of the heuristic. In particular, the main objective, z_1, is addressed by an iterative enumeration algorithm that produces a set of optimal configuration schemes. The minimization of total cabling, z_2, is addressed by the routing component of the algorithm, exploiting different metrics, the algorithm is able to compute different solutions for the giant TSP. Finally, different objectives are accounted in the clustering component, where the objective related to the mismatch loss estimation, z_3, is computed exploiting the meta-model presented earlier.

A configuration is defined by the subset of selected trackers $T' \subseteq T$ and for each selected tracker, $t \in T'$, by a feasible tracker configuration. In order to be feasible, a tracker configuration must not exceed the maximum number of modules, nm_t and must be composed by no more than ns_t strings all formed by at most um_t and at least lm_t modules. Optimal configurations are those that maximize the total number of connected modules. The set of optimal tracker configurations, G, can be computed by a recursive enumeration scheme. For each tracker, we iteratively consider ordered configurations. For a given selected configuration for tracker t, we ignore all configurations for tracker $t + 1$ with more modules than those associated with t. We therefore obtain all simple combinations and we avoid the brute force enumeration of all possible configurations and avoid configurations redundance. In order to explore some structurally different solutions, once we designed the set of optimal configurations we sample a subset of configurations with different variance in the number of assigned modules per tracker. In our experiments, we observed that there exist good solutions in which the configuration of trackers has low variance (all trackers are similarly configured) and good solutions in which the configuration of trackers has high variance (some trackers with many modules and some with few modules). Different types of optimal tracker configurations are therefore worth to be explored.

PV modules are assembled in a so called installation field, i.e. a bi-dimensional space with physical constraints for installation. In particular, as installation fields may be on rooftops, straight connection of any two modules is sometime impossible as rooftops can be non-convex or there may be obstacles along the path. Still, there may be a viable path that a cable can follow to connect any two pair of modules. To this purpose we developed a preprocessing component able to compute the cabling distance between any pair of modules in an installation field by defining a support graph in which nodes are represented by PV modules and vertices of obstacles and an edge between two nodes exists when a straight connection is feasible. For any pair of PV modules that does not have a straight connection we compute the shortest path in the support network with a standard Dijkstra algorithm according to a given distance norm: Euclidean norm, L1 norm and a problem-specific norm designed so that some module connections are preferred over others according to practical concerns related to installation. In practice, photovoltaic installations are composed of series of longitudinal tracks where modules are assembled and cables can be conveniently secured, therefore it is convenient to prefer connections between modules laying on the same track. The result of the pre-processing phase is a set of three distance matrices between any pair of modules.

Given a set of modules M and a set of distance matrices L computed in the preprocessing phase, according to the route first cluster second methodology (see Prins et al. [12]), we compute a set F of different so called giant TSP tours.

Given an optimal configurations $g \in G$ and an optimized giant tour $f \in F$, we build a solution to an instance for the PIDP splitting the tour in as many strings as the configuration g prescribes. Furthermore, we need to assign strings to trackers and decide the location of the junction point connecting the strings. We designed an iterated local search that produces an initial set of solutions with a constructive heuristic and then improves it with local exchanges.

5 Computational Results

We performed experiments on real-world instances proposed by our industrial partner. In this section, we compare solutions obtained with our methodology with those obtained by the company's best practices, currently a partially automated system. Results are reported in Table 2.

We have at hand 9 real-world instances. The smallest instance has 60 modules and an average of 10 modules per tracker, the largest instance has 1218 modules and an average of 305 modules per tracker. For the largest instance, the method converged in 192 s of computation. For all instances the method proposes different solutions in short computational time.

Savings related to cabling can be up to 31.76% in the largest instance and mismatch losses are reduced up to 45.20%. In some cases, the best solution with respect to one objective has a worse performance on the corresponding value of the other objective and this is expected. For example, in a 631 modules instance the best mismatch loss performance of 5.95 corresponds to a worsening of cable length by 3.27%. Anyway, for all instances, we were able to provide solutions improving both cabling and mismatch loss objectives. Finally, we remark that for the largest instance, the proposed method is able to connect all 1218 modules, while the reference solution connected one module less and still the overall cabling and mismatch loss are largely reduced.

Table 2 Computational results

Instance	Solution Type	Assigned Modules	Total Length	(%)	MML	(%)	Time(s)
instance_60_3_6	Reference	60	176.6		4.19		
	Best MML	60	200.9	−13.75	**3.74**	**10.74**	1.8
	Best length	60	**142.1**	**19.55**	4.40	−5.01	
	Best improvement	60	151.6	14.16	4.04	3.58	
instance_68_3_6	Reference	68	227.2		6.46		
	Best MML	68	226.1	0.50	**6.11**	**5.42**	2.1
	Best length	68	**158.4**	**30.30**	6.75	−4.49	
	Best improvement	68	185.2	18.49	6.43	0.46	
instance_210_5_10H	Reference	210	665.9		3.52		
	Best MML	210	582.7	12.50	**2.96**	**15.91**	6.9
	Best length	210	**548.4**	**17.65**	3.46	1.70	
	Best improvement	210	582.7	12.50	**2.96**	**15.91**	
instance_210_5_10L	Reference	210	665.9		2.21		
	Best MML	210	628.0	5.70	**1.99**	**9.95**	5.0
	Best length	210	**548.4**	**17.65**	2.19	0.90	
	Best improvement	210	555.2	16.62	2.14	3.17	
instance_577_6_12	Reference	577	2243.8		4.48		
	Best MML	577	2034.0	9.35	**3.49**	**22.10**	123.9
	Best length	577	**1765.1**	**21.33**	3.94	12.05	
	Best improvement	577	1785.5	20.43	3.87	13.62	
instance_631_3_3	Reference	631	2254.4		7.86		
	Best MML	631	2328.2	−3.27	**5.95**	**24.30**	79.9
	Best length	631	**1825.3**	**19.03**	7.52	4.33	
	Best improvement	631	2026.3	10.12	6.54	16.79	
instance_674_6_12	Reference	674	1884.2		2.97		
	Best MML	674	2251.0	−19.47	**2.60**	**12.46**	98.9
	Best length	674	**1719.1**	**8.76**	3.14	−5.72	
	Best improvement	674	1771.7	5.97	2.82	5.05	
instance_903_14_14	Reference	903	1979.3		3.48		
	Best MML	903	2072.6	−4.71	**2.87**	**17.53**	102.9
	Best length	903	**1736.3**	**12.28**	4.84	−39.08	
	Best improvement	903	1869.0	5.57	3.02	13.22	
instance_1218_4_4	Reference	1217	4957.2		10.93		
	Best MML	**1218**	3824.9	22.84	**5.99**	**45.20**	192.5
	Best length	**1218**	**3382.9**	**31.76**	9.45	13.54	
	Best improvement	**1218**	3592.0	27.54	6.15	43.73	

6 Conclusions

We have shown an efficient hybridization of machine learning and optimization to tackle a real-world problem. The Photovoltaic Installation Design Problem (PIDP) can be modelled as a location routing problem and solved with the arsenal of known OR methodologies. Results are encouraging: a set of non-dominated solutions is computed in a reasonable amount of time enabling the decision maker to compare among different solutions. The approach is shown to be applicable to a production environment. Further work should be concentrated in assessing the quality of the method with the computation of valid lower bounds.

References

1. Ahn, J., de Weck, O., Geng, Y., Klabjan, D.: Column generation based heuristics for a generalized location routing problem with profits arising in space exploration. Eur. J. Oper. Res. **223**(1), 47–59 (2012)
2. Beasley, J.: Route first cluster second methods for vehicle routing. Omega **11**(4), 403–408 (1983)
3. Di Dio, V., La Cascia, D., Miceli, R., Rando, C.: A mathematical model to determine the electrical energy production in photovoltaic fields under mismatch effect. In: 2009 International Conference on Clean Electrical Power, pp. 46–51. IEEE (2009)
4. Drexl, M., Schneider, M.: A survey of variants and extensions of the location-routing problem. Eur. J. Oper. Res. **241**(2), 283–308 (2015)
5. Fernández-Delgado, M., Cernadas, E., Barro, S., Amorim, D.: Do we need hundreds of classifiers to solve real world classification problems. J. Mach. Learn. Res. **15**(1), 3133–3181 (2014)
6. Friedman, J., Hastie, T., Tibshirani, R.: The Elements of Statistical Learning, vol. 1. Springer (2001)
7. Green, M.A.: Commercial progress and challenges for photovoltaics. Nat. Energy **1**, 15015 (2016)
8. Kang, M.H., Rohatgi, A.: Quantitative analysis of the levelized cost of electricity of commercial scale photovoltaics systems in the us. Solar Energy Mater. Solar Cells **154**, 71–77 (2016)
9. Laporte, G., Nobert, Y., Arpin, D.: An exact algorithm for solving a capacitated location-routing problem. Ann. Oper. Res. **6**(9), 291–310 (1986)
10. Nagy, G., Salhi, S.: Location-routing: issues, models and methods. Eur. J. Oper. Res. **177**(2), 649–672 (2007)
11. Peled, A., Appelbaum, J.: Enhancing the power output of pv modules by considering the view factor to sky effect and rearranging the interconnections of solar cells. Prog. Photovolt. Res. Appl. **25**(9), 810–818 (2017)
12. Prins, C., Lacomme, P., Prodhon, C.: Order-first split-second methods for vehicle routing problems: a review. Transp. Res. Part C Emerg. Technol. **40**(Supplement C), 179–200 (2014)
13. Prodhon, C., Prins, C.: A survey of recent research on location-routing problems. Eur. J. Oper. Res. **238**(1), 1–17 (2014)
14. Toro, E.M., Franco, J.F., Echeverri, M.G., Guimares, F.G.: A multi-objective model for the green capacitated location-routing problem considering environmental impact. Comput. Ind. Eng. **110**(Supplement C), 114–125 (2017)

Perspective Cuts for the ACOPF with Generators

Esteban Salgado, Claudio Gentile and Leo Liberti

Abstract The alternating current optimal power flow problem is a fundamental problem in the management of smart grids. In this paper we consider a variant which includes activation/deactivation of generators at some of the grid sites. We formulate the problem as a mathematical program, prove its **NP**-hardness w.r.t. activation/deactivation, and derive two perspective reformulations.

Keywords Power flow · Reformulation · NP-hardness

1 Introduction

The Alternating Current Optimal Power Flow (ACOPF) problem is one of the most important problems arising in the energy industry. It models the propagation of power flows in electrical grids. It is often used as second-level subproblem in bilevel problems modelling the decision of electricity prices subject to production and demands [14]. Multilevel problems with ACOPF at different time-scales are also considered [1]. The ACOPF received a lot of attention over the years, and specifically after smart grids were introduced [2].

E. Salgado
Ecole Polytechnique, 91128 Palaiseau, France
e-mail: easalgadovalenzuela@gmail.com

C. Gentile
IASI-CNR, Rome, Italy
e-mail: gentile@iasi.cnr.it

L. Liberti (✉)
CNRS LIX Ecole Polytechnique, 91128 Palaiseau, France
e-mail: liberti@lix.polytechnique.fr

© Springer Nature Switzerland AG 2018

P. Daniele and L. Scrimali (eds.), *New Trends in Emerging Complex Real Life Problems*, AIRO Springer Series 1,
https://doi.org/10.1007/978-3-030-00473-6_48

The ACOPF asks for the best power flow over an electrical network modelled by a digraph $\mathscr{D} = (\mathscr{N}, \mathscr{L})$, where $\mathscr{N}$ is the set of *buses* and $\mathscr{L}$ the set of *lines*. It is well known that the natural formulation can be simplified using only voltage variables [10]. The ACOPF is usually cast as a Mathematical Programming (MP) problem over the complex numbers (which make their appearance due to the cyclic nature of alternating currents). The standard ACOPF can be reformulated as a (larger) MP over the reals, by separating real and complex parts [17].

While the standard version of the ACOPF only has continuous variables, more realistic variants include binary variables which activate/deactivate various electrical components. In this paper we consider the possibility of activating/deactivating electrical generation at some of the buses. This defines an ACOPF variant which we call ACOPF with Generators (ACOPFG) [15].

Note that the ACOPF is **NP**-hard even without binary variables, as shown in [13]. Experimentally, however, it was found that many standard benchmarks, as well as randomly generated instances, can be solved efficiently. It is shown in [12] that this happens whenever the duality gap is zero. One might then question whether the ACOPFG is **NP**-hard simply because of the addition of the binary activation variables. The first contribution of this paper is to prove that this is indeed the case.

While ACOPF objective functions vary in the literature, it is common to consider quadratic objectives with respect to voltage. In this paper, however, we focus on a more general objective function, quadratic with respect to active power and quartic (without cubic terms) w.r.t. voltage [10]. The second contribution of this paper is the application of two perspective reformulations (PR) to the ACOPFG with the more general (quartic) objective [6, 8].

2 MP Formulation

We consider the network digraph $\mathscr{D}$ mentioned in Sect. 1. Let $n = |\mathscr{N}|$ and $\ell = |\mathscr{L}|$. We identify a subset $\mathscr{G}$ of *generator buses*, and let $n' = |\mathscr{G}|$. We note that, in modern smart grids, generators may produce *and* consume electricity. Because we are dealing with alternating currents, power is represented by a complex number. The real part is the *active power* while the complex part is *reactive*.

Notationwise, we use $[\underline{\alpha}, \overline{\alpha}]$ to denote lower/upper bounds to a quantity, and α^* to denote complex conjugate.

The parameters of our problem are as follows:

- $\forall b \in \mathscr{N}$ $S_b \in \mathbb{C}$ is the power demand at bus b;
- $\forall g \in \mathscr{G}$ $\mathfrak{S}_g = [\underline{S}_g, \overline{S}_g]$ is the (complex) interval where g can generate power if active;
- $\forall b \in \mathscr{N}$ $\mathfrak{v}_b = [\underline{v}_b, \overline{v}_b]$ is the (real) interval where the voltage magnitude at bus b can range;
- $\forall (a, b) \in \mathscr{L}$ $\overline{\imath}_{ab}$ is the maximum current which can flow through the line (a, b);

- Y is a complex $n \times n$ *bus admittance matrix* (it plays a role analoguous to the reciprocal of resistance in Ohm's law);
- Y^0, Y^1 are complex $\ell \times n$ *line admittance matrices* (they "encode" some electrical properties of the lines).

The decision variables are:

- $\forall g \in \mathscr{G}$ $s_g \in \mathbb{C}$ is the power generated at g;
- $\forall g \in \mathscr{G}$ $z_g \in \{0, 1\}$ denotes the deactivation (0) or activation (1) of generator g;
- $\forall b \in \mathscr{N}$ $v_b \in \mathbb{C}$ is the voltage at bus b;
- $\forall (a, b) \in \mathscr{L}$ $i_{ab} \in \mathbb{C}$ is the current on the line (a, b).

At each generator $g \in \mathscr{G}$, the injected complex power $s_g - S_g = v_g \sum_{(g,a)\in\mathscr{L}} i^*_{ga}$, and at each non-generator bus $b \in \mathscr{N} \smallsetminus \mathscr{G}$, we have $-S_b = v_b \sum_{(b,a)\in\mathscr{L}} i^*_{ba}$. Kirchoff's law and a generalized form of Ohm's law allow us to derive $i = Yv$, which implies that the RHS of the above equations can be reformulated to $v_b(Y^*v^*)_b = \sum_{(a,b)\in\mathscr{L}} v_b v^*_a Y^*_{ab}$ for each $b \in \mathscr{N}$ [17]. This allows us to express current in function of voltage and power. We obtain the following constraints:

$$\forall g \in \mathscr{G} \quad \sum_{(g,a)\in\mathscr{L}} Y^*_{ga} v_g v^*_a = s_g z_g - S_g \tag{1}$$

$$\forall b \in \mathscr{N} \smallsetminus \mathscr{G} \quad \sum_{(b,a)\in\mathscr{L}} Y^*_{ba} v_b v^*_a = -S_b \tag{2}$$

$$\forall (a, b) \in \mathscr{L}, \omega \in \{0, 1\} \quad \sum_{h\neq k\in\mathscr{N}} (Y^\omega_{abh})^* (Y^\omega_{abk})^* v^*_h v_k \leq \bar{\imath}_{ab} \tag{3}$$

$$\forall g \in \mathscr{G} \quad s_g \in \mathfrak{S}_g \tag{4}$$

$$\forall b \in \mathscr{N} \quad |v_b| \in \mathfrak{v}_b \tag{5}$$

$$\forall g \in \mathscr{G} \quad z_g \in \{0, 1\}. \tag{6}$$

We remark that complex power variables s only appear in Eqs. (1) and (4). We can eliminate them by replacing Eqs. (1) and (4) with the following inequalities:

$$\forall g \in \mathscr{G} \quad \underline{S}_g z_g \leq \sum_{(g,a)\in\mathscr{L}} Y^*_{ga} v_g v^*_a + S_g \leq \overline{S}_g z_g. \tag{7}$$

Moreover, if we define z over all of $\mathscr{N}$ and fix $z_b = 0$ for all $b \notin \mathscr{G}$, Eq. (7) quantified on $\mathscr{N}$ can also replace Eq. (2).

In the ACOPF literature [4, 11, 12, 15] we consider the following generation cost function, to be minimized:

$$f(s, z) = \sum_{g\in\mathscr{G}} z_g (c_{g2}(\mathsf{Re}(s_g))^2 + c_{g1}\mathsf{Re}(s_g) + c_{g0}). \tag{8}$$

Again we can replace s by v using Eq. (1) and removing constant terms in order to express Eq. (8) as a function of voltage: essentially, we obtain $f(v, z)$ from Eq. (8) by replacing s_g with $\sum_{(g,a)\in\mathscr{L}} Y^*_{ga} v_g v^*_a + S_g$.

Let F be the feasible subset of $\mathbb{C}^n$ defined by Eqs. (2)–(6) and (7). We call $\mathsf{ACOPFG}_{\mathbb{C}}$ the formulation $\min_{(v,z)\in F} f(v,z)$.

Finally, we can obtain a real formulation as follows:

1. replace each quadratic constraint $v^H M v \Diamond \alpha + j\beta$ (where $\Diamond \in \{=, \leq, \geq\}$ and $j = \sqrt{-1}$) by the pair of constraints
$$v^H M^+ v \Diamond \alpha + j\beta \wedge v^H M^- v \Diamond \alpha + j\beta,$$
where $M^+ = \frac{1}{2}(M + M^H)$ and $M^- = \frac{1}{2}(M - M^H)$;
2. replace each complex matrix M by the real matrix $\begin{pmatrix} \mathsf{Re}(M) & -\mathsf{Im}(M) \\ \mathsf{Im}(M) & \mathsf{Re}(M) \end{pmatrix}$;
3. replace each complex vector v by the real vector $(\mathsf{Re}(v)\ \mathsf{Im}(v))^\top$.

We call this reformulation $\mathsf{ACOPFG}_{\mathbb{R}}$.

3 Complexity

Assume $c_{g2} = 0$ for all $g \in \mathscr{G}$ in Eq. (8). By ignoring activation variables we obtain the ACOPF, which is a Quadratically Constrained Quadratic Program (QCQP). Since the ACOPF is **NP**-hard [3], it follows by inclusion that the ACOPFG is also **NP**-hard. On the other hand, it was shown in [12] that many practical ACOPF instances turn out to be easy rather than hard. We remark that "easy", in this setting, does not necessarily mean "in **P**", since the decision version of the QCQP is not known to be in **NP** (unless there are no quadratic constraints, in which case the problem class is known to be in **NP** [19]). The meaning of "easy" in this context is that global optima can be obtained by means of a local, rather than global, optimization procedures.

The question we answer in this section is whether the addition of the binary activation variables constitute an actual additional difficulty. To show that this need not necessarily be the case, we consider a linear system

$$\left.\begin{aligned} Ax &\leq b \\ x &\leq 1 \\ x &\in \mathbb{R}^n_+, \end{aligned}\right\} \quad (9)$$

where A is totally unimodular. Finding a feasible solution can obviously be done in polynomial time by the, say, interior point algorithm (irrespective of total unimodularity), and so this formulation is in **P**. If we add n additional binary activation variables $y_1, \ldots, y_n \in \{0, 1\}$ and n additional activation/deactivation constraints

$$\forall j \leq n \quad x_j \leq y_j, \quad (10)$$

then the new system has a constraint matrix:

$$\begin{pmatrix} A & 0 \\ I_n & 0 \\ I_n & -I_n \end{pmatrix},$$

which is easily seen to also be totally unimodular [20]. Therefore this new Mixed-Integer Linear Programming (MILP) formulation is in **P**. This provides an example where adding boolean activation variables does not make the underlying problem more difficult.

Having established that the question makes sense, we present a reduction of the weakly **NP**-complete SUBSET- SUM problem to a subclass of ACOPFG. Given an instance $(\sigma_1, \ldots, \sigma_n, S_0)$ of SUBSET- SUM, we must decide whether there is a subset $I \subseteq \{1, \ldots, n\}$ such that

$$\sum_{i \in I} \sigma_i = S_0. \tag{11}$$

This is equivalent to asking whether the following linear diophantine equation has a solution $x \in \{0, 1\}^n$:

$$\sum_{i \leq n} \sigma_i x_i = S_0. \tag{12}$$

We now show that we can naturally express Eq. (12) using the ACOPFG formulation of Sect. 2. We consider a simple network $\mathscr{D}$ with $\mathscr{G} = \{1, \ldots, n\}$ generators with demand $S_g = 0$ for $g \leq n$ and a single non-generator bus (indexed by 0) with demand S_0 (so that $\mathscr{N} = \{0, \ldots, n\}$). The set $\mathscr{L}$ of lines is $\{(g, 0) \mid 1 \leq g \leq n\}$, namely each generator is linked to the only non-generator bus. Each generator $g \in \mathscr{G}$ has generation interval $\mathfrak{S}_G = [\sigma_g, \sigma_g]$, i.e. each generator can either be inactive, or else, if active, must produce exactly σ_g. Then Eq. (7) becomes:

$$\forall g \leq n \quad Y^*_{g0} v_g v^*_0 = \sigma_g z_g. \tag{13}$$

Since we know σ_g is real and positive, we arrange Y^* so that the complex part of the LHS of Eq. (13) is zero; in particular, we arrange $Y^*_{g0} v^*_0$ to yield a $j^2 = -1$ coefficient (this can be easily done when we derive $\text{ACOPFG}_{\mathbb{R}}$). So we get:

$$\forall g \leq n \quad \mathsf{Re}(Y^*_{g0} v_g v^*_0) = -\sigma_g z_g. \tag{14}$$

Furthermore, Eq. (2) is:

$$\sum_{(g0) \in \mathscr{L}} Y^*_{g0} v_0 v^*_g = -S_0,$$

whence, by Eq. (14), we have:

$$\sum_{g\leq n}(-\sigma_g z_g) = -S_0,$$

which is exactly Eq. (12).

4 Perspective Reformulation

The objective function Eq. (8) can be restated using additional variables $p_g = \mathsf{Re}(v_g \sum_{(g,a)\in\mathscr{L}} v_a^* Y_{ga}^* + S_g)$. In practice use the convex constraints

$$\forall g \in \mathscr{G} \text{ s.t. } c_{g2} > 0 \quad p_g \geq \mathsf{Re}(v_g \sum_{(g,a)\in\mathscr{L}} v_a^* Y_{ga}^* + S_g), \tag{15}$$

which are justified by the objective function direction. We now reformulate Eq. (8) using these new variables:

$$f(p, z) = \sum_{g\in\mathscr{G}}(c_{g2}p_g^2 + c_{g1}p_g + c_{g0}z_g). \tag{16}$$

The reformulations proposed below can all be carried out on a per-generator basis. In the rest of this paper, we assume they are only applied to generators $g \in \mathscr{G}$ for which $c_{g2} > 0$.

The power p_g is subject to the following activation constraints:

$$p_g \leq \overline{P}_g z_g \quad \wedge \quad p_g \geq \underline{P}_g z_g \tag{17}$$

where $\overline{P}_g = \mathsf{Re}(\overline{S}_g)$ and $\underline{P}_g = \mathsf{Re}(\underline{S}_g)$. The PR reformulation [8] can be applied to (16) as follows:

$$\hat{f}(p, z) = \sum_{g\in\mathscr{G}}\left(c_{g2}\frac{p_g^2}{z_g} + c_{g1}p_g + c_{g0}z_g\right). \tag{18}$$

The function (18) can be optimized using the perspective cuts (PC) method [8], which works as follows: (i) first we add new variables t_g representing the nonlinear part of the cost in (18) by considering the following constraints

$$t_g \geq c_{g2}\frac{p_g^2}{z_g}, \tag{19}$$

and replacing (18) with $\tilde{f}(t, p, z) = \sum_{g\in\mathscr{G}}(t_g + c_{g1}p_g + c_{g0}z_g)$; (ii) then constraints (19) can be replaced by PCs:

$$t_g \geq c_{g2}(2\check{p}_g p_g - \check{p}_g^2 z_g), \tag{20}$$

where $\check{p}_g$ are fixed values of the real power p_g varying in the feasible interval $\underline{P}_g \leq \check{p}_g \leq \check{P}_g$ when $z_g = 1$. The addition of PCs does not add further difficulties in the problem formulation except for the condition that they should be generated iteratively as their number is not finite.

We can alternatively apply the AP2R technique [6, 7], which works in two phases. The first phase is a projection where the optimal value of z_g for the continuous relaxation of Eq. (18) subject to Eq. (17) is found depending on p_g. The second phase is a lifting where the variables z_g are lifted back. The resulting problem can be solved using an off-the-shelf MIP solver. This is equivalent to replacing (18) and (17) with:

$$\left.\begin{array}{lr} \min \sum\limits_{g\in\mathscr{G}} \left(z_g(p_g^{\text{int}})^2 + \check{f}_g(\pi_g + p_g^{\text{int}}) - \check{f}_g(p_g^{\text{int}}) + c_{g1}p_g + c_{g0}z_g\right) & \\ \forall g \in \mathscr{G} \qquad (\underline{P}_g - p_g^{\text{int}})z_g \leq \pi_g \leq (\overline{P}_g - p_g^{\text{int}})z_g & \\ \forall g \in \mathscr{G} \qquad \pi_g = p_g - p_g^{\text{int}} z_g, & \end{array}\right\} \tag{21}$$

where $\check{f}_g(x) = c_{g2}x^2$ and p_g^{int} is

$$p_g^{\text{int}} = \max(\underline{P}_g, \min(\sqrt{c_{g0}/c_{g2}}, \overline{P}_g)). \tag{22}$$

The final AP2R reformulation consists in Eqs. (21), (2)–(6), the complex part of Eqs. (7) and (15).

5 Computational Results

We tested PRs with 4 cuts and AP2R (implemented using AMPL [5]): both on the ACOPFG formulation $\text{ACOPFG}_{\mathbb{R}}$ in Sect. 2 (Table 1) and on the dual Diagonally Dominant Programming (DDP) outer-approximation proposed in [18] (Table 2) solved using CPLEX [9]. We compared these results with local optima of the ACOPF (all active generators) obtained by MatPower [21] and by solving the $\text{ACOPFG}_{\mathbb{R}}$ using Baron [16] to global optimality (within a limited CPU time of 1h). The test set includes small to medium scale instances taken from MatPower; results on one larger-scale instance are reported in Table 2. All results were obtained on an Intel i7 dual-core CPU at 2.1 GHz with 16 GB RAM.

In the "Perspective reformulation" columns we show: number of iterations, CPU seconds (limited to 1 h), PR objective value obtained on 1st iteration and final value, original objective function value at optimum, percentage of active generators at

Table 1 Results on $\mathsf{ACOPFG}_{\mathbb{R}}$ ('x': solution not found within time limit)

Instance	Perspective reformulation ($\mathsf{ACOPFG}_{\mathbb{R}}$)						AP2R ($\mathsf{ACOPFG}_{\mathbb{R}}$)				MATPOWER	Mi-Quartic			Solution's distances	
	It	Time	First value	Last value	% active	Real value	Time	% active	Value	Real value	Calue	Time	% active	Best value	Persp/Mi-Quartic	AP2R/Mi-Quartic
WB2	2	24	878.182	878.182	100	878.182	13	100	878.18	878.182	877.78	13	100	878.182	0	0
WB3	2	169	417.244	417.244	100	417.244	109	100	417.244	417.244	417.25	108	100	417.244	0	0
WB5	2	3600	947.056	947.056	100	947.056	2269	100	947.056	947.056	1082.33	2454	100	947.056	0	0
6ww	1	3600	2913.58	2913.58	x	2881.28	3600	100	10948.7	135.18	3134.35	3600	100	3135.18	0.1584	0
case9	2	3600	2062.65	5115.72	100	5430.38	3600	66.7	10948.74	7335.42	5296.69	3600	100	5296.69	0.2561	0.6234
case14	1	3600	5250.22	5250.22	x	5375.94	3600	80	6589.72	5287.72	8081.53	3600	60	5476.90	0.1370	0.8304
case30	1	3600	430.906	430.906	x	536.307	3600	66.7	503.508	503.508	576.89	3600	83.3	515.807	0.4234	0.3503

Table 2 Results on dual DDP [18] ('x': solution not found within time limit)

Instance	Perspective reformulation (dual DDP salgado3)						AP2R (dual DDP salgado3)				MATPOWER	Mi-Quartic			Solution's distances	
	It	Time	First value	Last value	% active	Real value	Time	% active	Value	Real value	Value	Time	% active	Best value	Persp/Mi-Quartic	AP2R/Mi-Quartic
WB2	2	0.001	876.923	876.923	100	876.923	0.001	100	876.923	876.923	877.78	13	100	878.182	0.0014	0.0014
WB3	2	0.002	398.443	398.443	100	398.443	0.001	100	398.443	398.443	417.25	108	100	417.244	0.0472	0.0472
WB5	2	0.005	677.688	677.688	100	677.688	0.008	100	677.688	677.688	1082.33	1200	100	947.056	0.3563	0.3992
6ww	4	0.02	2760.751	2838.693	66.7	2844.44	0.008	66.7	57287.6	2841.61	3134.35	1200	100	3135.177	0.6108	0.6108
case9	2	0.03	2012.135	5034.028	100	5430.38	0.031	66.7	10810.8	7197.45	5296.69	1200	100	5296.686	0.2422	0.8423
case14	3	0.15	5091.340	5390.800	60	5406.53	1200	80	4746.99	4746.99	8081.53	1200	60	5476.905	0.0572	0.0325
case30	2	3.28	398.427	509.554	83.3	518.42	21	66.7	492.232	492.232	576.89	1200	83.3	515.807	0.6031	0.4068
case89pegase	2	297.33	5730.152	5730.152	100	5730.15	286.484	100	5730.15	5730.15	5817.60	x	x	x	x	x

optimum. In the "AP2R" and "Mi-Quartic" columns we show CPU time, objective function value and percentage of active generators. In "Solution's distances" we report a scaled distance of the optima found by PR/AP2R w.r.t. Mi-Quartic, namely $\frac{\|p^{\omega}-p^{\text{MI-Quartic}}\|_1}{\|p^{\omega}\|_1}$, $\omega \in \{\mathsf{PR},\ \mathsf{AP2R}\}$.

While it is clear that the tests with $\mathsf{ACOPFG}_{\mathbb{R}}$ are inconclusive, those on the dual DDP approximation give very tight bounds in relatively little time.

Acknowledgements This paper has received funding from the European Union's Horizon 2020 research and innovation programme under the Marie Sklodowska-Curie grant agreement n. 764759. The second author is partially supported by MIUR PRIN2015 project no. 2015B5F27W. The last author (LL) is grateful for financial support by CNR STM Program prot. AMMCNT-CNR n. 80058 dated 05/12/2017.

References

1. Bacher, R.: Power system models, objectives and constraints in optimal power flow calculations. In: Frauendorfer, K., Glavitsch, H., Bacher, R. (eds.) Optimization in Planning and Operation of Electric Power Systems, pp. 217–264. Springer, Heidelberg (1993)
2. Bienstock, D.: Electrical Transmission System Cascades and Vulnerability: an Operations Research Viewpoint. Number 22 in MOS-SIAM Optimization. SIAM, Philadelphia (2016)
3. Bienstock, D., Verma, A.: Strong NP-hardness of AC power flows feasibility. Technical report (2015). arXiv:1512.07315
4. Chen, C., Atamtürk, A., Oren, S.: Bound tightening for the alternating current optimal power flow problem. IEEE Trans. Power Syst. **31**(5), 3729–3736 (2016)
5. Fourer, R., Gay, D., Kernighan, B.W.: AMPL: A Modeling Language for Mathematical Programming. AMPL Optimization LLC (2003)
6. Frangioni, A., Furini, F., Gentile, C.: Approximated perspective relaxations: a project&lift approach. Comput. Optim. Appl. **63**(3), 705–735 (2016)
7. Frangioni, A., Furini, F., Gentile, C.: Improving the approximated projected perspective reformulation by dual information. Oper. Res. Lett. **45**, 519–524 (2017)
8. Frangioni, A., Gentile, C.: Perspective cuts for a class of convex 0–1 mixed integer programs. Math. Program. **106**(2), 225–236 (2006)
9. IBM. ILOG CPLEX 12.6 User's Manual. IBM (2014)
10. Josz, C.: Application of polynomial optimization to electricity transmission networks. Ph.D. thesis, Univ. Paris VI (2016)
11. Kuang, X., Ghaddar, B., Naoum-Sawaya, J., Zuluaga, L.: Alternative LP and SOCP hierarchies for ACOPF problems. IEEE Trans. Power Syst. **32**(4), 2828–2836 (2017)
12. Lavaei, J., Low, S.: Zero duality gap in optimal power flow problem. IEEE Trans. Power Syst. **27**(1), 92–107 (2012)
13. Lehmann, K., Grastien, A., van Hentenryck, P.: AC-Feasibility on tree networks is NP-hard. IEEE Trans. Power Syst. **13**(1), 798–801 (2016)
14. Panciatici, P.: Private Communication (2016)
15. Ruiz, M., Maeght, J., Marié, A., Panciatici, P., Renaud, A.: A progressive method to solve large-scale AC optimal power flow with discrete variables and control of the feasibility. In: Proceedings of the Power Systems Computation Conference, PSCC, Piscataway, vol. 18. IEEE (2014)
16. Sahinidis, N.: BARON user Manual v. 17.8.9. The Optimization Firm LLC (2017)
17. Salgado, E.: Fast relaxations for alternating current optimal power flow. Master's thesis, LIX, Ecole Polytechnique (2017)

18. Salgado, E., Scozzari, A., Tardella, F., Liberti, L.: Alternating current optimal power flow with generator selection. In: Proceedings of ISCO 2018, vol. 10856, pp. 364–375. LNCS (2018)
19. Vavasis, S.: Quadratic programming is in NP. Inf. Process. Lett. **36**, 73–77 (1990)
20. Wolsey, L.A.: Integer Programming. Wiley, New York (1998)
21. Zimmermann, R., Murillo-Sanchez, C., Thomas, R.: MATPOWER: steady-state operations, planning, and analysis tools for power systems research and education. IEEE Trans. Power Syst. **26**(1), 12–19 (2010)

Coalitional Games in Evolutionary Supply Chain Networks

Laura Scrimali

Abstract We focus on the coalition formation in a supply chain network that consists of three layers of decision-makers, namely, suppliers, manufacturers, and retailers, with prices and shipments that evolve over time. We suppose that some partners in the chain vertically merge each other and act as one player to confront the other players that make their choices independently. In this model, the retailer is the dominant player and is a profit-maximizer. We present a non-cooperative approach to the coalitional game and provide the equilibrium conditions governing the model as well as an equivalent evolutionary variational inequality formulation.

Keywords Evolutionary variational inequality · Supply chain · Coalitions
Nash equilibrium

1 Introduction

This paper investigates the effects of coalition formation on the economic performance of a supply chain network in a time-dependent setting. In particular, we consider the vertical integration of suppliers, manufacturers, and retailers who make a joint venture to confront the other players. In this situation, the retailer is considered as the dominant player of the coalition and acts as a profit-maximizer.

Papers in the literature study different aspects of supply chain networks in both the static and time-dependent case. For instance, in [1], the authors study a three-echelon supply chain network and discuss the value of integrating a pair of partners in the chain. They study different scenarios, presenting and solving the related optimization

L. Scrimali (✉)
Department of Mathematics and Computer Science, University of Catania,
Viale Andrea Doria 6, Catania, Italy
e-mail: scrimali@dmi.unict.it

© Springer Nature Switzerland AG 2018

P. Daniele and L. Scrimali (eds.), *New Trends in Emerging Complex Real Life Problems*, AIRO Springer Series 1,
https://doi.org/10.1007/978-3-030-00473-6_49

problems in finite-dimensional spaces. In [3], the author considers a three-layer supply chain network with prices and shipments evolving over time and in the presence of excesses of production and demand. Equilibrium conditions and a time-dependent variational formulation of the complete supply chain are presented. In [6], the authors study the selection of chain partners, the formation of supply chains and outcome allocations. They formulate the static coalitional game as a variational inequality problem and provide an iterative diagonalization algorithm to determine the steady state for the game. In [2], the process of coalition formation is modeled as a non-cooperative bargaining game, where firms make offers and respond according to an exogenous rule of order, which establishes who is the first proposer and the order of reply. The equilibrium of the merger formation game for any possible asset structure is given, taking the coalitions' profit function resulting from Cournot competition.

In our paper, we depart from the partner selection problem and model the coalition formation game as an evolutionary variational inequality (EVI). Such tool well describes equilibrium problems and situations in which several decision-makers interact under technical and economic constraints. Variational inequality theory (see [4]) has been developed and applied to many research fields such as spatial equilibrium models, supply chain management, financial networks, transportation networks, electricity markets, and pollution. In addition, we give the equilibrium conditions governing the model and prove the equivalence with the EVI solution.

We emphasize that we study a continuous time model. Life unfolds continuously, thus suggesting that a continuous time framework is more realistic and allows one to follow the adjustment process through time. Consequently, our model does not provide static solutions, but curves of equilibria that describe the evolution of the whole system through time. The paper is organized as follows. In Sect. 2, we introduce the supply chain network model consisting of material providers, manufacturers, and retailers. In Sect. 3, we discuss the vertical integration of the partners of the chain and provide the equilibrium conditions. Finally, in Sect. 4, we present the variational inequality formulation of the coalition formation problem.

2 The Supply Chain Model

In this section, we present a supply chain network model that consists of three layers of decision makers: material providers, manufacturers, and retailers (Fig. 1).

We denote by $S = \{s_1, \ldots, s_h\}$ the set of suppliers operating in the market, with typical supplier denoted by s_i; $M = \{m_1, \ldots, m_l\}$ the set of manufacturers, with typical manufacturer denoted by m_j; and $R = \{r_1, \ldots, r_n\}$ the set of retailers, with typical retailer denoted by r_k. Manufacturers purchase raw material from suppliers and are involved in the production of a product, which can then be bought by retailers, who, in turn, make the product available to consumers. Suppliers are located at the top tier of nodes in the network; the manufacturing firms are located at the middle tier, whereas the retailers are located at the third or bottom tier. We assume that all players have complete information and no shortages are considered.

Fig. 1 A three-layer supply chain network

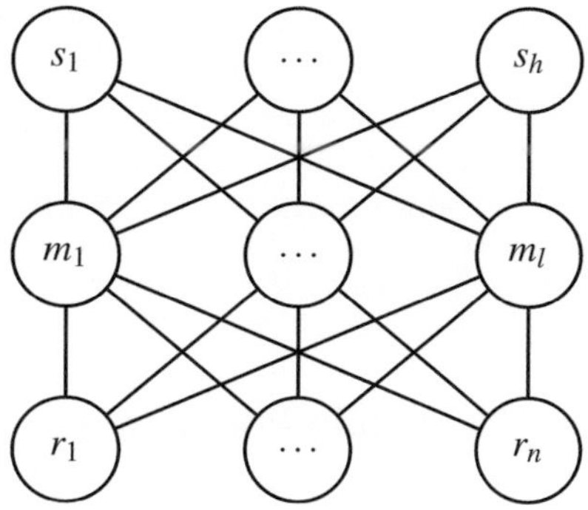

In the decentralized case, players in the same layer mutually compete, playing non-cooperative Nash games ([8, 9]). In other words, each supplier competes with the other suppliers; each manufacturer competes with the competitors in the related market, and each retailer competes with the other retailers. Thus, suppliers set their own price; manufacturers decide the order quantity and the wholesale price, while retailers charge the final price to consumers.

In our paper, we focus on the centralized model, in which the partners of some supply chains vertically merge and make a joint venture. The retailer of each integration is the dominant player and each coalition acts as one retailer to confront the other players that make their choices independently.

We now give the notation for the model.

- $q_{s_i}(t)$ is the nonnegative quantity of material produced by supplier s_i at time t.
- $q_{m_j}(t)$ is the nonnegative quantity of final product produced by manufacturer m_j at time t.
- $q_{r_k}(t)$ is the nonnegative quantity sold by retailer r_k at time t.
- $q(t) = (q_{r_k}(t), q_{r_{-k}}(t))$ is the the n-dimensional vector of the quantity sold by retailers, where $q_{r_{-k}}(t)$ denotes the $(n-1)$-dimensional vector of the quantities sold by competitors of retailer r_k.
- $p_{r_k}(t, q)$ is the demand function of retailer r_k at time t.
- $c_{s_i}(t, q_{s_i})$ is the nonnegative material production cost of supplier s_i at time t.
- $c_{m_j}(t, q_{m_j})$ is the nonnegative production cost of manufacturer m_j at time t.
- $c_{r_k}(t, q_{r_k})$ is the nonnegative holding cost of retailer r_k at time t.
- $\tau_{r_k}(t, q_{r_k})$ is the nonnegative transaction/transportation cost from retailer r_k to consumers at time t.

We also introduce the nonnegative production coefficient $\delta_{s_i m_j}$, which indicates the amount of raw material necessary to produce one unit of the final product.

We study the evolution of the model in the time interval $[0, \bar{t}]$, with $\bar{t} > 0$. We choose as our functional setting the space $L^2([0, \bar{t}], \mathbb{R}^N)$ (where N will be specified in the following) of square-integrable functions defined in the closed interval $[0, \bar{t}]$, endowed with the scalar product $\langle \cdot, \cdot \rangle_{L^2} = \int_0^{\bar{t}} \langle \cdot, \cdot \rangle dt$ and the usual associated norm $\| \cdot \|_{L^2}$. This choice is convenient since the space is very large, and allows us to consider both smooth and non-smooth functions. We note that some of the above functions are often piecewise constant, and this class of functions largely falls within the L^2 space. Moreover, in this space, we can prove the existence of solutions under minimal assumptions. Finally, we assume that all the price and cost functions are Carathéodory and continuously differentiable operators, and belong to the functional space.

3 The Chain Formation

In this section, we study the vertical integration of three partners of the supply chain network. We propose that a supplier, a manufacturer, and a retailer merge each other. In this scenario, the retailer is the dominant player and the coalition acts as the retailer. Thus, the revenue of the coalition will be the revenue generated by the products sold by the retailers. By vertically integrating supplier, manufacturer, and retailer, the coalition will affect costs and benefits of the members. The incurred costs include production costs, holding costs and transaction costs with consumers. We denote by $d_{s_i}(t)$ and $d_{m_j}(t)$ the disagreement profit of supplier s_i and manufacturer m_j, respectively, at time t which represents the profit in a competitive economy. The players that do not merge act in a non-cooperative fashion.

For the typical coalition (s_i, m_j, r_k), the retailer r_k is the dominant player and is a profit-maximizer. For any given quantity of the opponent retailers $q_{r_{-k}}(t)$, we define the chain value, or payoff of the coalition, as:

$$z_{s_i m_j r_k}(t, q_{s_i}, q_{m_j}, q_{r_k}, q_{r_{-k}}) = p_{r_k}(t, q_{r_k}, q_{r_{-k}})q_{r_k}(t) - c_{s_i}(t, q_{s_i}) - c_{m_j}(t, q_{m_j})$$
$$-c_{r_k}(t, q_{r_k}) - \tau_{r_k}(t, q_{r_k}) - d_{s_i}(t) - d_{m_j}(t). \tag{1}$$

The maximization profit problem is then given by:

$$\max \int_0^{\bar{t}} z_{s_i m_j r_k}(t, q_{s_i}, q_{m_j}, q_{r_k}, q_{r_{-k}})\mathrm{d}t, \text{ subject to}$$
$$(q_{s_i}, q_{m_j}, q_{r_k}) \in K_{s_i m_j r_k}, \tag{2}$$

where

$$K_{s_i m_j r_k} = \Big\{(q_{s_i}, q_{m_j}, q_{r_k}) \in L^2([0,\bar{t}], \mathbb{R}^3) : 0 \le q_{s_i}(t) \le Q^{max}_{s_i}(t),$$
$$0 \le q_{m_j}(t) \le Q^{max}_{m_j}(t), 0 \le q_{r_k}(t) \le Q^{max}_{m_j}(t), \text{ a.e. in } [0,\bar{t}]\Big\},$$

$Q^{max}_{s_i}(t)$ is the positive production capacity of supplier s_i at time t, and $Q^{max}_{m_j}(t)$ the positive production capacity of manufacturer m_j at time t.

We note that set $K_{s_i m_j r_k}$ is a bounded, convex, and closed subset of the space $L^2([0,\bar{t}],\mathbb{R}^3)$, then it is also weakly compact. Thus, if the payoff of the coalition is weakly upper semicontinuous problem (2) admits solutions (see [5]).

We now can give the following definition of equilibrium of the coalition formation problem.

Definition 1 The vector $(q^*_{s_i}, q^*_{m_j}, q^*_{r_k}) \in K_{s_i m_j r_k}$ is an equilibrium of the evolutionary coalition formation problem if and only if the following market conditions are verified a.e. in $[0,\bar{t}]$:

$$\frac{\partial z_{s_i m_j r_k}(t, q^*_{s_i}, q^*_{m_j}, q^*_{r_k}, q_{r_{-k}})}{\partial q_{s_i}} \begin{cases} \geq 0 \text{ if } q^*_{s_i}(t) = Q^{max}_{s_i}(t) \\ = 0 \text{ if } 0 < q^*_{s_i}(t) < Q^{max}_{s_i}(t) \\ \leq 0 \text{ if } q^*_{s_i}(t) = 0, \end{cases} \tag{3}$$

$$\frac{\partial z_{s_i m_j r_k}(t, q^*_{s_i}, q^*_{m_j}, q^*_{r_k}, q_{r_{-k}})}{\partial q_{m_j}} \begin{cases} \geq 0 \text{ if } q^*_{m_j}(t) = Q^{max}_{m_j}(t) \\ = 0 \text{ if } 0 < q^*_{m_j}(t) < Q^{max}_{m_j}(t) \\ \leq 0 \text{ if } q^*_{m_j}(t) = 0, \end{cases} \tag{4}$$

$$\frac{\partial z_{s_i m_j r_k}(t, q^*_{s_i}, q^*_{m_j}, q^*_{r_k}, q_{r_{-k}})}{\partial q_k} \begin{cases} \geq 0 \text{ if } q^*_{r_k}(t) = Q^{max}_{m_j}) \\ = 0 \text{ if } 0 < q^*_{r_k}(t) < Q^{max}_{m_j}(t) \\ \leq 0 \text{ if } q^*_{r_k}(t) = 0. \end{cases} \tag{5}$$

Condition (3) has the following meaning. If the quantity of raw material produced reaches the production capacity, then the marginal profit of supplier s_i is non negative, namely, the marginal revenue is greater than or equal to the marginal costs. If the quantity of raw material produced is positive and strictly less than the capacity $Q^{max}_{s_i}$, then market clears and the marginal revenue equals the marginal costs. Finally, there will be no production if the marginal profit is less than or equal to zero (the marginal revenue is less than or equal to the marginal costs). A similar meaning can be deduced for the conditions on the marginal profits of manufacturer m_j and retailer r_k.

4 Variational Inequality Formulation

In this section, we present a variational inequality approach to the coalition formation problem.

Theorem 1 *$(q^*_{s_i}, q^*_{m_j}, q^*_{r_k}) \in K_{s_i m_j r_k}$ is a solution of problem (2) if and only if it satisfies the following variational inequality:*

$$\int_0^{\bar{t}} \Bigg(-\frac{\partial z_{s_i m_j r_k}(t, q^*_{s_i}, q^*_{m_j}, q^*_{r_k}, q_{r_{-k}})}{\partial q_{s_i}} \Big(q_{s_i}(t) - q^*_{s_i}(t)\Big)$$
$$-\frac{\partial z_{s_i m_j r_k}(t, q^*_{s_i}, q^*_{m_j}, q^*_{r_k}, q_{r_{-k}})}{\partial q_{m_j}} \Big(q_{m_j}(t) - q^*_{m_j}(t)\Big)$$
$$-\frac{\partial z_{s_i m_j r_k}(t, q^*_{s_i}, q^*_{m_j}, q^*_{r_k}, q_{r_{-k}})}{\partial q_{r_k}} \Big(q_{r_k}(t) - q^*_{r_k}(t)\Big)\Bigg)\mathrm{d}t \geq 0,$$
$$\forall (q_{s_i}, q_{m_j}, q_{r_k}) \in K_{s_i m_j r_k}. \tag{6}$$

Proof We assume that $(q^*_{s_i}, q^*_{m_j}, q^*_{r_k}) \in K_{s_i m_j r_k}$ is a solution to problem (2). Then, for all $(q_{s_i}, q_{m_j}, q_{r_k}) \in K_{s_i m_j r_k}$ the function

$$G(\lambda) = \int_0^{\bar{t}} z_{s_i m_j r_k}(t, \lambda q^*_{s_i} + (1-\lambda) q_{s_i}, \lambda q^*_{m_j} - (1-\lambda) q_{m_j}, \lambda q^*_{r_k} + (1-\lambda) q_{r_k}, q_{r_{-k}})\mathrm{d}t,$$
$$\forall \lambda \in [0, \bar{t}],$$

admits a maximal solution at $\lambda = 1$ and $G'(1) \geq 0$. Therefore, we find

$$0 \leq G'(1) = \int_0^{\bar{t}} \Bigg(-\frac{\partial z_{s_i m_j r_k}(t, q^*_{s_i}, q^*_{m_j}, q^*_{r_k}, q_{r_{-k}})}{\partial q_{s_i}} \Big(q_{s_i}(t) - q^*_{s_i}(t)\Big)$$
$$-\frac{\partial z_{s_i m_j r_k}(t, q^*_{s_i}, q^*_{m_j}, q^*_{r_k}, q_{r_{-k}})}{\partial q_{m_j}} \Big(q_{m_j}(t) - q^*_{m_j}(t)\Big)$$
$$-\frac{\partial z_{s_i m_j r_k}(t, q^*_{s_i}, q^*_{m_j}, q^*_{r_k}, q_{r_{-k}})}{\partial q_{r_k}} \Big(q_{r_k}(t) - q^*_{r_k}(t)\Big)\Bigg)\mathrm{d}t,$$
$$\forall (q_{s_i}, q_{m_j}, q_{r_k}) \in K_{s_i m_j r_k},$$

which is variational inequality (6).

Conversely, we assume that $(q^*_{s_i}, q^*_{m_j}, q^*_{r_k}) \in K_{s_i m_j r_k}$ is a solution to (6). Since the function

$$\bar{z}_{s_i m_j r_k}(q_{s_i}, q_{m_j}, q_{r_k}, q_{r_{-k}}) = \int_0^{\bar{t}} z_{s_i m_j r_k}(t, q_{s_i}, q_{m_j}, q_{r_k}, q_{r_{-k}})\mathrm{d}t$$

is concave, then for all $(q^*_{s_i}, q^*_{m_j}, q^*_{r_k}) \in K_{s_i m_j r_k}$ the following estimate holds

$$-\bar{z}_{s_i m_j r_k}(\lambda q_{s_i} + (1-\lambda) q^*_{s_i}, \lambda q_{m_j} + (1-\lambda) q^*_{m_j}, \lambda q_{r_k} + (1-\lambda) q^*_{r_k}, q_{r_{-k}})$$
$$\leq -\lambda \bar{z}_{s_i m_j r_k}(q_{s_i}, q_{m_j}, q_{r_k}, q_{r_{-k}}) - (1-\lambda) \bar{z}_{s_i m_j r_k}(q^*_{s_i}, q^*_{m_j}, q^*_{r_k}, q_{r_{-k}}).$$

Thus, $\forall \lambda \in [0, 1]$ we find

$$\frac{1}{\lambda}\Big(- \overline{z}_{s_i m_j r_k}(q^*_{s_i} + \lambda(q_{s_i} - q^*_{s_i}), q^*_{m_j} + \lambda(q_{m_j} - q^*_{m_j}), q^*_{r_k} + \lambda(q_{r_k} - q^*_{r_k}))$$
$$+\overline{z}_{s_i m_j r_k}(q^*_{s_i}, q^*_{m_j}, q^*_{r_k}, q_{r_{-k}})\Big)$$
$$\leq -\overline{z}_{s_i m_j r_k}(q_{s_i}, q_{m_j}, q_{r_k}, q_{r_{-k}}) + \overline{z}_{s_i m_j r_k}(q^*_{s_i}, q^*_{m_j}, q^*_{r_k}, q_{r_{-k}}). \tag{7}$$

When λ tends to zero, the left-hand side of (7) converges to the left-hand side of variational inequality (6); hence

$$0 \leq -\overline{z}_{s_i m_j r_k}(q_{s_i}, q_{m_j}, q_{r_k}, q_{r_{-k}}) + \overline{z}_{s_i m_j r_k}(q^*_{s_i}, q^*_{m_j}, q^*_{r_k}, q_{r_{-k}}),$$

and $(q^*_{s_i}, q^*_{m_j}, q^*_{r_k})$ is an optimal solution to problem (2). □

We now prove that equilibrium conditions (3)–(5) are equivalent to variational inequality (6).

Theorem 2 *A vector $(q^*_{s_i}, q^*_{m_j}, q^*_{r_k}) \in K_{s_i m_j r_k}$ is an equilibrium of the evolutionary coalition formation problem if and only if it is solution to variational inequality (6).*

Proof Let $(q^*_{s_i}, q^*_{m_j}, q^*_{r_k}) \in K_{s_i m_j r_k}$ be an equilibrium solution. Therefore, a.e. in $[0, \overline{t}]$, we may write

$$-\frac{\partial z_{s_i m_j r_k}(q^*_{s_i}, q^*_{m_j}, q^*_{r_k}, q_{r_{-k}})}{\partial q_{s_i}}\Big(q_{s_i}(t) - q^*_{s_i}(t)\Big) \geq 0,$$
$$-\frac{\partial z_{s_i m_j r_k}(q^*_{s_i}, q^*_{m_j}, q^*_{r_k}, q_{r_{-k}})}{\partial q_{m_j}}\Big(q_{m_j}(t) - q^*_{m_j}(t)\Big) \geq 0,$$
$$-\frac{\partial z_{s_i m_j r_k}(q^*_{s_i}, q^*_{m_j}, q^*_{r_k}, q_{r_{-k}})}{\partial q_{r_k}}\Big(q_{r_k}(t) - q^*_{r_k}(t)\Big) \geq 0.$$

Summing up the above inequalities and integrating, we immediately find (6).

We now assume that $(q^*_{s_i}, q^*_{m_j}, q^*_{r_k}) \in K_{s_i m_j r_k}$ satisfies variational inequality (6) and choose $q_{m_j} = q^*_{m_j}, q_{r_k} = q^*_{r_k}$ in $[0, \overline{t}]$, which leads to

$$\int_0^{\overline{t}} -\frac{\partial z_{s_i m_j r_k}(q^*_{s_i}, q^*_{m_j}, q^*_{r_k}, q_{r_{-k}})}{\partial q_{s_i}}\Big(q_{s_i}(t) - q^*_{s_i}(t)\Big)\mathrm{d}t \geq 0, \forall q_{s_i}(t) : 0 \leq q_{s_i}(t) \leq Q^{max}_{s_i}(t).$$

By standard arguments, we find (3). Analogously, it is possible to deduce the other conditions. □

We note that in equilibrium, all the shipments between the levels have to coincide, namely,

$$q^*_{s_i}(t) = \delta_{s_i m_j} q^*_{m_j}(t) \text{ and } q^*_{m_j}(t) = q^*_{r_k}(t), \text{ a.e. in } [0, \overline{t}].$$

This means that, in equilibrium, the supplier will produce the quantity of raw material that the manufacturer demands, and the retailer will purchase from the manufacturer the quantity produced.

Thus, the optimization problem of coalition (s_i, m_j, r_k) becomes

$$\max \int_0^{\bar{t}} z_{s_i m_j r_k}(t, q_{r_k}, q_{r_{-k}})\mathrm{d}t, \text{ subject to } q_{r_k}(t) \in K_{r_k}, \tag{8}$$

where

$$\begin{aligned} z_{s_i m_j r_k}(t, q_{r_k}, q_{r_{-k}}) = \; & p_{r_k}(t, q_{r_k}, q_{r_{-k}}) q_{r_k}(t) - c_{s_i}(t, \delta_{s_i m_j} q_{r_k}) - c_{m_j}(t, q_{r_k}) \\ & -c_{r_k}(t, q_{r_k}) - \tau_{r_k}(t, q_{r_k}) - d_{s_i}(t) - d_{m_j}(t), \end{aligned} \tag{9}$$

$$K_{r_k} = \left\{ q_{r_k} \in L^2([0, \bar{t}]) : 0 \leq q_{r_k}(t) \leq Q_{m_j}^{max}(t), \text{ a.e. in } [0, \bar{t}] \right\}.$$

The profit-maximization model of a coalition (s_i, m_j, r_k) is then equivalent to the following evolutionary variational inequality:

$$\int_0^{\bar{t}} -\frac{\partial z_{s_i m_j r_k}(t, q^*_{r_k}, q_{r_{-k}})}{\partial q_{r_k}} (q_{r_k}(t) - q^*_{r_k}(t))\mathrm{d}t \geq 0, \; \forall q_{r_k} \in K_{r_k}. \tag{10}$$

If we consider a market in which $\bar{n} \in \{1, \ldots, n\}$ coalitions compete in a non-cooperative fashion, the equilibrium conditions for coalitions, simultaneously, can be expressed as the following variational inequality:

$$\sum_{k=1}^{\bar{n}} \int_0^{\bar{t}} -\frac{\partial z_{s_i m_j r_k}(t, q^*)}{\partial q_{r_k}} (q_{r_k}(t) - q^*_{r_k}(t))\mathrm{d}t \geq 0, \; \forall q \in \prod_{k=1}^{\bar{n}} K_{r_k}. \tag{11}$$

The existence of solutions is out of the scope of this paper; hence, we address the interested reader to [7].

Acknowledgements The research was partially supported by the research project "Modelli Matematici nell'Insegnamento-Apprendimento della Matematica" DMI, University of Catania. This support is gratefully acknowledged.

References

1. Amoozad Mahdiraji, H., Govindan, K., Zavadskas, E.K., Razavi Hajiagha, S.H.: Coalition or decentralization: a game-theoretic analysis of a three-echelon supply chain network. J. Bus. Econ. Manag. **15**(3), 460–485 (2014)
2. Bartolini, D.: Investment and merging as an endogenous coalition formation game. In: Discussion Paper n. 628, University of Essex (2011)

3. Daniele, P.: Evolutionary variational inequalities and applications to complex dynamic multi-level models. Transp. Res. Part E **46**, 855–880 (2010)
4. Facchinei, F., Pang, J.S.: Finite-Dimensional Variational Inequalities and Complementarity Problems. Vols. I and II. Springer, New York (2003)
5. Jahn, J.: Introduction to the Theory of Nonlinear Optimization. Springer, Berlin (1996)
6. Lin, C.C., Hsieh, C.C.: A cooperative coalitional game in duopolistic supply-chain competition. Netw. Spat. Econ. **12**(1), 129–146 (2012)
7. Maugeri, A., Raciti, F.: On existence theorems for monotone and nonmonotone variational inequalities. J. Convex Anal. **16**(3–4), 899–911 (2009)
8. Nash, J.F.: Equilibrium points in n-person games. Proc. Natl. Acad. Sci. **36**, 48–49 (1950)
9. Nash, J.F.: Non-cooperative games. Ann. Math. **54**, 286–295 (1951)

A Recent Approach to Derive the Multinomial Logit Model for Choice Probability

Roberto Tadei, Guido Perboli and Daniele Manerba

Abstract It is well known that the Multinomial Logit model for the choice probability can be obtained by considering a random utility model where the choice variables are independent and identically distributed with a Gumbel distribution. In this paper we organize and summarize existing results of the literature which show that using some results of the extreme values theory for i.i.d. random variables, the Gumbel distribution for the choice variables is not necessary anymore and any distribution which is asymptotically exponential in its tail is sufficient to obtain the Multinomial Logit model for the choice probability.

Keywords Random utility · Extreme values theory · Asymptotic approximation Multinomial Logit model

1 Introduction

In this paper we consider a discrete choice model where a decision maker needs to select an alternative among a finite set of mutually exclusive alternatives. Each alternative is characterized by a random utility. The decision maker will select the alternative with the greatest utility. Discrete choice models of this kind are called random utility models [13]. The aforementioned models are typical of several applications in operations management where decisions must be taken in advance with a limited knowledge of the alternatives, as in supply chain optimization, logistics, and transportation (see, e.g., [3, 4, 12, 15–17, 19, 22]).

R. Tadei (✉) · G. Perboli · D. Manerba
Department of Control and Computer Engineering, Politecnico di Torino,
Corso Duca degli Abruzzi 24, 10129 Turin, Italy
e-mail: roberto.tadei@polito.it

G. Perboli
e-mail: guido.perboli@polito.it

D. Manerba
e-mail: daniele.manerba@polito.it

© Springer Nature Switzerland AG 2018

P. Daniele and L. Scrimali (eds.), *New Trends in Emerging Complex Real Life Problems*, AIRO Springer Series 1,
https://doi.org/10.1007/978-3-030-00473-6_50

It is well known that the Multinomial Logit model (MNL) for the choice probability can be derived assuming that the random utilities are independent and identical distributed (i.i.d.) across alternatives and that their common distribution is a Gumbel function [1, 2, 6, 11].

In [9, 10] an asymptotic derivation of the MNL is given. Using some results of the extreme values theory for i.i.d. random variables [7], it is shown that the Gumbel distribution for the random variables is not necessary anymore. A distribution that is asymptotically exponential in its tail is just required to obtain the MNL model for the choice probability. Similar derivations are obtained in many applications of location, routing, loading, and packing [18–22].

In this paper we want to organize and summarize all the above existing results by presenting a very simple and intuitive random utility model.

The remainder of the paper is organized as follows. In Sect. 2, we recall the well-known derivation of the MNL model for the choice probability when the random variable distribution is a Gumbel function. In Sect. 3, we show that we can relax the Gumbel distribution assumption and still derive the MNL model for the choice probability. Finally, the conclusions of our work are reported in Sect. 4.

2 Derivation of the MNL Model When the Random Variable Distribution is a Gumbel Function

Let us consider

- $j = 1, \ldots, n$: mutually exclusive choice alternatives
- v_{ij}: deterministic utility of alternative j for decision maker i
- $\tilde{x}_{ij}$: random utility of alternative j for decision maker i.

The decision maker i assigns a total utility to each alternative as follows

$$\tilde{u}_{ij} = v_{ij} + \tilde{x}_{ij} \tag{1}$$

In the following, we recall the main results from the literature, where the MNL model is derived under the assumption that the random variables $\tilde{x}_{ij}$ are i.i.d. over alternatives and their common distribution is a Gumbel function.

Following [23], to derive the MNL model we first consider the density for each random component of utility $\tilde{x}_{ij}$

$$f(\tilde{x}_{ij}) = e^{-\tilde{x}_{ij}} e^{-e^{-\tilde{x}_{ij}}}. \tag{2}$$

Its cumulative distribution is

$$F(\tilde{x}_{ij}) = e^{-e^{-\tilde{x}_{ij}}} \tag{3}$$

which is a Gumbel function.

Following [14], the probability that decision maker i chooses alternative j is

$$\begin{aligned} p_{ij} &= Pr\{v_{ij} + \tilde{x}_{ij} > v_{ik} + \tilde{x}_{ik} \ \ \forall k \neq j\} = \\ &= Pr\{\tilde{x}_{ik} < v_{ij} - v_{ik} + \tilde{x}_{ij} \ \ \forall k \neq j\}. \end{aligned} \tag{4}$$

If $\tilde{x}_{ij}$ is given, this expression is the cumulative distribution for each $\tilde{x}_{ik}$ evaluated at $v_{ij} - v_{ik} + \tilde{x}_{ij}$, which, according to (3), is $exp(-exp(-(v_{ij} - v_{ik} + \tilde{x}_{ij})))$. Since the $\tilde{x}'s$ are independent, this cumulative distribution over all $k \neq j$ is the product of the individual cumulative distributions

$$p_{ij}|\tilde{x}_{ij} = \prod_{k \neq j} e^{-e^{-(v_{ij} - v_{ik} + \tilde{x}_{ij})}}. \tag{5}$$

Of course, $\tilde{x}_{ij}$ is not actually given, then the choice probability is the integral of $p_{ij}|\tilde{x}_{ij}$ over all values of $\tilde{x}_{ij}$ weighted by its density (2), i.e.

$$p_{ij} = \int_{-\infty}^{+\infty} \left[\prod_{k \neq j} e^{-e^{-(v_{ij} - v_{ik} + \tilde{x}_{ij})}} \right] e^{-\tilde{x}_{ij}} e^{-e^{-\tilde{x}_{ij}}} \, d\tilde{x}_{ij}. \tag{6}$$

After some manipulation of this integral one gets the following expression for the choice probability

$$p_{ij} = \frac{e^{v_{ij}}}{\sum_{k=1}^{n} e^{v_{ik}}} \tag{7}$$

where n is the total number of alternatives. Equation (7) is a MNL model.

3 Derivation of the MNL Model When the Random Variable Distribution is Not a Gumbel Function

We want to show that, under some mild conditions, the Gumbel distribution assumption for the i.i.d. random variables is unjustified to derive the MNL model for choice probability. Thus, we reformulate the random utility choice theory in terms of the asymptotic extreme values theory [7]. This theory deals with the properties of maxima (or minima) of sequences of random variables with a large number of terms. In particular, we will show that when the number of alternatives becomes large the Gumbel distribution assumption is not necessary anymore. In the following, we derive all the results for the maximization case, but the same results can be adapted to the minimization case.

We assume that $F(x)$ is asymptotically exponential in its right tail, i.e. there is a constant $\beta > 0$ such that

$$\lim_{y \to +\infty} \frac{1 - F(x + y)}{1 - F(y)} = e^{-\beta x}. \tag{8}$$

Our research question then becomes: Under condition (8) and the assumption that the number of alternatives is large, can we still asymptotically get the MNL model for the choice probability?

Let us consider a set J of $N = |J|$ alternatives. We assume that J is partitioned into n nonempty disjoint subsets J_j, $j = 1, \ldots, n$, called clusters, of $N_j = |J_j|$ alternatives. The partition into clusters of the alternatives is faced by many choice processes. For instance, when a household is looking for a dwelling, first of all it will select the district where to live (this is the cluster) and, then, inside that district, it will choose the actual dwelling among all the alternatives.

Let $\tilde{u}_{ij}^z$ be the utility for decision maker i for choosing alternative $z \in J_j$.

As already stated, in a random utility model we assume that $\tilde{u}_{ij}^z$ is the sum of a deterministic variable v_{ij} and a random variable $\tilde{x}_{iz}$, i.e.

$$\tilde{u}_{ij}^z = v_{ij} + \tilde{x}_{iz}. \tag{9}$$

The deterministic variable v_{ij} of the utility includes variables which represent attributes of the cluster and the decision context. The random variable $\tilde{x}_{iz}$ represents aspects of utility that the researcher does not observe, e.g., idiosyncrasies of decision maker i.

The decision maker i will choose among all alternatives the one with the maximum utility.

Let us define the distribution of the maximum utility for decision maker i among all alternatives z in all clusters j as

$$\tilde{u}_i = \max_{j=1,\ldots,n;\ z \in J_j} \tilde{u}_{ij}^z = \max_{j=1,\ldots,n} (v_{ij} + \max_{z \in J_j} \tilde{x}_{iz}) = \max_{j=1,\ldots,n} (v_{ij} + \tilde{x}_i^j), \tag{10}$$

where

$$\tilde{x}_i^j = \max_{z \in J_j} \tilde{x}_{iz}. \tag{11}$$

Let

$$G_i(x) = Pr\{\tilde{u}_i < x\} \tag{12}$$

be the distribution of $\tilde{u}_i$ and

$$P_{ij}(x) = Pr\{\tilde{x}_i^j < x\} \tag{13}$$

be the distribution of $\tilde{x}_i^j$.

By the i.i.d. assumption of the random variables, the distribution $P_{ij}(x)$ becomes

$$P_{ij}(x) = \prod_{z \in J_j} Pr\{\tilde{x}_{iz} < x\} = [F(x)]^{N_j}. \tag{14}$$

Now, because of (10), (11), and (14), Eq. (12) becomes

$$G_i(x) = Pr\{\tilde{u}_i < x\} = Pr\{\max_{j=1,\dots,n} (v_{ij} + \tilde{x}_i^j < x\} = \prod_{j=1,\dots,n} Pr\{v_{ij} + \tilde{x}_i^j < x\} =$$
$$= \prod_{j=1,\dots,n} Pr\{\tilde{x}_i^j < x - v_{ij}\} = \prod_{j=1,\dots,n} P_{ij}(x - v_{ij}) =$$
$$= \prod_{j=1,\dots,n} [F(x - v_{ij})]^{N_j}. \tag{15}$$

Following [18, 20], we will show that under assumption (8) the distribution $G_i(x)$ tends towards a Gumbel function as the total number of alternatives N becomes large. Then we will check if under these results the MNL model for the choice probability can be still derived.

First, let us consider that we can fix the origin for the utility scale arbitrarily, i.e., the choice probabilities are unaffected by a shift in the utility scale and any additive constant to the utilities can be ignored. Let us choose this constant as the root a_N of the equation

$$1 - F(a_N|N) = 1/N, \tag{16}$$

where we remind N is the total number of alternatives.

By replacing $\tilde{u}_i$ with $\tilde{u}_i - a_N$ in (15) one has

$$G_i(x|N) = \prod_{j=1,\dots,n} \left[F(x - v_{ij} + a_N|N)\right]^{N_j}. \tag{17}$$

Let us consider the ratio

$$\alpha_j = N_j/N \tag{18}$$

and assume that this ratio remains constant for each j while the values of $N = 1, 2, ...$ vary, as needed later to compute the asymptotic behavior while N increases.

Because of (18), Eq. (17) can be written as

$$G_i(x|N) = \prod_{j=1,\dots,n} \left[F(x - v_{ij} + a_N|N)\right]^{\alpha_j N}. \tag{19}$$

Let us assume that N is large enough to use $\lim_{N\to+\infty} G_i(x|N)$ as an approximation of $G_i(x)$. Then, the following theorem holds.

Theorem 1 *Under condition (8), the probability distribution $G_i(x)$ becomes the following Gumbel distribution*

$$G_i(x) = \lim_{N\to+\infty} G_i(x|N) = \exp\left(-A_i e^{-\beta x}\right) \tag{20}$$

where

$$A_i = \sum_{j=1,\dots,n} \alpha_j e^{\beta v_{ij}} \tag{21}$$

is the accessibility in the sense of Hansen [8] to the overall set of alternatives.

Proof By (17) and (18) one has

$$G_i(x) = \lim_{N \to +\infty} G_i(x|N) = \lim_{N \to +\infty} \prod_{j=1,\dots,n} \left[F(x - v_{ij} + a_N|N)\right]^{\alpha_j N} = \\ = \prod_{j=1,\dots,n} \lim_{N \to +\infty} \left[F(x - v_{ij} + a_N|N)\right]^{\alpha_j N}. \tag{22}$$

From (16), $\lim_{N \to +\infty} a_N = +\infty$ only if $x \to +\infty$, since $\lim_{N \to +\infty} 1/N = 0$ and $1 - F(x) = 0$, or $F(x) = 1$.

From (8) one obtains

$$\lim_{N \to +\infty} \frac{1 - F(x - v_{ij} + a_N|N)}{1 - F(a_N|N)} = e^{-\beta(x - v_{ij})}. \tag{23}$$

By (23) and (16), it holds that

$$\lim_{N \to +\infty} F(x - v_{ij} + a_N|N) = \lim_{N \to +\infty} \left(1 - [1 - F(a_N|N)]e^{-\beta(x - v_{ij})}\right) = \\ = \lim_{N \to +\infty} \left(1 - \frac{e^{-\beta(x - v_{ij})}}{N}\right) \tag{24}$$

and, by reminding that $\lim_{n \to +\infty}(1 + \frac{x}{n})^n = e^x$

$$\lim_{N \to +\infty} \left[F(x - v_{ij} + a_N|N)\right]^N = \lim_{N \to +\infty} \left(1 - \frac{e^{-\beta(x - v_{ij})}}{N}\right)^N = \\ = \exp\left(-e^{-\beta(x - v_{ij})}\right). \tag{25}$$

Substituting (25) into (22) and using (21), one finally has

$$G_i(x) = \prod_{j=1,\dots,n} \exp\left(-\alpha_j e^{-\beta(x - v_{ij})}\right) = \exp\left(-A_i e^{-\beta x}\right), \tag{26}$$

which is a Gumbel distribution. □

By (25), the following approximation holds for large values of N

$$F(x - v_{ij})^N = \exp\left(-e^{-\beta(x - v_{ij} - a_N)}\right), \tag{27}$$

where a_N is a constant. We want to prove that under Theorem 1 the MNL model for the choice probability still holds.

The choice probability p_{ij} for decision maker i to choose cluster j can be determined as follows. Decision maker i chooses cluster j if and only if

$$v_{ij} + \tilde{x}_i^j \geq v_{ik} + \tilde{x}_i^k, \quad \forall k \neq j \tag{28}$$

then

$$p_{ij} = Pr\{v_{ij} + \tilde{x}_i^j \geq v_{ik} + \tilde{x}_i^k, \quad \forall k \neq j\} = Pr\{v_{ij} + \tilde{x}_i^j \geq \max_{k=1,\ldots n;\ k \neq j} (v_{ik} + \tilde{x}_i^k)\}. \tag{29}$$

Since $\{\tilde{x}_i^k\}$ are independent, then

$$Pr\{\max_{k=1,\ldots n;\ k \neq j} (v_{ik} + \tilde{x}_i^k) < x\} = Pr\{\max_{k=1,\ldots n;\ k \neq j} \tilde{x}_i^k < x - v_{ik}\} =$$
$$= \prod_{k=1,\ldots,n;\ k \neq j} P_{ik}(x - v_{ik}) \tag{30}$$

and

$$Pr\{v_{ij} + \tilde{x}_i^j < x\} = Pr\{\tilde{x}_i^j < x - v_{ij}\} = P_{ij}(x - v_{ij}). \tag{31}$$

From the Total Probability Theorem [5], and the results in (30) and (31), Eq. (29) becomes

$$p_{ij} = \int_{-\infty}^{+\infty} \left[\prod_{k=1,\ldots n;\ k \neq j} P_{ik}(x - v_{ik}) \right] dP_{ij}(x - v_{ij}). \tag{32}$$

The following theorem holds

Theorem 2 *The choice probability p_{ij} for decision maker i to choose cluster j is given by*

$$p_{ij} = \frac{N_j e^{\beta v_{ij}}}{\sum_{k=1}^{n} N_k e^{\beta v_{ik}}}. \tag{33}$$

Proof From (32), by using (14) and (27), and setting $\gamma = e^{\beta a_N}$, one obtains

$$p_{ij} = \int_{-\infty}^{+\infty} \prod_{k=1,\ldots,n;\ k\neq j} [F(x - v_{ik})]^{\alpha_k N}\, d\left[F(x - v_{ij})\right]^{\alpha_j N} =$$

$$= \int_{-\infty}^{+\infty} \prod_{k=1,\ldots,n;\ k\neq j} \exp\left[-\alpha_k e^{-\beta(x-v_{ik}-a_N)}\right] d\exp\left[-\alpha_j e^{-\beta(x-v_{ij}-a_N)}\right] =$$

$$= \int_{-\infty}^{+\infty} \prod_{k=1,\ldots,n;\ k\neq j} \exp\left[-\gamma\alpha_k e^{-\beta(x-v_{ik})}\right] d\exp\left[-\gamma\alpha_j e^{-\beta(x-v_{ij})}\right] =$$

$$= \gamma\alpha_j e^{\beta v_{ij}} \int_{-\infty}^{+\infty} \beta e^{-\beta x} exp(-\gamma A_i e^{-\beta x})dx =$$

$$= \gamma\alpha_j e^{\beta v_{ij}} \int_{0}^{+\infty} e^{-\gamma A_i t} dt = \frac{\alpha_j e^{\beta v_{ij}}}{A_i} = \frac{N_j e^{\beta v_{ij}}}{\sum_{k=1}^{n} N_k e^{\beta v_{ik}}},$$

where $t = e^{-\beta x}$. □

Note that the choice probability in (33) still represents a MNL model.

4 Conclusions

This paper has summarized the usefulness of reinterpreting random utility models by means of the asymptotic theory of extremes, which allows to derive the Multinomial Logit model for the choice probability.

Most well-known results on extreme values statistics concern sequences of i.i.d. random variables. This independency assumption, even if it provides a very convenient form for the choice probability, could be considered too restrictive. In order to relax this assumption, the independency assumption could be replaced by the asymptotic independency for many results.

References

1. Ben-Akiva, M., Lerman, S.R.: Disaggregate travel and mobility choice models and measures of accessibility. In: Hensher, D., Stopher, P. (eds.) Behavioral Travel Modeling. Croom Helm, London (1979)
2. Ben-Akiva, M., Lerman, S.R.: Discrete Choice Analysis: Theory and Application to Travel Demand, vol. 9. MIT press (1985)
3. Beraldi, P., Bruni, M.E., Manerba, D., Mansini, R.: A stochastic programming approach for the traveling purchaser problem. IMA J. Manag. Math. **28**(1), 41–63 (2017)
4. Crainic, T.G., Gobbato, L., Perboli, G., Rei, W.: Logistics capacity planning: a stochastic bin packing formulation and a progressive hedging meta-heuristic. Eur. J. Oper. Res. **253**(2), 404–417 (2016)
5. DeGroot, M., Schervish, M.: Probability and Statistics, 3rd edn. Addison Wesley, Reading (2002)

6. Domencich, T., McFadden, D.: Urban Travel Dynamics: A Behavioral Analysis. North Holland, Amsterdam (1975)
7. Galambos, J.: The Asymptotic Theory of Extreme Order Statistics. Wiley, New York (1978)
8. Hansen, W.: How accessibility shapes land use. J. Am. Inst. Plan. **25**, 73–76 (1959)
9. Leonardi, G.: The structure of random utility models in the light of the asymptotic theory of extremes. In: Florian, M. (ed.) Transportation Planning Models, pp. 107–133. Elsevier (1984)
10. Leonardi, G.: Asymptotic approximations of the assignment model with stochastic heterogeneity in the matching utilities. Environ. Plan. A **17**, 1303–1314 (1985)
11. Luce, R.D.: Individual Choice Behavior: A Theoretical Analysis. Wiley, New York (1959)
12. Manerba, D., Mansini, R., Perboli, G.: The capacitated supplier selection problem with total quantity discount policy and activation costs under uncertainty. Int. J. Prod. Econ. **198**, 119–132 (2018)
13. Marschak, J.: Binary choice constraints on random utility indications. In: Arrow, K. (ed.) Stanford Symposium on Mathematical Methods in the Social Sciences, pp. 312–329. Stanford University Press (1960)
14. McFadden, D.: Conditional logit analysis of qualitative choice behavior. In: Zarembka, P. (ed.) Frontiers in Econometrics, pp. 105–142. Academic Press (1974)
15. Perboli, G., De Marco, A., Perfetti, F., Marone, M.: A new taxonomy of smart city projects. Transp. Res. Procedia **3**, 470–478 (2014)
16. Perboli, G., Ghirardi, M., Gobbato, L., Perfetti, F.: Flights and their economic impact on the airport catchment area: an application to the Italian tourist market. J. Optim. Theory Appl. **164**, 1109–1133 (2015)
17. Perboli, G., Gobbato, L., Maggioni, F.: A progressive hedging method for the multi-path traveling salesman problem with stochastic travel times. IMA J. Manag. Math. **28**, 65–86 (2015)
18. Perboli, G., Tadei, R., Baldi, M.: The stochastic generalized bin packing problem. Discrete Appl. Math. **160**, 1291–1297 (2012)
19. Perboli, G., Tadei, R., Gobbato, L.: The multi-handler knapsack problem under uncertainty. Eur. J. Oper. Res. **236**(3), 1000–1007 (2014)
20. Tadei, R., Perboli, G., Baldi, M.M.: The capacitated transshipment location problem with stochastic handling costs at the facilities. Int. Trans. Oper. Res. **19**(6), 789–807 (2012)
21. Tadei, R., Perboli, G., Perfetti, F.: The multi-path traveling salesman problem with stochastic travel costs. EURO J. Transp. Logist. **6**, 2–23 (2014). https://doi.org/10.1007/s13676-014-0056-2
22. Tadei, R., Ricciardi, N., Perboli, G.: The stochastic p-median problem with unknown cost probability distribution. Oper. Res. Lett. **37**, 135–141 (2009)
23. Train, K.E.: Discrete Choice Methods with Simulation. Cambridge University Press (2003)

The Optimal Tariff Definition Problem for a Prosumers' Aggregation

Antonio Violi, Patrizia Beraldi, Massimiliano Ferrara, Gianluca Carrozzino and Maria Elena Bruni

Abstract This paper deals with the problem faced by an aggregator in defining the optimal tariff structure for a group of prosumers aggregated within a coalition. The random nature of the main parameters involved in the decision process is explicitly accounted for by adopting the stochastic programming framework and, in particular, the paradigm of integrated chance constraints. Numerical experiments carried out on a realistic test case shows the efficacy of the proposed approach in providing more profitable rates for both consumers and producers with respect to the standard market alternatives.

Keywords Microgrid · Tariff definition · Chance constraints

A. Violi · P. Beraldi (✉) · G. Carrozzino · M. E. Bruni
DIMEG, University of Calabria, Rende (CS), Italy
e-mail: patrizia.beraldi@unical.it

G. Carrozzino
e-mail: gianluca.carrozzino@unical.it

M. E. Bruni
e-mail: mariaelena.bruni@unical.it

A. Violi
INNOVA s.r.l., Rome (RM), Italy
e-mail: antonio.violi@unical.it

M. Ferrara
Decision Lab, DIGIEC, Mediterranean University of Reggio Calabria, Reggio Calabria, Italy
e-mail: massimiliano.ferrara@unirc.it

M. Ferrara
ICRIOS, Bocconi University, Milan, Italy

© Springer Nature Switzerland AG 2018

P. Daniele and L. Scrimali (eds.), *New Trends in Emerging Complex Real Life Problems*, AIRO Springer Series 1,
https://doi.org/10.1007/978-3-030-00473-6_51

1 Introduction

The electricity market is facing a challenging transformation, slowly evolving from an electricity production system dominated by large centralized generating units to a system with an increasing amount of power production. In this new landscape, the position of the consumer is also changing. Household consumers, as well as small and large industrial consumers, are increasingly producing energy, for example, by installing solar panels, establishing a wind turbine on their property, or producing biogas from waste. In other words, they now act as prosumers, i.e. consumers that produce energy. The presence of several prosumers, representing the lower level of the power grid, modifies the energy chain, calling for its re-organization according to a bottom-up approach. Because of the existing entry barriers, imposed by national laws and technical issues, prosumers cannot act directly as suppliers, still keeping their passive role. In this new decentralized system, the creation of microgrid infrastructures, aimed at managing the electricity system locally, represents a valuable alternative. The aim of the microgrid is to create a cooperative system, aggregating a given number of agents (i.e. consumers, producers, prosumers) to act as a single entity when engaging in power system market (both wholesale and retail). The new system is coordinated by an entity, the aggregator, that is responsible of managing the energy exchanges between supplies and demands within the coalition and interacting with the power system with the aim of obtaining the maximum benefit from the grid. Nowadays, the fundamental value of aggregators in creating economies of scale and scope is widely recognized and different forms of aggregation are arising all over the world [1]. The management of aggregated systems poses new challenging problems, from the definition of new business models for the cooperation of the aggregated agents to decision problems related to the management of shared resources. These last problems can be classified according to the considered planning horizon. At a strategic level, considering a long-time horizon of several years, the main problems refer to the design of efficient energy solutions by investing, for example, in advanced technologies, in the efficiency of buildings, and in the adoption of storage systems [2]. Focusing on a shorter time horizon, typically one year, the main problems are related to the definition of the procurement plans by considering the optimal mix of the main supply sources (bilateral contracts, energy market, production from own plants). This problem has been widely investigated in the scientific literature. We refer, for example, to the recent paper [3, 4] (see also the references therein) where the authors proposed stochastic programming approaches to account for the uncertainty (prices, demands, production from renewable sources) that affect the decision problem. Over a shorter time horizon, typically a week, the main problem that the aggregator is called to solve is related to the satisfaction of the aggregated demand by properly managing the available resources. The problem involves scheduling decisions on the operation of the available conventional plants, the management of storage systems, the trading in the electricity market so to satisfy the aggregated demand with the final aim of maximizing the total wealth. The optimal management of aggregated systems, and more generally distributed energy resources has been widely studied and different contributions appeared in the recent scientific literature. We cite [5] (see also the references therein), where the authors propose a stochastic programming model with recourse, incorporating a risk measure to control potential losses caused by unfavorable events that may occur.

In this paper, we focus on the problem of the definition of the electricity tariffs (selling and buying side) for the prosumers aggregated within a coalition. The definition of pricing schemes represents an important issue that has been mainly investigated from the viewpoint of a retailer. As an intermediary between the wholesale electricity market and end-users, the retailer should determine the optimal selling price with the aim of maximizing the expected profit. Uncertainty affecting the pool market price, the demand, and eventually the supply from renewable sources should be explicitly accounted for. For this reason, the main contributions on the selling price determination rely on the adoption of stochastic optimization techniques. For example, Carrion et al. provides in [6] a stochastic programming methodology that allows an electricity retailer to engage in medium-term forward contracting and to optimally set selling prices to clients, with the final aim to maximize the expected profit given a pre-specified risk level on profit volatility. In [7] authors proposed a dynamic and flexible tariff structure for a distribution company that protects the consumers against the excessive fluctuations of the wholesales market prices, by means of a two-stage pricing scheme with a static and a dynamic component. More recently, Nojavan et al. have proposed in [8] a scenario based approach in a smart grid where the selling tariff is determined on the basis of a real-time pricing. In [9], the selling price problem has been addressed by the robust optimization approach.

In this paper, we analyze the problem of the tariff definition from the aggregator viewpoint. Differently from the retailer, the aggregator has to define both selling and buying tariffs to offer to the members of the coalition. A consumer finds it profitable to participate in the coalition if the offered price is lower than the price paid for buying energy from the market. The opposite happens for the selling side. Energy exchanged within the coalition only requires local distribution and the transmission savings can be shared between the local prosumers.

The rest of the paper is organized as follows. The following Sect. 2 introduces the mathematical formulation for the price determination problem. Section 3 presents and discusses the numerical results collected on a real case study. Concluding remarks and future research developments are discussed in Sect. 4.

2 The Mathematical Formulation

The definition of the electricity tariffs can be carried out by applying different pricing schemes [10]. The traditional Flat Pricing charges a time-invariant rate for each kWh. Critical Peak Pricing guarantees a constant price with the exception of critical time periods when the price is substantially raised. Real-time Pricing provides dynamic rates that track wholesale market prices. Finally, Time of Use (ToU) pricing offers electricity rates that vary with the time of the day. Typically, the hours of a day are divided in blocks (e.g., peak, intermediate, off-peak), that reflect the level of demand of the electricity network, attributing to the peak periods higher prices. It is evident that this flexible energy pricing scheme is able to offer economic incentives pushing towards the adoption of demand side management programs. We consider the ToU structure that reflects the current organization of the Italian Electricity market taken as reference model. We assume that the aggregator wants to define the selling and buying prices for a given time horizon, typically one year, divided into monthly time steps t. The hours of the different days of each month $t \in T$ are articulated into a set F of ToU blocks. We suppose that the aggregator has already defined, by using, for example, the model proposed in [11], the optimal procurement plan for the considered time horizon. In particular, for each time t and block f, we assume to know the power CP_{tf} purchased from bilateral contracts (at the fixed price PCP_{tf}), the amount MBP_{tf} taken from the market, the self production SP_{tf} from conventional plants and the amount MSP_{tf} to eventually sell to the market. The price determination problem, as the procurement one, is clearly an optimization problem under uncertainty, since the parameters involved in the decision process are unknown when the optimal tariffs should be determined. We deal with this more involved problem by adopting the stochastic programming framework, and we assume that the uncertain parameters are modeled as random variables defined on a given probability space $(\Omega, \mathscr{F}, \mathbb{P})$. We use letter in bold to denote the random variables. In defining the optimal tariffs, the aggregator should account for different issues. One the one hand, he should offer competitive tariffs: the more affordable the prices, the more attractive will be the coalition for the prosumers. On the other hand, it is important to guarantee that the eventual losses he incurs, if any, are very limited. We denote by ST_{tf} and BT_{tf} the selling (to the consumer) price and buying (from the producers) price for each time period t and block f. If bought on the market, the consumer would pay electricity at the market price $\mathbf{P_{tf}}$ (unknown when the tariff plan should be defined), increased by the margin of the distributor (Δ_A). Moreover, a producer would sell electricity at the price $\mathbf{W_{tf}}$, decreased by the distributor margin (Δ_V). Furthermore, for each time period t and block f, we denote by $\mathbf{D_{tf}}$ and $\mathbf{R_{tf}}$ the aggregated demand and the production from renewable sources, respectively. The aggregator should define the optimal tariff plan so to guarantee that under every circumstance that may occur the total monetary inflow should be greater than the total outflow:

$$\sum_{t\in T}\sum_{f\in F}\mathbf{D}_{tf}ST_{tf}+\sum_{t\in T}\sum_{f\in F}\mathbf{W}_{tf}MSP_{tf}\geq$$
$$\sum_{t\in T}\sum_{f\in F}\mathbf{P}_{tf}MBP_{tf}+\left(SP_{tf}+\mathbf{R}^{s}_{tf}\right)BT_{tf}+\sum_{t\in T}\sum_{f\in F}PCP_{tf}CP_{tf}+MC \quad \forall\omega\in\Omega \quad (1)$$

where the term MC represents the aggregator's margin that covers the management cost ed assures a given profit. Because of the presence of random parameters, constraint (1) is clearly stochastic. We deal with this constraint by adopting the paradigm of chance constraints (CC) in their integrated form (ICC) [12]. While the CC only measures the probability of shortage, imposing that the stochastic constraint should be satisfied with a given reliability value, the ICC uses the probability distribution to measure the expected magnitude of the shortage, thus, taking into account both quantitative and qualitative aspects of the shortage, whereas the CC only considers the qualitative side. In particular, we require that the average shortage is lower than a given threshold γ properly chosen by the decision maker:

$$\mathbb{E}[(\sum_{t\in T}\sum_{f\in F}\mathbf{P}_{tf}MBP_{tf} + \left(SP_{tf} + \mathbf{R}_{tf}\right)BT_{tf} + \sum_{t\in T}\sum_{f\in F}PCP_{tf}CP_{tf} + MC -$$
$$-\sum_{t\in T}\sum_{f\in F}\mathbf{D}_{tf}ST_{tf} - \sum_{t\in T}\sum_{f\in F}\mathbf{W}_{tf}MSP_{tf})_{+}] \leq \gamma \tag{2}$$

where $(a)_{+} = \max(0, a)$. In the following, we shall assume that the random variables follow a discrete distribution. We denote by S the set of possible realizations, scenarios, each occurring with a probability π_s and we shall use the superscript s to denote the s-th realization of the random parameters.

In this case, the ICC admits a nice reformulation, that requires the introduction of support variables. More specifically, for each scenario s, we introduce a nonnegative variable L_s and we replace (2) with

$$L_s \geq (\sum_{t\in T}\sum_{f\in F}(P^s_{tf}MBP_{tf} + \left(PQ_{tf} + R^s_{tf}\right)BT_{tf} + \sum_{t\in T}\sum_{f\in F}PCP_{tf}CP_{tf} + MC)$$
$$-\sum_{t\in T}\sum_{f\in F}(W^s_{tf}MSP_{tf} + D^s_{tf}ST_{tf}) \quad \forall s \tag{3}$$
$$\sum_{s\in S}\pi_s L_s \leq \gamma \tag{4}$$

It is interesting also to note the close connection of the ICC with the tradition Conditional Value at Risk measure [13]. Additional bound constraints are also included to mathematically represent the profitability of the coalition tariffs:

$$\sum_{t\in T}\sum_{f\in F}D^s_{tf}ST_{tf} \leq (1+\Delta_A)\sum_{t\in T}\sum_{f\in F}P^s_{tf}D^s_{tf} \qquad \forall s \tag{5}$$
$$\sum_{t\in T}\sum_{f\in F}(PP_{tf} + R^s_{tf})BT_{tf} \geq (1-\Delta_V)\sum_{t\in T}\sum_{f\in F}W^s_{tf}(PP_{tf} + R^s_{tf}) \qquad \forall s \tag{6}$$
$$LB_{ST} \leq ST_{tf} \leq UB_{ST} \qquad \forall t, \forall f \tag{7}$$
$$LB_{BT} \leq BT_{tf} \leq UB_{BT} \qquad \forall t, \forall f \tag{8}$$

Constraints (5)–(6) represent the economic advantage for the consumer and the producer, respectively, whereas conditions (7)–(8) bound the tariff values within specific lower and upper bounds computed on the basis of the expected value of the purchasing and selling market prices. In defining the optimal pricing, the aggregator aims at providing prosumers with more profitable conditions with respect to market. This goal can be achieved by pursuing different strategies. On the one hand, there is the need to define the lowest possible consumption rates, to encourage consumers (who should be the majority) to remain within the coalition. On the other hand, the desire to remunerate producers, from both traditional and renewable sources, in an appropriate way to promote the sale of energy within the coalition. In order to account for both the objectives, we consider a weighted objective function:

$$\max \lambda \sum_{s\in S}\sum_{t\in T}\sum_{f\in F} \pi_s D^s_{tf}((1+\Delta_A)P^s_{tf} - ST_{tf}) + \\ +(1-\lambda)\sum_{s\in S}\sum_{t\in T}\sum_{f\in F} \pi_s(PQ_{tf} + R^s_{tf})(BT_{tf} - (1-\Delta_V)W^s_{tf}) \quad (9)$$

Here the first term represents the advantage for "buying" measured in terms of total savings from buying the overall energy demand within the coalition w.r.t. the external market, while the second term is the gain from selling within the aggregation. A different choice of the parameter $\lambda \in [0, 1]$ can be used to calibrate the strategy of the aggregator, with a higher value of λ accounting for a greater attention devoted to the savings of consumers w.r.t. the gain of the producers.

3 Computational Experiments

In this section, we describe the computational experience carried out in order to validate the effectiveness of the proposed approach. The numerical code integrates MATLAB R2015a[1] for the scenario generation and parameters set-up phases and GAMS 24.5.2[2] as algebraic modeling system, with CPLEX 12.6.1[3] as solver for linear problems. All the test cases have been solved on a PC Intel Core I7 (2.80 GHz) with 16 GB of RAM. As testbed for the computational experience we have considered the "virtual" coalition described in [3]. We have considered a planning horizon of 12 months, starting from January 2017, and 3 time blocks (F_1,F_2,F_3) for each month, according to the Italian electricity market [14]. The expected value of coalition demand, production from the renewable plant and the market prices have been calculated by analyzing the available historical series of these data. We have considered a scenario set with 500 scenarios, generated by using the evolution model proposed in [15] and the scenario tree generation approach of [16, 17]. The procurement plan for the coalition has been already defined, in terms of energy procured from each one of the available sources, that are bilateral contracts (CP), production from traditional systems (SP) and amount to buy from the market (MBP), and eventually the amount to sell on the market (MSP). These data are reported in [18].

---

[1]www.mathworks.com.

[2]www.gams.com.

[3]https://www-01.ibm.com/software/commerce/optimization/cplex-optimizer/.

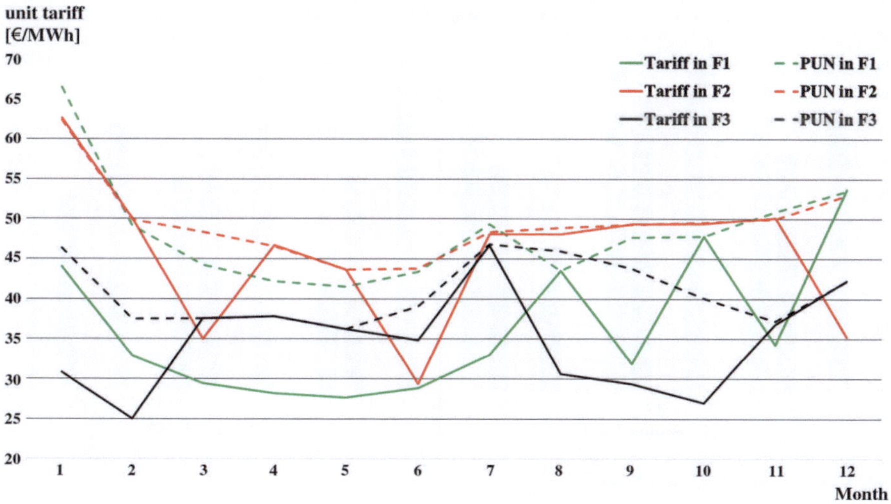

Fig. 1 Selling-side tariffs

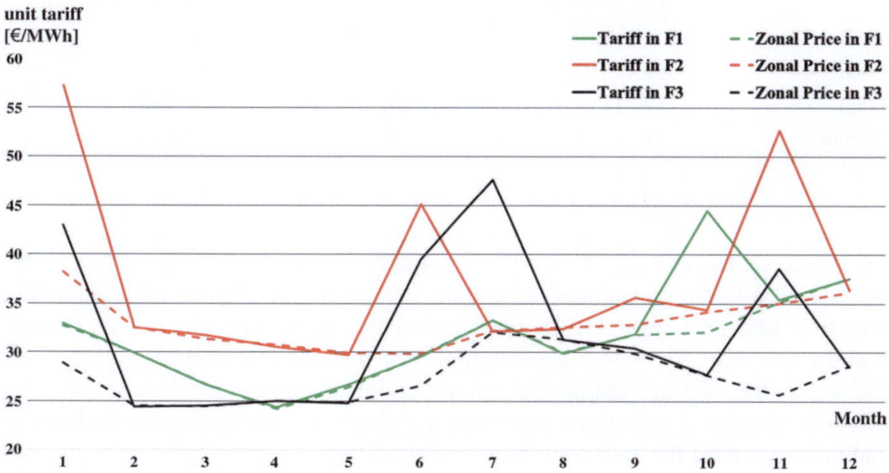

Fig. 2 Buying-side tariffs

The following Fig. 1 reports the optimal selling tariffs for the considered planning horizon compared with the market alternative (dashed line), obtained for a value of λ equal to 0.5 to balance the consumers' and producers' profitability.

Looking at the results, we may observe that the proposed tariff structure is profitable w.r.t. the market alternative, resulting in a substantial saving. A similar result has been obtained for buying side tariffs (see Fig. 2).

Additional experiments have been carried out to evaluate how the choice of λ impacts on the optimal solutions. Figure 3 reports the selling tariffs obtained with 3

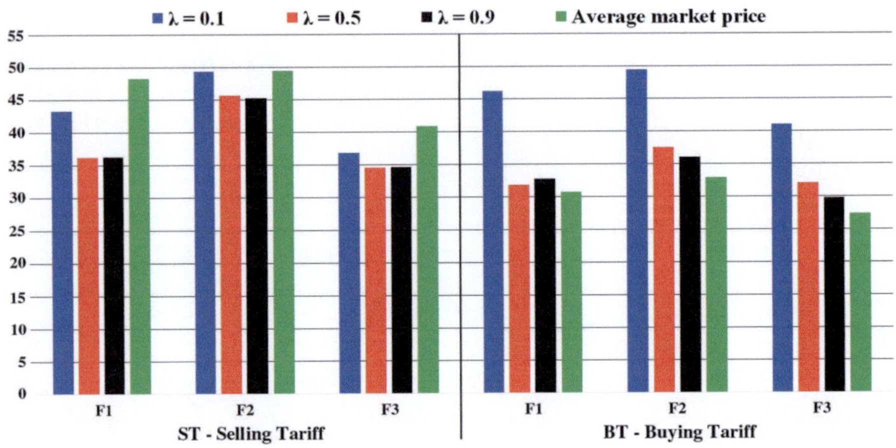

Fig. 3 Average selling tariffs as function of λ

Table 1 Solutions for different values of γ (€)

γ	Obj. function	Buying side benefit	Selling side benefit
0	46,213.10	84,627.18	7,799.03
1,000	53,425.94	99,708.24	7,143.64
5,000	58,317.29	109,456.78	7,177.80
10,000	61,529.94	115,545.84	7,514.03

different λ values, compared with the average market tariff. As expected, a higher attention devoted to consumers' economic advantage, represented by a value of λ close to 1, generates lower tariffs, both for buying and selling side: in fact, in order to allow low rates for energy consumption, the revenues for energy produced and sold within the coalition cannot be very profitable, even if better than the price for selling outside the aggregation. Similarly, a policy more oriented to production can lead to higher rates for both consumers and producers.

Finally, we have solved the model for different values of the threshold on potential losses γ. The results obtained with $\lambda = 0.5$ are reported in Table 1. It is evident that a higher value of threshold for potential average losses allows a greater overall benefit for both selling and buying side, even if a "risk-averse" approach ($\gamma = 0$) ensures good advantages anyway.

It is worthwhile noting that to the best of our knowledge no other decision approach for the tariff definition under similar conditions has been proposed in the literature. Nevertheless, we have empirically found that the straightforward approach based on the solution of a deterministic problem where the random parameters are replaced by the expected values performs very poorly.

4 Conclusions

The paper proposes a stochastic programming model based on the paradigm of integrated chance constraints for the problem faced by an aggregator that has to establish the optimal tariffs to apply to the prosumers of the coalition. A consumer finds profitable to participate in the coalition if the offered price is lower than the price paid for buying energy from the market. The opposite happens for the selling side. The numerical results carried out by considering a realistic case study have shown the validity of the proposed approach as a supporting tool for the aggregator in defining affordable tariffs for both sides. As future research, we plan to combine the procurement problem with the tariff definition into an unique model and to validate the effectiveness of the decision approaches with other policies. The resulting problem will present non linearity and specific approaches will be designed.

Acknowledgements This work has been partially supported by Italian Minister of Economic Development, Bando HORIZON 2020 PON I&C 2014–2020, with the grant for research project F/050159/01-03/X32 "Power Cloud: Tecnologie e Algoritmi nell'ambito dell'attuale quadro regolatorio del mercato elettrico verso un new deal per i consumatori e i piccoli produttori di energia da fonti rinnovabili".

References

1. Burger, S., Chaves-Ávila, J.P., Batlle, C., Pérez-Arriag, I.J.: A review of the value of aggregators in electricity systems. Renew. Sustain. Energy Rev. **77**, 395–405 (2017)
2. Min, D., Ryu, J., Choi, D.G.: A long-term capacity expansion planning model for an electric power system integrating large-size renewable energy technologies. Comput. Oper. Res. (2017) (in press)
3. Beraldi, P., Violi, A., Carrozzino, G., Bruni, M.E.: The optimal energy procurement: a stochatic programming approach. In: Springer Proceedings in Mathematics and Statistics, International Conference on Optimization and Decision Science ODS2017, vol. 217, pp. 357–365 (2017)
4. Beraldi, P., Violi, A., Bruni, M.E., Carrozzino, G.: A probabilistically constrained approach for the energy procurement problem. Energies **10**(12), Article no. 2179 (2017)
5. Beraldi, P., Violi, A., Carrozzino, G., Bruni, M.E.: A stochastic programming approach for the optimal management of aggregated distributed energy resources. Comput. Oper. Res. (2018) (in press)
6. Carrión, M., Conejo, A.J., Arroyo, J.M.: Forward contracting and selling price determination for a retailer. IEEE Trans. Power Syst. **22**(4), 2105–2114 (2007)
7. Triki, C., Violi, A.: Dynamic pricing of electricity in retail markets. 4OR **7**(1), 21–36 (2009)
8. Nojavan, S., Qesmati, H., Zare, K., Seyyedi, H.: Large consumer electricity acquisition considering time-of-use rates demand response programs. Arab. J. Sci. Eng. **39**(12), 8913–8923 (2014)
9. Fotouhi Ghazvini, M.A., Soares, J., Morais, H., Castro, R., Vale, Z.: Dynamic pricing for demand response considering market price uncertainty. Energies **10**, 1245 (2017)
10. Fridgen, G., Kahlen, M., Ketter, W., Rieger, A., Thimmel, M.: An empirical analysis of electricity tariffs for residential microgrids. Appl. Energ. **210**, 800–814 (2018)
11. Beraldi, P., Violi, A., Carrozzino, G., Bruni, M.E.: The optimal electric energy procurement problem under reliability constraints. Energy Procedia **136**, 283–289 (2017)

12. Ruszczyski, A., Shapiro, A.: Stochastic programming. In: Handbook in Operations Research and Management Science, vol. 672. Elsevier Science, Amsterdam (2003)
13. Rockafellar, R., Uryasev, S.: Optimization of conditional value-at-risk. J. Risk **2**, 21–41 (2000)
14. Beraldi, P., Conforti, D., Triki, C., Violi, A.: Constrained auction clearing in the Italian electricity market. **4 OR 2**(1), 35–51 (2004)
15. Menniti, D., Scordino, N., Sorrentino, N., Violi, A.: Short-term forecasting of day-ahead electricity market price. In: 2010 7th International Conference on the European Energy Market (2010)
16. Beraldi, P., De Simone, F., Violi, A.: Generating scenario trees: a parallel integrated simulation optimization approach. J. Comput. Appl. Math. **233**(9), 2322–2331 (2010)
17. Beraldi, P., Bruni, M.E.: A clustering approach for scenario tree reduction: An application to a stochastic programming portfolio optimization problem. TOP **22**, 1–16 (2013)
18. Beraldi, P., Violi, A., Ferrara, M., Carrozzino, G., Bruni, M.E.: A stochastic programming approach for microgrid tariff definition. Tech. Rep. TESEO Lab. **7**, (2018)

Impulse and Singular Stochastic Control Approaches for Management of Fish-Eating Bird Population

Yuta Yaegashi, Hidekazu Yoshioka, Koichi Unami and Masayuki Fujihara

Abstract Stochastic optimization serves as a central tool for effective population management. We present an impulse control model and a related singular control model for finding the cost-effective and sustainable population management policies of fish-eating birds, predators of fishery resources. The impulse control model considers the cost proportional to the amount of the killed bird and the fixed cost, while singular counterpart considers only the proportional cost. Their optimal controls are discussed from both qualitative and quantitative viewpoints.

Keywords Stochastic optimization · Impulse control · Singular control
Threshold-type population management

1 Introduction

Increasing feeding damage by fish-eating birds, such as *Phalacrocorax carbo*, to inland fishery resources, is one of the most urgent ecological problems to be resolved especially in Japan [3]. Recently, the authors have been approaching this issue from

Y. Yaegashi (✉) · K. Unami · M. Fujihara
Graduate School of Agriculture, Kyoto University, Kitashirakawa Oiwake-Cho, Sakyo-Ku, 606-8502 Kyoto City, Kyoto Prefecture, Japan
e-mail: Yaegashi.yuta.54s@st.kyoto-u.ac.jp

K. Unami
e-mail: unami@adm.kais.kyoto-u.ac.jp

M. Fujihara
e-mail: fujihara@kais.kyoto-u.ac.jp

Y. Yaegashi
Japan Society for the Promotion of Science, Tokyo, Japan

H. Yoshioka
Faculty of Life and Environmental Science, Shimane University, Nishikawatsu-Cho 1060, 690-8504 Matsue City, Shimane Prefecture, Japan
e-mail: yoshih@life.shimane-u.ac.jp

© Springer Nature Switzerland AG 2018
P. Daniele and L. Scrimali (eds.), *New Trends in Emerging Complex Real Life Problems*, AIRO Springer Series 1,
https://doi.org/10.1007/978-3-030-00473-6_52

the viewpoint of stochastic optimization where the bird population dynamics is considered [7]. The singular control [1, 8] was employed in our mathematical model, which leads to a threshold-type, cost-effective and sustainable population management policy under the stochastic population dynamics. In this paper, we propose an impulse control [2, 4] counterpart of the singular control model. These models employ similar performance indices, but have different admissible sets of controls with each other. The resulting optimal population management policies are compared and their practical implications are discussed. In addition, mathematical linkages between these models are analyzed focusing on parameter dependence of the solutions to the (quasi) variational inequalities.

2 Mathematical Models

We consider a management problem of a bird population in a habitat in an infinite horizon. The decision-maker, a manager of the bird population, can reduce the bird population through a countermeasure. We assume that the countermeasure is carried out in a much shorter timescale than that of the bird population dynamics.

2.1 *Impulse Control Model*

The bird population in the habitat at the time t is denoted as X_t, which is governed by the Itô's stochastic differential equation (SDE)

$$\begin{cases} \mathrm{d}X_t = X_t(\mu \mathrm{d}t + \sigma \mathrm{d}B_t), & \tau_i \leq t < \tau_{i+1} < \infty \\ X_{\tau_i} = X_{\tau_{i-}} - \zeta_i, & t = \tau_i \end{cases}, \; X_{0-} > 0, \tag{1}$$

where μ is the intrinsic growth rate, σ is the magnitude of stochastic fluctuation involved in the population dynamics, B_t is the 1-D standard Brownian motion [6], τ_i $(i = 0, 1, 2, \ldots, \tau_0 = 0 < \tau_1 < \tau_2, \ldots)$ is the time when the countermeasure is performed to reduce the population and $\zeta_i > 0$ is the magnitude of the killed bird at τ_i. Whenever the countermeasure is performed, the following cost is incurred:

$$K(\zeta_i) = k_1\zeta_i + k_0, \tag{2}$$

where $k_1 > 0$ is the cost proportional to ζ_i and $k_0 > 0$ is the fixed cost incurred regardless of ζ_i. In addition, if $\zeta_i = 0, k_i = 0$.The existence of the fixed cost distinguishes the impulse control model from the singular control counterpart [1, 8]. The performance index J_{imp} represents the expected net profit of the decision-maker:

$$J_{\text{imp}}(x;\eta) = \mathrm{E}\left[\int_0^{\infty} e^{-\delta s}\left(RX_s^M - rX_s^m\right)\mathrm{d}s - \sum_{i=0}^{\infty} e^{-\delta\tau_i} K(\zeta_i)\chi_{\tau_i}\right], \tag{3}$$

where δ is the discount rate, R, r, M and m are the positive constants, and χ_S is the indicator function for the subset S. In the right-hand side of (3), the term RX_s^M represents the ecological utility provided by the existence of the bird, the term $-rX_s^m$ represents the disutility by the existence of the bird, and the last summation term is the cost of the countermeasure. Here, η is the management policy defined by the timing and magnitude of the countermeasure

$$\eta = \{(\tau_i, \zeta_i)\}_{i \geq 0}, \tag{4}$$

and is called an admissible control if it satisfies the conditions stated in Definition 2.1 in Onishi and Tsujimura [4] and $X_t \geq 0$. Let $\mathfrak{A}$ be the set of the admissible controls. The goal of the decision-maker is to find the optimal control η^* such that

$$V_{\text{imp}}(x) = \sup_{\eta\in\mathfrak{A}} J_{\text{imp}}(x;\eta) = J_{\text{imp}}\left(x;\eta^*\right), \tag{5}$$

where $V_{\text{imp}}(x)$ is referred to as the value function. The dynamic programming principle [7] leads to its governing Quasi Variational Inequality (QVI) [2, 4] as

$$\max\left\{\mathcal{L}V_{\text{imp}}(x) + Rx^M - rx^m, \mathcal{M}V_{\text{imp}}(x) - V_{\text{imp}}(x)\right\} = 0,\ x > 0 \tag{6}$$

with the differential operator $\mathcal{L}$ and the intervention operator $\mathcal{M}$ defined as

$$\mathcal{L}W(x) = \frac{1}{2}\sigma^2 x^2 \frac{\mathrm{d}^2 W}{\mathrm{d}x^2} + \mu x \frac{\mathrm{d}W}{\mathrm{d}x} - \delta W,\ \mathcal{M}W(x) = \sup_{\zeta}[-K(\zeta) + W(x-\zeta)] \tag{7}$$

for generic sufficiently smooth function $W = W(x)$. The boundary condition for the QVI is $V_{\text{imp}}(0) = 0$, which means that no problem arises when there is no bird and the net profit of the decision-maker equals zero. Under the QVI controls (Definition 3.2 in [2]), the following control rule is the optimal management strategy with threshold values $\bar{x}_{\text{imp}}$ and $\underline{x}$ ($\bar{x}_{\text{imp}} > \underline{x}$):

1. If $X_{t-} < \bar{x}_{\text{imp}}$, then no countermeasure is performed and only if $X_{t-} = \bar{x}_{\text{imp}}$, the countermeasure is immediately performed and X_{t-} is reduced to $\underline{x}$ ($X_t = \underline{x}$).
2. If $X_{0-} > \bar{x}_{\text{imp}}$, then X_{0-} is immediately reduced to $\underline{x}$ ($X_0 = \underline{x}$) by the countermeasure, and follows the first rule.

A sample path of X_t following the above-mentioned rule is shown in Fig. 1.

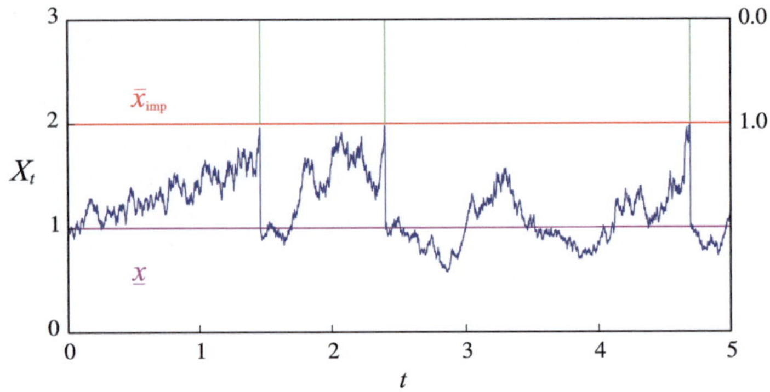

Fig. 1 The sample path of the impulse control with the thresholds $\bar{x}_{\text{imp}} = 2.0$ (red), $\underline{x} = 1.0$ (pink) and the magnitude of the killed bird ζ_i (green, from above)

2.2 *Singular Control Model*

In the singular control model, the population dynamics is described as

$$\mathrm{d}X_t = X_t(\mu \mathrm{d}t + \sigma \mathrm{d}B_t) - \mathrm{d}\omega_t,\ X_{0-} > 0, \tag{8}$$

where ω_t represents the decrease of the bird population by the countermeasure. This ω_t is admissible if it is the non-negative, non-decreasing, right-continuous and adopted process such that $X_t \geq 0$ [1, 7]. Let $\mathfrak{B}$ be the set of the admissible controls. The performance index in the singular control model is set as

$$J_{\text{sin}}(x;\omega) = \mathrm{E}\left[\int_0^\infty e^{-\delta s}\left(RX_s^M - rX_s^m\right)\mathrm{d}s - \int_0^\infty e^{-\delta s} k_2 \mathrm{d}\omega_s\right]. \tag{9}$$

where $k_2 > 0$ is the cost proportional to ω_s. The last term is the cost of the countermeasure. The value function in this case is

$$V_{\text{sin}}(x) = \sup_{\omega \in \mathfrak{B}} J_{\text{sin}}(x;\omega) = J_{\text{sin}}\left(x;\omega^*\right) \tag{10}$$

and the associated Variational Inequality (VI) that governs V_{sin} is derived as

$$\max\left\{\mathcal{L}V_{\text{sin}}(x) + Rx^M - rx^m, -\left(\frac{\mathrm{d}V_{\text{sin}}}{\mathrm{d}x} + k_2\right)\right\} = 0,\ x > 0 \tag{11}$$

with the boundary condition $V_{\text{sin}}(0) = 0$. The following control rule can be proved to the optimal management strategy with a threshold $\bar{x}_{\text{sin}}$ (Chap. 3 in [5]):

1. If the process is about to exceed the threshold $\bar{x}_{\text{sin}}$, the countermeasure is immediately performed and X_{t-} is reduced to $\bar{x}_{\text{sin}}$ ($X_t = \bar{x}_{\text{sin}}$). Otherwise, no countermeasure is performed.

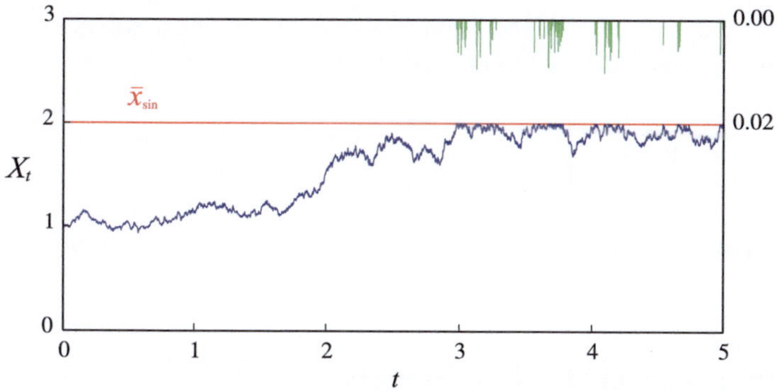

Fig. 2 The sample path of the singular control model with the threshold $\bar{x}_{\sin} = 2.0$ (red) and the magnitude of the killed bird ω_t (green, from above)

2. If $X_{0-} > \bar{x}_{\sin}$, then X_{0-} is immediately reduced to $\bar{x}_{\sin}$ ($X_0 = \bar{x}_{\sin}$) by the countermeasure, and follows the first rule.

A sample path of X_t following the above-mentioned rule is shown in Fig. 2.

3 Exact Solutions

The following assumptions are made for the model parameters:

$$\mu > \sigma^2/2,\ 0 < M < 1 < m < 2,\ \delta - \mu m - 0.5\sigma^2 m(m-1) > 0 \text{ and } \beta > m. \tag{12}$$

where β is given in (15). Then, based on the guessed solution method [4], the exact viscosity solution of the QVI (6) is found as

$$V(x) = \begin{cases} ax^\beta + A_1 x^M + A_2 x^m & (x < \bar{x}_{\mathrm{imp}}) \\ -k_1(x - \underline{x}) - k_0 + a\underline{x}^\beta + A_1\underline{x}^M + A_2\underline{x}^m & (x \geq \bar{x}_{\mathrm{imp}}) \end{cases}, \tag{13}$$

where

$$A_1 = R\left[\delta - \mu M - \frac{\sigma^2}{2}M(M-1)\right]^{-1} > 0,\ A_2 = r\left[\delta - \mu m - \frac{\sigma^2}{2}m(m-1)\right]^{-1} < 0, \tag{14}$$

$$\beta = \frac{1}{2}\left(1 - \frac{2\mu}{\sigma^2} + \sqrt{\left(\frac{2\mu}{\sigma^2} - 1\right)^2 + \frac{8\delta}{\sigma^2}}\right) > m, \tag{15}$$

and a, $\underline{x}$ and $\bar{x}_{\mathrm{imp}}$ are determined from a system of nonlinear equations. While the exact viscosity (classical) solution of the VI (11) is found as [8]

$$V(x) = \begin{cases} bx^{\beta} + A_1 x^M + A_2 x^m & (0 < x \le \bar{x}_{\mathrm{sin}}) \\ c - k_2 x & (x > \bar{x}_{\mathrm{sin}}) \end{cases}, \tag{16}$$

where b, c and $\bar{x}_{\mathrm{sin}}$ are determined from a system of nonlinear equations.

4 Comparisons of the Two Models

Based on a real case, the parameters are set as $\mu = 1.7\times10^{-1}$ (1/year), $\sigma = 5.3\times10^{-1}$ (1/year$^{1/2}$), $m = 2.0$ (–), $M = 0.5$ (–), $\delta = 1.0$ (1/year), $r = 5.0 \times 10^{-4}$ (–), $R = 1.0 \times 10^{-1}$ (–), and $k_0 = 1.0 \times 10^{1}$ (–) unless otherwise specified [8]. In addition, $k_1 = k_2 = 1.0$ is assumed without loss of generality. With these parameter values, the thresholds values are computed as $\bar{x}_{\mathrm{sin}} = 1052$ (–), $\bar{x}_{\mathrm{imp}} = 1312$ (–), and $\underline{x} = 869$ (–). The result indicates that the threshold of the impulse control model $\bar{x}_{\mathrm{imp}}$ is larger than that of the singular control model $\bar{x}_{\mathrm{sin}}$. With the impulse control model, the decision-maker should set a larger threshold to manage the bird population. Figure 3 shows the value functions and the corresponding thresholds, indicating that V_{sin} is slightly larger than V_{imp}. According to Table 1, as $k_0 \to 0$, namely if the fixed cost reaches 0, $\bar{x}_{\mathrm{imp}}$ monotonically decreases while $\underline{x}$ monotonically increases with $\underline{x} < \bar{x}_{\mathrm{sin}} < \bar{x}_{\mathrm{imp}}$. This result indicates that the impulse control model would approach the singular control model as the fixed cost k_0 decreases, although it should be mathematically investigated in future.

As shown in Fig. 2, for the singular control model, the decision-maker should continuously perform the countermeasure for suppressing increase of the bird population. In practice, it would be difficult to perform such a management policy. On the other hand, as shown in Fig. 1, for the impulse control model where the fixed cost k_0 is considered, the decision-maker should intermittently perform the countermeasure since the optimal policy is to decrease the finite amount of the population

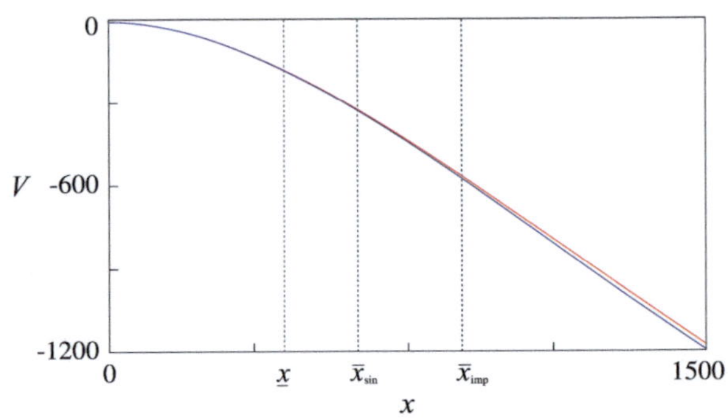

Fig. 3 The value functions V_{sin} (red) and V_{imp} (blue) and the thresholds $\underline{x}$, $\bar{x}_{\mathrm{sin}}$ and $\bar{x}_{\mathrm{imp}}$

Table 1 The thresholds $\bar{x}_{\mathrm{imp}}$ and $\underline{x}$ with changing k_0

	$k_0 = 1.0 \times 10^1$	$k_0 = 1.0 \times 10^{-1}$	$k_0 = 1.0 \times 10^{-3}$	$k_0 = 1.0 \times 10^{-5}$	$k_0 = 1.0 \times 10^{-7}$
$\bar{x}_{\mathrm{imp}}$	1312	1102	1068	1064	1063
$\underline{x}$	869	1006	1036	1040	1041

to the threshold value $\underline{x}$. Being different from the singular control, such a policy can be more easily implemented in real problems. In addition, in reality, the fixed cost such as implementation costs would arise whenever the countermeasure is performed regardless of its magnitude. The impulse control model thus seems to be more appropriate for the bird population management than the singular control model.

5 Conclusions

In this paper, an impulse control model and a singular control model for finding the cost-effective and sustainable management polices of fish-eating birds were proposed. The resulting optimal management policies were compared from both qualitative and quantitative viewpoints, and their feasibility in real problems was discussed. The threshold values for the optimal management policies of the two models were compared, showing that the decision-maker should set a larger threshold value with the impulse control model. In addition, numerical experiments implied that the singular control model can be seen as a limit of the impulse control model, which should be mathematically rigorously studied in future. Establishment of an efficient and stable numerical method for solving QVIs is an important research future research topic.

Acknowledgements This paper is partly funded by grants-in-aid for scientific research No. 16KT0018, No. 17J09125, and No. 17K15345 from the Japan Society for the Promotion of Science (JSPS) and Applied Ecology Research Grant No. 2016-02 from Water Resources Environment Center in Japan. The authors thank officers of Hirose Fisheries Cooperatives for their useful comments and suggestions.

References

1. Alvarez, L.H.: Singular stochastic control in the presence of a state-dependent yield structure. Stochastic Process. Appl. **86**(2), 323–343 (2000)
2. Cadenillas, A.: Optimal central bank intervention in the foreign exchange market. J. Econom. Theor. **87**, 218–242 (1999)
3. Kameda, K.: Population increase of the Great Cormorant *Phalacrocorax carbo* and measures to reduce its damage to the fisheries and forests of Lake Biwa. In: Lake Biwa: Interactions Between Nature and People, pp. 491–496. Springer (2012)
4. Ohnishi, M., Tsujimura, M.: An impulse control of a geometric Brownian motion with quadratic costs. Eur. J. Oper. Res. **168**(2), 311–321 (2006)

5. Øksendal, B.: Stochastic control problems where small intervention costs have big effects. Appl. Math. Optim. **40**(3), 355–375 (1999)
6. Øksendal, B.: Stochastic Differential Equations. Springer (2003)
7. Pham, H.: Continuous-Time Stochastic Control and Optimization with Financial Applications. Springer Science & Business Media. (2009)
8. Yaegashi, Y., Yoshioka, H., Unami, K., Fujihara, M.: Optimal policy of predator suppression for sustainable inland fishery management. In: 12th SDEWES Conference Archived Paper, 0309, p. 11 (2017)

When the Other Matters. The Battle of the Sexes Revisited

Asunción Zapata, Amparo M. Mármol, Luisa Monroy and M. Ángeles Caraballo

Abstract In this paper we address bimatrix games when the players take into account not only their own payoff, but they also show some concerns about the payoff of the other player. We propose a weighted Rawlsian representation of players' preferences which can accommodate the behaviours of different types of players, which are identified with different values of the parameters. The Battle of the Sexes game is analyzed in this extended setting where certain social interactions between the players determine their strategic behaviours. The best response correspondences are described depending on the relative importance that each player assigns to her own payoff and to the payoff of the other. This permits the identification of the corresponding sets of equilibria and the study of the changes produced with the variation of the parameters.

Keywords Battle of the sexes · Equilibria · Bi-matrix games · Rawlsian function

1 Bimatrix Games with Rawlsian Preferences

The so-called 2×2 bimatrix game is a non-cooperative game in normal form with two players, with only two pure strategies for each player. The consequences for each player of each possible combination of strategies are known and can be represented by a matrix.

A. Zapata · A. M. Mármol · L. Monroy
Departamento de Economía Aplicada III, IMUS, Universidad de Sevilla, Seville, Spain
e-mail: azapata@us.es

A. M. Mármol
e-mail: amarmol@us.es

L. Monroy
e-mail: lmonroy@us.es

M. Á. Caraballo (✉)
Departamento de Economía e Historia Económica, IUSEN, Universidad de Sevilla, Seville, Spain
e-mail: mcaraba@us.es

© Springer Nature Switzerland AG 2018

P. Daniele and L. Scrimali (eds.), *New Trends in Emerging Complex Real Life Problems*, AIRO Springer Series 1,
https://doi.org/10.1007/978-3-030-00473-6_53

Formally, if player 1 selects her i-th pure strategy and player 2 selects her j-th strategy, then the payoff of player 1 is denoted by a_{ij}, and the payoff of player 2 is denoted by b_{ij}. Thus, the payoffs can be described by a bimatrix (A, B), where $A = (a_{ij})$ and $B = (b_{ij})$, for $i, j = 1, 2$.

In this type of game the sets of mixed strategies (that is, the sets of distributions of probabilities on the pure strategies) for the players are $\{(p, 1-p), p \in [0,1]\}$, and $\{(q, 1-q), q \in [0,1]\}$ respectively. Thus, a mixed strategy for player 1 can be represented by $p \in [0,1]$, and a mixed strategy for player 2 by $q \in [0,1]$. The payoff functions, which represent the expected payoffs of the players, are defined on the combinations of mixed strategies. For $k = 1, 2$, $u_k : [0,1] \times [0,1] \to \mathbb{R}$, with $u_1(p,q) = (p, 1-p)A(q, 1-q)^t$ and $u_2(p,q) = (p, 1-p)B(q, 1-q)^t$.

Nash [3] proved that any finite game has at least an equilibrium in mixed strategies. In general, there could be any number of equilibria, with no compelling reason to choose among them. However, there are situations in which the result of the interaction may produce an undesirable payoff. When these situations are analysed, empirical evidence shows that social interactions between the players may affect individual choices in equilibrium. That is, players may have an incentive to act not entirely guided by their own interests. These circumstances have led us to incorporate different behaviours of the players in the analysis of the game. These behaviours emerge when the players take into account not only their own payoff function, but also they show some concerns about the payoff of the other player, and select their strategies accordingly.

We will specifically consider three types of players. Denote by $\nu^k : \mathbb{R}^2_+ \to \mathbb{R}$ the social preference function representing the evaluation of player k for each pair of expected payoffs. Player k is said to be *equanimous* if $\nu^k(u_1, u_2) = \nu^k(u_2, u_1)$, for all (u_1, u_2). She is said to be *pro-self* if for all $u_1 > u_2$, $\nu^k(u_1, u_2) \geq \nu^k(u_2, u_1)$, and she is said to be *pro-social* if for all $u_1 > u_2$, $\nu^k(u_1, u_2) \leq \nu^k(u_2, u_1)$. A pro-self player cares more for her payoff than for that of the other. The extreme case of a pro-self agent is an egoistic agent, who only cares for her own payoff, regardless what the other obtain. Similarly, a pro-social agent cares more for the payoff of the other. If a player is both pro-self and pro-social, then she is an equanimous player.

Monroy et al. [2] analysed these social attitudes in n-person non-cooperative games with an additive representation of the players' preferences. In this paper, we explore the implications of these attitudes in a setting where the preferences of the players are represented by weighted maxmin functions in the spirit of the egalitarian trend proposed by Rawls [4].

In our setting the social preference functions of the players are represented by weighted Rawlsian functions:

$$\omega^1_{\gamma^1}(u_1, u_2) = \min\left\{\frac{u_1}{\gamma^1}, \frac{u_2}{1-\gamma^1}\right\}, \quad \omega^2_{\gamma^2}(u_1, u_2) = \min\left\{\frac{u_1}{1-\gamma^2}, \frac{u_2}{\gamma^2}\right\},$$

with $\gamma^1, \gamma^2 \in [0, 1]$.

For $k = 1, 2$, the value γ^k represents the relative importance that player k assigns to her own payoff, and $1 - \gamma^k$ the relative importance that she assigns to the payoff of the other. In this model, the payoffs are not considered to be substitutes, they are, in fact, complementary, and the greater the value of the parameter γ^k, the more the value of u_k needs to grow in order to improve the value of v^k.

The different types of the players are characterised by the values of the parameters involved in their Rawlsian preference functions [5].

- $\gamma^k = 1/2$ if and only if player k is *equanimous*.
- $\gamma^k \geq 1/2$ if and only if player k is *pro-self*.
- $\gamma^k \leq 1/2$ if and only if player k is *pro-social*.

Apart from the individual behaviour of the players, we will also consider the relationship between the relative importances that both players assign to the payoffs. We say that the two players are *mirror* players when both players behave with respect to the other identically, that is, if $\gamma^1 = \gamma^2$. We say that the two players are *harmonized* players when they assign the same importance to the payoff of player 1 and the same importance to the payoff of player 2, that is, if $\gamma^1 = 1 - \gamma^2$.

In the next section we present an elaborate analysis of the well-known game *the Battle of the Sexes*. The calculations of the best responses and the equilibria for the Rawlsian case are exemplary for general bimatrix games.

2 Rawlsian Equilibria for the Battle of the Sexes

In this paper we consider the well-known game *Battle of the Sexes* introduced by Luce and Raiffa [1]. This game has two players, each of whom has the same two possible actions, one of which is preferred by each player. The players have some common interests, though, in that they would both prefer to choose the same strategy rather than doing different things. Thus, the players find themselves in a one-shot simultaneous-play game with the following payoff matrix:

$$(A, B) = \begin{pmatrix} (1,4) & (0,0) \\ (0,0) & (4,1) \end{pmatrix}$$

Let $(p, 1-p)$ with $0 \leq p \leq 1$ be a mixed strategy for the first player and $(q, 1-q)$ with $0 \leq q \leq 1$ a mixed strategy for the second player. The expected payoffs for both players are $u_1(p,q) = pq + 4(1-p)(1-q)$, and $u_2(p,q) = 4pq + (1-p)(1-q)$, respectively.

The best response correspondences of the players are:

$$r^1(\bar{q}) = \begin{cases} 0 & if\ \bar{q} < \frac{4}{5} \\ [0,1] & if\ \bar{q} = \frac{4}{5} \\ 1 & if\ \bar{q} > \frac{4}{5} \end{cases}, \quad r^2(\bar{p}) = \begin{cases} 0 & if\ \bar{p} < \frac{1}{5} \\ [0,1] & if\ \bar{p} = \frac{1}{5} \\ 1 & if\ \bar{p} > \frac{1}{5} \end{cases} \tag{1}$$

This game has two non-dominated Nash equilibria in pure strategies. The first one consists of the first strategy of each player with a payoff of $(1, 4)$ and the second one consists of the second strategy of each player with a payoff of $(4, 1)$. In addition a mixed-strategy equilibrium exists consisting of the mixed strategy $\left(\frac{1}{5}, \frac{4}{5}\right)$ for the first player, and $\left(\frac{4}{5}, \frac{1}{5}\right)$ for the second with expected payoff $\left(\frac{4}{5}, \frac{4}{5}\right)$.

In what follows we will analyse the equilibria of this game when the players show Rawlsian preferences.

Given $\gamma^1, \gamma^2 \in [0, 1]$, the Rawlsian functions for players 1 and 2 are

$$\omega^1_{\gamma^1}(p, q) = \min\left\{\frac{pq + 4(1-p)(1-q)}{\gamma^1}, \frac{4pq + (1-p)(1-q)}{1-\gamma^1}\right\} \tag{2}$$

$$\omega^2_{\gamma^2}(p, q) = \min\left\{\frac{pq + 4(1-p)(1-q)}{1-\gamma^2}, \frac{4pq + (1-p)(1-q)}{\gamma^2}\right\} \tag{3}$$

Equivalently, these functions can be written as:

$$\omega^1_{\gamma^1}(p, q) = \begin{cases} \frac{(5q-4)p+4(1-q)}{\gamma^1} & if\ ((5-10q)\gamma^1 + 5q - 4)p \leq (5\gamma^1 - 4)(1-q) \\ \frac{(5q-1)p+1-q}{1-\gamma^1} & if\ ((5-10q)\gamma^1 + 5q - 4)p \geq (5\gamma^1 - 4)(1-q) \end{cases} \tag{4}$$

$$\omega^2_{\gamma^2}(p, q) = \begin{cases} \frac{(5p-4)q+4(1-p)}{1-\gamma^2} & if\ ((10p-5)\gamma^2 - 5p + 1)q \leq (1 - 5\gamma^2)(1-p) \\ \frac{(5p-1)q+1-p}{\gamma^2} & if\ ((10p-5)\gamma^2 - 5p + 1)q \geq (1 - 5\gamma^2)(1-p) \end{cases} \tag{5}$$

2.1 *The Best Response Correspondences*

The Rawlsian preference function $\omega^1_{\gamma^1}$ depends on the payoff functions u_1 and u_2. Therefore, the best response of player 1 given the strategy $\bar{q}$ of player 2, denoted by $R^1_\gamma(\bar{q})$, will also depend on the best response functions of u_1 and u_2, $r^1(\bar{q})$ and $r^2(\bar{q})$.

First, note that $((5-10q)\gamma^1 + 5q - 4)p \geq (5\gamma^1 - 4)(1-q)$ can alternatively be written as $pq + 4(1-p)(1-q) \geq 5\gamma^1(pq + (1-p)(1-q))$. As a consequence, if $\gamma^1 \leq \frac{1}{5}$, then this inequality holds for any $p, q \in [0, 1]$, and $\omega^1_{\gamma^1}(p, \bar{q}) = \frac{u_2(p,\bar{q})}{1-\gamma^1}$. Therefore, $R^1_\gamma(\bar{q}) = r^2(\bar{q})$. With an analogous reasoning, when $\gamma^1 \geq \frac{4}{5}$ is $R^1_\gamma(\bar{q}) = r^1(\bar{q})$.

For $\frac{1}{5} < \gamma^1 < \frac{4}{5}$, in order to find the best reply of the first player to a fixed strategy of the second player, we consider different intervals for the values of $\bar{q}$. If $0 \leq \bar{q} < \frac{1}{5}$, then the best reply is $R^1_\gamma(\bar{q}) = 0$ since both individual best replies, r^1 and r^2, are null. Similarly, if $\frac{4}{5} < \bar{q} \leq 1$, then $R^1_\gamma(\bar{q}) = 1$.

For $\frac{1}{5} \leq \bar{q} \leq \frac{4}{5}$, consider $z(\bar{q}) = \frac{(5\gamma^1-4)(1-\bar{q})}{(5-10\bar{q})\gamma^1+5\bar{q}-4}$. Note that, for $\bar{q} = \frac{1}{5}$, if $((5-10\bar{q})\gamma^1 + 5\bar{q} - 4)p \leq (5\gamma^1 - 4)(1-\bar{q})$ then $\omega^1_{\gamma^1}(p, \bar{q}) = \frac{u_2(p,\bar{q})}{1-\gamma^1}$. Therefore, $R^1_\gamma(\frac{1}{5}) = [0, z(\frac{1}{5})]$, which coincides with the corresponding segment of r^2 (as shown in Fig. 1). Analogously, for $\bar{q} = \frac{4}{5}$, $R^1_\gamma(\frac{4}{5}) = [z(\frac{4}{5}), 1]$.

Finally, if $\frac{1}{5} < \bar{q} < \frac{4}{5}$, then $(5-10\bar{q})\gamma^1 + 5\bar{q} - 4 < 0$. As a consequence, if the strategy of player 1, p, is such that $p < z(\bar{q})$, then $\omega^1_{\gamma^1}(p, \bar{q}) = \frac{u_2(p,\bar{q})}{1-\gamma^1}$, and $\omega^1_{\gamma^1}(p + \varepsilon, \bar{q}) > \omega^1_{\gamma^1}(p, \bar{q})$. On the other hand, if $p > z(\bar{q})$, then $\omega^1_{\gamma^1}(p, \bar{q}) = \frac{u_1(p,\bar{q})}{\gamma^1}$, and $\omega^1_{\gamma^1}(p - \varepsilon, \bar{q}) > \omega^1_{\gamma^1}(p, \bar{q})$. It follows that $R^1_\gamma(\bar{q}) = z(\bar{q})$.

Formally, the explicit expression of the best response correspondence of player 1 for all the values of the parameter γ^1 is

(a) If $\gamma^1 \leq \frac{1}{5}$, then $R^1_\gamma(\bar{q}) = r^2(\bar{q})$,
(b) If $\frac{1}{5} < \gamma^1 < \frac{4}{5}$, then

$$R^1_\gamma(\bar{q}) = \begin{cases} 0 & if \quad \bar{q} < \frac{1}{5} \\ \left[0, \frac{4(4-5\gamma^1)}{15(1-\gamma^1)}\right] & if \quad \bar{q} = \frac{1}{5} \\ \frac{(5\gamma^1-4)(1-\bar{q})}{(5-10\gamma^1)\bar{q}+5\gamma^1-4} & if \ \frac{1}{5} \leq \bar{q} \leq \frac{4}{5} \\ \left[\frac{4-5\gamma^1}{15\gamma^1}, 1\right] & if \quad \bar{q} = \frac{4}{5} \\ 1 & if \quad \bar{q} > \frac{4}{5} \end{cases} \tag{6}$$

(c) If $\gamma^1 \geq \frac{4}{5}$, then $R^1_\gamma(\bar{q}) = r^1(\bar{q})$.

Analogously, the best response of player 2 to the strategy $\bar{p}$ of player 1, $R^2_\gamma(\bar{p})$, can be described.

(a) If $\gamma^2 \leq \frac{1}{5}$, then $R^2_\gamma(\bar{p}) = r^1(\bar{p})$,
(b) If $\frac{1}{5} < \gamma^2 < \frac{4}{5}$, then

$$R^2_\gamma(\bar{p}) = \begin{cases} 0 & if \quad \bar{p} < \frac{1}{5} \\ \left[0, \frac{4(5\gamma^2-1)}{15\gamma^2}\right] & if \quad \bar{p} = \frac{1}{5} \\ \frac{(1-5\gamma^2)(1-\bar{p})}{(10\gamma^2-5)\bar{p}+1-5\gamma^2} & if \ \frac{1}{5} \leq \bar{p} \leq \frac{4}{5} \\ \left[\frac{5\gamma^2-1}{15(1-\gamma^2)}, 1\right] & if \quad \bar{p} = \frac{4}{5} \\ 1 & if \quad \bar{p} > \frac{4}{5} \end{cases} \tag{7}$$

(c) If $\gamma^2 \geq \frac{4}{5}$, then $R^1_\gamma(\bar{p}) = r^2(\bar{p})$.

Figure 1 shows the best response correspondence when a player is pro-self (left-hand side), equanimous (center) and pro-social (right-hand side). Note that in these three cases the best response correspondences are decreasing for values of $\bar{q}$ between

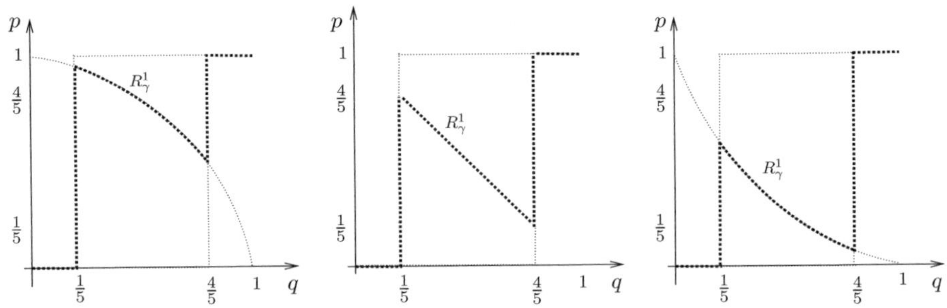

Fig. 1 Best responses correspondence for a pro-social, for an equanimous, and for a pro-self player

$\frac{1}{5}$ and $\frac{4}{5}$. They are concave or convex functions, depending on whether the relative importance that the player assigns to her own payoff is lower than or greater than the importance that she assigns to the payoff of the other player.

2.2 *The Equilibria*

An equilibrium is a pair of strategies that are mutually best replies, that is, (p, q) is an equilibrium if $p \in R^1_\gamma(q)$ and $q \in R^2_\gamma(p)$. If the players choose their strategies at equilibrium, no one can improve her valuation of the payoffs by deviating individually. For $\gamma^1, \gamma^2 \in [0, 1]$, we will denote the set of equilibria of the battle of the sexes game with a Rawlsian preference function as $E(\gamma^1, \gamma^2)$.

We obtain different sets of equilibria depending on the values of the parameters. The different cases are shown in Tables 1, 2 and 3. Note that the value of p where the best response R^1_γ switches to be represented by a decreasing curve, $\frac{4(4-5\gamma^1)}{15(1-\gamma^1)}$(in (6)), is in the interval $[\frac{1}{5}, \frac{4}{5}]$ if and only if $\frac{1}{2} \leq \gamma \leq \frac{13}{17}$. Similarly, the value of q where the best response R^2_γ switches to be represented by a decreasing curve, $\frac{4(5\gamma^2-1)}{15\gamma^2}$ (in (7)), is in the interval $[\frac{1}{5}, \frac{4}{5}]$ if and only if $\frac{4}{17} \leq \gamma \leq \frac{1}{2}$.

Figure 2 shows some cases of equilibria for pro-social mirror players. For values of γ lower than $\frac{4}{17}$ (left-hand side and center), three equilibria are obtained which coincide with the two equilibria in pure strategies in the standard game of the Battle of the Sexes, and a new equilibria in mixed strategies which consists of the mixed strategy $(\frac{4}{5}, \frac{1}{5})$ for player 1, and the mixed strategy $(\frac{1}{5}, \frac{4}{5})$ for player 2. The expected payoffs for the players at this equilibrium are $(\frac{4}{5}, \frac{4}{5})$. Observe that the expected payoffs coincide with the expected payoffs at the equilibria in the game when they play the game in a completely egoistic way ($\gamma = 1$). For $\gamma = \frac{4}{17}$ two more equilibria arise $(\frac{4}{5}, \frac{4}{5})$ and $(\frac{1}{5}, \frac{1}{5})$. As γ increases up to $\frac{1}{2}$ (right-hand side), two additional equilibria appear, where either p increases and q remains constant or q decreases and p doesn't change.

Table 1 Equilibria for mirror players

Mirror players $\gamma^1 = \gamma^2 = \gamma$		$E(\gamma^1, \gamma^2)$
Pro-social	$\gamma < \frac{4}{17}$	$\{(0,0), (\frac{4}{5}, \frac{1}{5}), (1,1)\}$
	$\gamma = \frac{4}{17}$	$\{(0,0), (\frac{1}{5}, \frac{1}{5}), (\frac{4}{5}, \frac{1}{5}), (\frac{4}{5}, \frac{4}{5}), (1,1)\}$
	$\frac{4}{17} < \gamma < \frac{1}{2}$	$\{(0,0), (\frac{1}{5}, \frac{1}{5}), (\frac{4(5\gamma-1)}{15\gamma}, \frac{1}{5}), (\frac{4}{5}, \frac{1}{5}),$ $(\frac{4}{5}, \frac{4-5\gamma}{15\gamma}), (\frac{4}{5}, \frac{4}{5}), (1,1)\}$
Pro-self	$\frac{1}{2} < \gamma < \frac{13}{17}$	$\{(0,0), (\frac{1}{5}, \frac{1}{5}), (\frac{1}{5}, \frac{4(4-5\gamma)}{15(1-\gamma)}), (\frac{1}{5}, \frac{4}{5}),$ $(\frac{5\gamma-1}{15(1-\gamma)}, \frac{4}{5}), (\frac{4}{5}, \frac{4}{5}), (1,1)\}$
	$\gamma = \frac{13}{17}$	$\{(0,0), (\frac{1}{5}, \frac{1}{5}), (\frac{1}{5}, \frac{4}{5}), (\frac{4}{5}, \frac{4}{5}), (1,1)\}$
	$\gamma > \frac{13}{17}$	$\{(0,0), (\frac{1}{5}, \frac{4}{5}), (1,1)\}$

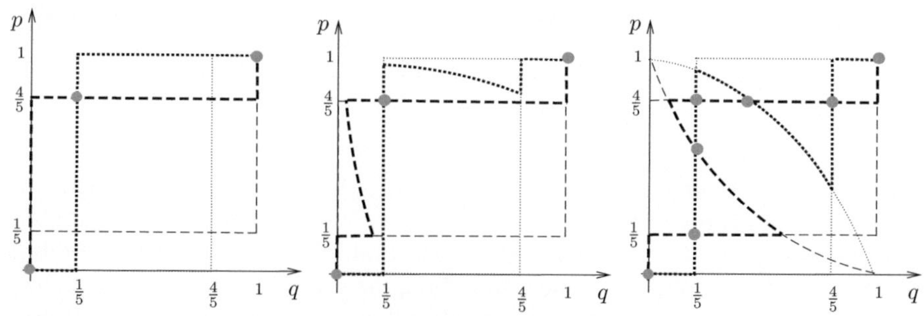

Fig. 2 Equilibria for mirror pro-social players: $\gamma < \frac{4}{17}$ and $\frac{14}{17} < \gamma < \frac{1}{2}$

Table 2 Equilibria for equanimous players

Equanimous players	
$\gamma^1 = \frac{1}{2} = \gamma^2$	$E(\gamma^1, \gamma^2) = \{(0,0), (1,1)\} \cup \{(p,q) :$ $p + q = 1, \frac{1}{5} \leq p \leq \frac{4}{5}\}$

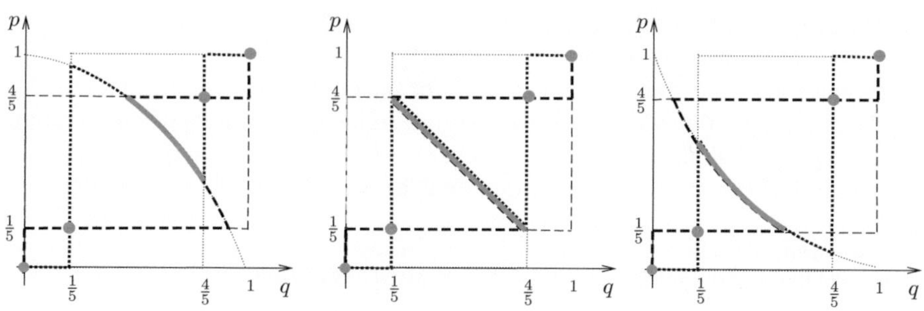

Fig. 3 Equilibria for harmonized players: player 1, pro-social and player 2, pro-self (left); both equanimous (center) and player 1, pro-self and player 2, pro-social (right)

Table 3 Equilibria for harmonized players

Harmonized players $\gamma^1 = 1 - \gamma^2$		$E(\gamma^1, \gamma^2)$
Player 1 pro-social	$\gamma^1 < \frac{4}{17}$	$\{(0,0), (\frac{1}{5}, \frac{1}{5}), (1,1)\}$
Player 2 pro-self	$\gamma^1 = \frac{4}{17}$	$\{(0,0), (\frac{1}{5}, \frac{1}{5}), (\frac{4}{5}, \frac{4}{5})(1,1)\}$
	$\frac{4}{17} < \gamma^1 < \frac{1}{2}$	$\{(0,0), (\frac{1}{5}, \frac{1}{5}), (\frac{4}{5}, \frac{4}{5}), (1,1)\} \cup \{(p,q) : p = \frac{(5\gamma^1-4)(1-q)}{(5-10\gamma^1)q+5\gamma^1-4}, \frac{4-5\gamma^1}{15\gamma^1} \leq q \leq \frac{4}{5}\}$
Player 1 pro-self	$\frac{1}{2} < \gamma^1 < \frac{13}{17}$	$\{(0,0), (\frac{1}{5}, \frac{1}{5}), (\frac{4}{5}, \frac{4}{5}), (1,1)\} \cup \{(p,q) : p = \frac{(5\gamma^1-4)(1-q)}{(5-10\gamma^1)q+5\gamma^1-4}, \frac{1}{5} \leq q \leq \frac{4(4-5\gamma^1)}{15(1-\gamma^1)}\}$
Player 2 pro-social	$\gamma^1 = \frac{13}{17}$	$\{(0,0), (\frac{1}{5}, \frac{1}{5}), (\frac{4}{5}, \frac{4}{5}), (1,1)\}$
	$\gamma^1 > \frac{13}{17}$	$\{(0,0), (\frac{4}{5}, \frac{4}{5}), (1,1)\}$

Figure 3 shows some examples of equilibria for harmonized players, including equanimous as a particular case. In these cases, the two pairs of pure strategies that are equilibria in the conventional game are still equilibria and a set of new equilibria emerge. Note that, in that set of new equilibria, the greater the relative importance for the own payoff, the lower the probability that the player plays her first pure strategy. Moreover, for equanimous players, each equilibrium in the segment provides the same expected payoff for both players $(5pq, 5pq)$. An equilibrium exists that dominates the others in the segment: $(\frac{1}{2}, \frac{1}{2})$ with expected payoffs $(\frac{5}{4}, \frac{5}{4})$.

3 Concluding Remark

The behavior based on self-interest which is assumed in standard bimatrix games is not evident in many real-world situations. We analyze these games considering agents that show concerns about the interest of the other. The approach proposed is based on the egalitarian model of Rawls. As one of the main results, when one of the players is pro-social, then the equilibrium in mixed strategies of the original game is not an equilibrium, regardless of the other player being pro-social or pro-self.

Acknowledgements The research of the authors is partially supported by the Spanish Ministry of Science and Innovation, Project ECO2015-68856-P (MINECO/FEDER).

References

1. Luce, R.D., Raiffa, H.: Games and Decisions: Introduction and Critical Survey. Wiley, New York (1957)
2. Monroy, L., Caraballo, M.A., Mármol, A.M., Zapata, A.: Agents with other-regarding preferences in the commons. Metroeconomica **68**, 947–965 (2017)
3. Nash, J.: Non-cooperative games. Ann. Math. **54**(2), 286–295 (1951)
4. Rawls, J.: A Theory of Justice, Revised edn. Harvard University Press, Cambridge (1971)
5. Zapata, A., Mármol, A.M., Monroy, L., Caraballo, M.A.: A Maxmin Approach for the Equilibria of Vector-Valued Games. Preprint (2018)